AF702137

NOVEL FRONTIERS IN THE PRODUCTION OF COMPOUNDS FOR BIOMEDICAL USE, VOLUME 1

FOCUS ON BIOTECHNOLOGY

Volume 1

Series Editors

MARCEL HOFMAN
Centre for Veterinary and Agrochemical Research Tervuren, Belgium

JOZEF ANNE
Rega Institute, University of Leuven, Belgium

COLOPHON

Focus on Biotechnology is an open-ended series of reference volumes produced for Kluwer Academic Publishers BV in co-operation with the Branche Belge de la Société de Chimie Industrielle a.s.b.l.

The initiative has been taken in conjunction with the Ninth European Congress on Biotechnology. ECB9 has been supported by the Commission of the European Communities, the General Directorate for Technology, Research and Energy of the Wallonia Region, Belgium and J. Chabert, Minister for Economy of the Brussels Capital Region.

NOVEL FRONTIERS IN THE PRODUCTION OF COMPOUNDS FOR BIOMEDICAL USE,
VOLUME 1

FOCUS ON BIOTECHNOLOGY

Volume 1

Series Editors
MARCEL HOFMAN
Centre for Veterinary and Agrochemical Research Tervuren, Belgium

JOZEF ANNE
Rega Institute, University of Leuven, Belgium

COLOPHON

Focus on Biotechnology is an open-ended series of reference volumes produced for Kluwer Academic Publishers BV in co-operation with the Branche Belge de la Société de Chimie Industrielle a.s.b.l.

The initiative has been taken in conjunction with the Ninth European Congress on Biotechnology. ECB9 has been supported by the Commission of the European Communities, the General Directorate for Technology, Research and Energy of the Wallonia Region, Belgium and J. Chabert, Minister for Economy of the Brussels Capital Region.

Novel Frontiers in the Production of Compounds for Biomedical Use, Volume 1

Edited by

ANNIE VAN BROEKHOVEN
FRED SHAPIRO
Innogenetics N.V., Ghent, Belgium

and

JOZEF ANNÉ
Catholic University of Leuven, Leuven, Belgium

KLUWER ACADEMIC PUBLISHERS
DORDRECHT / BOSTON / LONDON

A C.I.P. Catalogue record for this book is available from the Library of Congress.

ISBN 0-7923-6747-2

Published by Kluwer Academic Publishers,
P.O. Box 17, 3300 AA Dordrecht, The Netherlands.

Sold and distributed in North, Central and South America
by Kluwer Academic Publishers,
101 Philip Drive, Norwell, MA 02061, U.S.A.

In all other countries, sold and distributed
by Kluwer Academic Publishers,
P.O. Box 322, 3300 AH Dordrecht, The Netherlands.

Printed on acid-free paper

All Rights Reserved
© 2001 Kluwer Academic Publishers
No part of the material protected by this copyright notice may be reproduced or
utilized in any form or by any means, electronic or mechanical,
including photocopying, recording or by any information storage and
retrieval system, without written permission from the copyright owner.

Printed in the Netherlands.

EDITORS PREFACE

The present book entitled "Novel Frontiers in the Production of Compounds for Biomedical Uses" can perhaps be placed in its best perspective by the Shakespearean character in **The Tempest** who exclaimed" What's past is prologue". Indeed, this compilation of some of the outstanding presentations in the field of biomedicine made at the 9th European Congress on Biotechnology (Brussels, Belgium, July 11-15, 1999) not only reflects the achievements of the recent past, but provides a privileged glimpse of the biotechnology that is emerging in the first decade of the new Millennium.

It is becoming increasingly apparent that biotechnology is offering biomedicine novel approaches and solutions to develop a sorely needed new generation of biopharmaceuticals. This is all the more necessary because in recent years, new diseases have emerged with extraordinary lethality in all corners of the globe, while age-related chronic illnesses have filled the gap wherever biomedicine has made successful inroads. The rise of antibiotic resistance also poses major threats to public health. Thus, as disease patterns evolve, the rational development of new drugs is becoming urgent, not only for the clinical outcome of patients, but also in optimising the allocation of scarce health care resources through the use of cost-effective productions methods. It is in response to all these challenges that biotechnology offers new strategies that go beyond the more traditional approaches.

By the mid-1990's, the number of recombinant products approved annually for therapeutic use reached double digits. With the advent of the genomics revolution, coupled with proteomics, bioinformatics, high-throughput screening, and microelectronics, the number of potential bioproducts entering clinical trials will surely grow almost exponentially. The improved availability of structural and functional biochemical information, together with rational drug design, will also help fill the drug development pipeline. And it should certainly not be forgotten that the increasing demand for recombinant compounds necessitates improved production systems.

In order to better focus on the challenges at hand, the Editors have striven to adopt an integrative approach by selecting contributions linking genomics with bacterial resistance, antibiotics with improved production strategies, optimised production approaches with chemical development programs, chemical modifications with new antibody derivatives and biomaterials, the relation between bioartificial organs and xenotransplantation, apoptosis and process biotechnology, etc. It should be readily apparent from these chapters that the development of truly novel compounds for biomedical use not only requires leading-edge science and technology, but also curiosity, talent and hard work.

Finally, the editors wish to express their gratitude to Karen Vyaene for her outstanding secretarial editing skills. Without her spirited efforts, this book could not have seen the light of day.

Annie Van Broekhoven Fred Shapiro Jozef Anné

TABLE OF CONTENTS

Part 1

GENOMICS: THE NEW APPROACH TO THE DISCOVERY OF NEW COMPOUNDS

THE GENOMICS APPROACH: IS IT REALLY THE SOLUTION?

RAUL GOLDSCHMIDT AND KAREN BUSH
R. W. Johnson Pharmaceutical Research Institute
1000 Route 202Raritan, NJ 08869 USA
FAX: 908-526-3047 E-mail: kbush@prius.jnj.com

Summary

Antimicrobial drug discovery has undergone major changes over the past 25 years, with a shift in emphasis from natural product screening for whole cell inhibitory activity to biochemical approaches. However, a major hurdle has been to identify novel scaffolds for antimicrobial agents. The availability of genomic sequences from pathogenic organisms now offers the opportunity for creative identification of novel targets. Many challenges remain before a useful antimicrobial agent can be identified.

1.The Traditional Approach

Antimicrobial drug discovery began with the identification of natural products that had the ability to kill bacteria in agar-filled Petri dishes. The Golden Age of antibiotic discovery was synonymous with the Golden Age of natural product research. Chemists who could isolate and characterise the active antibacterial component in a fermentation broth were the driving force of antibiotic discovery research.

Mechanism of action considerations for antibiotics were not valued in the early days of antibiotic discovery. Antimicrobial spectrum of activity was the over-riding factor that determined the development of new agents. If a bacterial infection was cured after treatment, the agent was considered a success. As a result, antibiotic discovery was heavily dependent upon whole cell screens capable of identifying novel agents from fermentation broth. Broths with antimicrobial activity were examined and the most active compound(s) isolated. The spectrum of activity was established and efficacy in animal models was determined. Only after many years did scientists then determine how the agents acted upon bacterial targets in order to effect killing.

However, even after fifty years of therapeutic antibiotic use, very few drug discovery programs can claim success in identifying new agents using approaches that differ from those used in the 1950s and 1960s. Chemists still base their structure-activity relationships (SAR) upon data obtained from whole cell assays. Attempts to make these

A. Van Broekhoven et al. (eds.), Novel Frontiers in the Production of Compounds for Biomedical Use, 15–22.
© 2001 *Kluwer Academic Publishers. Printed in the Netherlands.*

analyses more quantitative have not been fully successful, due to the multiple interacting factors that are involved in targeting a specific compound to its site of action. Even when a specific target is known for an antibiotic, such as the penicillin-binding proteins for the β-lactams, enzyme-binding data are useful only as a first approximation for eventual antimicrobial activity. Other factors, such as solubility, stability, permeability into bacteria, and pharmacokinetic parameters that contribute to the eventual utility of an agent, often take precedent over the inherent activity of the agent against the primary target.

After the identification of thienamycin (Kahan et al., 1979), the olivanic acids (Butterworth et al., 1979), the clavams (Brown et al., 1976), and the monobactams (Imada et al., 1981, Sykes et al., 1981), in the late 1970s, natural products research did not provide the pharmaceutical industry with the wealth of structures previously identified. Although many companies retained a natural product screening effort, the results have been disappointing. In the past twenty years no major novel class of antibiotic with useful commercial utility has been identified from a natural source, in spite of attempts to improve the quality of the screening samples, such as expansion of the natural product libraries to include novel fermentation conditions and sampling from unusual sites such as marine sources and tropical forests.

As a result, the focus within the last decade has been to identify novel enzyme targets for selective anti-bacterial agents. Screening for enzyme inhibitors, rather than for compounds that work against whole cells, has been deemed a better strategy by some, as enzyme assays can be more quantitative and more sensitive than whole cell assays. Enzyme assays also allow chemists to develop a more predictive SAR based on inhibition of a single target, an approach that has been notably successful in the identification of novel antiviral agents. Following this lead, "novel target" programs have become an integral part of antibacterial discovery programs.

2. The Biochemical Genetics Approach

In their search for novel targets, antibacterial programs began to devote increasing attention to bacterial components that are essential for either survival or pathogenesis. Characterisation of such components is never a simple chore but has been greatly facilitated by the development of genetic and molecular tagging techniques, cloning methodology, and methods to deliver recombinant DNA into a variety of microorganisms. Consequently, over 100 essential microbial genes have been identified. As a result of advances in protein overexpression technology, a number of essential bacterial proteins of well-characterised enzymatic activities have been used in high throughput assays to screen for inhibitors. None of these assays has yet yielded antibacterials that have reached clinical trials. However, it is premature to evaluate the success of this approach, since the number of examined targets is only a small fraction of the number of available targets.

3. The Genomics Approach

3.1 TARGET IDENTIFICATION

Antimicrobial target selection has taken a very different slant with the ready availability of databases of microbial genomic sequence and of a number of computational tools that allow processing that information. The genomics approach permits one to gain information relatively quickly about potential antimicrobial targets within clinically important microorganisms (Moir et al., 1999). Algorithms that measure the degree of similarity of nucleic acids or protein sequences with homologous sequences of known function in other microorganisms allow one to assign presumed functions to a significant proportion of the proteins encoded by the ORFs of newly sequenced bacterial genomes. Furthermore, other analyses such as comparisons of the genetic organisation of genes within conserved loci, or the presence of particular motifs within encoded proteins, allows further conjectural assignment of functions to the genes. By these means, it is possible to assign a function and annotate 40 to 50% of the genes in a newly sequenced bacterial genome.

As mentioned earlier, essentiality of novel targets has been valued by many as a requirement for a therapeutically useful agent. For some of the targets identified through genomics, essentiality is known to exist from published data. For others with known or novel unknown function, extensive molecular approaches (see above) are necessary to demonstrate that a gene must be functional for bacterial survival. Unlike earlier studies in which essentiality was evaluated only in vitro, it is now accepted that this essential nature may be expressed either *in vitro* or *in vivo*. In fact, many technological approaches now being developed to demonstrate that a novel gene function is required for replication in a mammalian host.

3.2. COMPARATIVE GENOMICS

There are at present 15 completed microbial sequences in the public domain (www-fp.mcs.anl.gov/~gaasterland/genomes.html; and www.tigr.org/tdb/mdb/mdb.html), of which ten correspond to pathogenic bacteria (Table 1), and two to microbes that albeit non-pathogenic (*Bacillus subtilis*, *Saccharomyces cerevisiae*) bear close genomic similarity to clinically important microorganisms (Goffeau et al., 1996, Kunst et al., 1997). There are additional 67 microbial genomes being sequenced (www.tigr.org/tdb/mdb/mdb.html), of which 37 correspond to pathogenic microbes. Of these, 35 partly sequenced genomes are public (Table 2) and 33 can be accessed for homology searches either directly or indirectly:

- www.ncbi.nlm.nih.gov/BLAST/ouacgtbl.html;
- www.tigr.org/tdb/mdb/mdb.html;
- www.sanger.ac.uk/DataSearch/.

There are also a few restricted databases available for commercial subscribers.

Table 1. Completed publicly available genomes of pathogenic *bacteria*[a]

Genome	Strain	Size (Mb)	Institution	Reference
Borrelia burgdorferi	B31	1.44	TIGR	(Fraser et al., 1997)
Campylobacter jejuni	NCTC 11168	1.7	Sanger Centre	(Unpublished)
Chlamydia trachomatis	serovar D (D/UW-3/Cx)	1.05	UC Berkeley/Stanford	(Stephens et al., 1998)
Escherichia coli	K-12	4.60	Univ. Wisconsin	(Blattner et al., 1997)
Haemophilus influenzae Rd	KW20	1.83	TIGR	(Fleischmann et al., 1995)
Helicobacter pylori	J99	1.64	Astra /GTC	(Alm et al., 1999)
H. pylori	26695	1.66	TIGR	(Tomb et al., 1997)
Mycobacterium tuberculosis	H37Rv (lab strain)	4.40	Sanger Centre	(Cole et al., 1998)
Mycoplasma genitalium	G-37	0.58	TIGR	(Fraser et al., 1995)
Mycoplasma pneumoniae	M129	0.81	Univ. Heidelberg	(Himmelreich et al., 1996)
Rickettsia prowazekii	Madrid E	1.10	Univ. Uppsala	(Andersson et al., 1998)
Treponema pallidum	Nichols	1.14	TIGR/ Univ. Texas	(Fraser et al., 1998)

[a]Data compiled from www-fp.mcs.anl.gov/~gaasterland/genomes.html and www.tigr.org/tdb/mdb/mdb.html and www.sanger.ac.uk/Projects/.

Table 2. Incomplete publicly available genomes of microbial pathogens[a]

Genome	Strain	Size (Mb)	Institution	Anticipated Completion
Actinobacillus actinomycetemcomitans	HK1651	2.2	Univ. Oklahoma	
Aspergillus nidulan[b]		29	Cereon Genomics	
Bordetella bronchiseptica	RB50	4.9	Sanger Centre	
Bordetella parapertussis		3.9	Sanger Centre	
Bordetella pertussis	Tohama I	3.88	Sanger Centre	
Candida albicans[b]	1161	15	Sanger Centre	
C. albicans[b] (1.5X)	SC5314		Stanford	
Chlamydia pneumoniae	CWL029	1.23	UC Berkeley	
C. pneumoniae		1.00	TIGR/Univ. Manitoba	
Chlamydia trachomatis	MoPn	1.00	TIGR/Univ. Manitoba	
Clostridium difficile	630	4.4	Sanger Centre	
Enterococcus faecalis	V583	3.00	TIGR	1999
Giardia lamblia[b,c]	WB	12	Marine Biological Lab.	
Francisella tularensis[c]	schu 4	2.00	European & North American Consortium	
Leishmania major[b] Chr3	Friedlin		SBRI	
Chr4	Friedlin	0.5	Sanger Centre	

Table 2. Incomplete publicly available genomes of microbial pathogens[a]

Genome	Strain	Size (Mb)	Institution	Anticipated Completion
Chr5,13,14,19,21,23	Friedlin		Sanger Centre/European Consortium	
L. major Chr27,35			SBRI	
Mycobacterium avium	104	4.70	TIGR	2000
Mycobacterium bovis	AF2122/97	4.4	Sanger Center	
Mycobacterium leprae		2.8	Sanger Centre	
Mycobacterium tuberculosis	CSU#93 (clinical isolate)	4.40	TIGR	1999
Neisseria gonorrhoeae		2.20	Univ. Oklahoma	
Neisseria meningitidis	MC58	2.30	TIGR	
	serogroup A strain Z2491	2.30	Sanger Centre	
Plasmodium falciparum[b]				
Chr1,3,4,5,6,7,8,9,13	3D7	0.8	Sanger Centre	
Chr12	3D7	2.4	Stanford University	
Chr14	3D7	3.4	TIGR/NMRI	
Porphyromonas gingivalis	W83	2.20	TIGR/ Forsyth Dental Center	1999
Pseudomonas aeruginosa	PAO1	5.90	Univ. Washington PathoGenesis	
Pseudomonas putida		5.00	TIGR	
Salmonella typhi		4.5	Sanger Centre	
Salmonella typhimurium	SGSC1412	4.80	Washington Univ. Consortium	
Schizosaccharomyces pombe			14 Sanger Centre	
Shewanella putrefaciens	MR-1	4.50	TIGR	
Staphylococcus aureus	COL	2.8	TIGR	
	8325	2.8	Univ. Oklahoma	
Streptococcus mutans	UAB159	2.20	Univ. Oklahoma	
Streptococcus pneumoniae	type 4	2.20	TIGR	1999
Streptococcus pyogenes	M1GAS	1.98	Univ. Oklahoma	
Trypanosoma brucei Chr I			Sanger Centre	
Trypanosoma b. rhodesiense	TREU 927/4	35	TIGR	
Vibrio cholerae	serotype O1, Biotype El Tor, strain N16961	2.50	TIGR	1999
Yersinia pestis	CO-92 Biovar Orientalis	4.38	Sanger Centre	

[a]Data from www.tigr.org/tdb/mdb/mdb.html, www-fp.mcs.anl.gov/~gaasterland/genomes.html, and www.sanger.ac.uk/Projects/[b]Eukaryote[c]Annotated ORFS and sequences available for reading and downloading, but no homology searches at the site.

It is the use of comparative genomics that has been most useful for antimicrobial drug discovery. For example, the sequences from both pathogenic and non-pathogenic bacteria have been evaluated by identifying homologous genes shared amongst these bacteria (Arigoni et al., 1998) and absent from eukaryotic genomes, represented by *Saccharomyces cerevisiae* (Koonin et al., 1997). The hypothesis is that highly conserved genes should make good targets for broad-spectrum agents. A similar comparison can also be useful for identifying targets that may be selective for narrow-spectrum agents. If genes from a single organism such as *Helicobacter pylori* or *Mycobacterium tuberculosis* could be identified with low homology to genes in other microbial genomes, it might be possible to identify inhibitors that were selective for the target bacteria and that did not have effects on commensal bacteria.

4. Perspectives for an Integrated Approach

Genomic sequence information is now being mined in an exponential fashion. Most major pharmaceutical companies have invested in some type of bioinformatics technology, with some companies making this a major focus of their antimicrobial efforts. Traditional drug screening is becoming less frequent as genomic targets are identified; many of the resulting screening techniques are now based on enzyme inhibition assays rather than whole cell assays. However, it is necessary for the identified inhibitors to function in whole cells and *in vivo* before an antibacterial agent can be developed.

As important as the genomics approach has become for antimicrobial discovery research, it is equally important that we recognise it as a tool, and as such, that we become familiar with its strengths and limitations. It is a tool to guide the evaluation of targets according to the degree of similarity found with homologous prokaryotic and eukaryotic proteins. Nevertheless, the range of parameters to guide us in such a selection is still largely unknown and may be target dependent. For example, it is desirable to select as targets bacterial proteins that show little or no homology with eukaryotic proteins. However, proteins such as DNA gyrase, RNA polymerase and dihydrofolate reductase show moderate homology with eukaryotic proteins and yet they have been used successfully as targets for antimicrobials. The increasing number of available three-dimensional protein structures (www.ncbi.nlm.nih.gov/Structure/) should ultimately allow one to refine the approaches to selectivity, by taking into account the nature of structural homologies among the bacterial proteins, and between bacterial and eukaryotic proteins.

Furthermore, from a biochemical perspective, the essentiality of a gene, tested by totally eliminating its function, may not necessarily be a good predictor for the effects of inhibitors on enzymes. However, at present we do not yet have any assessment of how important this may be. Although genomic approaches at this time have not resulted in the identification of novel agents, it is just four years since the first complete bacterial genomic sequences became available. The proof of principle of the approach will require screening a much larger number of targets. Through the use of genomics it should be possible to make more educated choices and avoid targets marred by redundancy. With the numbers of resources being directed into this area, eventually a successful drug candidate will be identified based on a clever, or lucky, use of genomics information. From a somewhat cynical perspective, the absence among the pipelines of pharmaceutical companies of antimicrobials against genomics-defined targets, may be a blessing of sorts. Indeed, at the present time neither the pharmaceutical industry nor the health service sector may be able accommodate multiple antimicrobials with novel modes of action entering the market simultaneously.

It is a somewhat mischievous twist that the genomics approach has brought us back to a situation not unlike that of the Golden Age. Indeed, mechanisms of action are once again of diminished importance when screening targets of unknown function identified by the genomics approach. At the same time, the ultimate discriminator is the ability of the identified compounds to prevent replication of the organism *in vivo* (if not also *in vitro*), and this is still a formidable challenge. However, in this process, our perception of the available targets has changed fundamentally, and we can anticipate a time in which all essential gene products from bacteria of clinical importance will have been screened for inhibitors.

References

Alm, RA, LS Ling, DT Moir, BL King, ED Brown, PC Doig, DR Smith, B Noonan, BC Guild, BL deJonge, et al. 1999. Genomic-sequence comparison of two unrelated isolates of the human gastric pathogen *Helicobacter pylori*. Nature (London). 397:176-80.

Andersson, SG, A Zomorodipour, JO Andersson, T Sicheritz-Ponten, UC Alsmark, RM Podowski, AE Naslund, AS Eriksson, HH Winkler, and CG Kurland. 1998. The genome sequence of *Rickettsia prowazekii* and the origin of mitochondria. Nature. 396:133-140.

Arigoni, F, F Talabot, M Peitsch, MD Edgerton, E Meldrum, E Allet, R Fish, T Jamotte, ML Curchod, and H Loferer. 1998. A genome-based approach for the identification of essential bacterial genes. Nature Biotechnology. 16:851-856.

Blattner, FR, GI Plunkett, CA Bloch, NT Perna, V Burland, M Riley, J Collado-Vides, JD Glasner, CK Rode, GF Mayhew, et al. 1997. The complete genome sequence of *Escherichia coli* K-12. Science. 277:1453-1474.

Brown, AG, D Butterworth, M Cole, G Hanscomb, JD Hood, C Reading, and GN Rolinson. 1976. Naturally occuring ß-lactamase inhibitors with antibacterial activity. J. Antibiot. 29:668-669.

Butterworth, D, M Cole, G Hanscomb, and GN Rolinson. 1979. Olivanic acids, a family of beta-lactam antibiotics with beta-lactamase inhibitory properties produced by *Streptomyces* species. I. Detection, properties and fermentation studies. J. Antibiotics. 32:287-294.

Cole, ST, R Brosch, J Parkhill, T Garnier, C Churcher, D Harris, SV Gordon, K Eiglmeier, S Gas, CEr Barry, et al. 1998. Deciphering the biology of *Mycobacterium tuberculosis* from the complete genome sequence. Nature (London). 393:537-544.

Fleischmann, RD, MD Adams, O White, RA Clayton, EF Kirkness, AR Kerlavage, CJ Bult, J-F Tomb, BA Dougherty, JM Merrick, et al. 1995. Whole-genome random sequencing and assembly of *Haemophilus influenzae* Rd. Science. 269:496-512.

Fraser, CM, S Casjens, WM Huang, GG Sutton, R Clayton, R Lathigra, O White, KA Ketchum, R Dodson, EK Hickey, et al. 1997. Genomic sequence of a Lyme disease spirochaete, *Borrelia burgdorferi*. Nature (London). 390:580-586.

Fraser, CM, JD Gocayne, O White, MD Adams, RA Clayton, RD Fleischmann, CJ Bult, AR Kerlavage, G Sutton, JM Kelley, et al. 1995. The minimal gene complement of *Mycoplasma genitalium*. Science. 270:397-403.

Fraser, CM, SJ Norris, GM Weinstock, O White, GG Sutton, R Dodson, M Gwinn, EK Hickey, R Clayton, KA Ketchum, et al. 1998. Complete genome sequence of *Treponema pallidum*, the syphilis spirochete. Science. 281:375-388.

Goffeau, A, BG Barrell, H Bussey, RW Davis, B Dujon, H Feldmann, F Galibert, JD Hoheisel, C Jacq, M Johnston, et al. 1996. Life with 6000 genes. Science. 274:546-567.

Himmelreich, R, H Hilbert, H Plagens, E Pirkl, BC Li, and R Herrmann. 1996. Complete sequence analysis of the genome of the bacterium *Mycoplasma pneumoniae*. Nucl. Acids Res. 24:4420-4449.

Imada, A, K Kitano, K Kintaka, M Muroi, and M Asai. 1981. Sufazecin and isosulfazecin, novel b-lactam antibiotics of bacterial origin. Nature. 289:590-591.

Kahan, JS, FM Kahan, R Goegelman, SA Currie, M Jackson, EO Stapley, TW Miller, AK Miller, D Hendlin, S Mochales, et al. 1979. Thienamycin, a new beta-lactam antibiotic. I. Discovery, taxonomy, isolation and physical properties. J. Antibiotics. 32:1-12.

Koonin, EV, AR Mushegian, MY Galperin, and DR Walker. 1997. Comparison of archaeal and bacterial genomes: computer analysis of protein sequences predicts novel functions and suggests a chimeric origin for the archaea. Molec. Microbiol. 25:619-637.

Kunst, F, N Ogasawara, I Moszer, AM Albertini, G Alloni, V Azevedo, MG Bertero, P Bessieres, A Bolotin, S Borchert, et al. 1997. The complete genome sequence of the Gram-positive bacterium *Bacillus subtilis*. Nature (London). 390:249-256.

Moir, DT, KJ Shaw, RS Hare, and GF Vovis. 1999. Genomics and antimicrobial drug discovery. Antimicrob. Agents Chemother. 43:439-446.

Stephens, RS, S Kalman, C Lammel, J Fan, R Marathe, L Aravind, W Mitchell, L Olinger, RL Tatusov, Q Zhao, et al. 1998. Genome sequence of an obligate intracellular pathogen of humans: *Chlamydia trachomatis*. Science. 282:754-759.

Sykes, RB, CM Cimarusti, DP Bonner, K Bush, DM Floyd, NH Georgopapadakou, WH Koster, WC Liu, WL Parker, PA Principe, et al. 1981. Moncyclic ß-lactam antibiotics produced by bacteria. Nature. 291:489-491.

Tomb, J-F, O White, AR Kerlavage, RA Clayton, GG Sutton, RD Fleischmann, KA Ketchum, HP Klenk, S Gill, BA Dougherty, et al. 1997. The complete genome sequence of the gastric pathogen *Helicobacter pylori*. Nature (London). 388:539-547.

THE CONTRIBUTION OF GENOMICS TO THE DISCOVERY OF NEW ANTIBIOTICS

DAVID J. HOLMES, JOHN P. THROUP, NICOLA G. WALLIS, MARTIN K. R. BURNHAM, MAGDALENA ZALACAIN, SANJOY BISWAS, ALISON F. CHALKER, KAREN A. INGRAHAM, ANDREA MARRA, ALEX BRYANT, GARY WOODNUTT, PATRICK V. WARREN, JAMES R BROWN, MARTIN ROSENBERG

Anti-Infectives Research Department, SmithKline Beecham Pharmaceuticals, Upper Providence, PA 19426, USA

Abstract

The emergence of common bacterial pathogens that are resistant to multiple antibiotics, coupled with the failure of traditional methods to yield new anti-infective agents, threatens current paradigms of therapeutic intervention (Omura, 1992; Shlaes *et al.*, 1991; Tenover and Hughes, 1996). The current focus has been on improving existing antibiotic classes while little progress has been made in discovering chemically novel anti-infective agents. This unmet medical need may now be addressed if we can successfully exploit the new wealth of genomic sequence data to devise novel strategies for drug discovery (Moir *et al.*, 1999). Whole genome comparative sequence analysis now allows the identification of genes/gene products that are common to many or all pathogenic bacteria of clinical importance. bioinformatics-based gene homology and motif analyses allow rapid phylogenetic comparisons to be made and, in addition, predict functional information critical to target selection. Putative targets can then be assessed rapidly using gene essentiality testing methods which allow analyses both *in vitro* and in models of the infection state. Sensitive and direct *in vivo* expression analyses confirm that the expression of the target gene is relevant to the establishment and maintenance of infection. Ultimately, the gene products must be screened against novel chemical libraries and natural product banks derived from a wide bio-diversity in order to identify lead compounds with potential antibiotic activities that will be developed to provide the next generation of therapeutic agents. Those companies which can adapt and implement these genomic-based technologies, derive significant competitive advantage in creating product portfolios that will address the unmet clinical need.

A. Van Broekhoven et al. (eds.), Novel Frontiers in the Production of Compounds for Biomedical Use, 23–31.
© 2001 *Kluwer Academic Publishers. Printed in the Netherlands.*

1. Introduction

Traditional methods for identifying antibiotics have involved challenging laboratory cultivated bacteria with novel compounds and assaying their bacteristatic or bactericidal qualities. Once an antibacterial entity had been isolated, mode of action studies would determine which essential cellular function was being inhibited by that compound. This approach was highly successful for several decades, but since the 1970's, no new classes of antibiotics have been identified (Fig 1). It is perhaps a reflection of the method, that currently available antibiotics inhibit a relatively small number of cellular processes involved in DNA, RNA, protein and cell wall biosynthesis (Gale *et al.,* 1981).

Although antibiotic discovery may have been on the decline, the same cannot be said for the ability of pathogenic bacteria to discover ways to become resistant. The appearance of methicillin-resistant *Staphylococcus aureus* (MRSA) and vancomycin resistant *Enterococcus faecium* (VREF) as well as multiple drug resistance in clinically important bacteria (Table 1) has made it essential that new antibiotics be found (Swartz, 1994). Given the recent advances in molecular target screening technologies, assays are typically designed to be performed rapidly in high throughput using minimal volumes, thereby using smaller amounts of reagents. Consequently, an enzyme target, for example, can be screened against hundreds of thousands of compounds in a short timeframe to identify chemicals which functionally interact with the isolated protein.

3. The Properties of an Antibacterial Target

In order to make a good candidate target for therapeutic intervention, a gene product should possess a number of attributes. New antibacterial compounds will be more likely obtained by screening against diverse, novel proteins that are not the target of known antibiotics. With access to multiple genome sequences, it is possible to determine the bacterial spectrum of any given gene, and consequently, make predictions about the profile (broad/narrow spectrum) of a novel antibiotic compound. Of course, to be a feasible target, a gene must be essential for cell viability, and expressed during infection. Precise knowledge of every gene sequence allows radically new approaches to studying transcription during infection and gene essentiality. Finally, discovery of lead compounds results from high-throughput screening of chemical libraries that require robust and reproducible assays.

3.1 NOVELTY

If new classes of chemical entities having antibiotic activity are to be found, hitherto unexploited target enzymes involved in vital cellular process such as DNA replication, cell division, protein synthesis, cell wall biosynthesis, intermediary metabolic pathways and signal transduction must be identified and their specific function determined (Apfel *et al.,* 1999). Genome sequences supply all of the possible targets for therapeutic intervention. All of the genes involved in essential cell processes are there to be identified and exploited and in many cases this has already been done. In addition, whole genome information permits the classification of novel genes whose products will be, for example, new biosynthetic enzymes, transcription factors, or structural proteins. Such novel protein targets with new enzymatic activities are being identified and are likely to be inhibited by different chemotypes than those already in service for antibiotic therapy.

3.2 SPECTRUM/SELECTIVITY

The current unmet clinical need is for a broad-spectrum antibiotic to complement or replace current chemotherapies such as methicillin, amoxycillin and vancomycin. Comparison of multiple bacterial genome sequences that are now available (table 1) allows the identification of targets that are present in all clinically relevant pathogens and, equally important, have no homologue in the higher eukaryotes and certainly not in humans (Tatusov, *et al.*, 1997; Arigoni *et al.*, 1998). Moreover, the use of multiple genome sequences also permits targeted chemotherapy. With genomics and bioinformatics the spectrum of a potential target can be determined before inhibitors are identified rather than afterwards, as has been the case traditionally. Thus, not only broad spectrum, but specifically Gram positive or Gram negative specific genes can be targeted. Indeed, in some cases, such as *Helicobacter pylori* it may be desirable to screen a species-specific protein (Tomb *et al.*, 1997). In this manner, specific disease states could be treated without necessarily affecting the entire bacterial flora.

3.3 EXPRESSION DURING INFECTION

If a gene product is to be a target relevant for antibacterial therapeutic intervention, then it must be expressed during infection. The growth environment of a bacterium on agar plates or in liquid culture differs radically from the more natural surroundings of an animal host. However, studies of gene expression have historically been performed on *in vitro* grown cultures. While this has yielded valuable information, it does not give us a true picture of bacterial physiology during infection. It is not sufficient to assay the activity of a novel antibiotic on agar plates if it is not known whether the gene product being inhibited under laboratory conditions is also expressed in infection. Now, with genomic technology it is possible not only to get data of global gene expression in animal models of infection (Swartley *et al.*, 1998; DeRisi, *et al.*, 1997), but also to obtain those data at the relevant time during infection when treatment would commence. In order to determine this, infected tissue is isolated from an animal model and the total RNA extracted. Since the entire genome has been sequenced, selective reverse transcriptase PCR (RT-PCR) can be used to determine global gene expression using specific primer pairs to every gene in the genome. With suitable controls to ensure primer efficiency and that no bacterial chromosomal DNA is present in the sample, comparative studies can be made of global transcription between agar grown cells (*in vitro* grown) and cells from an infection model (*in vivo* grown). Moreover, it is possible to contrast the transcription levels during the early stages of infection (establishment of infection) with those in a late stage infection (maintenance of infection).

3.4 ESSENTIAL FOR CELL VIABILITY

Any potential target must be essential for cell viability during infection. A variety of techniques have been developed for testing essentiality both *in vitro* and in some cases in infection models. For example, random mutants can be generated in which colony formation is conditional for *in vitro* growth. In one case, a library of temperature sensitive mutants can be generated and each selected mutant complemented with a wild-type gene library to identify the mutated gene. In another case, a tagged transposon-based system (STM) has been developed to identify genes that are essential for infection (Hensel, *et al.*,1995; Mei, *et al.*, 1997). Only mutated genes dispensable for growth *in vitro* are identified and thus most are virulence factors. Due to the paucity of genetic information in pathogenic bacteria, these random mutagenic methodologies were the only way to identify genes vital for cell viability.

The availability of entire pathogen genome sequences, now allows a systematic gene-by-gene analysis of essentiality. A number of methods are available for knocking out genes in a directed fashion. The simplest form is plasmid insertion mutagenesis (Apfel *et al.*, 1999). In this method, a small (500 bp) internal fragment of the target gene is cloned in a non-replicative plasmid harbouring an antibiotic resistance determinant that is expressed in the host organism. This plasmid is introduced into the bacterium and drug-resistant colonies identified. Maintenance of the plasmid DNA requires homologous recombination into the target gene, thereby disrupting it. The method is quick and simple, but has caveats. Since most bacteria require approximately 500 bp for efficient recombinational insertion, significant portions of the "disrupted" gene may remain intact. The method is particularly inappropriate for small genes, and also raises the spectre of partial gene activity being expressed from the intact portions of the disrupted open reading frame. Partial activity of an essential gene may be sufficient for cell viability and thus single crossover insertional inactivation may miss essential genes (i.e. false negative analysis). Moreover, insertional mutagenesis requires integration of the entire vector into the chromosome and this may well effect the expression of genes downstream of the target (commonly referred to as polarity effects). As a result, a gene may be thought essential when in fact it is some distal gene that is actually vital for cell growth (i.e. false positive analysis).

A better approach requires removal of the gene to assure its inactivation and replacement with a small resistance determinant designed not to produce polarity effects (Norgren *et al.*, 1995; Hynes *et al.*, 1995; Buckley *et al.*, 1995;O'Connell *et al.*, 1993). While these experiments cannot prove that a gene is essential, they can certainly eliminate those genes which are not, thereby focusing attention on the potential antibacterial targets.

3.5 AMENABLE TO HIGH-THROUGHPUT SCREENING

The purpose of all of these efforts is to identify targets which can be screened in order to discover new chemical entities that will inhibit bacterial growth. In an attempt to analyse a given target for inhibition by hundreds of thousands of synthetic compounds and natural product extracts, the target protein must perform in a biochemical or cell-based assay that is robust and amenable to very high-throughput. Therefore an assayable activity of the novel target protein must be known (although knowledge of the exact cellular function is not necessarily required) and reagents readily available.

Below we will describe two kinds of targets which exemplify the ways in which genomics has effected anti-infective drug discovery.

4. Aminoacyl tRNA synthetases

Translation of messenger RNA into protein requires precise recognition of the 64 triplets by the appropriate tRNA charged with the cognate amino acid. This charging is performed by the aminoacyl tRNA synthetases in a two step reaction shown empirically below.

Step 1: Enzyme recognition and activation
E + aa + ATP -> E(aa-AMP) + PPi
Step 2: Enzyme transfer
E(aa-AMP) + tRNA -> aa-tRNA + AMP

The first step involves the formation of an enzyme-aminoacyladenylate complex in which the amino acid reacts with ATP, releasing pyrophosphate (step 1). In the second step, the 3' terminal adenosine of the cognate tRNA is linked to the aminoacyladenylate in an esterification reaction that releases AMP (step 2) (Meinnel *et al.*, 1995).

In Gram negative bacteria there is a specific enzyme for every amino acid which performs an esterification reaction between the tRNA molecule and it's cognate amino acid. Gram positive bacteria possess only 19 aminoacyl-tRNA synthetases. They lack the glutaminyl-tRNA synthetase (Lapointe *et al.*, 1986; Schon *et al.*, 1988). Instead, $tRNA^{gln}$ is glutamylated by the GluRS enzyme. The misacylated tRNA is then the substrate for a transamidase which converts the glutamyl adduct $tRNA^{Gln}$ (Schon *et al.*, 1988). Each of the tRNA synthetases are essential for cell viability and indeed one of them, the isoleucyl-tRNA synthetase is the target of pseudomonic acid, a valuable topical antibiotic with potent Gram positive activity including MRSA (Bactroban™).

Prior to 1994, aminoacyl-tRNA synthetases had only been well defined in *E. coli* K12, and little was known about them in Gram positive bacteria. Indeed, only 2 such enzymes (isoleucyl tRNA synthetase (Grundy *et al.*, 1997) and lysyl-tRNA synthetase (Green and Vold, 1994)) had been identified in *S. aureus* and none had been isolated from *S. pneumoniae.* Given the knowledge that there are a limited number of aminoacyl-tRNA synthetases makes the task of isolating each one of them by standard biochemical and genetic methods using appropriate DNA probes a plausible task. However, this would be time consuming and labour intensive whereas knowledge of the entire genome sequence allowed us to immediately identify all 19 of the aminoacyl-tRNA synthetase genes from both *S. aureus* and *S. pneumoniae in silico* by comparison with those from *E. coli.* In fact, with the considerable advances in sequencing technology that have reduced the cost of large-scale sequencing programs, it is a practical economic option. Using standard research methods for gene expression all of these proteins were made available as tools and relevant high throughput assays could be developed. The enzymes were screened against file compounds, combinatorial chemistry libraries and natural product extracts and lead molecules identified.

5. Two Component Signal Transduction Systems

Arguably the most widespread mechanism used by bacteria to sense and respond to environmental stimuli is through the use of two component signal transduction systems (TCSTS) (Parkinson and Kofoid, 1992; Stock *et al.*, 1989). Generally composed of a histidine kinase sensor protein and a phosphorylatable (DNA- binding) response regulating protein, TCSTS are known to be involved in a number of physiological processes including competence (Havarstein *et al.*, 1996, Magnisson *et al.*, 1994; antibiotic production (Chang *et al.*, 1996), chemotaxis (Falke *et al.*, 1997), osmoregulation (Russo and Silhavy, 1991) and sporulation (Tzeng *et al.*, 1998) in a wide range of bacteria, although they are apparently absent from higher eukaryotes. In response to specific stimuli, the sensor protein autophosphorylates at a specific histidine residue. The high-energy phosphate is then transferred to a conserved aspartate residue which lies within a cognate response regulator protein. This transfer activates the response regulator which is then able to mediate changes in gene expression or protein function (Egger *et al.*, 1997). In this manner, TCSTS control the expression of whole subsets of genes and operons, some of which may code for regulatory proteins. Thus, through this mechanism, a complex cascade of genetic control mechanisms can result from the recognition of a single environmental stimulus.

When pathogenic bacteria infect a host, they are confronted with dramatic changes in environmental conditions (oxygen/pH/nutrient stress) as well as being challenged in most cases by the host defence systems. It is reasonable to suggest that disruption of the TCSTS signalling systems would impair the bacteria's capacity to respond to it's environment affecting both the viability of the cell and it's ability to establish and maintain infection. Thus, the TCSTS of bacteria have the potential to be appropriate targets, the inhibition of which would impede the infection process.

Both histidine kinases and response regulators can be readily identified by sequence comparisons, and it is possible to isolate heterologous TCSTS gene pairs using appropriate DNA fragments as probes. However, before the advent of whole genome sequencing, it would have been impossible to know whether all of the TCSTS had been cloned from a particular bacterial species. Unlike aminoacyl-tRNA synthetases, it is difficult to estimate the number of TCSTS required by any particular microorganism. Examination of the entire genome sequence of *Streptococcus pneumoniae* 0100993 identified 14 TCSTS. Only three of these had been previously characterised and shown to be involved in the process of competence for DNA uptake (Guenzi *et al.*, 1994; Pestova *et al.*, 1996; Novak *et al.*, 1999). We have been able to identify 11 novel TCSTS and have investigated their requirement for *in vitro* cell viability as well as their role in the establishment and maintenance of infection. Each gene couple was examined and allelic replacement experiments performed (table 3). In certain cases individual genes were deleted, while in others the gene pair was deleted and replaced by the resistance marker *ermAM*. In three cases, after multiple attempts it was impossible to obtain a deletion mutant. Of these three TCST systems, one was shown to be essential for the viability of *S. pneumoniae in vitro.* When the response regulator was supplied on a plasmid vector it proved possible to delete the gene copy in the chromosome. This gene belongs to a system that is highly homologous to the *yycFG* TCSTS recently shown to be essential for *in vitro* cell growth in *Bacillus subtilis* (Fabret and Hoch, 1998). Of the 21 deletion mutants that were generated, 13 were tested in a murine respiratory tract infection model. Of these, a further four mutants were severely attenuated in a respiratory tract infection model (table 3). Thus, we have identified one novel TCSTS of *S. pneumoniae* that is essential for *in vitro* growth and a further two that may be essential. Moreover, there are 4 additional systems that, although cells grow *in vitro,* cannot cause disease in an RTI model.

6. Discussion

Antibiotic drug discovery in the genomic era clearly differs from the traditional methods of screening and mode of action studies. The wealth of sequence data is allowing us to identify new genes, assess their importance in the infection process and their essentiality for cell viability. Armed with a knowledge of the genes that are essential for growth, we can devise biochemical assays for the gene products that can be run in a high throughput format. Given the advances in screening technologies and the use of robotics, an enormous number of chemical entities can be assessed for their inhibitory activity against any given target protein. With so many targets to choose from, we can afford to select molecules with excellent profiles of spectrum, and that do not effect eukaryotic cells.

However, turning lead compounds into antibiotic medicines of use to the community remains a long process. Molecular screening technologies define inhibitory compounds in an *in vitro* assay. Intensive chemistry is then performed to establish structure-activity relationships with the target in order to improve potency. To have antibiotic activity, these molecules (often) must also penetrate the bacteria to reach their target and cause cell death. Consequently, active lead compounds are immediately profiled for their antibiotic activity against a range of pathogens. Having obtained an antimicrobial compound that specifically and potently inhibits a known target, the development process begins. Consequently, it is years before pharmacokinetics, pharmacodynamics, toxicology, and clinical testing are completed and a new antibiotic is demonstrated to be safe and effective in man. So while we have been able to expedite the identification of novel antibacterial targets and discover new lead compounds through high throughput screening, the development process remains long and arduous. Meanwhile, the march of bacterial antibiotic resistance continues.

References

Apfel, C.M., Takacs, B., Fountoulakis, M., Stieger, M., and Keck, W. (1999) Use of genomics to identify bacterial undecaprenyl pyrophosphate synthetase: cloning, expression, and characterization of the essential *uppS* gene. *J Bacteriol 181(2)* 483-492.

Arigoni, F., Talabot, F., Peitsch, M., Edgerton, M.D., Meldrum, E., Allet, E., Fish, R., Jamotte, T., Curchod, M.L., and Loferer, H. (1998) A genome-based approach for the identification of essential bacterial genes. *Nat Biotechnol 16(9)*851-856.

Buckley, N.D., Lee, L.N., and LeBlanc, D.J. (1995) Use of a novel mobilizable vector to inactivate the *scrA* gene of *Streptococcus sobrinus* by allelic replacement. *J Bacteriol 177(17)* 5028-5034.

Chang, H.M., Chen, M.Y., Shieh, Y.T., Bibb, M.J., and Chen, C.W. (1996) The cutRS signal transduction system of *Streptomyces lividans* represses the biosynthesis of the polyketide antibiotic actinorhodin. *Mol Microbiol 21(5)* 1075-1085.

DeRisi, J.L., Iyer, V.R., and Brown, P.O. (1997) Exploring the metabolic and genetic control of gene expression on a genomic scale. *Science 278(5338)* 680-686.

Egger, L.A., Park, H., and Inouye, M. (1997) Signal transduction via the histidyl-aspartyl phosphorelay. *Genes Cells 2(3)* 167-184.

Fabret, C., and Hoch, J.A. (1998) A two-component signal transduction system essential for growth of *Bacillus subtilis*: implications for anti-infective therapy. *J Bacteriol 180(23)* 6375-6383.

Falke, J.J., Bass, R.B., Butler, S.L., Chervitz, S.A., and Danielson, M.A. (1997) The two-component signaling pathway of bacterial chemotaxis: a molecular view of signal transduction by receptors, kinases, and adaptation enzymes. *Annu Rev Cell Dev Biol 13* 457-512.

Gale E.F., Cundliffe, E., Reynolds, P.E., Richmond, M.H., and Waring, M.J. (1981) The molecular basis of antibiotic action. 2nd ed. John Wiley and Sons, London, United Kingdom.

Green,C.J. and Vold,B.S. (1994) An unusual rRNA-tRNA gene organization in *Staphylococcus aureus*. GenBank Accession number L36472

Grundy, F.J., Haldeman, M.T., Hornblow, G.M., Ward, J.M., Chalker, A.F. and Henkin, T.M. (1997) The *Staphylococcus aureus ileS* gene, encoding isoleucyl-tRNA synthetase, is a member of the T-box family. *J Bacteriol 179(11)* 3767-3772.

Guenzi, E., Gasc, A.M., Sicard, M.A., and Hakenbeck, R. (1994) A two-component signal-transducing system is involved in competence and penicillin susceptibility in laboratory mutants of *Streptococcus pneumoniae*. *Mol Microbiol 12(3)* 505-515.

Havarstein, L.S., Gaustad, P., Nes, I.F., and Morrison, D.A. (1996) Identification of the streptococcal competence-pheromone receptor. *Mol Microbiol 21(4)* 863-869.

Hensel, M., Shea, J.E., Gleeson, C., Jones, M.D., Dalton, E., and Holden, D.W. (1995) Simultaneous identification of bacterial virulence genes by negative selection. *Science 269(5222)* 400-403.

Hynes, W.L., Hancock, L., and Ferretti, J.J. (1995) Analysis of a second bacteriophage hyaluronidase gene from *Streptococcus pyogenes*: evidence for a third hyaluronidase involved in extracellular enzymatic activity. *Infect Immun 63(8)* 3015-3020.

Lapointe, J., Duplain, L., and Proulx, M. (1986) A single glutamyl-tRNA synthetase aminoacylates tRNAGlu and tRNAGln in *Bacillus subtilis* and efficiently misacylates *Escherichia coli* tRNAGln1 *in vitro*. *J Bacteriol 165(1)* 88-93.

Magnuson, R., Solomon, J., and Grossman, A.D. (1994) Biochemical and genetic characterization of a competence pheromone from *B. subtilis*. *Cell 77(2)* 207-216

Mei, J.M., Nourbakhsh, F., Ford, C,W., and Holden D.W. (1997) Identification of *Staphylococcus aureus* virulence genes in a murine model of bacteraemia using signature-tagged mutagenesis. *Mol Microbiol 26(2)* 399-407.

Meinnel, T., Mechulam, Y., and Blanquet, S. (1995) tRNA: Structure, biosynthesis and Function. ed. Soll, D., and RajBhanbary, U. ASM Washington, DC, USA.

Moir, D.T., Shaw, K.J., Hare, R.S., and Vovis, G.F. (1999) Genomics and antimicrobial drug discovery. *Antimicrob Agents Chemother 43(3)* 439-446.

Norgren, M., Caparon, M.G., and Scott, J.R. (1989) A method for allelic replacement that uses the conjugative transposon Tn916: deletion of the *emm6.1* allele in *Streptococcus pyogenes* JRS4. *Infect Immun (12)* 3846-3850.

Novak, R., Cauwels, A., Charpentier, E,, and Tuomanen, E. (1999) Identification of a *Streptococcus pneumoniae* gene locus encoding proteins of an ABC phosphate transporter and a two-component regulatory system. *J Bacteriol 181(4)* 1126-1133.

O'Connell, C., Pattee, P.A., and Foster, T.J. (1993) Sequence and mapping of the *aroA* gene of *Staphylococcus aureus* 8325-4. *J Gen Microbiol 139* 1449-1460.

Omura, S. (1992) Thom Award Lecture. Trends in the search for bioactive microbial metabolites. *J Ind Microbiol* 10*(3-4)* 135-156.

Parkinson, J.S., and Kofoid, E.C. (1992) Communication modules in bacterial signaling proteins. *Annu Rev Genet 26* 71-112.

Pestova, E.V., Havarstein, L.S., and Morrison, D.A. (1996) Regulation of competence for genetic transformation in *Streptococcus pneumoniae* by an auto-induced peptide pheromone and a two-component regulatory system. *Mol Microbiol 21(4)* 853-862.

Russo, F.D., and Silhavy, T.J. (1991) EnvZ controls the concentration of phosphorylated OmpR to mediate osmoregulation of the porin genes. *J Mol Biol 222(3)* 567-580.

Schlaes, D., Levy, S., and Archer G. (1991) Antimicrobial resistance: new directions. *ASM news 57* 455-458.Tatusov, R.L., Koonin, E.V., and Lipman, D.J. (1997) A genomic perspective on protein families. *Science 278(5338)* 631-637.

Tenover, F.C., and Hughes, J.M. (1996) The challenges of emerging infectious diseases. Development and spread of multiply-resistant bacterial pathogens. *JAMA 275(4)* 300-304.

Schon, A., Kannangara, C.G., Gough, S., and Soll, D. (1988) Protein biosynthesis in organelles requires misaminoacylation of tRNA. *Nature 331(6152)* 187-190.

Stock, J.B., Ninfa, A.J., and Stock, A.M. (1989) Protein phosphorylation and regulation of adaptive responses in bacteria. *Microbiol Rev 53(4)* 450-490.

Swartley, J.S., Liu, L.J., Miller, Y.K., Martin, L.E., Edupuganti, S., and Stephens, D.S. (1998) Characterization of the gene cassette required for biosynthesis of the (alpha1-->6)-linked N-acetyl-D-mannosamine-1-phosphate capsule of serogroup A *Neisseria meningitidis. J Bacteriol 180(6)* 1533-1539.

Swartz, M.N. (1994) Hospital-acquired infections: diseases with increasingly limited therapies. *Proc Natl Acad Sci U S A 91(7)* 2420-2427.

Tomb, J.F., White, O., Kerlavage, A.R., Clayton, R.A., Sutton, G.G., Fleischmann, R.D., Ketchum, K.A., Klenk, H.P., Gill, S., Dougherty, B.A., Nelson, K., Quackenbush, J., Zhou, L., Kirkness, E.F., Peterson, S., Loftus, B., Richardson, D., Dodson, R., Khalak, H.G., Glodek, A., McKenney, K., Fitzegerald, L.M., Lee, N., Adams, M.D., Hickey, E.K, Berg, D.E, Gocayne, J.D., Utterback, T.R, Peterson, J.D., Kelley J.M., Cotton, M.D., Weidman, J.M., Fujii, C., Bowman, C., Watthey, L., Wallin, E., Hayes, W.S., Borodovsky, M., Karp, P.D., Smith, H.O., Fraser, C.M. and Venter, J.C. (1997) The complete genome sequence of the gastric pathogen *Helicobacter pylori. Nature 388(6642)* 539-547.

Tzeng, Y.L., Zhou, X.Z., and Hoch, J.A. (1998) Phosphorylation of the Spo0B response regulator phosphotransferase of the phosphorelay initiating development in *Bacillus subtilis. J Biol Chem 273(37)* 23849-23855.

Part 2

ANTIBIOTICS

THE ANTIBIOTIC GALLIDERMIN - EVOLUTION OF A PRODUCTION PROCESS

MARKUS KEMPF, UWE THEOBALD AND
HANS-PETER FIEDLER

1.The antibiotic gallidermin

In 1983 Devrise *et al.* isolated a *Staphyloccocus* strain from the skin of chicken that was named *Staphylococcus gallinarum.* Five years later Kellner *et al* (1988) investigated cultures of *S. gallinarum* Tü 3928 and found the peptide antibiotic gallidermin (Figure 1).

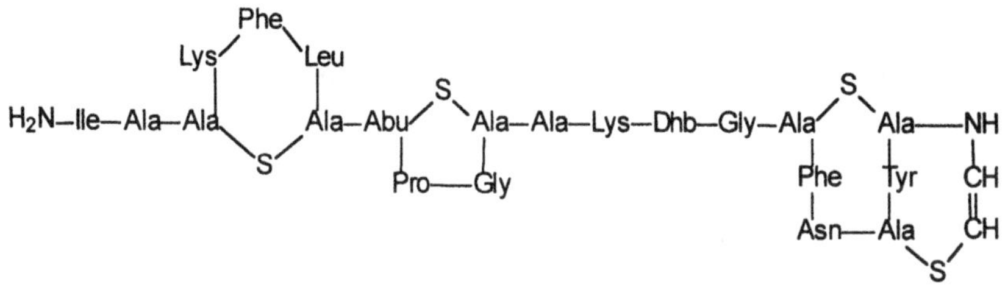

Figure 1: Structure of the antibiotic gallidermin

As shown in Figure 1 gallidermin, as a result of posttranslantional enzymatic modifications contains some unusual amino acid residues such as lanthionine, β-methyllanthionine or α,β-didehydroamino acids. These amino acids building intramolecular thioether bridges are used for the chemical characterisation of a group of peptide antibiotics called lantibiotics (Jung, 1991). Other renowned members of this group are e.g. nisin, subtilin, pep5, epidermin, mersacidin or actagardin (Jack *et al.*, 1998).

According to major structural variations and to different modes of action, lantibiotics are subdivided into an A-type and a B-type. Gallidermin as a type-A lantibiotic is of an elongated structure and its mode of action is the formation of pores into the cytoplasmatic membrane of Gram-positive bacteria (Benz *et al.*, 1991). Type-B lantibiotics in contrast show a more globular structure and do not form pores. Mersacidin and the closely related actagardin are two examples of type-B lantibiotics that inhibit the peptidoglycan biosynthesis at the level of transglycosylation (Börtz *et al.*, 1997). Both antibiotics exhibit a strong activity against the global spreading multi-drug resistant Staphylococcus aureus strains (MRSA) and thus are important candidates for a

A. Van Broekhoven et al. (eds.), Novel Frontiers in the Production of Compounds for Biomedical Use, 35–55.
© 2001 *Kluwer Academic Publishers. Printed in the Netherlands.*

probable new generation of clinically used antibiotics. The mode of action of mersacidin and actagardin is similar to that of the " last-defence-line-antibiotic" vancomycin (Börtz *et al.*, 1998). Although these antibiotics also are active against vancomycin resistant pathogens they target the same cellular structure and thus resistant strains probably will evolve faster than by the use of an antibiotic with a completely different mode of action. Gallidermin in contrast shows a bactericidal anti-MRSA activity (including vancomycin resistance) based upon its pore forming ability. Due to this bactericidal effect combined with the uncommon mode of action gallidermin is a very promising candidate for the treatment of several other crucial infections such as endocarditis, abscesses or the human acne disease.

These pharmacological properties of gallidermin led to extensive investigations in the last decade to elucidate the biosynthetic pathway of gallidermin formation (*Bierbaum et al.*, 1996; Sahl *et al.*, 1995) and to optimise the production process of this drug for industrial purposes (Hörner *et al.*, 1990; Kempf *et al.*, 1999a/b; Kempf *et al.*, 2000).

2. The improvement of the production process

From its discovery until 1995 the gallidermin concentrations during fermentation could be increased mainly by a variation of the production medium from initially 50 mg/l to 220 mg/l (Brecke *et al*, 1995). At this stage of optimisation three major disadvantages were identified:

- a low reproducibility of the entire production process that seriously interfere with further optimisation steps
- a very cost intensive production medium based on meat extract
- a non-optimised scale-up procedure into the dimension of industrial scales.

In order to eliminate these disadvantages further research investigations were focused on this topics.

3. Stabilisation of product-formation

A production process in general has to comply with some major prerequisites. One of the most important is the reproducibility of the whole process. The critical points that may alter the quantity or quality of the final product in such systems are transitions of any kind (e.g. a change from solid to liquid cultivation or each scale-up procedure; Humphrey, 1998). So, the whole process could be divided into several consecutive smaller sections where the quality of each input has a big influence upon the results obtained at the end. Particularly the very early stages of e.g. a fermentative metabolite production, should be well optimised to avoid an accumulation of disadvantages over the whole production period.

In the case of gallidermin, a creeping loss in its formation was observed in liquid cultures in-between a period of time of about half a year. These liquid cultures were inoculated from agar slants used for stock cultures which could be identified to be the reason of the unstable gallidermin production. This negative effect finally could be traced back to the medium composition of the stock culture medium.

In order to eliminate such a fluctuating productivity and thus to avoid problems during a scale-up procedure, a medium development had to be carried out.

4. Stability test

To investigate the suitability of a certain medium composition for stock cultures, agar slants were inoculated and incubated for 24 h. After this incubation time the next plate of the same medium was inoculated from this plate. Such passages were repeated up to 12 times. Each incubation period was followed by the inoculation of a liquid culture containing the same production medium (an illustration of the procedure is given in Figure 2).

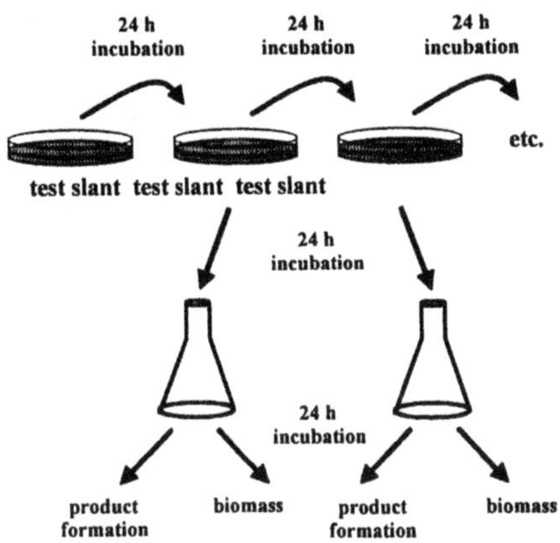

Figure 2: Experimental set-up of the suitability test (for details see text)

To facilitate the handling of the huge amount of data generated in those experiments, a parameter designated as hs-value (high and stable product formation) could be established to analyse the effect of the tested medium formulations on product formation and reproducibility in liquid cultures (Theobald & Kempf, 1998).

5. Hs-value

The total number of passages carried out with one medium composition depended on the obtained product concentrations and yields of the inoculated liquid cultures. Media resulting in low product concentrations (< 200 mg/l) or in dropping C_P or $Y_{P/X}$ values with a repeated loss in the average range of >5 % of the values obtained from the passages before were disregarded (at least 4 passages have been made with every medium). Liquid cultures from stock culture media creating values better than those threshold values were observed up to 12 passages.

To facilitate analysis of the generated data of in all 53 different media formulations the hs-value was developed that reduces the huge amounts of data to one single number for each medium composition:

$$hs - value = \left(\overline{Y_{P/X}} + \overline{\Delta Y_{P/X}}\right) \cdot \overline{C_P} \text{ [mg}^2\text{g}^{-1}\text{l}^{-1}\text{]} \tag{1}$$

$\overline{Y_{P/X}}$ overall average of product yields $Y_{P/X}$ [mg/g]
$\overline{\Delta Y_{P/X}}$ overall average of the difference between two consecutive $Y_{P/X}$ values [mg/g]
$\overline{C_P}$ overall average of maximum product concentration [mg/l].

For each medium composition one single hs-value for all passages made was calculated using the corresponding data at 24 hours of fermentation.

Due to the aim of reaching high product concentrations, the most prominent parameter in this formula is $\overline{C_P}$ the over all average of maximum product concentration obtained after all passages. A reduction of product formation during a few passages will reduce the value of $\overline{C_P}$ and thus the assessment of a medium composition would be lowered as it would happen by a low over all production.

In general, high $\overline{C_P}$ values could be reached in two different ways (or a mixture of both):

a) a drastically increased biomass (C_X) formation without any change in product yields ($Y_{P/X}$)
b) increased product yields ($Y_{P/X}$) without any change in biomass formation (C_X).

The first case (a) should be avoided to prevent an optimisation of biomass formation instead of product formation and thus to channel the available energy of the liquid production medium into product formation. To transform the mentioned into mathematical terms $\overline{C_P}$ was multiplied with the parameter $\overline{Y_{P/X}}$. $\overline{Y_{P/X}}$ is the overall average of $Y_{P/X}$ (the product yield) for one medium determined throughout all passages. Thus, $Y_{P/X}$ is the best available parameter to qualify the physiological condition of a culture. $\overline{Y_{P/X}}$ was modified by the addition of $\overline{\Delta Y_{P/X}}$. $\overline{\Delta Y_{P/X}}$ is the overall average of the difference of consecutive $Y_{P/X}$ values indicating a rising, dropping or equal tendency of during all passages. Cultures reaching low but during all passages stable $Y_{P/X}$ values will be assessed better than cultures with decreasing $Y_{P/X}$ values although both may produce the same amounts of product. So, a change in biomass formation as origin of stable or rising product concentrations could be uncovered this way.

The range of the hs-values obtained with the tested media was found to be at any value between zero mg^{2}g^{-1}l^{-1}, indicating no growth and/or no production at all, and positive values depending upon the product and biomass concentrations gained with a certain medium. The tested medium compositions reached values of nearly 4.000 $mg^2g^{-1}l^{-1}$ for the most suitable formulations. It should be mentioned that the unit $mg^2g^{-1}l^{-1}$ has no correspondence in physiological terms, thus being a completely artificial construct.

A comparison of gallidermin production and the corresponding hs-value of some media differing in the major components is shown in Figure 3.

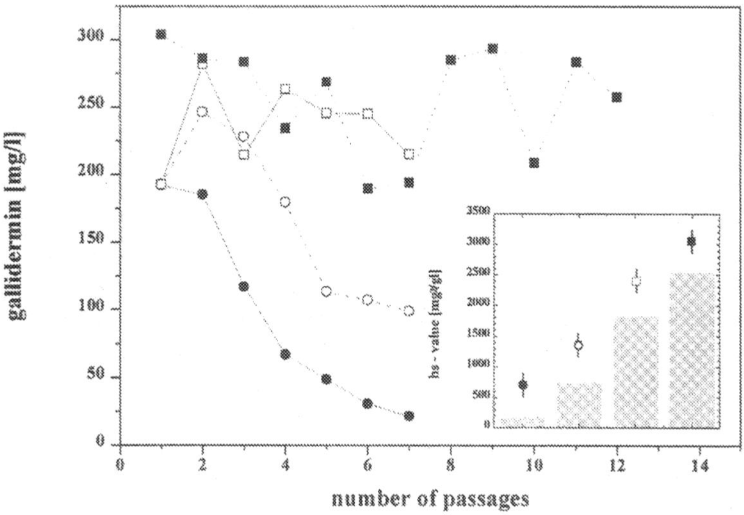

Figure 3: Gallidermin concentrations and corresponding hs-values obtained from a suitability test shown for four different stock culture media during several passages. -○-: medium 8 (BM, the former medium used for stock cultures), -●-: medium 15 (containing yeast extract Ohly KAT at a concentration of the production medium), -□-: medium 31 (containing yeast extract Ohly KAT at a low concentration), -■-: medium 49 (containing yeast extract Hy-Yest 455 from Quest)

Figure 3 illustrates the time course of gallidermin concentrations over several passages found in the corresponding liquid cultures after a fermentation period of 24 hours. Biomass formation was almost the same in all four cases (data not shown). Medium 8 and medium 15 could be identified as unsuitable due to a continuously dropping gallidermin concentration and thus very low hs-values. The production pattern of media 31 and 49 can not be distinguished clearly. In contrast, a clear difference between both media could be seen regarding the corresponding hs-values. In this case priority was given to medium 49.

If a certain ingredient of the medium is of a dominant influence on product formation, mainly acting independently from other components, even a comparison of hs-values concerning the concentrations of one single medium component is possible. Table 1 shows the average hs-values gained with different concentrations of maltose.

Table 1: Average hs-values gained with different concentrations of maltose without any respect to other ingredients.

Maltose concentration [g/l]	hs-value [mg^2/gl]
0	1028
0.25	898
0.5	855
1.0	640
2.0	314

The table demonstrates the tendency to obtain higher hs-values with lower concentrations of maltose.

Glucose, in contrast the carbon source of the former medium mostly led to higher hs-values although not showing a clear tendency (data not shown). So, maltose was disregarded in favour of glucose.

A comparison of different medium compositions and the corresponding concentrations of certain ingredients led to an optimised medium that does not show any product fluctuations over a long period of time (Figure 4).

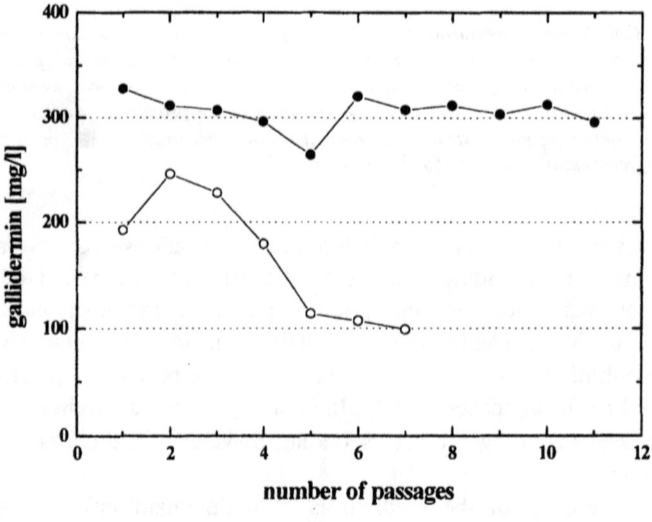

Figure 4 Gallidermin concentration obtained from a suitability test of agar slants with -○-: the former medium (medium 8 in figure 2), -●-: the optimised medium

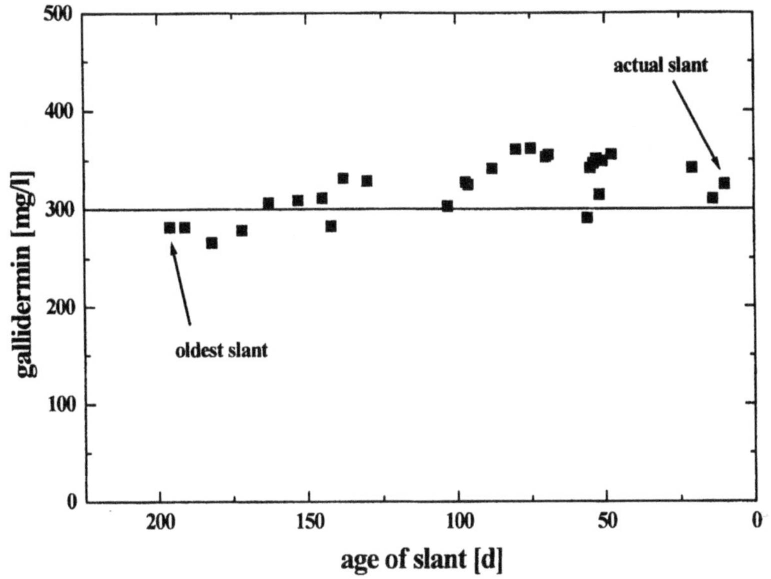

Figure 5: Gallidermin concentrations obtained from a suitability test of agar slants (optimised medium) of different ages. The agar slants were inoculated consecutively from the respective slant before-■-: gallidermin concentration.

As shown in figure 4 the optimised stock culture medium led to a stable product formation over all passages. Such stability was observed too, if agar slants were inoculated consecutively within 1 to 14 days during a time period of 200 days (Figure 5).

Each square in figure 4 represents the maximum gallidermin concentration obtained in liquid cultures inoculated from slants of the corresponding age. It could be seen that product formation is stable and a trend to higher gallidermin concentrations could be observed in comparison to the best yields obtained with the formally used medium (indicated by the line at 300 mg/l in Figure 5).

6. Economic improvement of the production process

For optimisation, major efforts were focused on economical aspects of the medium used for gallidermin production since it contains a very cost intensive complex component. Although meat extract was thought to be the only suitable complex medium component that led to appropriate product yields (Hörner *et al.*, 1990) several other complex ingredients were tested in order to replace it. Table 2 summarises cell growth and product formation obtained with different complex nutrient sources (Kempf *et al.*, 1999a).

6.1 NUTRIENT SOURCES

Table 2. Comparison of growth and gallidermin concentrations obtained in liquid cultures with different complex nutrient sources.

medium [a]	Growth [g/l]	gallidermin [mg/l]	optimal conc. of complex component [g/l]
ME 1 (original)[b]	26	239	150
ME 2[c]	25	199	80
YE 1[c]	30	249	50
YE 2[c]	28	150	50
YE 3[c]	27	204	50

[a] ME 1: (original Medium) 150 g/l meat extract (Lab Lemco Powder, Oxoid), 38 g/l malt extract and 30 g/l $CaCl_2$; ME 2: 150 g/l meat extract (Marcor C, Hartge Ingredients), 2.5 g/l maltose, 30 g/l $CaCl_2$; YE 1: different concentrations of yeast extract (Ohly KAT, Deutsche Hefewerke), 2.5 g/l maltose, 30 g/l $CaCl_2$; YE 2: different concentrations of yeast extract (Difco), 2.5 g/l maltose, 30 g/l $CaCl_2$; YE 3: different concentrations of yeast extract (Hy-Yest 455, Quest International), 2.5 g/l maltose, 30 g/l $CaCl_2$; YE 4: (optimised medium) 50 g/l yeast extract (Ohly KAT), 5 g/l maltose, 45 g/l $CaCl_2$.
[b] Malt extract (in ME1) was generally replaced by 2.5 g/l maltose for medium development.
[c] The concentration of the complex component in each medium was varied between 10 – 100 g/l. Only the concentrations leading to the highest yields are given.

The data given in Table 2 indicate that it is possible to replace meat extract in the formally used medium by a yeast extract (Ohly KAT). Moreover, higher yields were obtained with this nutrient compared to meat extract.

Four other commonly used media (Table 3) were used to compare the results of Table 2 in order to insure the specificity of media ME 1-2 and YE 1-3 for gallidermin production.

Basic medium (BM) is a specific medium for selective isolation of staphylococci from soils. NL 19 in contrast, is a medium commonly used for cultivation of Actinomycetes, main producers of secondary metabolites (Sanglier *et al.*, 1993). It contains an elevated nitrogen level due to the soybean meal used as complex nutrient source. Antibiotic medium 3 and fermentation broth are two commercial media of general purpose.

Table 3. Comparison of growth and gallidermin concentrations obtained in liquid culture with some media of more general purpose.

Medium [a]	growth [g/l]	gallidermin conc. [mg/l]
BM	1.65	30
NL 19	0	0
antibiotic medium 3	0	0
Fermentation broth	4.3	45

[a] *BM: casein hydrolysate 10 g/l, yeast extract 5 g/l, maltose 2 g/l, K_2HPO_4 1 g/l; NL 19: soybean meal 20 g/l, mannitol 20 g/l; antibiotic medium 3 and fermentation broth were purchased from Difco.*

The gallidermin production in liquid cultures of those media given in table 3 could not reach the results obtained with ME 1-2 or YE 1-3, indicating that the media ME 1-2 and YE 1-3 are specific for gallidermin production. Thus, further optimisation experiments were carried out on the basis of YE 1 medium that led to the highest product yields (Table 2).

Since Staphylococcus species are very halotolerant in general and the former production media for gallidermin contained large amounts of salts (Hörner *et al.*, 1990; Kempf *et al.*, 1997), the influence of different concentrations of salts as $CaCl_2$ and NaCl were considered in detail. To optimise this parameter, different concentrations up to 50 g/l of both salts were tested.

As shown in Figure 6, there is an optimum in $CaCl_2$ concentration of about 30 - 40 g/l whereas the NaCl concentration is only of slight influence. Biomass formation (data not shown) is not influenced by NaCl in any way but shows the same dependency towards $CaCl_2$ as product formation.

After this rough medium adjustment, the final quantitative optimisation was carried out with a computer program using genetic algorithms (Goldberg, 1989; Holland, 1975) for multiple parameter optimisation. The program varied simultaneously each concentration in fixed steps in order to maximise both biomass (optical density) and product formation (gallidermin concentration). The concentration of yeast extract was varied from zero to 100 g/l in steps of 10 g/l, $CaCl_2$ from zero to 50 g/l in steps of 5 g/l and maltose from zero to 15 g/l in steps of 0.5 g/l, respectively. After five iterative optimisation steps (each consisted of 8 different medium variations), an optimum for all three medium components was found. In contrast to common optimisation procedures (variation of one parameter after another) only 40 different combinations out of 3751 possible were necessary. The resulting medium was designated as medium YE4 and consisted of yeast extract (Ohly KAT) 50 g/l, $CaCl_2$ 45 g/l and maltose 5 g/l. YE4 medium led to a cell dry weight of about 37 g/l (OD 578nm = 82 at 24 hours of incubation) and a product concentration of almost 300 mg/l gallidermin in the culture broth. Figure 7 compares the time courses for gallidermin concentrations observed with the original production medium ME1 (Hörner *et al.*, 1990) and the optimised medium YE4.

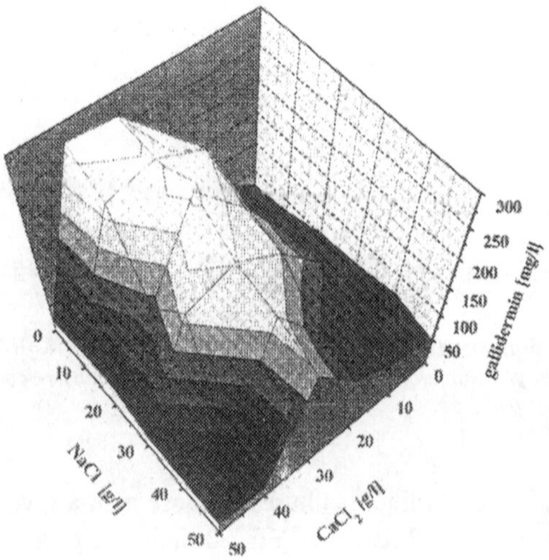

Figure 6: Maximal gallidermin concentrations in the culture broth of Staphylococcus gallinarum TÜ 3928 in dependence of different NaCl/CaCl2 proportions in medium YE1.

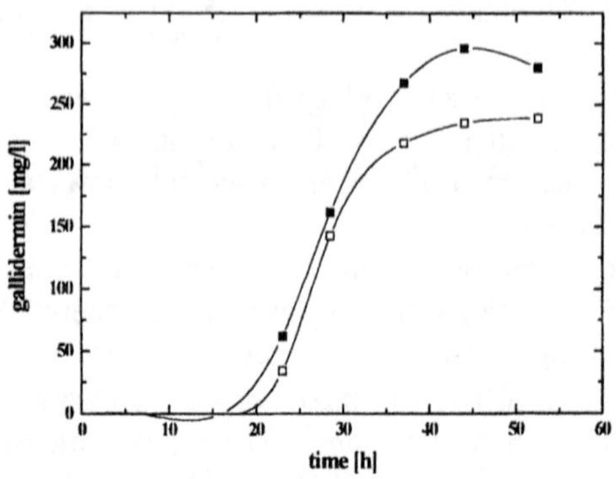

Figure 7: Comparison of gallidermin formation by Staphylococcus gallinarum TÜ 3928 grown in the previously used medium ME1 (-□- ; containing 150 g/l meat extract) and the new developed production medium YE4 (-■- ; containing 50 g/l yeast extract). Both liquid cultures were inoculated from the same BM-agar plate.

Figure 7 demonstrates, that beside the replacement of meat extract by a more cost effective complex component the optimisation led to an increase in gallidermin formation of approximately 50 mg/l. The total economic outcome of this medium optimisation was a 15-fold cost reduction. The reduction of medium costs and the increase in gallidermin concentration during the three optimisation steps (replacement of ingredients, rough adjustment of salt concentrations and final optimisation using a computer program) are summarised in Figure 8.

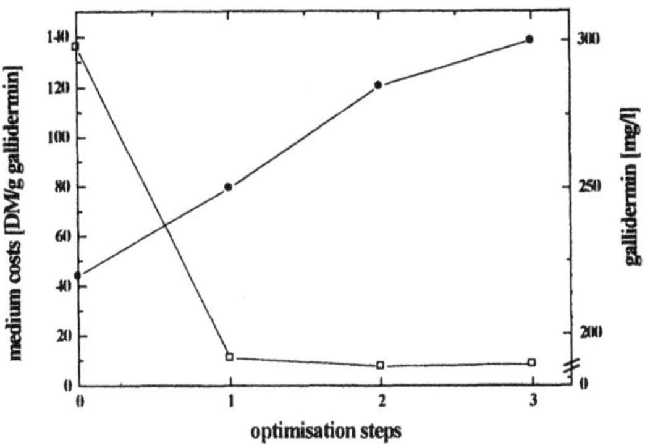

Figure 8:. Evolution of medium costs (-□ -) and product yields (- ■ -) during three optimisation steps (see text).

As demonstrated in Figure 8 medium optimisation led to a reduction of medium costs by 93 % compared to the previously used medium. Additionally higher product yields were achieved.

A scale-up of the process based on the new medium was tested with several fermentation modes. In preliminary experiments with such non-optimised fermentation processes best results were obtained with batch or fed-batch processes. So, further investigations were focused on fermentation modes of that kind.

6.2 DEVELOPMENT OF SCALE-UP PROCEDURE

6.2.1 Development of a fed-batch process

For gallidermin production *Staphylococcus gallinarum* was grown on an optimised production medium (Kempf *et al.* 1999a). Nearly 75 % of the complex nutrient source of this medium (yeast extract) consists of amino acids (free form) or as peptides and small proteins. Thus, an influence of amino acids on the formation of the peptide antibiotic gallidermin is very probable. Analysis of the amino acid consumption (Kempf *et al.*, 1999b) in a 2 litre bioreactor revealed that aspartate (Asp), glutamic acid (Glu), asparagine (Asn), glycine (Gly), serine (Ser), threonine (Thr), alanine (Ala) and arginine (Arg) are consumed in the same phase of fermentation as growth and gallidermin formation was observed (Figure 9).

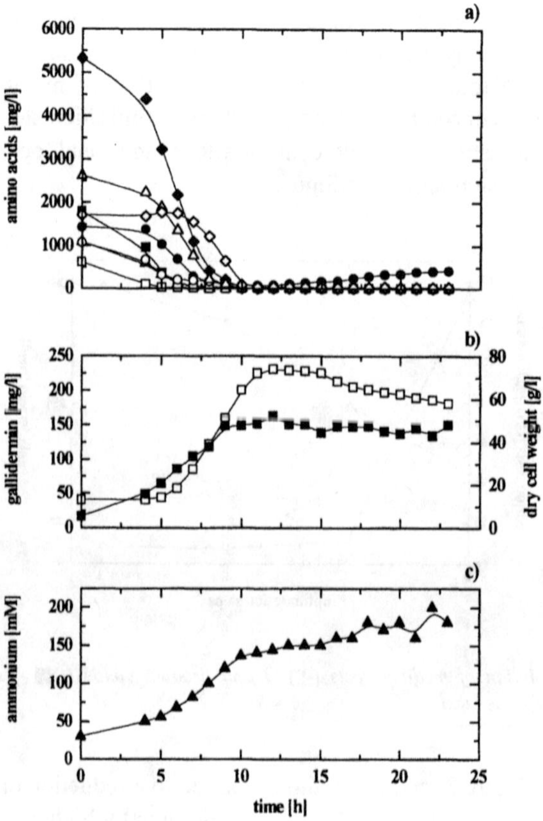

Figure 9: Investigations in a 2-litre bioreactor. The symbols represent the time course of concentrations of a) the metabolised amino acids: -■- Glutamic acid, -●- aspartate, -▲- asparagine, -◆- glycine, -%- serine, -O- threonine, -Δ- alanine and -◇- arginine; b) -%- gallidermin and the -■- cell dry weight; c) -▲- (NH4)⁺.

All other amino acids are either not constituents of the used yeast extract, were not metabolised in significant amounts or are not detectable (Cys, Pro) by means of the analytical system used for amino acid determination (OPA - derivatisation method). To investigate a possible stimulation of amino acids on gallidermin formation, pulse experiments were carried out with single amino acids in shake flasks. Only those amino acids which were metabolised during the production phase of gallidermin in batch fermentation were analysed in detail. The experiments were realised as single pulses at incubation times of 0, 24 or 30 hours, representing the different phases of fermentation like lag phase, exponential phase and stationary phase during cultivation in shake flasks. It could be shown that a single dose of either Glu, Gly, Ser or Thr at 24 to 30 hours of fermentation could increase gallidermin production significantly (Figure 10).

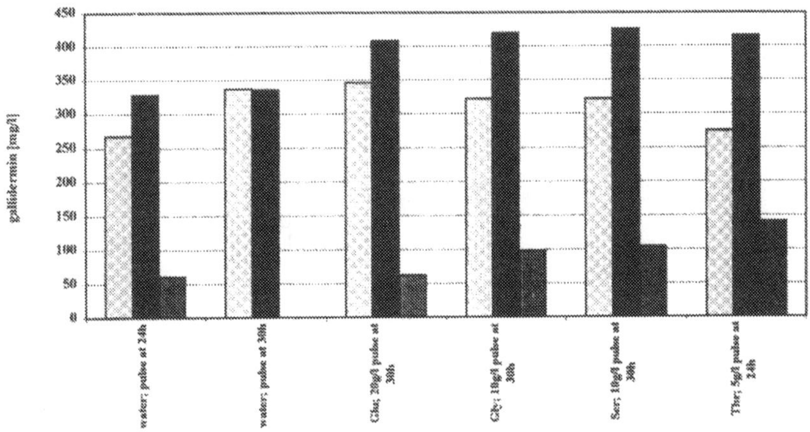

Figure 10: Increase of gallidermin production obtained in pulse experiments with the corresponding amino acids. The control experiments were carried out as a pulse of sterile water solution (10 % v/v) at a fermentation time of: see brackets in Figure 10. Glutamic acid (final conc. 20 g/l), glycine and serine (final conc. 10 g/l) were pulsed at a fermentation time of 30 hours; threonine (final conc. 5 g/l) at 24 hours. The given parameters led to the best results with the corresponding amino acid. Colours symbolise: white: gallidermin conc. before the pulse; grey: max. gallidermin conc. after the pulse; light grey: difference of both conc..

Each of these amino acids could increase the product concentration to about 400 mg/l representing an increase of 25 % compared to batch experiments. With respect to economical aspects, glutamic acid was chosen for further investigations in a 2-litre bioreactor. In order to maintain a better control of the process in this scale a feeding of glutamic acid was preferred instead of a single pulse.

Furthermore, an on-line HPLC-system for amino acid determination was implemented to the process to determinate the appropriate time for an optimal glutamic acid feed indicated by depletion of the amino acids. For this measurement, the same analytical system was used as for the generation of off-line data (Figure 9) but coupled to the bioreactor with a cross-flow filtration module to obtain cell free culture filtrate. The module was attached to the bioreactor with a by-pass where the retentate (containing biomass) was recycled into the bioreactor. The volume of the fermenter was kept constant by addition of the feeding solution. Thus, the flow rate of the feeding solution was controlled by the flow rate across the membrane of the module. The experimental set-up of the entire system is given in Figure 11.

The results obtained with this fed-batch process are illustrated in Figure 12. In general it is difficult to compare the results of a fermentation in batch with those obtained in a fed-batch process due to dilution effects caused by the addition of the feeding solution in the fed-batch process. So, the best way is a comparison of the absolute product amounts (volume independent) of the product in both processes. As shown in Figure 12 about 650 mg gallidermin could be produced in the fed-batch process whereas 460 mg (= 2 litres x 230mg/l) were achieved in the batch process. Thus it could be concluded that the production phase of gallidermin could be successfully prolonged by a feed of glutamic acid, although productivity was not increased

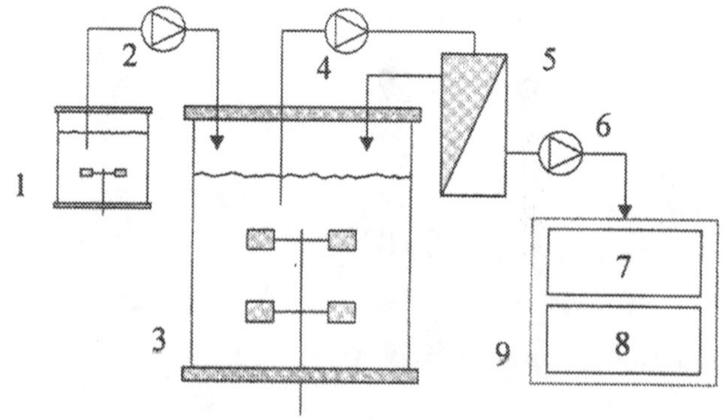

Figure 11: Experimental system of a fed-batch process. Amino acid concentrations are monitored by on-line HPLC-system. Numbers symbolise: 1: feeding solution; 2: feeding pump; 3: bioreactor; 4: by-pass pump; 5: cross-flow filtration unit; 6: filtrate pump; 7: automatic probing, dilution and pre-column derivatisation (autosampler); 8: amino acid determination (HPLC); 9: HPLC-system and autosampler.

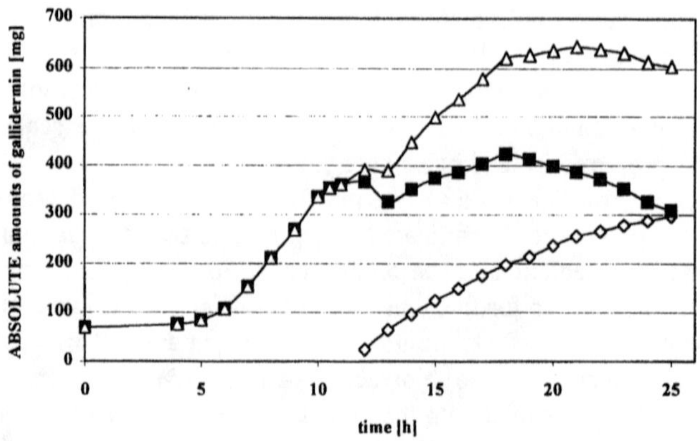

Figure 12: Absolute amounts of gallidermin during a fed-batch process in a 2-litre bioreactor. During the feeding (11 to 25 hours of fermentation), 1.7 litre of cell free culture filtrate was replaced by a feeding solution (glutamic acid; 20 g/l). The symbols represent the absolute amounts of gallidermin found: -■- in the bioreactor (3 in Figure 3); -◇- in the filtrate (6 in Figure 3); -Δ- total amount found in the entire system (addition of bioreactor and filtrate).

However, the positive effect of an amino acid supplementation raised the question whether the amino acids were metabolised as carbon or as nitrogen source (or both). A characteristic time course of a batch fermentation (Figure 9) indicated no limitation of ammonia. These results could be confirmed by addition of different $(NH4)^+$ salts and of $Mg_3(PO_4)_2$. $Mg_3(PO_4)_2$ was used for the complexation of $(NH4)^+$ (Ōmura & Tanaka 1984) and thus served as a negative control. In both experiments, addition and reduction of the $(NH4)^+$ concentration during cultivation in shake flasks did not show any effect on the production of gallidermin (data not shown). These experiments suggested an influence of the amino acids on the product formation due to the additional carbon rather then the nitrogen source. This suggestion is consistent with a significant increase of cell dry weight found in the fed-batch process (max. about 70 g/l) compared to a batch fermentation (max. about 50 g/l), although this was not observed in a scale of shake flasks (data not shown).

Consequently, a pulse experiment with maltose was performed instead of an addition of amino acids in a shake flask scale. This experiment led to a gallidermin concentration of about 400 mg/l after an addition of maltose at 24 hours of cultivation, too.

6.2.2 Investigations of batch processes

A scale-up procedure in general is divided into three steps: (i) cultivation in small scale at a volumetric range of millilitres, (ii) fermentation in pilot scale at ranges of a few litres to some hundred litres, and (iii) production in plant scale with large, mostly industrial installations.

A change in fermentation scale may be associated with a loss in productivity. The most critical step usually is found in the transfer from small to pilot scale, due to the equipment needed to handle larger volumes. The most dramatic changes of the physical, environmental conditions of a culture such as vessel geometry, mixing or aeration occurs at this scale-up level (Reuss and Bajpai, 1991) .

In pilot scale, the consecutive next scale-up step, most factors that may have negative influence on production, such as mixing time delay, mass transfer or sheer stress are expected to be already present. In plant scale, these negative influences finally may also be observed but should not be of completely new origin. So, a pilot scale fermentation could be considered as a good simulation for an optimisation of a large plant scale production process.

The improvement of the stock culture medium together with an optimised production medium led to a reproducible production of about 300 mg/l gallidermin in the scale of shake flasks. Despite those optimisations, a 10 to 20 % lower product concentration was observed during a scale-up process in a non-optimised pilot scale fermentation.

So, the aim was to eliminate or to compensate this loss in gallidermin formation during a scale-up process.

As described by Hosobuchi and Yoshikawa (1999) most of the factors that influence the efficiency of a scale-up procedure can be determined by a simulation of the scale-up in small scale.

The basis of investigations covered in this section is the assumption that a scale-up procedure also could be simulated by transitions of consecutive liquid cultures of the same scale. So, investigations could be made without an enlargement of volume and thus being less time and cost intensive.

6.2.3 Small scale fermentations.

To obtain information about the parameters involved in the loss of product formation during scale-up, several experiments were carried out in shake flasks, simulating a scale-up procedure by inoculation of several consecutive liquid cultures of the same volume (Figure 13).

6.2.4 Variation of parameters in shake flask experiments

In these experiments, three individual subjects were investigated: the cultivation conditions for *S. gallinarum* on solid medium in Petri dishes (A), the cultivation conditions of pre-cultures (first liquid culture, B) and the cultivation of the consecutive scale-up steps (simulated by D-H). The effect of any variation in A or B was assessed by the product concentration in flask C, whereas cultures D-H were used to examine the stability of product formation in the simulated steps of scale-up.

The experiments indicated that any variation of the incubation time of the Petri disk (A) or the first liquid culture (B) always resulted in unaffected maximal gallidermin concentrations in the consecutive liquid culture (C). In this context one exception must be made: a pre-culture (B) that was incubated longer than 32 hours led to a significant decrease in gallidermin formation in the following culture (C). This results suggested that a culture used for inoculation of any other liquid culture in a scale-up process should be transferred into the next cultivation volume before reaching the stationary growth phase, when gallidermin concentrations are found to be maximal.

Although the amount of biomass inoculated from Petri disks (A) into the pre-culture (B) was varied by a factor of more than thousand, no effect in the resulting product concentration was observed. So, the amount of cells could be negligible.

If different volumes were used for the inoculation of culture C some surprising results were obtained: an "inoculum" of 50 % (v/v) led to a slightly (but reproducibly) increase of gallidermin formation in consecutive liquid cultures. Such an "inoculum" is equivalent to a fed-batch procedure where the same volume of fresh medium is added to a culture. In contrast, other experiments in which medium concentrations in liquid cultures C-D were doubled or tripled always revealed a negative influence on product formation in this flasks. Thus, it could be assumed that production medium added at a certain time of the production phase could probably increase or prolong antibiotic formation. Since no other modifications concerning the steps described in Figure 13 were found in order to increase product formation further investigations were focused on feeding strategies during the phase of product formation.

As described in the section above several amino acids and maltose could be identified to increase gallidermin formation in pulse experiments. This positive effect from parts of the medium was observed in the scale of shake flasks. Investigations in 2 litres and 20 litres bioreactors revealed that only a pulse with maltose could successfully increase product formation in both larger scales. This findings were independent from several other parameters like aeration, agitation, cultivation conditions of pre-cultures or the production phase when certain amino acids were added to the culture broth of a

bioreactor (data not shown). So, a maltose pulse was favoured to develop a scale-up procedure.

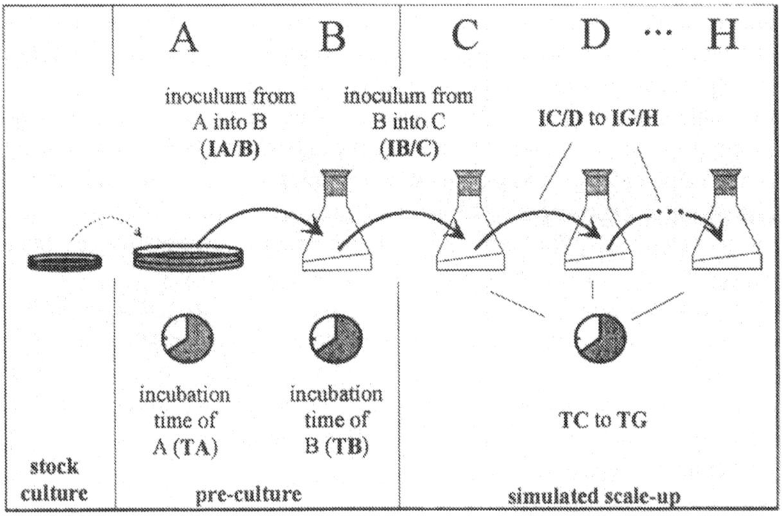

Figure 13: Experimental set-up of small-scale parameter optimisation: To obtain sufficient amounts of biomass for the inoculation of B (the first liquid culture) an Petri disk (A) was inoculated from a stock culture. The second liquid culture (C) was inoculated from B, etc. . This transitions from consecutive liquid cultures were repeated up to 7 times, simulating a scale-up procedure. Abbreviations: A: Petri disk for inoculation of B; B to H: first to seventh consecutive liquid culture; T: Incubation time of a culture before inoculating the next consecutive culture; TA to TH: Incubation time T of cultures A to H; I: Inoculation volume (%) into the consecutive culture; IA/B to IG/H: Inoculation volume I from A into B to G into H.
Incubation times of Petri disk A (TA): up to 168 hours at alternating steps of 8 and 16 hours. Incubation time of the first liquid culture B (TB): 8, 16, 24, 32, 40 and 48 hours. Incubation times of consecutive liquid cultures D to G (TD to TG), was 24 hours. Inoculation from Petri disk A into the first liquid culture B (IA/B) with a suspension of cells in sterile water. Inoculation was performed with different volumes and cell densities: 800µl, OD578nm 123; 500µl, OD578nm 78; 500µl, OD578nm 16; 100µl, OD578nm 1.6. Inoculation of consecutive liquid cultures C to E (IB/C to ID/E) were carried out with 1%, 5%, 10%, 25% and 50% (v/v); F to H (IE/F to IG/H): 5% and 25%.

6.2.5 Pilot scale fermentations

As described for shake flasks cultivation in small scale, the transition from a 20 litres to a 200 litres bioreactor was first simulated by consecutive batches of 20 litres (Figure 14) For these investigations the cultivation conditions of first steps were kept constant: agar Petri disks used for the inoculation of pre-cultures were incubated for 24 hours. According to the findings at the shake flask level not to exceed 32 hours of cultivation, 2 litres of a 24 hours old pre-culture were inoculated (10 % (v/v)) into the first 20 litres bioreactor. The culture broth of this bioreactor finally was used to inoculate the second (consecutive) 20 litres fermenter simulating the 20 to 200 litres transition (Figure 14).

.Agitation, aeration, the cultivation time of the first bioreactor before inoculation, the inoculum into the (consecutive) second reactor and the growth phase of fermentation when maltose was supplemented, were varied.
This results of about 20 different fermentations revealed that neither the cultivation time of the first bioreactor nor the inoculum into the second reactor were of any significant influence on product formation (data not shown).
A positive outcome concerning gallidermin production was observed when at the transition point between 20 and 200 litres fermentations the cell dry weight reached a characteristic value of 20 g/l and gallidermin concentration was about 100 mg/l.

Results of pulse experiments with pilot scale reactors were found to be best if the shift from the batch to the fed-batch phase of fermentation were started (by the addition of maltose) at maximal gallidermin concentrations (250-270 mg/l).

Thc typical concentration of cell dry weight reached before the addition of maltose was about 40 g/l. Generally, an addition of maltose during the late phase of product formation (second bioreactor) always led to an increase of the same magnitude of nearly 100 mg gallidermin per litre. So, it can be concluded that the maximal concentration reached in a pulsed process depends solely on the actual gallidermin concentration in the bioreactor before maltose is added.

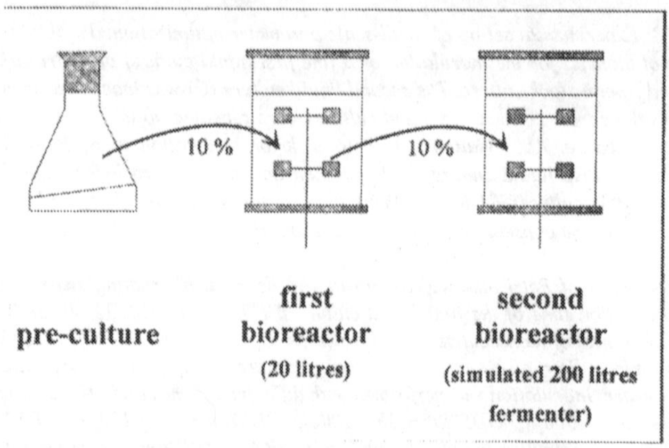

Figure 14: Experimental set-up of scale-up investigations in a 20 litres scale, simulating the transition from a pre-culture to 200 litres bioreactor.

6.2.6 An optimised scale-up procedure for the production of gallidermin

On the basis of data generated in this simulation experiments a scale-up strategy was developed and transferred to a 200 litres bioreactor. Essential details for each scale-up step are summarised in Table 4.
Figure 15 shows the concentrations of gallidermin and cell dry weight obtained with such a scale-up procedure in a 200 litres bioreactor.

The maximal gallidermin concentration was found to be about 330 mg/l in the 200 litres bioreactor. Compared to the concentrations found in a non-pulsed 200 litres fermentation (250-270 mg/l) this is an increase of 20 – 30 % (Kempf *et al.*, 2000).

Table 4: Cultivation conditions during scale-up leading to a reproducible antibiotic formation in large-scale fermentations.

Scale	parameter	value
pre-culture	inoculum	not relevant
500 ml	incubation	24 hours (not exceeding growth/production phase)
pre-fermenter	inoculum	10 % (v/v)
20 litres	oxygen supply	allowing moderate growth (reduced growth rate)
	cultivation [a]	until c_P > 100 mg/l and $c_X \sim$ 20 g/l
	inoculum	10 % (v/v)
pilot-scale	oxygen supply [b]	allowing moderate growth (reduced growth rate)
reactor	batch phase [a]	until $c_P \sim$ 250-270 mg/l and $c_X \sim$ 40 g/l
200 litres	pulse	15g/l maltose (final conc.); volume 10% (v/v)

[a] c_P: concentration of gallidermin; c_X: concentration of cell dry weight; [b] oxygen supply: regulated by aeration and agitation

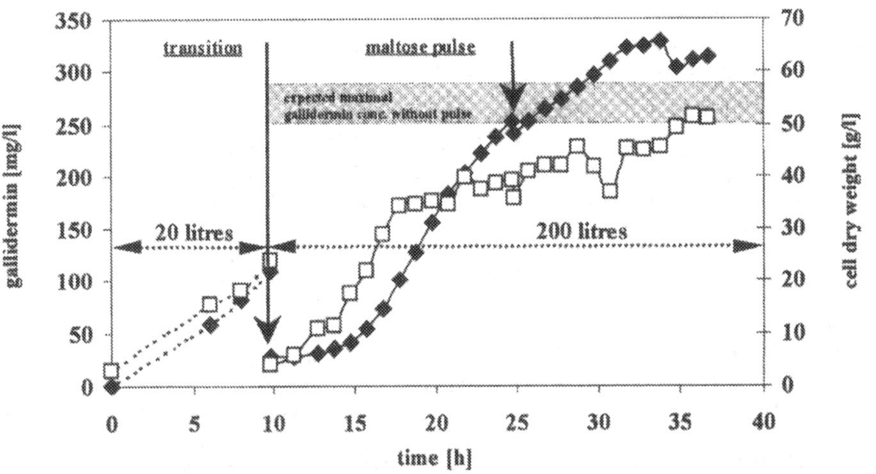

Figure 15: Gallidermin (-◆-) and cell dry weight (-□-) concentrations observed during a scale-up procedure (parameters see text). Concentrations found in a 20 and the consecutive 200 litres bioreactor. The arrow marks the transition from 20 to 200 litres and the time when maltose is added into the 200 litres bioreactor; the grey area shows the range of the gallidermin concentration (including errors) expected in a 200 litres bioreactor without maltose pulse.

References

Benz, R., Jung, G. and Sahl, H.-G. (1991) Mechanism of channel-formation by lantibiotics in black lipid membranes in Jung, G. and Sahl, H.-G. (eds.), *Nisin and Novel Lantibiotics* , ESCOM, Leiden, pp. 359-372.

Börtz, H., Bierbaum, K., Reynolds, P.E. and Sahl, H.-G. (1997) The lantibiotic mersacidin inhibits peptidoglycan biosynthesis at the level of transglycosylation, *Eur. J. Biochem. 246*, 193-199.

Börtz, H., Bierbaum, G., Leopold, K., Reynolds, P.E. and Sahl H.-G. (1998) The lantibiotic mersacidin inhibits peptidoglycan synthesis by targeting lipid II, *Antimicrob. Agents Chemother. 42*, 154-160.

Bierbaum, G., Götz, F., Peschel, A., Kupke, T., van de Kamp, M. and Sahl, H.-G. (1996) The biosythesis of the lantibiotics epidermin, gallidermin, Pep5 and epilancin K7, *Antonie von Leeuwenhoek 69*, 119-127.

Breckel, A., Harder, M., Fiedler, H.-P. and Zähner, H. (1995) Production of gallidermin by *Staphylococcus gallinarum* Tü3928, in Schmid, R.D. (Ed.), *Biochemical Engeneering 3*, Kurz, Stuttgart, pp 62-66.

Devrieses, L.A., Poutrel, B., Kilpper-Bälz, R. and Schleifer, K.H. (1983) *Staphylococcus gallinarum* and *Staphylococcus caprae*, two new species from animals, *Int. J. Syst. Bacteriol. 33*, 480-486.

Goldberg, D.E. (1989) *Genetic Algorithms in Search, Optimization and Machine Learning*, Addison-Welsley Publishing, Reading, Massachusetts.

Hörner, T., Ungermann, V., Zähner, H., Fiedler, H.-P., Utz, R., Kellner, R. and Jung G. (1990) Comparative studies on the fermentative production of lantibiotics by staphylococci, *Appl. Microbiol. Biotechn. 32*, 511-517.

Holland, J.H. (1992) Genetische Algorithmen, *Spekt. Wiss. 9.*

Hosobuchi. M. and Yoshikawa. H. (1999) Scale-up of microbial processes, in Demain, A.L. and Davies, J.E. (eds.), *Manual of Industrial Microbiology and Biotechnology*, 2nd edn., ASM Press, Washington, pp 236-239.

Humphry, A. (1998) Shake flask to fermentor: what have we learned? *Biotechno.* Prog. 14, 3-7.

Jack, R.W., Bierbaum, G. and Sahl, H.-G. (1998) Lantibiotics and Related Peptides, Springer-Verlag, Heidelberg.

Jung, G. (1991) Lantibiotics: a survey, in Jung, G. and Sahl, H.-G. (eds.), Nisin and Novel Lantibiotics, ESCOM, Leiden, pp. 1-34.

Kellner, R., Jung, G., Hörner, T., Zähner, H., Schnell, N., Entian, K.-D. and Götz, F. (1988) Gallidermin: a new lanthionine-containing polypeptide antibiotic, Eur. J. Biochem. 177, 53-59.

Kempf, M., Theobald, U. and Fiedler H.-P. (1999a) Economic improvement of the fermentative production of gallidermin by Staphylococcus gallinarum, Biotechnology Letters 21, 663-667.

Kempf, M., Theobald, U. and Fiedler H.-P. (1999b) Correlation between the consumption of amino acids and the production of the antibiotic gallidermin by Staphylococcus gallinarum, Biotechnology Letters 21, 959-963.

Kempf, M., Theobald, U. and Fiedler H.-P. (2000) Production of the antibiotic gallidermin by *Staphylococcus gallinarum* – development of a scale-up procedure, *Biotechnology Letters 22*, 123-128.

Kempf, M., Theobald, U. and Fiedler, H.-P. (1997) Influence of dissolved oxygen on the fermentativ production of gallidermin by *Staphylococcus gallinarum*, *Biotech Letters 19*, 1063-1065.

Ōmura, S. and Tanaka, Y, (1984) Control of ammonium ion level in antibiotic fermentation, in Ortiz-Ortiz, L., Bojalil, L.F. and Yakoleff, V. (eds.), *Biological, Biochemical and Biomedical Aspects of Actinomycetes*, Academic Press, Orlando, p. 367.

Reuss, M. and Bajpai, R. (1991) Stirred tank models, in Rehm. H.-J. and Reed. G. (eds.), *Biotechnology Vol. 4*, 2nd edn., VCH, Weinheim, pp 299-348.

Sahl, H.-G., Jack, R., Bierbaum, G. (1995) Biosynthesis and biological activities of lantibiotics with unique post-translational modifications, *Eur. J. Biochem. 230*, 827-853.

Sanglier, J.J., Wellington, E.M.H., Behal, V., Fiedler, H.P., Ellouz Ghorbel, R.,

Finance, C., Hacene, M., Kamoun, A., Kelly, C., Mercer, D.K., Prinzis, S. and
Trigo, S. (1993) Novel bioactive compounds from actinomycetes, *Res. Microbiol. 144*, 661-663.
Theobald, U. and Kempf M. (1998) A novel tool for medium optimization and characterization in the early stages of a metabolite production process, *Biotechnology Techniques 12*, 893-897.

RESISTANCE TO β-LACTAMS, A SELF-REGENERATING PROBLEM

JOZSEF ASZODI* AND ANDRÉ BRYSKIER,
Hoechst Marion Roussel, *102 Route de Noisy, F-93235 Romainville Cedex, France*

1. Resistance to Beta-Lactams

Due to the short generation time and the simple structure, bacteria can adjust rapidly to environmental changes. The development of resistance is such an adaptation to a new, hostile environment. This phenomenon is truly natural for these microorganisms as they can modify their structure and metabolic pathways in order to be able to survive in a wide variety of conditions. Most of today's antibiotics are natural products or derivatives thereof and they have been used by the producing microorganisms for their own defence. In these conditions, bacteria having the appropriate defence mechanism could more easily survive.

In 1945, at the beginning of the extensive clinical use of penicillin G, most Gram positive bacteria were sensitive, but due to lack of penetration, the compound did not kill Gram negative rods. The extensive clinical use of this substance rapidly selected those strains which produced an enzyme, a penicillinase, capable of hydrolysing the beta lactam nucleus. This was only a first step of a long lasting race between man and bacteria in the development of new molecules, followed by the rapid appearance of new resistant strains.

Today, after more than 50 years, the family of beta-lactams still represents the most widely used, the most efficient and safest class of antibiotics. This is partly due to the fact that their target is extracellular, and also that the irreversible inhibition contributes to their broad spectrum and high efficacy. Combined with a good safety profile, these unique properties made penicillin, and later other molecules bearing a β-lactam nucleus, particularly attractive targets for semisynthetic modifications.

2. Mode of Action of β-Lactams

Bacteria are surrounded by a solid murein sacculus in order to maintain the shape, the rigidity and the osmotic stability of the cell. This peptidoglycan is built up from a mono or multilayer two-dimensional polymeric meshwork, consisting of a glycan chain formed by repeating disaccharides (N-acetyl glucosamine and N-acetyl muramic acid). These linear chains are cross-linked by a short peptide consisting of L and D amino

A. Van Broekhoven et al. (eds.), Novel Frontiers in the Production of Compounds for Biomedical Use, 57–83.
© 2001 *Kluwer Academic Publishers. Printed in the Netherlands.*

acids, linked to the lactyl function of the muramate. This reticulation is essential for the solidity of the mesh. The cross-linking takes place usually between a D-alanine terminus of a tetrapaptide (formed by the cleavage of the terminal D-alanine of a pentapeptide) and the Ω-amino group of a lysine or diaminopimelic acid of a neighbouring chain. In staphylococci, there is an additional pentaglycin unit, inserted between the D-Ala on one chain, and the lysin of another

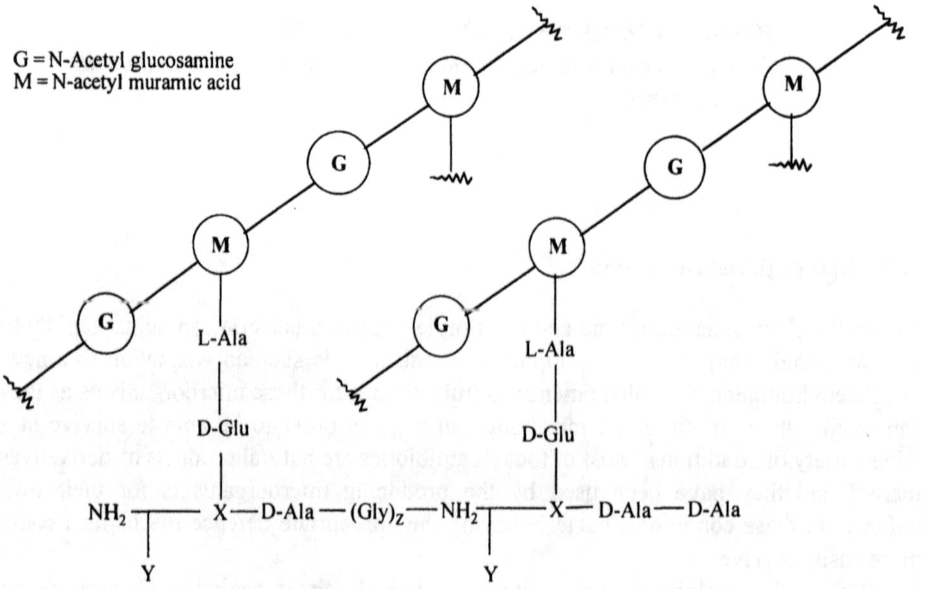

X = Diaminopimelic acid, Y = CO_2H, Z=0

(In *Staphylococcus* X = L-Lysine, Y = H, Z=5)

Polymerisation is carried out by the so-called penicillin binding proteins (1). The common feature amongst the members of this family is that they all covalently bind radiolabelled penicillin. This is not a homogeneous family of enzymes. Some of them are transpeptidases, others possess endo- or carboxypeptidase activity. The role of each member of the family is not clearly understood, and often there is only indirect evidence about the possible function.

The most important from our point of view are those with high molecular weight, which are often essential for the cell survival. They can be divided into two groups. Those of the class A are bifunctional. The amino terminal part does not bind penicillin and possesses a transglycosylase activity. It is responsible for the elongation of the glycan chains by catalysing the formation of a glycosidic linkage between the sugar units. The carboxy terminal part, conserved both in class A and B PBPs, catalyse the formation of a peptide bond and thereby is responsible for the reticulation. The PBPs of the class B can only carry out this transpeptidation reaction, the non penicillin binding domain does not possess any known enzymatic activity (2).

During the reticulation phase, the terminal alanine of the pentapeptide is cleaved off and a covalent acyl enzyme intermediate is formed. In the next step, this acyl residue is transferred to the terminal amino group of a diaminopimelic acid (or lysine in staphylococci) receptor. These proteins are therefore transpeptidases. Beta-lactams have a different structure from the D-Ala-D-Ala terminus of the pentapeptide, but similar enough to be recognised to a certain extent. Due to the high reactivity of the beta lactam ring compared to a simple peptide bond, the first reaction can readily take place. However, the active site of the enzyme being fully occupied, the transfer of the acyl group to the amine receptor cannot occur and the enzyme remains blocked.

The low molecular weight PBPs with carboxy- or endopeptidase activity play a role in the remodelling of the peptidoglycan. They are not essential *in-vitro* for the survival of bacteria, the absence or inhibition of these proteins do not alter the viability of the cells but can influence the polymerisation (3).

3. Beta Lactamases

The most common resistance mechanism is the hydrolysis of the β-lactam nucleus by β-lactamases (4). Derived historically from PBPs, β-lactamases play a protective role for bacteria, by hydrolysing the antibiotic (5). The enzymes produced by Gram positive bacteria are mostly secreted into the environment and can hydrolyse the penicillins before they can reach their targets. The consequence was that by the end of the 1940's in some hospitals, 59% of bacteria became resistant to penicillin G (6).

The discovery of cephalosporins solved the problem only for a short while, as cephalosporinases appeared rapidly. This hydrolysis represents the first line of defence by bacteria.

During the 1970's, bacteria exhausted all natural sources for new beta lactamases, and adopted a more active strategy. By successive mutations, a continuous race started between these microorganisms and the chemists. The tens of thousands of alterations of penicillins and cephalosporins using diverse substituents have been accompanied by specific changes in the specificity of the enzymes by successive amino acid replacements in the active site.

Today even the carbapenems, having the reputation of being particularly stable to β-lactamases, are hydrolysed. Only very few compounds can resist the new metallo β-lactamases (MBLs). If they are not frequent today, the fact that some of them are plasmid mediated may give them more importance in the future.

β-Lactamases function much like carboxypeptidases with their natural substrates. The difference with PBPs is that after opening of the β-lactam ring, the acyl-enzyme intermediate is unstable, and is rapidly hydrolysed. In the MBLs, there is no serine in the active site and one or two zinc atoms catalyse the hydrolysis of the β-lactam nucleus.

There is a direct relationship between the amount of enzyme produced constitutively and the MICs. This is also seen in the inoculum effect, where the MICs rise with an increased size of inoculum of *S. aureus*. Despite the presence of the outer membrane which is supposed to contain the β-lactamases of Gram negative bacteria, they also show reduced susceptibility with increased inoculum size (7).

Several attempts have been made to classify the ~200 known β-lactamases (8) according to their structure and biochemical properties (9) In a commonly accepted scheme proposed by Ambler (10), serine based enzymes belong to the classes A,C and D, while the class B contains the metallo enzymes.

3.1. CLASS A β-LACTAMASES

The class A β-lactamases represents by far the highest diversity and they are the best studied. The structure of 5 of them has been determined by X-ray crystallography (4). The enzymes belonging to this class are either plasmid mediated or chromosome encoded, and hydrolyse both penicillins and cephalosporins. Enzymes conferring resistance to third generation cephalosporins and carbapenems are also members of this group, together with those resistant to β-lactamase inhibitors (9). The catalytic properties and primary structures of this class differ considerably (11)) and although only nine residues are strictly conserved (12) they validate the original classification by Ambler (10). The catalytic mechanism of theses enzymes is to some extent related to the one generally accepted for serine proteases, including the presence of an "oxyanion hole". However, the nature of the specific residue that enhances the nucleophilicity of the active site serine is still a matter of debate (4).

Third generation cephalosporins, by the time of their introduction into clinical use, were particularly stable to these enzymes. However, selective pressure rapidly leads to the emergence of resistant strains. These bacteria produce β-lactamases with only a tiny difference from the sensitive ones. The large number of TEM and SHV extended spectrum β-lactamases (13) attests to the fantastic capability of bacteria for adaptation. By only a few mutations, the enzymes completely changed their specificity and the modified mutants are capable of specifically hydrolysing newer compounds. The mutations mostly lead to an expansion of the spectrum. Interestingly however, in some cases the new enzymes can specifically hydrolyse third generation cephalosporins without destroying older molecules. TEM 46, which contains mutations similar to TEM-3 on one hand and TEM-5 on the other confers resistance to Escherichia coli against ceftazidime, but the bacterium remains sensitive to cephalotin (14). Although the emergence of these mutants seems to be very simple, recent studies show that it cannot be explained by a series of single mutations - selection events (15).

Due to their high resistance to serine β-lactamases, imipenem and other carbapenems have become quite popular in hospitals and are often used as antibiotics of last resort against infections caused by gram-negative bacteria resistant to other ß-lactams. These molecules used to be resistant to serine β-lactamases, due to the hydroxyethyl side chain. X-ray studies and molecular modelling provided evidence that the steric hindrance of this substituent results in a poor interaction between the carbonyl of the primary acyl–enzyme and the oxyanion hole leading to a decreased hydrolysis rate. The relative stability of this species offers the possibility to the system to undergo a conformational change which further stabilise the enzyme-inhibitor complex (16). The recent emergence of carbapenem hydrolysing β-lactamases however casts a shadow over this idyllic picture.

The β-lactamases Sme-1 and NMC-A from *Serratia marcescens* (17, 18) and *Enterobacter cloacae* (19, 20) respectively efficiently hydrolyse carbapenems. Molecular modelling predicted an unusual disulfide bridge near the active site between the residues 69 and 283 (21). These results have been also confirmed in the high resolution X-ray structure of NMC-A (22). The presence of the disulfide bridge modifies the topology of the substrate binding site in a way that it can easily accommodate the carbapenem, however the position of the critical active site residues remains unchanged within a r.m.d.s. of 0.3 Å. Also the position of Asn-132 is different from usual class A enzymes, providing additional space in a critical area for the accommodation of sterically hindered compounds.

Inhibitors of β-lactamases have been developed to prolong the life of clinically useful β-lactams by inhibiting their destruction. However, as PBPs learned how to escape beta lactam antibiotics, β-lactamases adopted the same strategy. *E. coli* can resist β-lactam - inhibitor combinations by overproduction of TEM-1 (23). This effect is further enhanced if penetration is impaired (24). More recently, inhibitor resistant β-lactamases have emerged (25, 26). By amino acid mutations, inhibitor resistant (IRT) β-lactamases evolved from TEM-1 in *E. coli* and from TEM-2 in *Klebsiella pneumoniae* and *Proteus mirabilis*. They display reduced sensitivity to clavulanic acid, sulbactam and tazobactam (27).

3.2. CLASS C β-LACTAMASES

The class C enzymes are mostly chromosomal cephalosporinases containing a serine in the active site. They also hydrolyse penicillins, albeit with a low turnover. From a structural point of view they are more similar to DD-peptidases than to class A β-lactamases (28). They are produced by *Enterobacteriaceae, Pseudomonas* and *Aeromonas*. This is a more homogenous group than the smaller class A enzymes and it displays more than 35% sequence homology within the class (4). When produced in large quantities, they confer resistance to third generation cephalosporins (29,161). The catalytic mechanism is less controversial than for class A enzymes, and has been established, based on X-ray structure (30).

Originally they were chromosome encoded, however, more recently plasmid mediated forms are emerging (4), presenting a serious clinical threat.

Imipenem and newer carbapenems have long been resistant to β-lactamases, and therefore been considered as drugs of choice against gram-negative bacteria, resistant to cephalosporins. Although AmpC slowly hydrolyses carbapenems, it only contributes to resistance in the case of impaired penetration due to the lack of the D3 porin (31) or more rarely, in porin deficient enterobacteria (32).

3.3. CLASS D β-LACTAMASES

With 31 kDa, the size of class D β-lactamases falls between that of classes A and C. Produced by Gram negative bacteria, they are much less known than the other three categories, mainly due to the lack of an X-ray structure. They are mostly, but not exclusively plasmid mediated. Nine primary structures have been determined so far (33). Both the substrate profile and the kinetics of hydrolysis of the class D enzymes are different from those of classes A and C. They are often called oxacillinases, as they can efficiently hydrolyse both oxacillin and cloxacillin (34). Also, they exhibit a biphasic "burst" kinetics (35).

3.4. CLASS B β-LACTAMASES

Contrary to serine β-lactamases, these metallo enzymes contain one or two zinc atoms in the active site. The second Zn atom, however, does not necessarily play a role in the hydrolysis mechanism (36). In *Bacteroides fragilis*, the second Zn atom is not only dispensable for the catalysis, but can even be detrimental for the efficiency. The second binding site may be needed if a translocation of the metal ion occurs during the catalytic cycle. The fact that two Zn atoms are found in purified samples may only be due to the high concentration of Zn^{2+} during the enzyme production and the monocomplexed form could be predominant *in-vivo* (37).

The mechanism of action is different from serine β-lactamases, and they are not inhibited by usual β-lactamase inhibitors (38). Historically, they have not played an important role in the resistance and their presence was limited to *B. cereus* until carbapenem hydrolysing chromosomal enzymes appeared (39,40). Only a few years later, plasmid born enzymes emerged in *Pseudomonas aeruginosa* and *Bacteroides fragilis* (41,42).

The metallo enzymes belong to three groups according to the sequence homology. They are differentiated by the Zn binding amino acids (43).

The IMP-1 metallo-β-lactamase produced by *Serratia marcescens* or *Pseudomonas aeruginosa* exhibits a broad-spectrum activity profile. It hydrolyses not only penicillins and cephalosporins, including cephamycins, but also carbapenems. Monobactams are resistant to this enzyme (44,159).

4. PBPs

The high molecular weight penicillin binding proteins, those having a transpeptidase activity, are the primary targets of β-lactams. The inhibition of sometimes one but more often several of them leads to bacterial death. The PBPs of *E. coli* are numbered from 1 to 7. PBPs 1A and 1B are complementary, and only a double knockout or inhibition of both proteins lead to bacterial death. These enzymes are important for the formation of an intact peptidoglycan, and their selective inhibition leads to cell lysis. The PBPs 2 and 3 play a role in cell elongation and septum formation respectively. All these enzymes have the same substrate, but play a different role, and display different sensitivity to beta lactams.

The structure of the staphylococcal cell wall is slightly different from that of Gram negative bacteria. It is a multilayer peptidoglycan, the cell is round shaped and the cross-linking involves a lysine together with a pentaglycin bridge. Therefore, the PBPs found in this bacterium are slightly different too.

Beta lactams are poorly recognised by the PBPs, but due to the irreversible nature of inhibition, they can block a wide variety of PBPs of different origin.

4.1. NATURALLY RESISTANT PATHOGENS

The various members of the beta lactam antibiotic family can inhibit one, two or several PBPs in one single cell, however the inhibited enzymes are not necessarily essential for the bacterial survival.

Enterococci, essentially *E. faecium and E. faecalis*, possibly due to their living conditions in the gastrointestinal tract, can survive in particularly harsh environments and are intrinsically resistant to a variety of antibacterial agents, including β-lactams, particularly cephalosporins. *E. faecium* has 3 essential PBPs, 1-3. PBPs 4 and 5 are not essential for growth. Although present in small quantities, the latter, acting alone can produce a stable peptidoglycan in various enterococcal species (45,46,47) providing a protection to the cell against β-lactams, due to their low affinity to these antibiotics. The level of resistance can be further increased by the overproduction of the enzyme resulting from the inactivation of the repressor gene (46). Mutations in the C-terminal part of the protein lead to very high level of resistance (48). PBP5 proteins display 54-77 % sequence homology between different enterococcal strains (49). The homology is even higher for the C terminal transpeptidase domain.

In *E. hirae* there is a second, low affinity PBP, the PBP3r (50). The gene coding for this protein is born on a large plasmid. *Enteroccus faecium* could be the origin of this PBP3r as it possesses a 99% sequence identity with the chromosomal pbp5fm of the latter (51).

For obvious reasons, β-lactam producing microorganisms are naturally resistant to these compounds. *Streptomyces clavuligerus*, contains a protein, the PcbR, which seems to be responsible for the resistance to penicillin. This protein could well be a resistant PBP as it shares structural similarity with both class A and B penicillin binding proteins (52,53).

4.2. INTRINSIC RESISTANCE THROUGH ACQUISITION OF RESISTANT PBPS

The introduction of methicillin, a penicillinase stable penicillin to the market was rapidly followed in by the emergence of resistant strains, the so-called methicillin resistant *S. aureus*, MRSA (54). These staphylococci acquired a new protein, of a still unknown origin, the so-called PBP 2a or 2', which is insensitive to β-lactams. It can substitute for all the other PBPs, although the level of cross-linking is relatively low when all other Penicillin Binding Proteins are inhibited (55).

The mec determinant, coding for PBP 2a (56,57,58), confers resistance to all families of β-lactams, including penicillins, cephalosporins and carbapenems. The reason for the resistance is still unknown at the molecular level. MRSA is often resistant to non β-lactam antibiotics as well. Since the transfer of vancomycin resistance from *Enterococcus* to *Staphylococcus* in laboratory conditions and the first isolation of a *S. aureus* with diminished susceptibility to vancomycin in Japan, the Methicillin Resistant Staphylococci represent the most important danger in the treatment of bacterial infections.

Four PBPs are normally present in staphylococci, one of them being essential. PBP 2a is an additional protein. It has been cloned by Matsuhashi (59), and shown to cause methicillin resistance to usual *Staphylococcus aureus* strains (156,157). Coagulase negative staphylococci are thought to be the origin of methicillin resistance (60). This protein also displays sequence homology with both low affinity PBPs 3r and 5 of *E. hirae* (61).

The protein displays sequence similarity with major portions of the PBP2 and PBP3 of *E. coli*, and also sequence homology between the N-terminal part and a staphylococcal penicillinase suggesting that the N-terminal portion may arise from a β-lactamase and the C-terminal from a PBP (62).

Although PBP2a is considered as the primary cause of methicillin resistance, inactivation of the natural PBP2 leads to decreased methicillin resistance (63) suggesting that even in the presence of β-lactams, the other PBPs still show some level of activity.

The expression of the resistance at a high level was well correlated with the IC_{50}'s against PBP2a. On the other hand, in staphylococci with low level of resistance the MIC correlates both with the activity on PBP2a and PBP4 possibly due to a reduction in the cell wall crosslinking by the inhibition of PBP4 (64). Also some synergy was observed against MRSA between β-lactams having different binding activities to PBP2a and other PBPs (65).

The level of methicillin resistance is further increased in two ways. Point mutations in the mecA gene result in even lower susceptibility of the coded protein (66). Overproduction of PBP4 also increases the level of resistance to β-lactams (67). This is in line with the observation that PBP2 knock out decreases the resistance.

The expression of PBP2a is regulated via the mecR1-mecA regulatory system. MecR1 is a transmembrane β-lactam sensor and *mecI* is a repressor of the *macA* transcription. The protein sequences are similar to those of *BlaR1* and *BlaI* respectively (68). There is also crosstalk between the two systems. The β-lactamase regulatory BlaR1-BlaI system can also induce expression of *mecA* (69). *MecA* is essential for the expression of the resistance but the relationship between the amount of *mecA* transcription and the production of PBP2a and the level of resistance depend on the genetic background of *Staphylococcus aureus*. These results show that probably additional factors are also involved in the regulation of methicillin resistance (70).

In a number of clinical isolates, the regulatory system is inactivated by point mutation, allowing a constitutive expression of PBP2a and further increasing the level of methicillin resistance (71).

A series of other staphylococcal genes, the so-called fem factors (factors essential for methicillin resistance) regulate the level of resistance to β-lactams (72). The *femAB* operon codes for proteins which are essential for the building up of the pentaglycin chain. Inactivation of *femAB* leads to monoglycin side chains and to a dramatic (>4000 fold) reduction in methicillin MICs. Also single knock out of *femA* causes decrease in MICs, although to a lesser (512 fold) extent (73,74,75). These knockout experiments also show that, in the assay conditions, these genes are not essential for the cell survival, but are necessary to the expression of methicillin resistance.

The precise role of femA and femB is unknown. They tentative function could be the successive ligation of the glycin units, but as no enzymatic activity has been shown so far *in-vitro*, they also could play an indirect role in the biosynthesis of the pentapeptide crossbridge.

FemA and *FemB* are specific and cannot substitute for each other (76). It is unknown, why these deletions decrease methicillin resistance, but it is possible that the side chain with one single glycin is not recognised by PBP2a (77). Interestingly, the inactivation of *femAB* in the absence of PBP2a also increases the susceptibily not only to β-lactams but also to different classes of antibiotics i.e. fosfomycin, fluoroquinolones and macrolides (78,73), showing that the cell is much less viable. This is in agreement with the observation that femAB mutants are not able to cause infections *in-vivo*.(79).

4.3. MOSAIC GENES

The fact that numerous bacteria have penicillin binding proteins insensitive to beta lactams was unnoticed for a long period of time, simply because lots of them are not pathogens. This is the case with some streptococcal strains which possess also PBPs with low sensitivity to beta lactams. They can share these characteristics with other strains of streptococci. Instead of donating a whole gene, they exchange only some fragments. This process yields mosaic PBPs where large parts are identical to the natural PBP but some inserts are different.

The low affinity PBP2 of *Neisseria* is a homologue of the *E. coli* PBP3 (40% sequence similarity). The difference from the sensitive enzyme is located only in the transpeptidase region (80). The pathogens *N. gonorrhoeae* and *N. meningitidis* are capable of exchanging parts of their PBP genes with proteins less susceptible to penicillin, coming from the non pathogenic *N. flavescens* or *N. cinerea* (81,82)

Streptococcus pneumoniae shows increasing resistance to antibiotic treatment.

Three genes PBP1a, 2b and 2x with reduced susceptibily to penicillin, isolated from penicillin resistant *S. pneumoniae* have also a mosaic structure (83,84,85). The same phenomenon is observed in *H. influenzae.*

The origin of these mosaic genes in naturally transformable strains is an interspecies recombination event, permitting the exchange of DNA fragments of the host's original PBPs with an allele of a resistant PBP acquired by transformation from a related species from the same environment. The alteration of PBP1 and PBP2 of gonococcus leads only to a low level of resistance *in-vitro*. However the modification of both enzymes results in a 1,000 decrease in penicillin susceptibility, a situation similar to the one observed in clinical isolates.

Different PBPs play different roles depending on the β-lactam. PBP2x acts as the primary PBP target in cefotaxime-resistant mutants, whereas PBP2b is the primary target in piperacillin-resistant mutants. Depending on the mutations in PBP2x, Pneumococcus can resist only to cefotaxime or also to penicillins (86, 166).

The reduced susceptibility of PBP2X is explained by the lack of a water molecule in the active site cavity of PBP2X as evidenced by X-ray crystallography and site directed mutagenesis (87). Via a local structural modification, the partial or complete loss of this water molecule could result in the reduction of the acylation efficiency of PBP2x substrates. Cell wall synthesis would still occur, but the sensitivity to β-lactams would be reduced.

Sequence comparison of PBP1A from penicillin resistant Pneumococci suggests that mutation of Thr-371 would contribute to the development of penicillin resistance in penicillin resistant *S. pneumonia* with altered PBP2x and PBP2b genes (88).

4.4. POINT MUTATIONS

Single mutations can not only increase the resistance level of PBPs of low affinity to β-lactams, i.e. PBP5 of enterococci or PBP2a of *S. aureus*, but also confer resistance to normally sensitive proteins.

A simple insertion of an extra amino acid into PBP2 can cause resistance in *N. gonorrhoeae* (89).

Also a single amino acid change Thr550→Ala in PBP 2x of laboratory mutants confer first-level β-lactam resistance to *S. pneumoniae* against cefotaxime (90). This mutation is observed in a highly resistant clinical isolate, but confers only low level of resistance in the laboratory strain. Interestingly the same mutation induces hypersensitivity to penicillin (66). An additional modification of Gly550 further decreases the sensitivity. The transformation Thr446→Ala0 in PBP 2b confers a low level of resistance to piperacillin.

These single mutations are of little clinical relevance. They are, however, important in the stepwise development of highly resistance strains containing multiple alterations of several PBPs, and also in the better understanding of the interaction between β-lactams and PBPs.

In *S. aureus,* mutations in the natural PBP2 lead to a low level of resistance through the decrease of affinity to β-lactams.

These point mutations were detected near the conserved penicillin-binding motifs in BB255R and CDC6, two low-level methicillin-resistant strains. Kinetic data suggest that in both strains the binding of penicillin to PBP2 is altered due to both lower binding affinity and more rapid release of bound drug (66).

In clinical isolates of *Enterococcus faecium* the highest resistance to β-lactams was observed when low expression of PBP5 was combined with mutations in positions 485, 499, 629 and 667. In the most resistant isolates to ampicillin, an additional serine was inserted after Ser-466 (91).

There are also laboratory mutants of *E. coli* with reduced susceptibility to mecillinam and cephalexin due to reduced affinities for PBPs 2 and 3 respectively. However, these sorts of mutants have not been found in a clinical setting, possibly due to their reduced viability. The β-lactamase mediated resistance remains more efficient in Gram negative bacteria (92).

5. Penetration Barrier

Gram negative bacteria are surrounded by a second protective layer, the outer membrane, in addition to the peptidoglycan. This membrane can be crossed by small hydrophobic molecules (93), but not by hydrophilic compounds. This penetration barrier of the outer membrane has long been considered as an important hurdle in Gram negative bacteria for β-lactams to reach their targets and as being the major cause of resistance in *Pseudomonas aeruginosa.*

The inner leaflet of this membrane is composed of phospholipids, the major constituents of the outer leaflet are lipopolysaccharides. The outer membrane of enteric bacteria contains approximately 50% of proteins (94). A part of these proteins, the porins, form hydrophilic channels enabling small hydrosoluble substances of <600 Da to reach the periplasmic space, but exclude hydrophobic ones (95). *E. coli* K12, expresses two porins with no particular specificity, OmpF and OmpC (96). The relative expression depends on the osmolarity of the medium regulated by the proteins ompR and the sensor envZ (97). The structures of the porins have largely been investigated by X-ray crystallography and electron microscopy (167).

In *Escherichia coli,* β-lactams utilise predominantly this porin pathway (98). OmpF and OmpC are non-specific and allow β-lactams and other substances to penetrate in a passive way.

Pseudomonas aeruginosa lacks the high permeability porins present in most other Gram negative bacteria. The low efficiency porins of this organism allow a hundred fold lower diffusion rate compared to usual porins (99). Because of this small diffusion rate Pseudomonas is intrinsically resistant to a wide variety of antibiotics.

The penetration barrier in Gram negative bacteria is not simply a yes or no question, and is rarely a reason for resistance alone (100). For example, by using *Salmonella typhimurium* cells lacking an efflux pump or by inhibiting the efflux by de-energetisation with CCCP it has been shown that nafcillin, a lipophilic penicillin, can quite easily cross the outer membrane of this Gram negative rod and the drug displays a reasonable MIC of 8 µg/ml (101).

To be active, β-lactams have to reach a stationary concentration in the periplasmic space in order to be able to inhibit PBPs. Several factors, however, tend to decrease this concentration. Due to the short generation time, the periplasmic space is steadily growing. This phenomenon tends to permanently decrease the antibiotic concentration as do some efflux proteins. In addition, β-lactamases also continuously deactivate the antibiotic. Consequently the decreased penetration becomes only a resistance factor when combined with other effects. Using permeability coefficients determined in reconstituted liposome systems it was possible to calculate the external concentration (MIC) necessary to reach the minimal periplasmic concentration sufficient for the inhibition of PBPs (102). An improved procedure (103) allowed the correct prediction of MICs in 61 out of 65 assays. Also, with very slow penetration rates, the MIC tends to be underestimated, possibly because of some "leakage" of the outer membrane, allowing the slow penetration of solutes when the ability to cross the porins is very low (100).

Laboratory mutants of *Escherichia coli* K12, showing moderate resistance to carbenicillin were isolated. They are deficient in the ompF porin (104), but retained the ompC protein, which is a narrower channel. The MIC only increased with those antibiotics which originally displayed low penetration rate. Similar mutants have also been isolated in other laboratories. Even though decreased penetration can lead to increased resistance in laboratory strains, it is less frequent in clinical settings as mutants with less efficient porins may be disadvantaged *in-vivo* (100).

The less efficient the protection, the more frequent the selection of permeability mutants, as shows the case of *S. marcescens* in presence of broad spectrum cephalosporins, which are stable to Serratia's chromosomal β-lactamase (100).

Although most β-lactams use passive diffusion through non-specific porins, there are also specific transporters in bacteria. Imipenem is one of the few β-lactams that can inhibit *Pseudomonas aeruginosa* in clinically useful concentrations. This is partly due to its intrinsic stability to β-lactamases but also to a specific transport into the periplasmic space through the D2 protein. This transporter is specific for basic amino acids and because of structural similarity helps also some carbapenems crossing the outer membrane. Imipenem resistant Pseudomonas mutants are deficient in the D2 protein (105). There is however evidence that while imipenem requires the presence of the D2 porin, meropenem, a related compound may use other pathways (106).

6. Antibiotic Efflux

New or altered PBPs and β-lactamases are the major cause for emergence of resistance in bacteria but limited access to the target protein can also impair the efficacy of β-lactam antibiotics. Naturally, Gram negative rods display more intrinsic resistance due the presence of an outer membrane. Until recently, decreased penetration across this external bacterial membrane has been considered as the only limitation to the antibiotic to have free access to the penicillin binding proteins. More recently, it has been shown that the efflux mechanisms cannot only pump out unwanted molecules from the cytoplasm but also from the periplasmic space, where the targets of β-lactams are located.

Several genes have been reported to be directly or indirectly involved in multidrug resistance in bacteria, including resistance to β-lactams (164): AcrRAB and marRAB in *Salmonella typhimurium* and *Escherichia coli*, ramA in *Klebsiella pneumoniae*, pqrA in *Proteus vulgaris*, mtrRCDE in *Neisseria gonorrhoeae*, Cje in *Campylobacter jejuni* (101,107). *Pseudomonas aeruginosa* possesses three efflux systems, but only the *mexAB-oprM* seems to be constitutively expressed in normal laboratory conditions.

The *mexA*, *-mexB*, *-oprM* operon, involved in the pyoverdine transport in *P. aeruginosa* (108,109) efficiently contribute to the resistance to β-lactams. MecA and oprM null mutants show increased susceptibility to various β-lactams, indicating that these molecules may be substrates of the *MexA-MexB-OprM* pump. *MexB* functions as an efflux pump across the cytoplasmic membrane, and is linked to outer membrane component *OprM* through the lipoprotein *MexA* (110). The *MexA-MexB-OprM* system confers resistance to meropenem, cephems and quinolones.

Two other systems *mexCD-oprJ* and *mexEF-oprN* are also present in *Pseudomonas aeruginosa* with different substrate specificity (111). They recognise new cephalosporins i.e. cefpirom and cefozopran and carbapenems i.e. imipenem, respectively.

Cloning the *MexA-MexB-OprM* operon in *E. coli*, deficient in the endogenous multidrug efflux system, conferred resistance to several β-lactams among other antibiotics (112). Interestingly, despite the periplasmic location of the β-lactams, expression of the outer membrane component OprM alone is not sufficient to provide resistance to this family of antibiotics (158). The same experience with the *mexCD-oprJ* operon resulted in resistance to various antibiotics but not to β-lactams.

The presence of β-lactamases can potentiate the effect of the efflux pumps, even with those molecules which are only slowly hydrolysed (113). *Pseudomonas aeruginosa* lacking the expression of both the AmpC β-lactamase or that of the *MexAB-OprM* pump display hypersensitivity compared to those strains where either one or both of the defence systems are present.

β-Lactamase inhibitors are also substrates of *mexAB-oprM* mediated efflux. Direct evidence has been obtained taking advantage of the (weak) intrinsic antibacterial activity of Cloxacillin, Clavunate and BRL42715 (114). The mechanism has also been confirmed indirectly by the increased synergistic activity of Cloxacillin and BRL42715 with ampicillin and cephaloridin in pump deficient *P. aeruginosa* (114).

Although β-lactams carry out their activity in the periplasmic space the intact tripartite system is necessary for the efflux. The outer membrane component OprM does not participate in the recognition of the β-lactam and this role is assured by the inner membrane-associated component (115). Consequently, the *MexAB-OprM* must be able to accommodate its substrate from the outer site of the inner membrane. According to this model β-lactams would be picked up from the outer leaflet and cytoplasmic drugs from the inner leaflet of the cytoplasmic membrane (116).

This hypothesis is supported by the observation that the AcrAB system in *Salmonella typhimurium* was more efficient with β-lactams containing a more lipophilic side chain, as they could be at least partially partitioned into the lipid bilayer of the inner membrane and thereby be captured by the pump (101).

A very subtle equilibrium between influx - efflux - hydrolysis by β-lactamases is needed for the activity. A mutant cell of PAO1, with no functional mexAB-oprM efflux is 64 fold more sensitive against carbenicillin than the parent strain (117). Knocking out the AmpC gene, however, has no effect on the activity, confirming the resistance of this substance to the AmpC enzyme. Amoxicillin behaves in the opposite way. The β-lactamase negative mutant becomes sensitive to this antibiotic, but the absence of the efflux system has no effect on the MICs.

Meropenem is stable to AmpC and is a poor substrate of the efflux pump. Therefore only the double knock out improves the activity tenfold.

The double mutation has only a four-fold effect on imipenem, however, when the penetration rate is limited because of the absence of the D2 channel, the absence of AmpC leads to a 64-fold decrease in activity (117).

7. The future of β-Lactams

β-lactams have certainly been the most successful family of antibiotics and still today a lot of effort is devoted to the improvement of their properties in order to overcome resistance (118). There are essentially two ways that have been exploited. The first is, of course, the synthesis of new derivatives which are not susceptible to the known resistance phenomena. The second is the development of compounds, to be used in combination with known β-lactams, interfering with the expression of the resistance.

7.1. NEW FAMILIES

Since the discovery of penicillin tens of thousands of new β-lactam derivatives have been prepared and more than one hundred are in clinical use. The first significant event was the synthesis of methicillin, which solved the problem of penicillin resistance as it was stable to penicillinases. However, this relief was only of short duration and new strains (MRSA) rapidly emerged resistant to methicillin by another mechanism. The discovery of cephalosporins was also a significant event to escape the destructive action of penicillinases, but these compounds only selected another class of β-lactamases. Next, the advent of carbapenems was followed by the appearance of metallo enzymes. Monocyclic β-lactams, in particular those having a tetrazole substituent, are stable to cephalosporinases, but having sacrificed the activity against staphylococci, their clinical use was limited. The successive discovery of new families i.e. cephalosporins, carbapenems, led to a significant jump in the antibacterial spectrum but selected shortly new resistance phenomena.

7.2. NEW GENERATIONS

The introduction of various substituents in the proximity of the cephem nucleus i.e. the α-methoxy group of cephamycins, or the syn-oxyimino group of third generation cephalosporins temporarily decreased the susceptibility to β-lactamases, due to steric hindrance (119,120) or interference with the catalytic water molecule (121), but mutations in some penicillinases led to the development of the extended spectrum enzymes (122).

Carbapenems are naturally stable to classical β-lactamases, although metallo enzymes and new serine-carbapenemases are able to hydrolyse these compounds. The research was focussed on the improvement of their antibacterial spectrum and *in-vivo* behaviour (123).

With the aim of broadening the antibacterial spectrum and fight against β-lactamases, cephalosporins, but also penems and carbapenems, have been combined in one single molecule with antibacterials with a different mode of action (124).

The latest major development in the cephalosporin and to a lesser extent in the penicillin field was the use of an iron transport system, to make compounds enter more easily into *Pseudomonas aeruginosa* (125) and the development of new molecules active against MRSA.

Pseudomonas aeruginosa has long been a particularly challenging target. Active compounds i.e. quaternary ammonium cephalosporins and in particular ceftazidime can diffuse through the outer membrane across non-specific porins. Some carbapenem such as imipenem penetrate via the D2 channel, specific for the transport of basic amino acids. However, resistance is emerging against both classes of compounds. In 1982, E-0702, shown to possess potent activity *against Pseudomonas aeruginosa* (126,127) turned out to be the prototype of a series of catechol containing β-lactams. This was followed in 1984 by a catechol containing penicillin, BRL 36650.

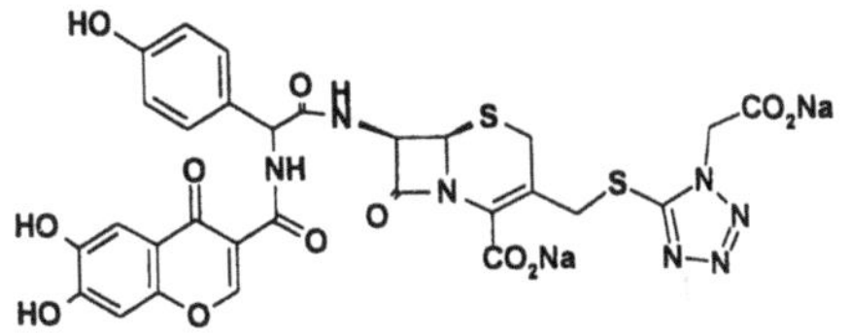

E-0702

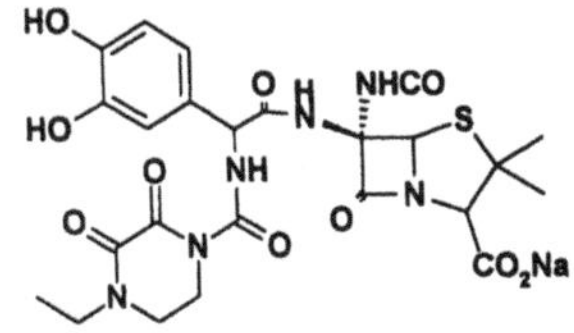

BRL 36650

Since the observation that catechol containing β-lactams can diffuse through a siderophor transport system (128,129) a wide variety of molecules have been produced containing a catechol or a surrogate in different positions of the molecules (125).

As the resistance is multifactorial, the most active compounds combine high penetration rate with good β-lactamase stability.

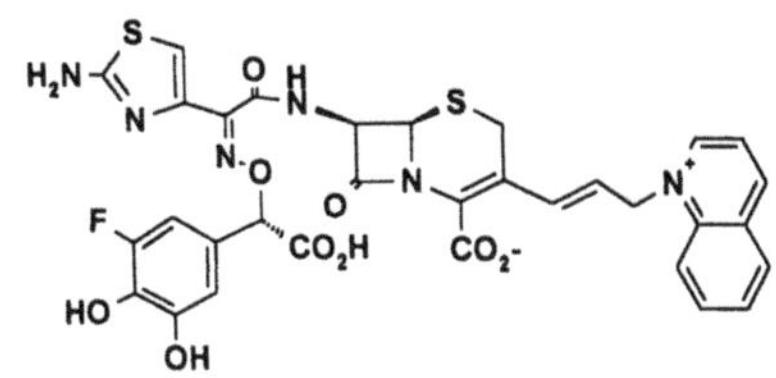

RU 59863

For example, RU 59863 (130) has been optimised to this end. This compound is perfectly stable to most chromosomal cephalosporinases and extended spectrum β-lactamases. The compound is highly active against both *Pseudomonas aeruginosa* and staphylococci. MRSA, however, remain resistant.

Another recent development in the field, targets methicillin resistant *Staphylococcus aureus*. Based on molecular modelling using crystal structures of class A β-lactamase as a model for PBPs, cis penems have been prepared. These compounds show improved activity towards MRSA, however, they tend to be less stable to β-lactamases (131).

Most of the work has been done by introducing large lipophilic substituents in the 3 position of the cephalosporins or in the 2 position of carbapenems. Early molecules were characterised by poor solubility and high protein binding, deleterious for *in-vivo* activity. Newer compounds usually contain a quaternary ammonium or a basic amino function which is protonated at physiological pH. The double ammonium substituent of L-786,392 affords derivatives with increased activity against resistant Gram positive bacteria (132). J-111,225 (133) is a well tolerated compound which is active *in-vivo* both against MRSA and *Pseudomonas aeruginosa* (134).

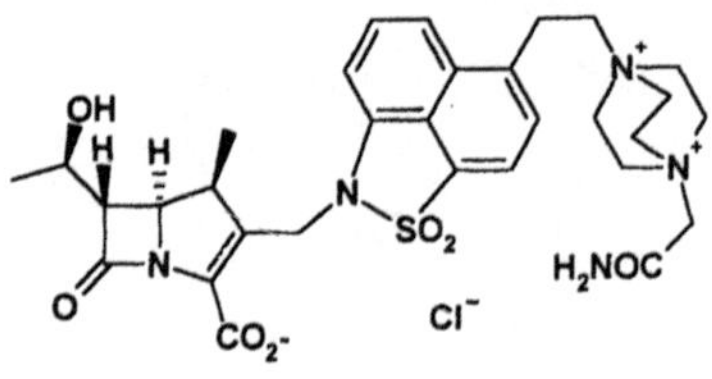

L-786,392

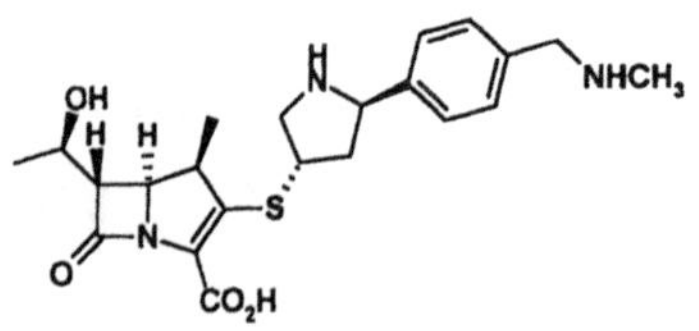

J-111,225

The 7-side chain of the new cephems is often a hydroxyimino version of the most common aminothiazolyl side chain of third generation cephalosporins, well known for its good performance in general against Gram positive cocci. MC-02,479, a new member of a series of anti-MRSA cephalosporins has increased water solubility compared to previous compounds from the same family. It displays 10 times higher activity against MRSA than imipenem (135) but it is inactive against *Enterobacteriaceae*. Ro 63-9141 inhibits PBP2a at submicromolar level and therefore it is active against MRSA (136). The compound also shows cefepime like potency against Gram negatives.

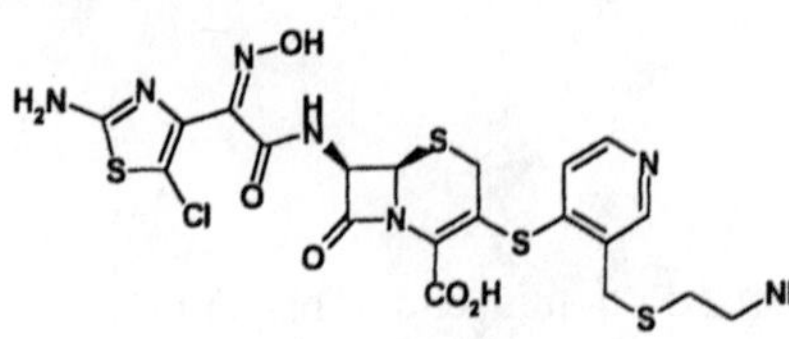

MC-02,479

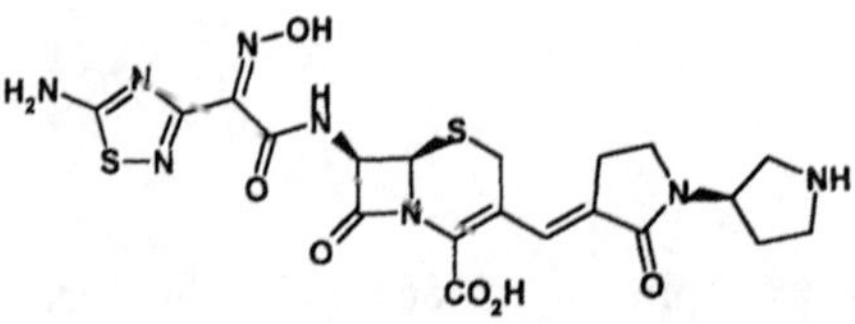

Ro 63-9141

As the reason for methicillin resistance is still unknown at the molecular level, these modifications have been made based on experiments and observation. Moreover, the highly lipophilic nature of these anti-MRSA β-lactams result often in poor solubility, antibacterial spectrum limited to Gram positive cocci and high protein binding, accompanied by a poor performance in animal disease models.

In order to find completely new compounds by screening, a Scintillation Proximity Assay has been set up to screen a small library compounds which compete with the binding of radiolabelled penicillin to PBP 2a. Compounds found positive in the assay possess weak antibacterial activity and are synergistic with β-lactams (160). Another assay is based on the nitrocephin induced precipitation of soluble PBP2a and is amenable to high throughput screening (137).

7.3. POTENTIATORS OF β-LACTAMS

The most well known class of compounds is constituted by the β-lactamase inhibitors which have been in clinical use for almost twenty years. These are mostly irreversible inhibitors, with the exception of the Beta Lactamase Inhibitory Protein (BLIP), which is a small, 17.5 kDa protein produced by *Streptomyces clavuligerus* (138). The same strain produces Clavulanic acid, an irreversible inhibitor, which is a component of Augmentin, a combination of a penicillin with a β-lactamase inhibitor. Two synthetic derivatives are also commercially available, sulbactam and tazobactam. The vast majority of (semi)synthetic β-lactamase inhibitors made in the research laboratories are penem derivatives. The difference with classical β-lactams is that after reaction with the active site serine a rearrangement takes place thus stabilising the acyl enzyme complex (139). They are highly active against class A enzymes, but with the exception of two compounds, BRL 42715 and tazobactam, (139), they are mostly hydrolysed by class C β-lactamases and are therefore ineffective (140, 141). This happens when the hydrolysis of the acyl-enzyme intermediate is faster than the rearrangement.

A large number of β-lactamase inhibitors have been synthesised by the modification of the usual (oxa)penem and cephem nuclei (142).

From the understanding of the catalytic mechanism, with the help of X-ray structures of the *C. freundii* β-lactamase with aztreonam, another paradigm has recently been implemented. This is based on the rigidity of the acyl enzyme which needs no rearrangement after ring opening (143). A series of bridged monobactams displayed good inhibitory activity against class C enzymes and potentialised the activity of ceftriaxone against *Citrobacter freundii.* Ro 48-1256 reduced imipenem MICs from 1-2 mg/l to 0.25-0.5 for *Pseudomonas aeruginosa* isolates (144).

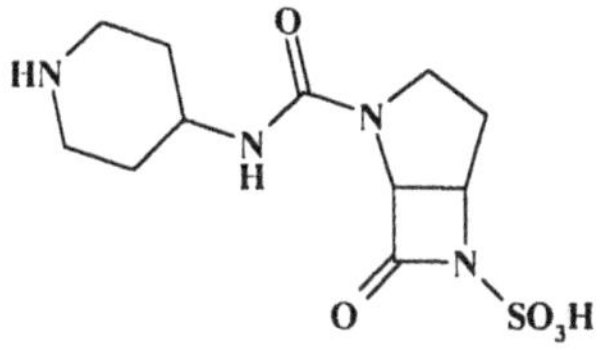

Ro 48-1256

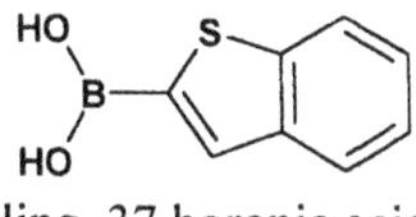

With the help of molecular modelling, 37 boronic acid derivatives have been designed as transition state analogues. The most active compound, despite of the very simple structure, displays an affinity (Ki) to *E. coli* AmpC β-lactamase of 27 nM (145).

Inhibitors of metallo β-lactamases of imipenem resistant *Bacteroides fragilis* have been identified by combination of screening and molecular modelling. A series of diphenyl tetrazoles i.e. L-161,189 has been identified. An X-ray structure shows that the tetrazole moiety directly interacts with one of the two Zn atoms in the active site. The presence of these inhibitors increased the zone of inhibition of imipenem on an Agar plate. Some of these compounds also displayed weak antibacterial activity on their own (165).

Also from a screening of metallo protease inhibitors non β-lactam "penicillin like" thiol containing compounds have recently been identified. They display synergistic effect with β-lactams (146).

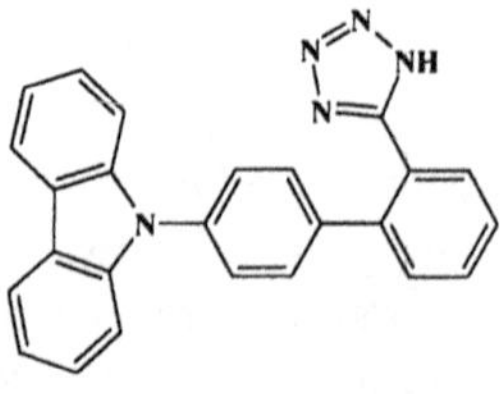

L-161-189

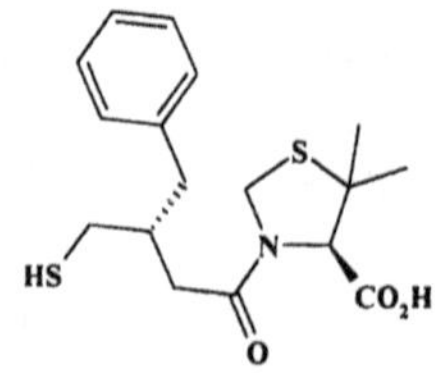

SB-224043

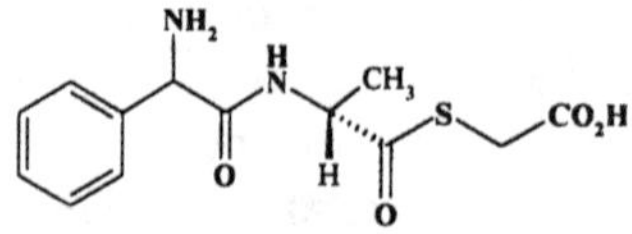

SB-216271

Thiol esters such as SB 216271, are poor inhibitors of metallo β-lactamases but introduce an interesting new concept. They are mechanism based irreversible inhibitors and not simple complexing agents. Upon cleavage of the thiol ester bond, a disulfide bridge would form with the active site cysteine in the aerobic conditions of the assay (162, 163).

As done with β-lactamase inhibitors, another logical approach is the inhibition of the expression of methicillin resistance. Compounds could either interfere with the expression of PBP2a, or they could inhibit it selectively compared to other PBPs. Inhibitors of the Fem A/B pathway should also suppress methicillin resistance.

It was suggested that some antibiotics, acting on cell wall biosynthesis, display synergistic activity with β-lactams against *E. hirae*, through the inhibition of PBP synthesis (147).

Although the antibacterial activity of the tea (*Camellia sinensis*) has been known for 90 years, some fractions of the tea extract, with no antibacterial activity, can potentiate the activity β-lactams against MRSA by interfering with the expression of PBP2a (148).

LY 301621, an unnatural tripeptide, potentiates the activity of methicillin against MRSA. Alone, it is only a very weak inhibitor of bacterial growth (149). From another screening assay MC-200,612 and MC-207,252 have been identified (150). The compounds potentiate the activity of imipenem against MRSA by 500 fold. Both molecules seem to modify the expression of PBP2a, the latter also increases the potency of β-lactams *in-vivo* (151).

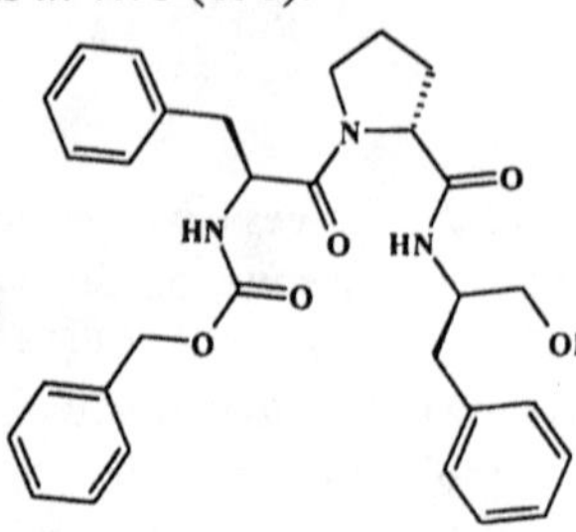

LY 391621

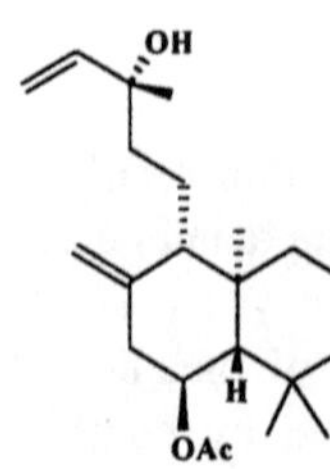

MC 200,616

8. Conclusions

Resistance to β-lactams emerged in parallel with the increased clinical use of these antibiotics. The discovery of new families of natural products i.e. penams, cephems, carbapenems or monobactams was followed by the rapid emergence of new resistance phenomena specific to the new classes of compounds. Extensive chemical modifications of existing compounds led to the selection of variants of existing resistance mechanisms capable of protecting bacteria from the new drugs.

Resistance is most often multifactorial. New PBPs confer resistance to old compounds but this resistance is further enhanced by additional mutations or overexpression of the proteins. The efficiency of β-lactamases is reinforced by a decreased penetration rate of the drug into the periplasmic space. The penetration barrier of Gram negative bacteria also acts in synergy with efflux pumps and can render even highly β-lactamase resistant drugs inefficient (152). The problem of altered PBPs and β-lactamases is specific to the β-lactam class of antibiotics. However, the penetration barrier and even more, the presence of efflux pumps conferring multidrug resistance, will remain a challenge in the future for whatever new classes of antibiotics, even acting on completely new targets.

β-lactam producing organisms may possess still unknown mechanisms of resistance. Resistance to carbapenems from the producing strain *Erwinia carotovora* could already be transferred to *Escherichia coli* in laboratory conditions (153).

For over 50 years, β-lactam research (154) has been able to solve most of the resistance problems associated with β-lactams, while expanding the antibacterial spectrum. However, because of MRSA and increasingly specific and efficient β-lactamases, more focussed solutions may be needed in the future.

β-Lactamases play an ubiquitous role in resistance not only in association with penetration barriers and efflux pumps but also in the borderline methicillin resistance. One can foresee that perfectly stable molecules, possibly with novel mechanisms of action should, to a large extent, solve this problem.

A tremendous amount of knowledge about β-lactam resistance has been accumulated over the last 40 years. However, it is fair to say that up to now pharmaceutical research and bacteria have mainly used the same error and trial approach to respectively overcome and produce resistance by random screening and derivatisation of inhibitors on one hand and random mutation of genes on the other. This traditional pharmaceutical approach, however, has shown its limitation, yielding molecules with higher and higher complexity, and narrower antibacterial spectrum. Knowledge about resistance played little role in drug design.

In the recent years, a more rational approach has been emerging taking advantage of the structural information about PBPs and β-lactamases (155). New β-lactamase inhibitors with novel mechanism of action and a better understanding of the role of efflux versus penetration barrier pave the way for rational design of more powerful penicilloyl serine transferase inhibitors.

References

1, Ghuysen, J.-M., Charlier, P., Coyette, J., Duez, C., Fonzé, E., Fraipont, C., Goffin, C., Joris, B., Nguyen-Distéche, M. (1996), Penicillin and beyond : evolution, protein fold, multimodular polypeptides and multiprotein complexes. *Microb. Drug. Resist.*, *2*, 163-175.

2, Adam, M., Fraipont, C., Rhazi, N., Nguyen-Disteche, M., Lakaye, B., Frere, J.-M., B Devreese, Van Beeumen, Van Heijenoort, Y., Van Heijenoort, J., Ghuysen, J.-M., (1997), The bimodular G57-V577 polypeptide chain of the class B penicillin- binding protein 3 of Escherichia coli catalyzes peptide bond formation from thiolesters and does not catalyze glycan chain polymerization from the lipid II intermediate, *J. Bacteriol., 179*, 6005-6009.

3, Benz R., Beacterial Cell Wall, Matsuhashi, M., R., Elsevier 1994, pp. 55-71.

4, Matagne A., Dubus A., Galleni M., Frère J.-M., (1999), The β-lactamase cycle: a tale of selective pressure and bacterial ingenuity, *Nat. Prod. Rep., 16*, 1-19.

5, Massova, I., Mobashery, S., (1998), Kinship and diversification of bacterial penicillin-binding proteins and β-lactamases, *Antimicrob. Agents Chemother., 42(1)*, 1-17.

6, Levy, S.B., (1992), The antibiotic paradox, Plenum Press, pp. 1-12.

7, Livermore, D.M. (1998), β-lactamases: quantity and resistance, *Clinical Microbiology and Infection, 3*, 4S10-4S19.

8, Philippon, A., Dusart, J., Joris, B., Frère, J.-M., (1998), The diversity, structure and regulation of β-lactamases, *Cell. Mol. Life Sci., 54*, 341-346.

9, Bush, K., Jacoby, G.A., Medeiros, A.A., (1995), A functional classification scheme for β-lactamases and its correlation with molecular structure, *Antimicrob. Agents Chemother., 39(6)*, 1211-1233.

10, Ambler, R.P., (1980), The structure of β-lactamases, *Philos. Trans. R. Soc. Lond. (Biol), 289*, 321-331

11, Waley, S.G., (1992), in The chemistry of β-lactams, ed. M.I. Page, pp. 198-228, A. & P. Blackie, London.

12, Matagne, A., Lamotte-Brasseur, J., Frère, J.-M., (1998), Catalytic properties of class A β-lactamases: efficiency and diversity, *Biochem. J., 330*, 581-598.

13, Jacoby, G., Bush, K., Amino Acid Sequences for TEM, SHV and OXA Extended-Spectrum and Inhibitor Resistant ß-Lactamases *http://www.lahey.org/studies/webt.htm*

14, Chanal-Claris, C., D. Sirot, L. Bret, P. Chatron, R. Labia, and J. Sirot. (1997), Novel extended-spectrum TEM-type ß-lactamase from an Escherichia coli isolate resistant to ceftazidime and susceptible to cephalothin, *Antimicrob. Agents Chemother. 41*, 715-716.

15, Bastarrachea, F., (1998), On the origin of plasmid-borne, extended-spectrum, antibiotic resistance mutations in bacteria, *J. Theor. Biol., 190*, 379-387.

16, Maveyraud, L., Mourey, L., Kotra, L.P., Pedelaq, J.-D., Guillet, V., Mobashery, S., Samama, J.-P., (1998), Structural basis for clinical longevity of carbapenem antibiotics in the face of challenge by the common class A β-lactamases from the antibiotic resistant bacteria, *J. Am., Chem., Soc., 120*, 9748-9752.

17, Yang, Y., Wu, P., Livermore, D.M., (1990), Biochemical characterization of a β-lactamase that hydrolyses penems and carbapenems from two *Serratia marcescens* isolates, *Antimicrob. Agents Chemother., 34*, 755-758.

18, Nass, Y., Vandel, L., Sougakoff, W., Livermore, D.M., Nordmann, P., (1994), Cloning and sequence analysis of the gene for a carbapenem-hydrolysing class A β-lactamases, Sme-1, from *Serratia marcescens* S6, *Antimicrob. Agents Chemother., 38*, 1262-1270.

19, Nordmann, P., Mariotte, S., Naas, T., Labia, R., Nicolas, M.H., (1993), Biochemical properties of a carbapenem-hydrolysing β-lactamase from *Enterobacter cloacae* and cloning of the gene into *Escherichia coli, Antimicrob. Agents Chemother., 37*, 939-946.

20, Naas, T., Nordmann, P., (1994), Analysis of a carbapenem-hydrolysing class A β-lactamase from *Enterobacter cloacae* and of its Lys-R type regulatory protein, *Proc. Natl. Acad. Sci., USA, 91*, 7693-7697.

21, Raquet, X., Lamotte-Brasseur, J., Bouillenne, F., Frère, J.-M., (1997), A disulfide bridge near the active site of carbapenems-hydrolysing class A β-lactamases might explain their unusual substrate profile. *Proteins, Structure, function and genetics, 27*, 47-58.

22, Swaren P, Maveyraud L, Raquet X, Cabantous S, Duez C, Pedelacq JD, Mariotte-Boyer S, Mourey L, Labia R, Nicolas-Chanoine MH, Nordmann P, Frere JM, Samama JP, (1998) X-ray analysis of the NMC-A beta-lactamase at 1.64-A resolution, a class A carbapenemase with broad substrate specificity, *J Biol Chem, 273*, 26714-26721

23, Wu. P.J., Shannon., K.S., Philips, I., (1994), Effect of hyperproduction of TEM-1 β-lactamase on in vitro susceptibility of *Echerishia coli* to β-lactam antibiotics, *Antimicrob. Agents Chemother., 38*, 494-498.

24, Reguera, J.A., Baquero, A.F., Perez-Diaz, J.C., Martinez, J.L., (1991), Factors determining resistance to β-lactam combined with β-lactamase inhibitors in *Echerishia coli, J. Antimicrob. Chemother., 27*, 569-575.

25, Stapleton, P., Wu, P.J., King, A., Shannon, K., French, G., Phillips, I., (1995), Incidence and mechanism of resistance to the combination of amoxicillin and clavulanic acid in *Echerishia coli*, *Antimicrob. Agents Chemother.*, *39*, 2478-2483.
26, Henquell, C., Chanal, C., Sirot, D., Labia, R., Sirot J., (1995), Molecular characterisation of nine different types of mutants among 107 inhibitor resistant TEM β-lactamases from clinical isolates of *Echerishia coli*, *Antimicrob. Agents Chemother.*,*39*, 427-430.
27, Nicholas-Chanoine, M.H. (1997) Inhibitor-resistant β-lactamases, *J. Antimicrob. Chemother.*, *40*, 1-3
28, Knox, J.R., Moews, P.C., Frère J.-M., (1996), Molecular evolution of bacterial β-lactam resistance, *Chem. Biol.*, *3*, 937-947.
29, Lindberg, F., Normark, S., (1986), Contribution of chromosomal beta-lactamases to beta-lactam resistance in enterobacteria, *Rev. Infect. Dis.*, *Suppl. 3*, S292-304.
30, Oefner, C., Darcy, A., Daly, J.J., Gubernator, K., Charnas, R.L., Heinze, I., Hubshcwerlen, C., Winkler, F.K., (1990), Refined crystal structure of beta-lactamase from *Citrobacter freundii* indicates a mechanism for beta-lactam resistance, *Nature*, *343*, 284-288.
31, Livermore, D.M., (1992), Interplay of impermeability and chromosomal β-lactamase in imipenem resistant *Pseudomonas aeruginosa*, *Antimicrob. Agents Chemother. 36*, 2046-2048.
32, Bradford, P.A., Urban, C., Mariano, N., Projan, S.J., Rahal, J., Bush, K., (1996), Imipenem resistance in *Klebsiella pneumoniae* is associated with the combination of ACT-1, a plasmid-mediated Ampc β-lactam, and the loss of an outer membrane protein, *Antimicrob. Agents Chemother.*, *41*, 563-569.
33, Sanschagrin, F., Couture, F., Levesque, R.C., (1995) Primary structure of OXA-3 and phylogeny of oxacillin-hydrolyzing class D beta-lactamases *Antimicrob. Agents Chemother.*, *39*, 887-893.
34, Ledent, P., Raquet, X., Joris, B., Van Beeumen, J., Frère, J.-M., (1993), A comparative study of class-D beta-lactamases. *Biochem. J.*, 292, 555-562.
35, Couture, F., Lachapelle, J., Levesque, R.C., (1992), Phylogeny of LCR-1 and OXA-5 with class A and class D beta-lactamases, *Mol. Microbiol.*, *12*, 1693-1705.
36, Fabiane, S.M., Sohi, M.K., Wan, T., Payne, D.J., Bateson, J.H., Mitchell, T., Sutton, B.J., (1998), Crystal structure of the zinc-dependent beta-lactamase from *Bacillus cereus* at 1.9 Å resolution: binuclear active site with features of mononuclear enzyme. *Biochemistry*, *37*, 12404-11.
37, Paul-Soto, R., Hernandez-Valladares, M., Galleni, M., Bauer, R., Zeppezauer, M., Frère, J.-M., Adolph H.-W., (1998), Mono and binuclear Zn-β-lactamase from Bacteroides fragilis: catalytic and structural roles of the Zn ions, *FEBS Lett.*, *438*, 137-140.
38, Bush K., (1998), Metallo-beta-lactamases: A class apart, *Clin. Infect. Diseases*, *27(suppl. 1)* S48-53
39, ShannonK., King A., Phillips I., (1986), β-Lactamases with high activity against imipenem and SCH 34343 from *Aeromonas hydrophila*, *J. Antimicrob. Chemother.*, 17, 45-50.
40, Cuchural G.J. Jr, Malamy M.H., Tally F.P., (1986), β-Lactamase-mediated imipenem resistance in *Bacteroides fragilis*, *Antimicrob. Agents Chemother.*, *30*, 645-648.
41, Watanabe M. Iyobe S., Inoue M., Mitsuhashi S., (1991), Transferable imipenem resistance in *Pseudomonas aeruginosa*, *Antimicrob. Agents Chemother.*, *35*, 147-151.
42, Bandoh K., Watanabe K., Muto Y., Tanaka Y., Kato N., Ueno K., (1992), Conjugal tyransfer of imipenem resistance in *Bacteroides fragilis*, *J. Antibiot.*, *145*, 542-547.
43, Rasmussen B.A., Bush K., (1997), Carbapenem hydrolysing β-lactamases, *Antimicrob. Agents Chemother.*, *41*, 223-232.
44, Laraki, N., Franceschini, N., Rossolini, G. M., Santucci, P., Meunier, C., de Pauw, E., Amicosante, G., Frère, J. M., Galleni, M. (1999), Biochemical Characterization of the *Pseudomonas aeruginosa* 101/1477 Metallo-beta-Lactamase IMP-1 Produced by *Escherichia coli*, *Antimicrob. Agents Chemother. 43*, 902-906.
45, Canepari, P., Del Mar Lleo, M., Fontana, R., Satta, G. (1981), *Streptococcus faecium* mutants that are temperature sensitive for cell growth and show alterations in penicillin-binding proteins, *J. Bacteriol.*, 169, 2432-2439.
46, Fontana R., Ligozzi M., Pittaluga F., Satta G., (1996), Intrinsic penicillin resistance in enterococci. *Microb. Drug Resist.*, *2*, 209-213.
47, Signoretto, C., Boaretti, M., Canepari, P., (1998), Peptidoglycan synthesis by *Enterococcus faecali* penicillin binding protein 5. *Arch. Microbiol.*, *170*, 185-190.
48, Zorzi, W., Zhou, X.Y. Dardenne O., Lamotte, J., Raze D., Pierre J., Gutmann L., Coyette J., (1996), Structure of the Low Affinity Penicillin-Binding Protein 5 PBP5fm in Wild-Type and High Penicillin Resistant Strains of *Enterococcus faecium*, *J. Bacteriol.*, *178(16)*, 4948-57.

49, Ligozzi, M., Pittaluga, F., Fontana, R., (1996), Modification of penicillin binding protein 5 associated with high level Ampicillin resistance in *Enterococcus faecium*, *Antimicrob. Agents Chemother.*, *40*, 354-357.

50, Piras, G., El Kharroubi, A., Van Beeumen, J., Coeme, E., Coyette, J., Ghuysen, J.-M. (1990), Characterization of an *Enterococcus hirae* Penicillin-binding protein 3 with low penicillin affinity, *J. Bacteriol.*, 172, 6856-6862.

51, Raze, D., Dardenne, O., Hallut, S., Martinez-Bueno, M., Coyette, J., Ghuysen, J.-M. (1998), The gene encoding the low-affinity penicillin-binding protein 3r in *Enterococcus* hirae S185R is borne on a plasmid carrying other antibiotic resistance determinants, *Antimicrob. Agents, Chemother.*, *42*, 534-539.

52, Paradkar, A.S., Aidoo, K.A., Wong, A., and Jensen, S.E., (1996), Molecular analysis of a beta-lactam resistance gene encoded within the cephamycin gene cluster of Streptomyces clavuligerus, *J. Bacteriol*, *178*, 6266-6274.

53, Goffin, C., Ghuysen, J.-M., (1998), Multimodular panicillin-binding proteins: An enigmatic family of orthologs and paralogs, *Microbiol. Mol. Biol. Rev.*, *62(4)*, 1079-1093.

54, Jevons, M.P. (1961), "Celbenin" -resistant Staphylococci, *Br. Med. J.* 124-125.

55, Boudewijn de Jonge, L.M., Tomasz, A. (1993), Abnormal peptidoglycan produced in a methicillin-resistant strain of *Staphylococcus aureus* grown in the presence of methicillin: functional role for penicillin-binding protein 2A in cell wall synthesis, *Antimicrob. Agents, Chemother.*, *37*, 342-346.

56, Beck, W.D., Berger-Bächi, B., Kayser, F.H., (1986), Additional DNA in methicillin-resistant *Staphylococcus aureus* and molecular cloning of *mec*-specific DNA, *J. Bacteriol.*, *165*, 373-378.

57, Archer, G.L., Niemeyer, D.M., (1994), Origin and evolution of DNA associated with resistance to methicillin in staphylococci, *Trends Microbiol.*, *2*, 343-347.

58, Ubukata, K., Nonoguchi, R., Matsuhashi, M., Konno, M., (1989), Expression and inducibility in *Staphylococcus aureus* of the *mecA* gene, which encodes a methicillin-resistant *Staphylococcus aureus* specific penicillin binding protein, *J. Bacteriol.*, *171*, 2882-2885.

59, Matsuhashi, M., Song, M.D., Ishino, F., Wachi, M., Doi, M., Inoue, M., Ubukata, K., Yamashita, N., Konno, M., J. (1986), Molecular cloning of the gene of a penicillin-binding protein supposed to cause high resistance to β-lactam antibiotics in *Staphylococcus aureus*, *J. Bacteriol.* *167*, 975-980.

60, Archer, G.L., Niemeyer, D.M. (1994), Origin and evolution of DNA associated with resistance to methicillin in staphylococci, *Trends Microbiol*, *2*, 343-347.

61, Piras, G., Raze, D., el Kharroubi, A., Hastir, D., Englebert, S., Coyette, J., Ghuysen, J.-M., (1993), Cloning and sequencing of the low-affinity penincillin binding protein 3r ancoding gene of *Enterococcus hirae*: modular design and structural organization of the protein, *J. Bacteriol.*, *175*, 2844-2852.

62, Song, M.D., Wachi, M., Doi, M., Ishino, F., Matsuhashi, M. (1987) Evolution of an inducible penicillin-target protein in methicillin-resistant *Staphylococcus aureus* by gene fusion, *FEBS Lett*, *221*, 167-171.

63, Pinho, M.G., Ludovice, A.M., Wu, S., De Lencastre, H. (1997), Massive reduction in methicillin resistance by transposon inactivation of normal PBP2 in a methicillin-resistant strain *Staphylococcus aureus*, *Microb. Drug Resistance-Mechanisms, Epidemiology and Disease*, *3*, 409-413.

64, Higashi, Y. Wakabayashi, A., Matsumoto, Y., Watanabe, Y., Ohno, A. (1999), Role of inhibition of penicillin binding proteins and cell wall cross-linking by beta-lactam antibiotics in low- and high-level meticillin resistance of *Staphylococcus* aureus, *Chemotherapy*, *45*, 37-47.

65, Matsuda K., Asahi Y., Sanada M., Nakagawa S., Tanaka N., (1991), In-vitro activity of imipenem combined with beta-lactam antibiotics for methicillin-resistant *Staphylococcus aureus*. *J. Antimicrob. Chemother.*, *27*, 809-815.

66, Hackbarth, C.J., Kocagoz, T., Kocagoz, S., Chambers, H.F. (1995), Point mutations in *Staphylococcus aureus* PBP 2 gene affect penicillin-binding kinetics and are associated with resistance, Antimicrob. Agents Chemother., *39*, 103-106.

67, Henze, U.U., Berger-Bächi , B. (1996), Penicillin-binding protein 4 overproduction increases β-lactam resistance in *Staphylococcus aureus*, *Antimicrob. Agents, Chemother.*,*40*, 2121-2125.

68, Hiramatsu, K., Asada, K., Suzuki, E., Okonogi, K., Yokota, T. (1992), Molecular cloning and nucleotide sequence determination of the regulator region of *mecA* gene in methicillin-resistant *Staphylococcus aureus* (MRSA*)*, *FEBS Lett.*, *298*, 133-136.

69, Ryffel, C., Kayser, F.H., Berger-Bächi, B. (1992), Correlation between regulation of *mecA* transcription and expression of methicillin resistance in staphylococci, *Antimicrob. Agents, Chemother.*, *36*, 25-31.

70, Niemeyer, D.M., Pucci, M.J., Thanassi, J.A., Sharma, V.K., Archer G.L., (1996), Role of *mecA* transcriptional regulation in the phenotypic expression of methicillin resistance in *Staphylococcus aureus, J. Bacteriol. 178*, 5464-5471.

71, Suzuki, E., Kuwahara-Arai, K., Richardson, J.F., Hiramatsu, K. (1993), Distribution of *mec* regulator genes in methicillin-resistant *Staphylococcus* clinical strains, *Antimicrob. Agents, Chemother., 37*, 1219-1226

72, Ehlert K., (1999), Methicillin-resistance in *staphylococcus aureus* - molecular basis, novel targets and antibiotic therapy, *Current Pharm. Design, 5*, 45-55.

73, Ling, B., Berger-Bächi, B. (1998), Increased overall antibiotic susceptibility in *Staphylococcus aureus femAB* null mutants, *Antimicrob. Agents Chemother., 42*,936-938.

74, Stranden, A.M., Ehlert, K., Labishinski, H., Berger-Bächi, B., (1997), Cell wall monoglycine cross-bridges and methicillin hypersusceptibility in a *femAB* null mutant of methicillin-resistant *Staphylococcus aureus, J. Bacteriol. 179*, 9-16.

75, Kopp, U., Roos, M., Wecke, J., Labischinski, H., (1996), Staphylococcal peptidoglycan interpeptide bridge biosynthesis: a novel antistaphylococcal target? *Microb Drug Resist., 2*, 29-41.

76, Ehlert, K., Schroder, W., Labishinski, H. (1997), Specificities of FemA and FemB for different glycine residues: FemB cannot substitute for FemA in staphylococcal peptidoglycan pentaglycine side chain formation, *J. Bacteriol., 179*, 7573-7576.

77, De Lencastre, H., Boudewijin de Jonge, L.M., Mattews, P.R., Tomasz, A. (1994), Molecular aspects of methicillin resistance in *Staphylococcus aureus, J. Antimicrob. Chemother., 33*, 7-24.

78, Labishinski, H., Ehlert, K., Berger-Bächi, B. (1998), The targeting of factors necessary for expression of methicillin resistance in staphylococci , *J. Antimicrob. Chemother., 41*, 581-584.

79, Mei, J.M., Nourbaskhsh, F., Ford, C.W., Holden D.W. (1997), Identification of *Staphylococcus aureus* virulence genes in a murine model of bacteraemia using signature-tagged mutagenesis, *Mol. Microbiol., 26*, 399-407.

80, Spratt, B.G. (1988), Hybrid penicillin-binding proteins in penicillin-resistant strains of *Neisseria gonorrhoeae, Nature, 332*, 173-176.

81, Spratt, B.G., Bowler L.D., Zhang, Q.-Y., Zhou, J., Smith, J.M. (1992), Role of interspecies transfer of chromosomal genes in evolution of penicillin resistance in pathogenic and commensal *Neisseria* species, *J. Mol. Evol., 34*, 115-125.

82, Maynard Smith, J., Dowson, C.G., Spratt, B.G. (1991), Localized sex in bacteria, *Nature, 349*, 29-31.

83, Martin, C., Sibold, C., Hackenbeck, R. (1992), Relatedness of penicillin-binding protein 1a genes from differnt clones of penicillin-resistant *Streptococcus pneumoniae* isolated in South Africa and Spain, *EMBO J., 11*, 3831-3836.

84, Dowson, C.G., Hutchinson, A., Spratt, B.G. (1989), Extensive re-modelling of the transpeptidase domain of penicillin-binding protein 2B of a penicillin-resistant South Africain isolate of *Streptococcus pneumoniae, Mol. Microb., 3*, 95-102.

85, Laible, G., Spratt, B.G., Hekenbeck, R. (1991), Interspecies recombinational events during the evolution of altered PBB 2x genes in penicillin-resistant clinical isolates of *Sreptococcus pneumoniae, Mol. Microb., 5*, 1993-2002.

86, Krauss, J., Van Der Linden, M., Grebe, T., Hakenbeck, R., (1996), Penicillin-binding proteins 2x and 2b as primary PBP targets in *Strepococcus pneumoniae. Microb. Drug Resist. 2(2)*, 183-186.

87, Mouz, N., Gordon, E., Di Gulmi, A.-M., Petit, I., Petillot, Y., Dupont, Y., Hakenbeck, R.., Vernet, T., Dideberg, O. (1998), Identification of a structural determinant for resistance to β-lactam antibiotics in Gram-positive bacteria, *Proc. Natl. Acad. Sci. U. S. A., 95*, 13403-13406.

88, Asahi, Y., Ubukata, K. (1998), Association of a Thr-371 substitution in a conserved amino acid motif of penicillin-binding protein 1A with penicillin resistance of *Streptococcus pneumoniae., Antimicrob. Agents Chemother., 42*, 2267-2273.

89, Brannigan J.A., (1990), Insertion of an extra amino acid is the main cause of the low affinity of penicillin-binding protein 2 in penicillin-resistant strains of Neisseria gonorrhoeae, *Mol Microbiol., 4*, 913-919.

90, Grebe, T., Hakenbeck, R. (1996), Penicillin-binding proteins 2b and 2x of *Streptococcus pneumoniae* are primary resistance determinants for different classes of β-lactam antibiotics, *Antimicrob. Agents Chemother., 40(4)*, 829-834.

91, Rybkine, T., Mainardi, J.L., Sougajoff, W., Collatz, E., Gutmann, L., (1998), Penicillin-binding protein 5 sequence alteration in clinical isolates of *Enterococcus faecium* with different levels of beta-lactam resistance, *J. Infect. Dis.*, *178*, 159-163.

92, Georgopapadakou, N. H. (1993), Penicillin-binding proteins and bacterial resistance to β-lactams, *Antimicrob. Agents Chemother.*, *37*, 2045-2053.

93, Nikaido, H. (1976), Outer membrane of *Salmonella typhimurium* transmembrane diffusion of some hydrophobic substances, *Biochim. Biophys. Acta* ,*433*, 118-132.

94, Benz R., Beacterial Cell Wall, Ghuysen J.M., Hackenbeck R., Elsevier 1994, page 397-423.

95, Schulz G., (1993), Bacterial porins: structure and function., *Curr. Opin. Cell Biol.*, 5, 701-707.

96, Van Alphen, W., Lutenberg, B. (1977), Influence of osmolarity of the growth medium on the outer membrane protein pattern of *Escherichia coli*, J. Bacteriol., *131*, 623-630.

97, Mizuno, T., Wurtzel, E.T., Inouye, M. (1982), Osmoregulation of gene expression, *J. Biol. Chem.*, *257*, 13692-13698.

98, Yoshimura, F., Nikaido, H. (1985), Diffusion of β-lactam antibiotics through the porin channels of *Escherichia coli* K-12, *Antimicrob. Agents Chemother.*, *27*, 84-92.

99, Nikaido, H. (1994), Prevention of drug access to bacterial targets: permeability barriers and active efflux, *Science*, *264*, 382-387.

100, Nikaido, H. (1989), Outer membrane barrier as a mechanism of antimicrobial resistance, *Antimicrob. Agents Chemother.*, *33*, 1831-1836.

101, Nikaido, H., Basina, M., Nguyen, V.Y., Rosenberg E.Y. (1998), Multidrug efflux pump AcrAB of *Salmonella typhimurium* excretes only those β-lactam antibiotics containing lipophilic side chains, *J. Bacteriol*, *180*, 4686-4692.

102, Nikaido, H., Normark S., (1987), Sensitivity of *Escherichia coli* to various beta-lactams is determined by the interplay of outer membrane permeability and degradation by periplasmic beta-lactamases: a quantitative predictive treatment, *Mol. Microbiol.*, *1*, 29-36.

103, Frère, J.-M. (1989), Quantitative relationship between sensitivity to β-lactam antibiotics and β-lactamase production in gram-negative bacteria_I *Biochem. Pharmacol.*, *38*, 1415-1426.

104, Harder K.J., Nikaido H., Matsuhashi M. (1981), Mutants of Escherichia coli that are resistant to certain beta-lactam compounds lack the ompF porin. *Antimicrob. Agents Chemother.*, *20*, 549-552.

105, Trias,J., Dufresne, J., Levesque, R.C., Nikaido, H. (1990), Decreased outer membrane permeability in Imipenem-resistant mutants of *Pseudomonas aeruginosa*, *Antimicrob. Agents Chemother.*, *33*, 1201-1206.

106, Garcia de Lomas, J., Navarro, D., Gimeno, C., (1997), Penetration of meropenem in Gram-negative bacilli. Differences in the activity with imipenem, *Enferm. Infecc. Microbiol. Clin.*, *Suppl 1*, 14-19.

107, George A.M. (1996), Multidrug resistance in enteric and other Gram-negative bacteria, *FEMS Microbiology Letters*, *139*, 1-10.

108, Poole K., Krebes K., McNally C., Neshat S. (1993), Multiple antibiotic resistance in pseudomonas aeriginosa: evidence for involvement of an efflux operon, *J. Bacteriol*, *175*, 7363-7372.

109, Li X-.Z., Nikaido H., Poole K. (1995), Role of MexA-MexB-OprM in antibiotic efflux in *pseudomonas aeruginosa*, *Antimicrob. Agents Chemother.*, *39*, 1948-1953.

110, Ma D., Cook D.N., Hearst J.E., Nikaido H. (1994) Efflux pumps and drug resistance in Gram-negative bacteria,*Trends Microbiol*, *2*, 489-493.

111, Masuda, N., Sakagawa, E., Ohya, S. (1995), Outer membrane proteins responsible for multiple drug resistance in *pseudomonas aeruginosa*, *Antimicrob. Agents Chemother.*, *39*, 645-649.

112, Srikumar, R., Kon, T., Gotoh, N., Poole, K. (1998), Expression of *pseudomonas aeruginosa* multidrug efflux pumps MexA-MexB-OprM and MexC-MexD-OprJ in a multidrug-sensitive *Escherichia coli* strain, *Antimicrob. Agents Chemother.*, *42*, 65-71.

113, Masuda, N., Gotoh, N., Ishii, C., Sakagawa E., Ohya S., Nishino T. (1999), Interplay between chromosomal β-lactamase and the MexAB-OprM efflux system in intrinsic resistance to β-lactams in *Pseudomonas aeruginosa*, *Antimicrob. Agents Chemother.*, *43*, 400-402.

114, Li, X-.Z, Zhang, L., Srikumar, R., Poole, K. (1998), β-lactamase inhibitors are substrates for the multidrug efflux pumps of *pseudomonas aeruginosa*, *Antimicrob. Agents Chemother.*, *42*, 399-403.

115, Srikumar, R., Li, X-.Z, Poole, K. (1997), Inner membrane efflux components are responsible for β-lactam specificity of multidrug efflux pumps in *pseudomonas aeruginosa*, *J. Bacteriol.*, *179)*, 7875-7881. 178(20), 5853-5859.

116, Nikaido, H. (1996), Multidrug efflux pumps of gram-negative bacteria, *J. Bacteriol*,*178*, 5853-5859.

117, Masuda, N., Gotoh, N., Ishii, C., Sakagawa, E., Ohya, S., Nishino, T. (1999), Interplay between chromosomal β-lactamase and the MexAB-OprM efflux system in intrinsic resistance to β-lactams in *Pseudomonas aeruginosa, Antimicrob. Agents Chemother., 43*, 400-402.

118, Burton G., Osborne N.F., Pearson M.J., Southgate R., *in Burger's Medicinal Chemistry and Drug Discovery* (Fifth Edition), ed. Wolff M.E., 1998, vol. 4: Therapeutic Agents, pp. 277-363, John Wiley & Sons, Inc.

119, Raquet, X., Lamotte-Brasseur, J., Fonzè, E., Goussard, S., Courvalin, P., Frère, J.-M., (1994), *J. Mol. Biol., 244*, 625-639.

120, Saves, I., Burlet-Schiltz, O., Maveyraud, L., Samama, J.P., Promé, J.C., Masson, J.M., (1995), Mass spectral kinetic study of acylation and deacylation during the hydrolysis of penicillins and cefotaxime by β-lactamase TEM-1 and the G238S mutant, *Biochem., 34*, 11660-11667.

121, Matagne, A., Lamotte-Brasseur, J., Dive, G., Knox, J.R., Frère, J.-M., (1993), Interactions between active-site serine beta-lactamases and compounds bearing methoxy side chain on the alpha-face of the beta-lactam ring: Kinetic and molecular modeling studies. *Biochem. J., 293*, 607-611.

122, Davis J., (1994), Inactivation of antibiotics and the dissemination of resistance genes, *Science, 264*, 375-382.

123 Coulton S., Hunt E. (1996), Recent advances in the chemistry and biology of carbapenem antibiotics, *Progress in Med. Chem., Ed. Ellis G.P Luscombe D.K., Elsevier Science B.V., 33*, 99-145.

124, Bryskier, A. (1997), Anti-infectives Dual β-lactam-fluoroquinolone compounds: a novel approach to antibacterial treatement, *Exp. Opin. Invest. Drugs, 6(10)*, 1479-1499.

125, Chang, M.H., Koh, H.Y., Cho, Y.S., Choi, K.I. (1997), Novel anti-pseudomonas cephalosporins, *Current Pharm. Design,3*,209-226.

126, Katsu K., Kitoh K., Inoue M., Mitsuhashi S., (1982), In vitro antibacterial activity of E-0702, a new semisynthetic cephalosporin, *Antimicrob. Agents Chemother., 22*, 181-185.

127, Neu, H., Labthavikul, P. (1983), In vivo antibacterial activity and β-lactamase stability of E-0702, a new cephalosporin, *Antimicrob. Agents Chemother, 24*, 313-320.

128, Watanabe, N., Nagasu, T., Katsu, K., Kitoh, K. (1987), E-0702, a novel cephalosporin, is incorporated into *Escherichia coli* cells via the *tonB*-dependent iron transport system, *Antimicrob. Agents Chemother., 31*, 497-504.

129, Nikaido, H., Rosenberg, E.Y. (1990), Cir and Fiu proteins in the outer membrane of *Escherichia coli* catalyze transport of monomeric catechols: study with β-lactam antibiotics containing catechol and analogous groups, *J. Bacteriol., 172*, 1361-1367.

130, Aszodi J., Bonnefoy A., Chantot J.F., RU 59863, A Novel Catechol-substituted Vinylogous Cephalosporin : *In-vivo* Antibacterial Activity, 34th Interscience Conference on Antimicrobial Agents and Chemotherapy, 1994, F-101.

131, Ishiguro, M., Tanaka, R., Namikawa, K., Nasu, T., Inoue, H., Nakatsuka, T., Oyama, Y., Imajo, S. (1997), 5,6-cis-penems: broad-spectrum anti-meticillin-resistant *Staphylococcus aureus* β-lactam antibiotics, *J. Med. Chem., 40*, 2126-2132.

132, Wilkening R.R., Ratcliffe R.W., Wildonger K.J., Waddell S.T., Santorelli G.M., Parker D.L., Morgan J., Blizzard T.A., Cama L.D., Sun W., Leone J.F., Hammond M.L., Discovery of a novel PBP2a-binding, anti-MRSA 1,8-naphtosultamyl pharmacophore, 38th Interscience Conference on Antimicrobial Agents and Chemotherapy, 1998, Abs. N° F032.

133, Shimizu A., Imamura H., Sato H., Sakuraba S., Sugimoto Y., Kiyonaga H., Ogawa M., Nakano M., Ohtake N., Ushijima R., Nakagawa S., Suzuki C., Hashizume T., Morishima H., J-111,225, a novel trans - 3,5-disubstituted pyrrolidinylthio 1β-methylcarbapenem; synthesis and physicochemical properties, 38th Interscience Conference on Antimicrobial Agents and Chemotherapy, 1998, Abs. N° F061.

134, Shibata, K., Nagano, R., Hashizume, T., Morishima, H. (1998), F-057 *In-vivo* efficacy of J-111,225 in mouse experimental infection models including mixed infection with MRSA and *Pseudomonas aeruginosa*, 38th Interscience Conference on Antimicrobial Agents and Chemotherapy, Abs. N° F057

135, Glinka T., Cho I.S.,Zhang J.S., et al., Discovery of MC-02,479 and Analogs, New Cephalosporins Active Against Resistant Gram-Positive Bacteria, 37th Interscience Conference on Antimicrobial Agents and Chemotherapy, 1997, Abs. N° F176.

136, Hebeisen P., Angehrn P., Heinze-Krauss I., Kansy M., Page M., Runtz V., Weiss U., Synthesis and SAR of pyrrolidinone-3-ylidenmethyl cephems against methicillin resistant *staphylococcus aureus*, 38th Interscience Conference on Antimicrobial Agents and Chemotherapy, 1998, Abs. N° F021.

137, Roychoudhury S., Kaiser R.E., Brems D.N., Yeh W.-K., (1996), Specific interaction between beta-lactams and soluble penicillin-binding protein 2a from methicillin-resistant *Staphylococcus aureus*: development of a chromogenic assay, *Antimicrob. Agents Chemother.*, *40*, 2075-2079.

138, Doran, J.L., Leskiw, B.K., Aippersbach, S., Jensen, S.E., (1990), Isolation and characterisation of a β-lactamase-inhibitory protein from *Streptomyces clavuligerus* and cloning and analysing of the corresponding gene. *J. Bacteriol.*, *172*, 4909-4918.

139, Massova, I., Mobashery S., (1997), Molecular basis for interactions between β-lactam antibiotics and β-lactamases, *Acc. Chem. Res.*, *30*, 162-168.

140, Richter, H.G.F., Angehrn, P., Hubschwerlen, C., Kania, M., Page, M.G.P., Specklin, J.L., Winkler F.K. (1996), Design, synthesis, and evaluation of 2β-Alkenyl penam sulphone acids as inhibitors of β-lactamases, *J. Med. Chem.*, *39*, 3712-3722.

141, Pratt R.F., Beta lactamase inhibition. In : The chemistry of beta-lactams, Page M.I. Ed Blackie, Academic & Professional, London, 229-271.

142, Maiti, S.N., Phillips, O.A., Micetich, R.G., Livermore, D.M. (1998), β-lactamase inhibitors: agents to overcome bacterial resistance, *Current Med. Chem.*, *5*, 441-456.

143, Heinze-Krauss I., Angehrn P., Charnas R.L., Gubernator K., Gutknecht E.-M., Hubschwerlen C., Kania M., Oefner C., Page M.G.P., Sogabe S., Specklin J.-L., Winkler F., *J. Med. Chem.*, *1998*, *41*, 3961-3971.

144, Livermore, D.M., Chen, H.Y. (1997), Potentiation of β-lactams against *Pseudomonas aeruginosa* strains by Ro 48-1256, a bridged monobactam inhibitor of AmpC β-lactamases, *J. Antimicrob. Chemother.*, *40*, 335-343.

145, Weston, G.S., Blazquez, J., Baquero, F., Shoichet, B.K. (1998), Structure-based enhancement of boronic acid-based inhibitors of AmpC β-lactamase, *J. Med. Chem.*, *41*, 4577-4586.

146, Payne D.J., Broad spectrum inhibitors of metallo β-lactamases, 3rd Internat. Symposium on Antibacterial Drug Discovery & Development Princeton March 8-9, 1999.

147, Grossato,A., Sartori, R., Fontana, R. (1991), Effect of non-β-lactam antibiotics on penicillin-binding protein synthesis of *Enterococcus* hirae ATCC 9790, *J. Antimicrob. Chemother.*, *27*, 263-271.

148, Yam, T.S., Hamilton-Miller, J.M.T., Shah, S. (1998), The effect of a component of tea *Camellia sinensis* on methicillin resistance, *PBP2* synthesis, and β-lactamase production in *Staphylococcus aureus*, *J. Antimicrob. Chemother.*, *42* , 211-216.

149, Eid, C.N.,Halligan, N.G., Nicas, T.I., Mullen, D.L., Butler, T.F., Loncharich, R.J., Paschal, J.W. (1997), Tripeptide LY301621 and its diastereomers as methicillin potentiators against methicillin resistant *Staphylococcus aureus*, *J. Antibio.*, *50*, 283-285.

150, Chamberland S., Blais J., Boggs A.F., Bao Y., Maloin F., Hecker S.J., Lee V.J., MC-207,252 Abolishes PBP 2a Mediated β-Lactam Resistance in Staphylococci, 35th Interscience Conference on Antimicrobial Agents and Chemotherapy, 1995, Abs. N° F-144.

151, Griffith D., Annamatai T., Williams R., Parr Jr. T.R., MC-207,252 Potentiates the Activity of β-Lactams Against Methicillin-Resistant *Staphyloccus aureus* in a Murine Soft Tissue Model, 35th Interscience Conference on Antimicrobial Agents and Chemotherapy, 1995, Abs. N° F-143.

152, Lakaye, B., Dubus, A., Lepage S., Groslambert, S., Frère J.-M., (1999), When drug inactivation rebders the target irrelevant to antibiotic resistance: a case story with β-lactams, *Mol. Microbiol.*, *31*, 89-101.

153, McGowan, S.J., Sebaihia, M., O'Leary, S., Hardie, K.R., Williams, P., Stewart G.S.A.B., Bycroft, B.W., Salmond, P.C., (1997), Analysis of the carbapenem gene cluster of *Erwinia carotovora*: definition of the antibiotic biosynthetic genes and evidence for a novel β-lactam resistance mechanism, *Mol. Microbiol.*, *26*, 545-556.

154, Rolinson, G.N. (1998), Forty years of β-lactam research, *J. Antimicrob. Chemother.*, *41*, 589-603.

155, Ghuysen, J.-M., (1994), Molecular structures of penicillin-binding proteins and β-lactamases, *Trends. Microbiol.*, *10*, 372-380.

156, Hartman, B., Tomasz, A. (1984), Low-affinity penicillin-binding protein associated with β-lactam resistance in *Staphylococcus aureus*, *J. Bacteriol*, *158*, 513-516.

157, Ubukata, K., Nonoguchi, R., Matsuhashi, M., Konno, M. (1989), Expression and inducibility in *Staphylococcus aureus* of the *mec*A gene, which encodes a methicillin-resistant S. *aureus*-specific penicillin-binding protein, *J.Bacteriol.*, *171*, 2882-2885.

158, Wong, K.K.Y., Poole, K., Gotoh, N., Hancock R.E.W. (1997), Influence of OprM expression on multiple antibiotic resistance in *Pseudomonas aeruginosa*, *Antimicrob. Agents Chemother.*, *41*, 2009-2012.

159, Osano, E., Arakawa, Y., Wacharotayankun, R., Ohta, M., Horii, T., Ito, H., Yoshimura, F., Kato, N., (1994), Molecular characterization of an enterobacterial metallo beta-lactamase found in a clinical isolate of Serratia marcescens that shows imipenem resistance *Antimicrob. Agents Chemother.*, *38*, 71-78.

160, McCall, J.O., Xuey, X., Sabet, R., Discovery of Compounds Reversing β-Lactam Resistance in MRSA (Methicillin-Resistant *Staphylococcus aureus*) Using a High-throughput Screen Based on SPA (Scintillation Proximity Assay) Technology, 37th Interscience Conference on Antimicrobial Agents and Chemotherapy, 1997, F-233.

161, Livermore, D.M., (1987), Clinical significance of beta-lactamase induction and stable derepression in gram-negative rods, *Eur. J. Clin. Microbiol. 6(4),* 439-445.

162, Payne, D.J., Bateson, J.H., Gasson, B.C., Proctor, D., Khushi, T., Farmer, T.H., Tolson, D.A., Bell, D., Skett, P.W., Marshall, A.C., Reid, R., Ghosez, L., Combret, Y., Marchand-Brynaert, J., (1997), Inhibition of metallo-beta-lactamases by a series of mercaptoacetic acid thiol ester derivatives, *Antimicrob Agents Chemother., 41,* 135-140.

163, Payne, D.J., Bateson, J.H., Gasson, B.C., Khushi, T., Proctor, D., Pearson, S.C., Reid, R., (1997), Inhibition of metallo-beta-lactamases by a series of thiol ester derivatives of mercaptophenylacetic acid, *FEMS Microbiol. Lett.*, *157,* 171-5.

164, Lawrence L.E., Barrett, J.F. (1998), Efflux pumps in bacteria: overview, clinical relevance, and potential pharmaceutical target, *Exp. Opin. Invest. Drugs, 7,* 199-217.

165, Toney, J.H., Fitzgerald, P.M.D., Grover-Sharma, N., Olson, S.H., May, W.J., Sundelof, J.G., Vanderwall, D.E., Cleary, K.A., Grant, S.K., Wu, J.K., Kozarich, J.W., Pompliano, D.L., Hammond, G.G. (1998), Antibiotic sensitization using biphenyl tetrazoles as potent inhibitors of *Bacteroides fragilis* metallo-β-lactamase, *Chem. & Biol., 5,* 185-196.

166, Coffey, T.J., Daniels, M., McDougal, L.K., Dowson, C.G., Tenover, F.C., Spratt, B.G. (1995), Genetic analysis of clinical isolates of *Streptococcus pneumoniae* with high-level resistance to expanded-spectrum cephalosporins, *Antimicrob. Agents Chemother.*,*39*, 1306-1313.

167, Jap, B.K., Wallan, P.J. (1996), Structure and functional mechanism of porins, *Physiological Reviews, 76*, 1073-1088.

Resistance to aminoglycoside antibiotics: Function meets structure

GERARD D. WRIGHT AND ALBERT M. BERGHUIS

Antimicrobial Research Centre, Department of Biochemistry, McMaster University, 1200 Main St W., Hamilton, Ontario, Canada, L8N 3Z5

Abstract

The aminoglycoside antibiotics find use as important broad-spectrum antibacterial agents. The clinical utility of these compounds is threatened by the emergence of resistance in both Gram-positive and Gram-negative organisms. The principle route of resistance to the aminoglycoside antibiotics is through the expression and action of enzymes that covalently modify the compounds. Chemical modification takes the form of phosphorylation, acetylation, or adenylylation. The mechanism of action of some of these enzymes has been investigated and now three-dimensional crystal structures have been solved for representative members catalyzing each type of reaction. The structural and functional information, which is now available, provides the basis for the design of enzyme inhibitors that have the potential to reverse aminoglycoside antibiotic resistance

1. Introduction

The discovery of first streptomycin and then neomycin by Selman Walksman and colleagues over 50 years ago ushered in a golden era of aminoglycoside antibiotic discovery, which resulted in the isolation, and clinical implementation of compounds such as kanamycin, spectinomycin, and gentamicin. These water-soluble compounds showed a broad-spectrum antibacterial activity and in most cases were bactericidal which made them highly useful and important clinical agents, despite some unwelcome side effects such as ototoxicity for some compounds. However, already by the mid-1950s, microbial resistance to the aminoglycosides was well documented. Streptomycin was the first anti-tubercular drug of significance, but resistance to single agent therapy rapidly evolved. Resistance to kanamycin was first described in early 1960s and associated with chemical modification of the drug (Umezawa et al., 1967).

The theme of chemical modification resulting in aminoglycoside resistance was to become familiar over the next decades. This prompted the search for naturally occurring aminoglycosides that were not susceptible to resistance such as tobramycin, and spurred the chemical modification of known aminoglycosides for the semi-

A. Van Broekhoven et al. (eds.), Novel Frontiers in the Production of Compounds for Biomedical Use, 85–98.
© 2001 *Kluwer Academic Publishers. Printed in the Netherlands.*

synthesis of agents that were less susceptible to resistance such as amikacin and isepamicin. However, despite these efforts, aminoglycoside resistance has continued to be a significant problem and has contributed to the reduced use of these antibiotics and the fact that no new aminoglycosides have been brought to market in North America for many years.

Resistance to the aminoglycosides can occur in five ways: decreased transport into the cell, efflux, mutation at the ribosome which results in lower affinity of the antibiotics, enzymatic modification (methylation) of the ribosome, and the production of intracellular aminoglycoside modifying enzymes. The first three mechanisms are problematic only in a limited number of cases, e.g. ribosomal mutation in *Mycobacterium tuberculosis*. Ribosomal methylation is associated with high-level aminoglycoside resistance, but confined to aminoglycoside producing bacteria. On the other hand, the production of aminoglycoside modifying enzymes is the most clinically relevant and prevalent mechanism of aminoglycoside resistance. Modification of aminoglycosides can occur through three routes: *O*-phosphorylation, *O*-nucleotidylation, and *N*-acetylation. The enzymes that catalyze these reactions are the aminoglycoside phosphotransferases (APH), nucleotidyltransferases (ANT), and acetyltransferases (AAC) (Fig. 1).

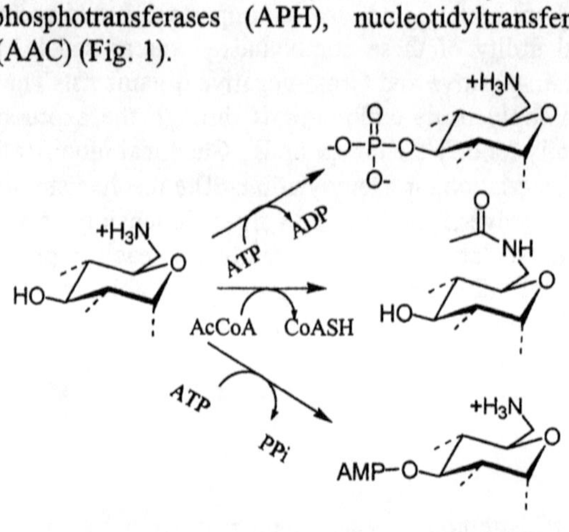

Figure 1. Enzymatic modification of aminoglycoside antibiotics

2. Aminoglycoside modifying enzymes

The aminoglycoside modifying enzymes show a broad regioselectivity of group transfer and thus there are many distinct APHs which modify hydroxyls at position 3' and 2", and AACs which specifically modify N1, N3, N2' and N6' for example (Fig. 2). Furthermore, there exist a large number of genes which have been identified as encoding aminoglycoside modifying enzymes and these have been found in plasmids, transposons, and within the chromosomes of bacteria (Shaw et al., 1993). It is in part the large number of resistance elements that has discouraged research into new aminoglycosides. However, despite the abundance of resistance enzymes and their different substrate and regioselectivities, most share amino acid sequence homology within their class of enzymes (APH, ANT, AAC), indicating similarity in structure and function. Thus enzyme function and structure studies performed on one enzyme should be broadly applicable to others within the family.

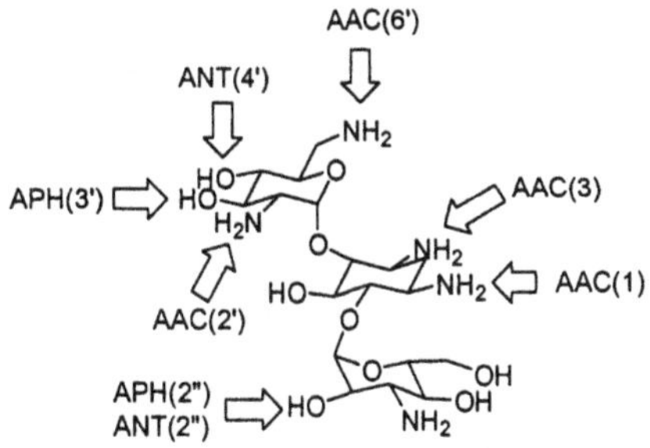

Figure 2. Sites of enzymatic modification of kanamycin B

Research on the mechanisms and structure of aminoglycoside modifying enzymes has benefited greatly from the recent descriptions of the 3D-structures of members of all three classes of aminoglycoside modifying enzymes. The results of these studies have demonstrated unexpected structural similarity with various eukaryotic proteins and raise several questions not only for enzyme mechanism and inhibitor studies, but also regarding the origins of antibiotic resistance.

2.1. ANT

Aminoglycoside nucleotidyltransferases, are especially relevant to clinical resistance in Gram-negative organisms where ANT(2")-Ia is frequently associated with resistance to gentamicin and tobramycin (Miller et al., 1997). This enzyme has been purified and enzymatically characterized by Northop's group. The 31.5 kDa enzyme shows broad substrate specificity for both aminoglycosides and nucleotides (Gates and Northrop, 1988b) and it has been determined that the rate limiting step in aminoglycoside modification is release of the adenylated aminoglycoside (Gates and Northrop, 1988a). Adenylation of aminoglycosides occurs with inversion of the configuration at the α-phosphate group suggesting that the adenylation reaction occurs by direct nucleophilic attack of the aminoglycoside hydroxyl on the nucleotide α-phosphate (Van Pelt et al., 1986).

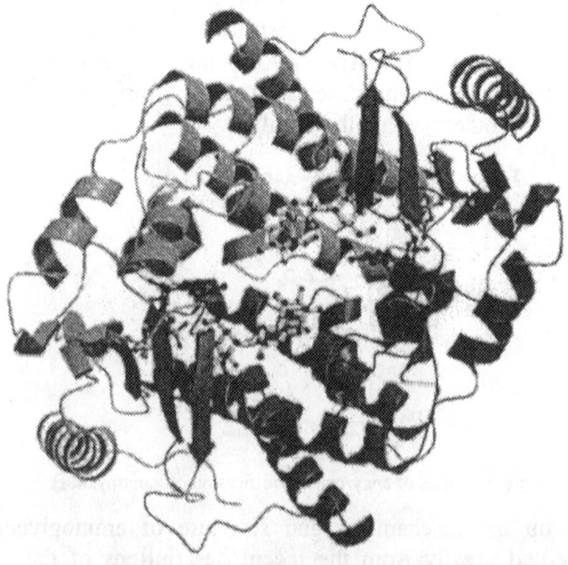

Figure 3. Structure of the ANT(4')-Ia dimer

While the 3D-structure of ANT(2")-Ia is not known, the crystal structure of 4'-kanamycin nucleotidyltansferase, ANT(4')-Ia, has been determined, both in the absence and presence of substrates, by Holden and colleagues (Perdersen et al., 1995; Sakon et al., 1993). The structure reveals that the active enzyme exists as an intimate dimer with two heterodimeric catalytic sites (Fig. 3.). Each catalytic site is composed of residues originating from both monomers, with residues from one monomer predominantly contributing to the binding site of ATP and magnesium, and residues from the other monomer principally participating in aminoglycoside binding (Fig. 4). An intriguing feature of the aminoglycoside binding site is the overwhelming predominance of acidic residues which, under physiological conditions, generate a negatively charged surface patch (Fig. 5). Since aminoglycoside antibiotics are invariably positively charged compounds, it is clear that presence of a negatively charged patch of ANT(4') will aid in guiding the aminoglycoside substrate to the active site and facilitate binding. It can also be argued that the abundance of negatively charged hydrogen acceptor groups can accommodate a broad spectrum of substrates which are rich in positively charged and/or hydrogen donating groups (i.e. NH_3^+, NH_2, and OH functional groups). Thus, the nature of the ANT(4')-Ia aminoglycoside binding site provides insight into the substrate promiscuity observed for this class of enzymes. As will be discussed below, the presence of acidic groups in the aminoglycoside binding site appears to be a recurring theme in aminoglycoside modifying enzymes.

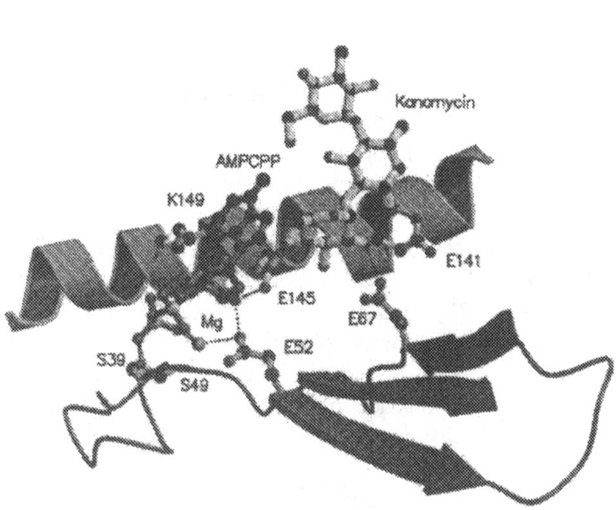

Figure 4. Active site region of ANT(4')-Ia

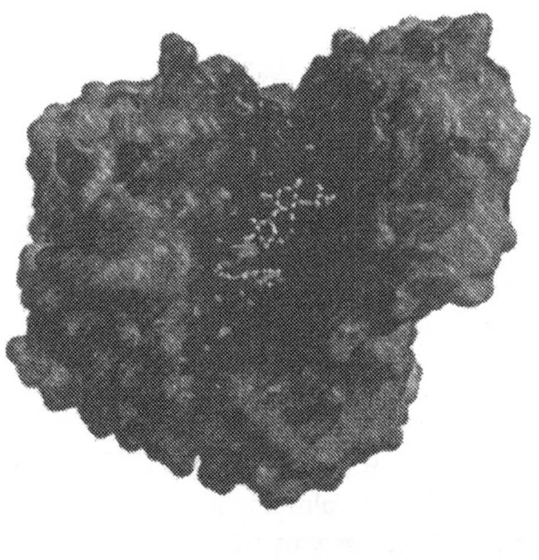

Figure 5. Negative surface charge density of ANT(4')-Ia

The crystal structure of ANT(4') in complex with kanamycin and the non-hydrolyzable ATP analogue AMPCPP is most relevant for understanding the mechanism of action for the nucleotidyltransferase class of aminoglycoside modifying enzymes. In fact, this is currently the only structure of an aminoglycoside modifying enzyme which has an antibiotic bound in its active site. The ANT(4')•AMPCPP•Kanamycin structure strongly suggests that a glutamic acid (Glu145) functions as the general base which deprotonates the incoming aminoglycoside thereby increasing the nucleophilicity of the attacking 4' hydroxyl group. A positively charged residue, Lys149, interacts with the α-phosphate group of ATP. This has the capacity to increase electrophilicity of this phosphate, making it more susceptible to nucleophilic attack (Fig. 4). Such a direct attack mechanism is consistent with the results of stereochemical analysis of the ANT(2")-Ia reaction (Van Pelt et al., 1986). However, care should be taken in further interpreting this crystal structure. As Pedersen *et al.* mention in their description of the ternary complex structure, the electrondensity for kanamycin is not unambiguous, and the authors do not rule out that the kanamycin can also be oriented in the opposite orientation, i.e. the 4" hydroxyl is directed towards the α-phosphate group. This ambiguity in orientation is not unexpected, given that ANT(4') is also capable of modifying the 4" hydroxyl group (Santanam and Kayser, 1978). Furthermore, the 4' (or 4") hydroxyl group is approximately 5Å away from the reactive phosphate group, which is somewhat distant. Structure determination of this conformation may have occurred since crystals of ANT(4') were grown in the presence of kanamycin, and after completion of the crystallization process AMPCPP was soaked into these crystals. Note that for the nucleotidyltransfer reaction to occur, ATP is required to bind first to ANT(4') and the aminoglycoside substrate must be second, i.e. the reverse of what occurred during the generation of ANT(4')•AMPCPP•Kananmycin crystals.

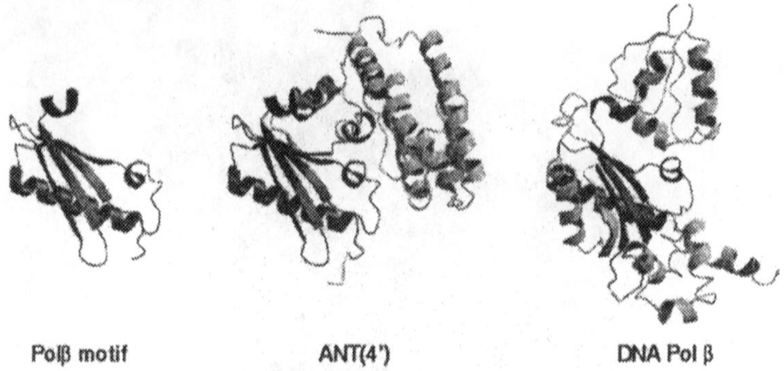

Figure 6. Strucutral similarity between ANT(4')-Ia and rat DNA polymerase β

Analyses of similarities in the fold of different protein by Holm and Sanders revealed that part of the ANT(4') fold was also present in the crystal structure of eukaryotic DNA polymerase β, an enzyme which also catalyses *O*-nucleotidyl transfer, despite minimal sequence similarity (Fig. 6) (Holm and Sander, 1995). This structural similarity implies that these two NMP transferases are evolutionary related, suggesting the existence of a DNA polymerase β-like nucleotidyltransferase superfamily (polβ superfamily). Using sequence alignment tools (e.g. PSI_BLAST), Aravind and Koonin have recently exploited the structural similarity between ANT(4') and DNA Polβ for identifying other members of the polβ superfamily (Aravind and Koonin, 1999). Their analysis suggests that the polβ superfamily encompasses enzymes involved in antibiotic resistance (e.g. ANT(4')-Ia), DNA replication and repair (e.g. DNA Polβ), chromatin remodeling (e.g. yeast Trf4p), and perhaps signal transduction (e.g. GlnD). Furthermore, this superfamily spans all three kingdoms of life, with members in bacteria, archaea and eukaryotes.

2.2. AAC

Aminoglycoside acetyltransferases (AACs) catalyze the acetylCoA dependent *N*-acetylation of the antibiotics. Individual aminoglycoside molecules carry many amine functional groups, and families of enzymes which modify amino groups on both the 6-aminohexose ring (AAC(6'), AAC(2')) and the 2-deoxystreptamine ring (AAC(1), AAC(3)) are known. Both AAC(6') and AAC(3) are common resistance elements in Gram-negative bacteria (Miller et al., 1997). In Gram-positive bacteria, AAC(6')-Ie is the major acetyltransferase associated with clinically important gentamicin resistance (Miller et al., 1997). AAC(6')-Ie forms the N-terminal portion of a bifunctional enzyme, AAC(6')-APH(2") (Ferretti et al., 1986), and is found in Gram-positive cocci such as Staphylococci, Enterococci, and Streptococci. This enzyme also has the unusual distinction to be the only AAC yet reported with the capacity to direct *O*-acetyltransfer as well (Daigle et al., 1999). *Enterococcus faecium* also harbor the chromosomally encoded AAC(6')-Ii (Costa et al., 1993).

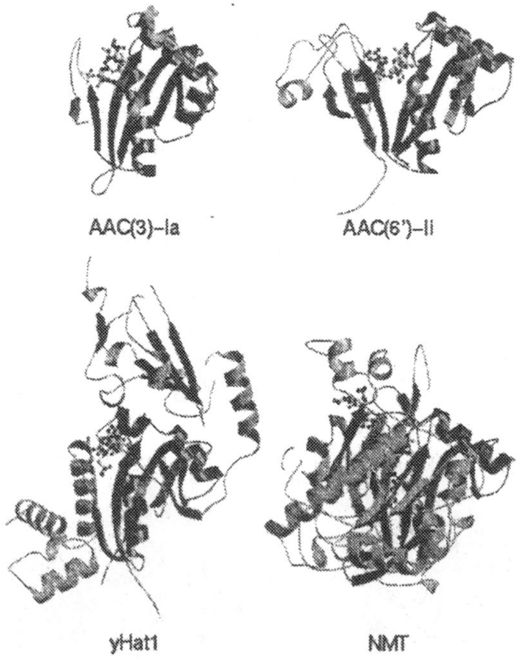

Figure 7. Structures of AAC and other acyl-transferases

Two aminoglycoside *N*-acetyltransferases with different regiospecificities have been crystallized and their structures determined by X-ray crystallographic methods. The first is AAC(3)-Ia which confers resistance to gentamicin and is found in Enterobacteriacea (Wolf et al., 1998), and the second is AAC(6')-Ii from *E. faecium*, which is chromosomally encoded (Wybenga-Groot et al., 1999). The structure of

AAC(3)-Ii was determined in complex with Coenzyme A, while AAC(6')-Ii was determined in complex with Acetyl-CoA. However, a crystal structure of any of these enzymes in complex with an aminoglycoside is not yet available.

The two AAC crystal structures reveal extensive similarity in their 3-dimensional fold despite very low (<15%) amino acid sequence homology (Fig. 7). Therefore, it could be assumed that all members of the AAC enzyme class share this similar folding motif. However, based on the observed diversity in primary structure present within the AAC family, this assumption may not necessarily be true. On the other hand, as will be discussed below, dissimilarity in primary structure may not be a good indication of dissimilarity in ternary structure for AAC enzymes.

While a crystal structure for the ternary complex of an AAC enzyme with bound co-factor and aminoglycoside is not available, some information has been derived from NMR experiments. The conformation of the aminoglycosides isepamicin and butirosin A when bound to AAC(6')-Ii have been examined (DiGiammarino et al., 1997). These studies suggest that the 4,6-disubstituted aminoglycoside isepamicin can bind in two alternative conformations, one of them being perhaps a non-productive complex, and that the 4,5-disubstituted aminoglycoside butirosin A possess only one binding mode, which differs from those observed for isepamicin. The conclusion that 4,5- and 4,6-disubstituted aminoglycosides bind differently to AAC(6')-Ii is in agreement with steady state kinetic measurements (Wright and Ladak, 1997) .

Additional insight into aminoglycoside binding to AAC enzymes can be derived from examination of the two available crystals structures. The position and orientation of the CoA and AcCoA groups for AAC(3)-Ia and AAC(6')-Ii, respectively, delineate the region where the antibiotic should be bound to form a productive ternary complex. Examination of the area adjacent to the CoA sulfur atom reveals a negatively charged surface patch in both AAC(3)-Ia and AAC(6')-Ii, analogous to the aminoglycoside binding region of ANT(4')-Ia (Fig. 8). It is therefore reasonable to assume that the two AAC enzymes use an identical strategy to attract and bind a broad spectrum of aminoglycoside substrates, as suggested for the ANT family of aminoglycoside resistance enzymes.

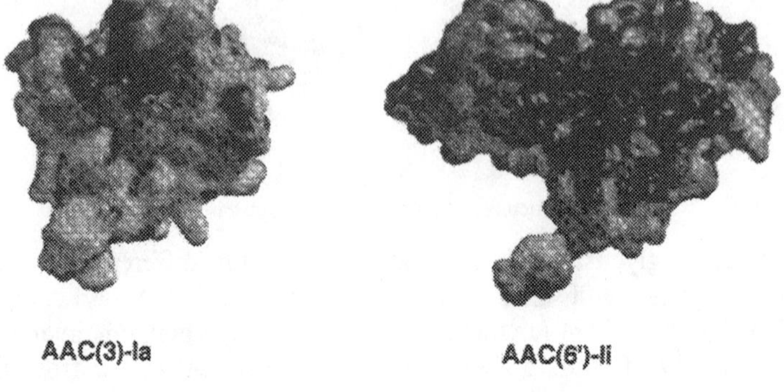

Figure 8. Negative surface charge density on AAC(3)-Ia and AAC(6')-Ii

As can be seen in Fig. 7, the common structural motif found in AAC(3)-Ia and AAC(6')-Ii is also present in other enzymes. Specifically, two enzymes have been identified that contain the same structural motif, yeast Histone Acetyltransferase Hat1 (yHat1) and the fungal N-myristoyltransferase (Nmt) (Bhatnagar et al., 1998; Dutnall et al., 1998; Weston et al., 1998). This observation confirms that both AAC(3)-Ia and AAC(6')-Ii belong to the GCN5-related N-acetyltransferase (GNAT) superfamily. The first members of the GNAT superfamily to be identified were a class of histone acetyltransferase enzymes, typified by the yeast MAK3 and GCN5 histone acetyltransferase, which are involved in regulation of gene expression (Tercero et al., 1992). Using a multiple alignment and database-search procedure, additional *N*-acetyltransferases, present in bacteria, archaea and eukaryotes, were classified as members of the GNAT superfamily (Neuwald and Landsman, 1997). With the structure determination of Nmt, it is now realized that the GNAT superfamily is not restricted to acetyltransferases, but also encompasses certain acyltransferases (Modis and Wierenga, 1998).

When comparing the members of the GNAT superfamily whose structures have been determined, and closely related members, a surprising observation can be made; sequence conservation is essentially not observed within the GNAT superfamily. In fact, a structure based sequence alignment reveals that there are no conserved catalytic residues, i.e. the only residues that are functionally or chemically conserved are hydrophobic in nature (Wybenga-Groot et al., 1999). This observation must imply that the reaction mechanism employed by GNAT superfamily members does not involve a covalent acyl-enzyme intermediate. Apparently, the sole role of the enzyme is to bring the AcCoA and substrate in sufficiently close proximity with the proper orientation so that direct acyl transfer can take place. An additional consequence of the absence of primary sequence conservation within the GNAT superfamily is that even with the low sequence homology observed within the AAC family of enzymes, these proteins may share a common folding motif.

2.3. APH

Aminoglycoside phosphotransferases (APHs) catalyze the ATP-dependent *O*-phosphorylation of aminoglycoside hydroxyl groups. These aminoglycoside kinases are found in both Gram-negative and Gram-positive bacteria. As noted above, the bifunctional enzyme AAC(6')-APH(2") is the prevalent mechanism of gentamicin resistance in the Gram-positive cocci. The C-terminal APH(2") activity of this enzyme has a broad substrate specificity and also has the capacity to phosphorylate aminoglycosides on the 6-aminohexose and 3-aminohexose rings.

The most abundant class of APH enzymes are the APH(3')s which modify both the kanamycin and neomycin class of aminoglycosides. One of these, APH(3')-IIIa, has been studied extensively from the perspectives of both a function and structure. The enzyme has broad substrate specificity with the capacity to modify many aminoglycosides, though the clinically important gentamicin C is not a substrate. Phosphorylation occurs exclusively on the 3'-hydroxyl of the kanamycin class of aminoglycosides, but can occur in both the 3'- and 5"-positions of the neomycin class of antibiotic. Intriguingly, some of these are readily di-phosphorylated. ADP-release is the rate limiting step of the reaction and the structural basis for this became apparent when the crystal structure of APH(3)-IIIa was solved as the nucleotide binding site is buried at the 'bottom'of the enzyme active site.

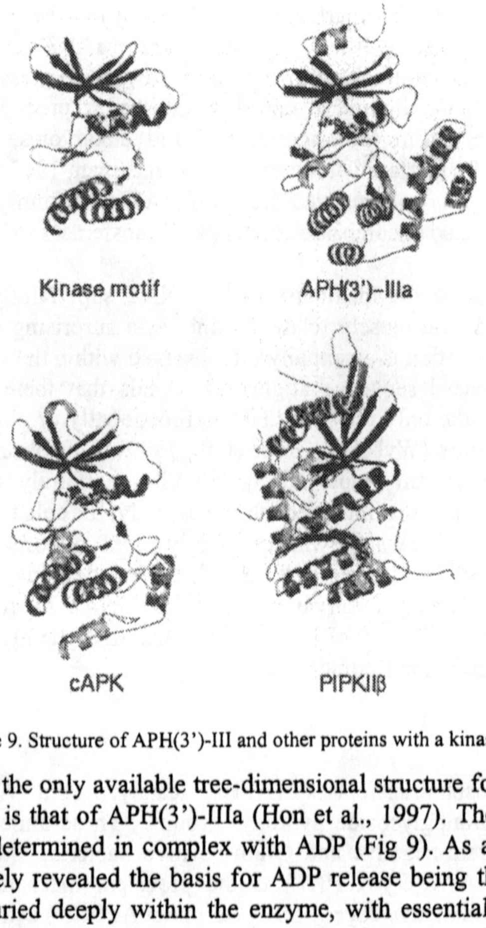

Figure 9. Structure of APH(3')-III and other proteins with a kinase motif

Currently, the only available tree-dimensional structure for an aminoglycoside phosphotransferase is that of APH(3')-IIIa (Hon et al., 1997). The crystal structure of APH(3')-IIIa was determined in complex with ADP (Fig 9). As alluded to above, the structure immediately revealed the basis for ADP release being the rate-limiting step. The co-factor is buried deeply within the enzyme, with essentially only the terminal phosphate group being exposed to the solvent. Furthermore, ADP forms numerous interactions, such as hydrogen bonds and van der Waals contacts with the enzyme. The location and orientation of the bound nucleotide also suggested the area where an aminoglycoside should bind for phosphoryl transfer to occur. Again, analogous to what has been described above for the ANT and AAC classes of enzymes, an analysis of the residues lining this putative aminoglycoside binding area, reveals a preponderance of acidic residues (Fig. 10). Thus, employing electrostatics for guiding and binding a broad spectrum of aminoglycosides appears to be a strategy employed by all aminoglycoside modifying enzymes. The crystal structure also unveiled the structural basis for the order of substrate binding and product release observed for APH(3')-IIIa, i.e. ATP binds prior to the aminoglycoside, and release of phosphorylated aminoglycoside precedes that of ADP, since binding of an aminoglysocide necessarily obstructs the entrance to the nucleotide binding pocket.

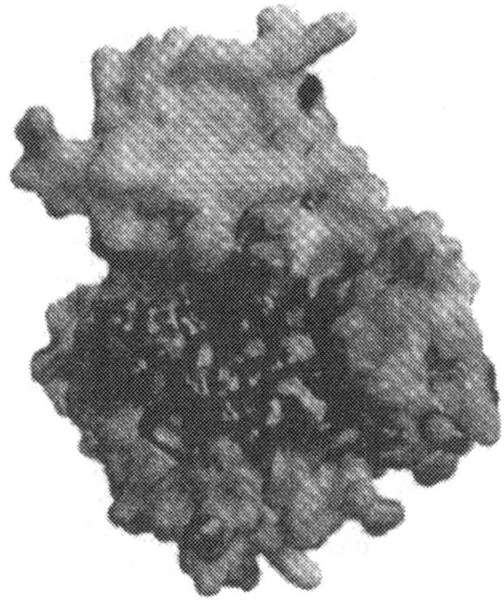

Figure10. Negative surface charge density on APH(3')-IIIa

Unfortunately, no crystallographic data has yet become available for the APH(3')-IIIa enzyme in complex with an antibiotic. However, using NMR, site-directed mutagenesis and computational simulation methods, much insight has been gained in what the three-dimensional structure of such a complex may look like (Cox et al., 1996; Cox and Serpersu, 1997; Thompson et al., 1999). In brief, the data suggests that there are multiple binding modes for aminoglycoside antibiotics. This variety in substrate binding geometry allowed by APH(3')-IIIa provides an explanation for the broad substrate specificity observed for this enzyme.

In addition to providing much insight into the mechanisms of action of APH(3')-IIIa, the crystal structure for this enzyme has also shed light on other APH enzymes. Specifically, since sequence similarity within the APH family is high, it has been possible to identify a several absolutely and functionally conserved residues, and explain much of the mutational data previously collected (Wright et al., 1998). However, the value of the structural data became obvious when the similarity in fold to eukaryotic protein kinases was noted. Despite the essential absence of amino acid sequence similarity (<3%), nearly 50% of the enzyme adopts a fold identical to that of protein kinases (Fig. 9) (Taylor et al., 1992). The structural homology is specifically centered about the nucleotide-binding pocket, and of the only seven residues that are conserved six are located in the part of the enzyme. It is noteworthy, that recently a second structural homologue of eukaryotic protein kinases was determined, namely type IIβ phosphatidylinositol phosphate kinase (PIPKIIβ; Fig. 9) (Rao et al., 1998). However, judged by the structural similarity, this enzyme is more distantly related to the protein kinases than is APH(3')-IIIa.

The structural similarities with protein kinases have been paralleled by similarities in mechanism, sensitivity to inhibitors, and capacity to phosphorylate peptides and proteins. Furthermore, these similarities largely extend to the AAC(6')-APH(2") kinase domain demonstrating the universality of structure and function. Site directed mutagenesis studies have supported the critical role for two APH(3')-IIIa Asp residues, Asp190 and Asp208, in catalysis, both of which are conserved in protein kinases. The first is thought to play a role in activation of the aminoglycoside substrate, possibly acting as an active site base and the second is an essential Mg^{2+} ligand (Wright and Thompson, 1999). Inhibitors of protein kinases such as the flavanoids quercein and genestein and the isoquinoline sulfonamides are also inhibitors of aminoglycoside kinases. These compounds are competitive inhibitors of ATP and noncompetitive inhibitors of aminoglycoside substrates in the aminoglycoside and protein kinases (Daigle et al., 1997). Furthermore, it has been recently determined that aminoglycoside kinases can phosphorylate known substrates of protein kinases such as myelin basic protein (Daigle et al., 1998). Phosphoamino acid analysis demonstrated that phosphoryation was occurring on Ser residues. Thus aminoglycoside kinases share many characteristics with protein kinases which suggests an evolutionary link between these classes of proteins.

3. Conclusions

The availability of 3D-structures for all three classes of aminoglycoside modifying enzymes has permitted great insight into the mechanism of resistance to this important class of antibiotics. The merger of both structure and functional data should facilitate the rational design of novel inhibitors of these enzymes, and the initial steps in this direction have begun with the identification of the APH sensitivity to protein kinase inhibitors. Potent and selective inhibitors of aminoglycoside modifying enzymes have the potential to block aminoglycoside resistance in vivo and therefore could find use in extending the clinical utility of many aminoglycosides. Of special challenge in this scenario will be inhibition of the AAC(6')-APH(2") bifunctional enzyme which is the predominant source of gentamicin resistance in clinically important Gram-positive cocci. Furthermore, the structural data that is emerging from the aminoglycoside resistance area is pointing to unexpected and exciting relationships with enzymes important in eukaryotic cell life.

Acknowledgements
GDWE is the recipient of an MRC Scholarship, and AMB is the recipient of an MRC/PMAC-HRF Research Career Award in the Health Sciences.

References

Aravind, L., and Koonin, E. V. (1999). DNA polymerase β-like nucleotidyltransferase superfamily: identification of three new families, classification and evolutionary history. Nucleic Acids Res. *27*, 1609-1618.

Bhatnagar, R. S., Futterer, K., Farazi, T. A., Korolev, S., Murray, C. L., Jackson-Machelski, E., Gokel, G. W., Gordon, J. I., and Waksman, G. (1998). Structure of *N*-myristoyltransferase with bound myristoylCoA and peptide substrate analogs. Nat. Struct. Biol. *5*, 1091-1097.

Costa, Y., Galimand, M., Leclercq, R., Duval, J., and Courvalin, P. (1993). Characterization of the chromosomal *aac(6')-Ii* gene specific for *Enterococcus faecium*. Antimicrob. Agents Chemother. *37*, 1896-1903.

Cox, J. R., McKay, G. A., Wright, G. D., and Serpesu, E. H. (1996). Arrangement of substrates at the active site of an aminoglycoside antibiotic 3'-phosphotransferase as determined by NMR. J. Am. Chem. Soc. *118*, 1295-1301.

Cox, J. R., and Serpersu, E. H. (1997). Biologically important conformations of aminoglycoside antibiotics bound to an aminoglycoside 3'-phosphotransferase as determined by transferred nuclear Overhauser effect spectroscopy. Biochemistry, *36*, 2353-2359.

Daigle, D. M., Hughes, D. W., and Wright, G. D. (1999). Prodigious substrate specificity of AAC(6')-APH(2"), an aminoglycoside antibiotic resistance determinant in enterococci and staphylococci. Chem. Biol. *6*, 99-110.

Daigle, D. M., McKay, G. A., Thompson, P. R., and Wright, G. D. (1998). Aminoglycoside phosphotransferases required for antibiotic resistance are also Serine protein kinases. Chem. Biol. *6.*, 11-18.

Daigle, D. M., McKay, G. A., and Wright, G. D. (1997). Inhibition of aminoglycoside antibiotic resistance enzymes by protein kinase inhibitors. J. Biol. Chem. *272*, 24755-24758.

DiGiammarino, E. L., Draker, K. a., Wright, G. D., and Serpesu, E. H. (1997). Solution studies of isepamicin and conformational comparisons between isepamicin and butirosin A when bound to an aminoglycoside 6-*N*-acetyltransferase determined by NMR spectroscopy. Biochemistry, *37*, 3638-3644.

Dutnall, R. N., Tafrov, S. T., Sternglanz, R., and Ramakrishnan, V. (1998). Structure of the histone acetyltransferase Hat1: a paradigm for the GCN5-related *N*-acetyltransferase superfamily. Cell, *94*, 427-438.

Ferretti, J. J., Gilmore, K. S., and Courvalin, P. (1986). Nucleotide sequence analysis of the gene specifying the bifunctional 6'-aminoglycoside acetyltransferase 2"-aminoglycoside phosphotransferase enzyme in *Streptococcus faecalis* and identification and cloning of gene regions specifying the two activities. J. Bacteriol. *167*, 631-638.

Gates, C. A., and Northrop, D. B. (1988a). Determination of the rate-limiting segment of aminoglycoside nucleotidyltransferase 2"-I by pH- and viscosity-dependent kinetics. Biochemistry, *27*, 3834-3842.

Gates, C. A., and Northrop, D. B. (1988b). Substrate specificities and structure-activity relationships for the nucleotidylation of antibiotics catalyzed by aminoglycoside nucleotidyltransferase 2"-I. Biochemistry, *27*, 3820-3825.

Holm, L., and Sander, C. (1995). DNA polymerase β belongs to an ancient nucleotidyltransferase superfamily. Trends Biol. Chem. *20*, 345-347.

Hon, W. C., McKay, G. A., Thompson, P. R., Sweet, R. M., Yang, D. S. C., Wright, G. D., and Berghuis, A. M. (1997). Structure of an enzyme required for aminoglycoside resistance reveals homology to eukariotic protein kinases. Cell, *89*, 887-895.

Miller, G. H., Sabatelli, F. J., Hare, R. S., Glupczynski, Y., Mackey, P., Shlaes, D., Shimizu, K., Shaw, K. J., and Groups, a. A. R. S. (1997). The most frequent aminoglycoside resistance mechanisms--changes with time and geographic area: a reflection of aminoglycoside usage patterns? Clin. Infect. Dis. *24*, S46-S62.

CoA-dependent enzymes. Structure *6*, 1345-1350.

Neuwald, A. F., and Landsman, D. (1997). GCN5-related histone *N*-acetyltransferases belong to a diverse superfamily that includes the yeast SPT10 protein. Trends Biochem. Sci. *22*, 154-155.

Perdersen, L. C., Benning, M. M., and Holden, H. M. (1995). Structural investigation of the antibiotic and ATP-binding sites in kanamycin nucleotidyltransferase. Biochemsitry, *34*, 13305-13311.

Rao, V. D., Misra, S., Boronenkov, I. V., Anderson, R. A., and Hurley, J. H. (1998). Structure of type IIβ phosphatidylinositol phosphate kinase: a protein kinase fold flattened for interfacial phosphorylation. Cell, *94*, 829-839.

Sakon, J., Liao, H. H., Kanikula, A. M., Benning, M. M., Rayment, I., and Holden, H. M. (1993). Molecular structure of kanamycin nucleotidyl transferase determined to 3 Å resolution. Biochemistry, *32*, 11977-11984.

Santanam, P., and Kayser, F. H. (1978). Purification and characterization of an aminoglycoside inactivating enzyme from *Staphylococcus epidermidis* FK109 that nucleotidylates the 4'- and 4''-hydroxyl groups of the aminoglycoside antibiotics. J. Antibiot. *31*, 343-351.

Shaw, K. J., Rather, P. N., Hare, R. S., and Miller, G. H. (1993). Molecular genetics of aminoglycoside resistance genes and familial relationships of the aminoglycoside-modifying enzymes. Microbiol. Rev. *57*, 138-163.

Taylor, S. S., Knighton, D. R., Zheng, J., Ten Eyck, L. F., and Sowadski, J. M. (1992). Structural framework for the protein kinase family. Annu. Rev. Cell Biol. *8*, 429-62.

Tercero, J. C., Riles, L. E., and Wickner, R. B. (1992). Localized mutagenesis and evidence for post-transcriptional regulation of MAK3. A putative *N*-acetyltransferase required for double-stranded RNA virus propagation in Saccharomyces cerevisiae. J. Biol. Chem. *267*, 20270-20276.

Thompson, P. R., Schwartzenhauer, J., Hughes, D. W., Berghuis, A. M., and Wright, G. D. (1999). The COOH terminus of aminoglycoside phosphotransferase (3')-IIIa is critical for antibiotic recognition and resistance. J. Biol. Chem. *274*, 30697-30706.

Umezawa, H., Okanishi, M., Kondo, S., Hamana, K., Utahara, R., Maeda, K., and Mitsuhashi, S. (1967). Phosphorylative inactivation of aminoglycopside antibioics by *Escherichia coli* carrying R factor. Science, *157*, 1559-1561.

Van Pelt, J. E., Iyengar, R., and Frey, P. A. (1986). Gentamicin nucleotidyltansferase. Stereochemical inversion at phosphorus in enzymatic 2'-deoxyadenylyl transfer to tobramycin. J. Biol. Chem. *261*, 15995-15999.

Weston, S. A., Camble, R., Colls, J., Rosenbrock, G., Taylor, I., Egerton, M., Tucker, A. D., Tunnicliffe, A., Mistry, A., Mancia, F., de la Fortelle, E., Irwin, J., Bricogne, G., and Pauptit, R. A. (1998). Crystal structure of the anti-fungal target *N*-myristoyl transferase. Nat. Struct. Biol. *5*, 213-221.

Wolf, E., Vassilev, A., Makino, Y., Sali, A., Nakatani, Y., and Burley, S. K. (1998). Crystal structure of a GCN5-related *N*-acetyltransferase: *Serratia marcescens* aminoglycoside 3-*N*-acetyltransferase. Cell, *94*, 439-449.

Wright, G. D., Berghuis, A. M., and Mobashery, S. (1998). Aminoglycoside antibiotics: Structure, function and resistance. In Resolving the Antibiotic paradox: Progress in Drug Design and Resistance, B. P. Rosen and S. Mobashery, eds. (New York: Plenum Press), pp. 27-69.

Wright, G. D., and Ladak, P. (1997). Overexpression and characterization of the chromosomal aminoglycoside 6'-*N*-acetyltransferase from *Enterococcus faecium*. Antimicrob. Agents Chemother. *41*, 956-960.

Wright, G. D., and Thompson, P. R. (1999). Aminoglycoside phosphotransferases: proteins, structure, and mechanism. Front. Biosci. *4*, D9-D21.

Wybenga-Groot, L., Draker, K. a., Wright, G. D., and Berghuis, A. M. (1999). Crystal structure of an aminoglycoside 6'-*N*-acetyltrasnferase: defining the GCN5-related *N*-acetyltransferase superfamily fold. Structure, *7*, 497-507.

THE GENETICS AND BIOCHEMISTRY OF RESISTANCE TO GLYCOPEPTIDE ANTIBIOTICS

Glycopeptide resistance mechanisms

P. E. REYNOLDS
Department of Biochemistry, University of Cambridge, Tennis Court Road, Cambridge, CB2 1QW, U.K.

Summary

Enterococci and staphylococci have evolved different resistance mechanisms to counteract glycopeptide antibiotics. High- and low-level resistance in enterococci results from new pathways of peptidoglycan biosynthesis together with elimination of peptidoglycan precursors with high affinity for glycopeptides. In staphylococci the main mechanism appears to involve sequestration of the antibiotic in the outer layers of the cell wall before it reaches the vital target sites.

1. Action of Glycopeptides: Vancomycin and Teicoplanin

Glycopeptide antibiotics have a unique mechanism of action. They bind to a specific arrangement of amino acids (acyl-D-Ala-D-Ala) in the cell wall peptidoglycan and its precursors and inhibit, probably by steric hindrance, a transglycosylase enzyme that catalyses synthesis of nascent peptidoglycan chains from membrane-bound lipid intermediates (Barna and Williams, 1984; Reynolds, 1989, Arthur *et al.*, 1996b). Two aspects of this mechanism have influenced the length of time it has taken for bacteria to acquire resistance to glycopeptides.

Firstly, although glycopeptides bind to acyl-D-Ala-D-Ala, the reaction inhibited does not involve the peptide region of the lipid intermediate precursor or nascent peptidoglycan. In the transglycosylation reaction the bonds that are broken and formed involve the carbohydrate moieties of the precursors. Computer simulations of the space occupied by the glycopeptide bound to acyl-D-Ala-D-Ala indicate that the relatively bulky, but compact, glycopeptide does not extend as far as the sugar residues of the precursors (Jeffs and Nisbet, 1988): consequently the presence of the glycopeptide does

A. Van Broekhoven et al. (eds.), Novel Frontiers in the Production of Compounds for Biomedical Use, 99–115.
© 2001 *Kluwer Academic Publishers. Printed in the Netherlands.*

not directly shield the substrate from its enzyme but appears to prevent the active site of the transglycosylase from being positioned correctly in order to catalyse the reaction.

Secondly, even if the transglycosylase reaction was not inhibited, the binding of vancomycin and teicoplanin would inhibit the subsequent transpeptidation reaction that links newly-synthesised nascent peptidoglycan to the existing cell wall. This, and other peptidoglycan cross-linking reactions, which occur outside the cytoplasmic membrane where ATP is unavailable, rely on the hydrolysis of an existing peptide bond and retention of the energy liberated, by formation of a covalent ester bond between the carbonyl of a D-Ala residue and the hydroxyl of a serine in the active site of the transpeptidase enzyme (Frère and Joris, 1985). The unique conformation of the acyl-D-Ala-D-Ala terminus of the pentapeptide chain of the peptidoglycan precursor is essential for the process. If this arrangement was altered in a potential resistance mechanism to hinder or prevent binding of glycopeptides, then the transpeptidase would probably be inactive unless the change was relatively minor. Development of glycopeptide resistance would be dependent on a decreased affinity of the glycopeptide for the altered substrate.

2. Potential Mechanisms of Glycopeptide Resistance

These two aspects of glycopeptide action limit the possible mechanisms for resistance. One obvious potential mechanism involves inactivation. However, although the glycopeptide structure is based on a heptapeptide, the oxidative reactions that link together the phenolic R groups on amino acids 2, 4, and 6, and 5 and 7 in vancomycin give rise to an atypical peptide that adopts a bracelet conformation and no enzyme has yet been detected that modifies or destroys the activity of vancomycin or teicoplanin. Secondly, the possibility of restriction of access of the glycopeptide to its target site by synthesis of a 'barrier' is unlikely as glycopeptides do not penetrate the cytoplasmic membrane but target peptidoglycan precursors as they emerge through the membrane. The cell wall is permeable to large molecules and does not restrict the accessibility of glycopeptides to their essential target sites. Alternatively, overproduction of non-essential target molecules either in the outer region of the cell wall or by synthesis and release into the medium might sequester free antibiotic as a complex in the wall or growth medium and prevent it reaching the limited number of crucial target sites on the outside surface of the cytoplasmic membrane (Reynolds, 1989). The fourth possibility for a resistance mechanism is for the pathway of peptidoglycan synthesis to be changed so that glycopeptides either do not bind to the precursor or bind with a greatly reduced affinity: there are severe constraints on structural alterations that can be tolerated as discussed above. The basic structure to which glycopeptides bind is assembled inside the cell and so is inaccessible. A glycopeptide resistance mechanism which alters the metabolic pathway must involve not only new enzymes for a new pathway but other enzymes to eliminate the susceptible pathway. It is probably for this reason that acquired glycopeptide resistance emerged only after prolonged clinical use, compared with the usual short period of 1 - 3 years for the majority of antibiotic resistance mechanisms.

3. Glycopeptide Resistance in Enterococci

The critical enzyme that catalyses synthesis of the D-Ala-D-Ala dipeptide before its addition to UDP-MurNAc-tripeptide (tripeptide) is D-Ala:D-Ala ligase (Ddl). It is the specificity of this enzyme coupled with the availability of D-Ala or other potential substrates that determines the nature of the residue incorporated at the carboxy-terminus of the peptidoglycan precursors (Table 1). One of the earliest findings in glycopeptide-resistant enterococci was the identification of a gene encoding a second dipeptide ligase (Dutka-Malen *et al.*, 1990a); the normal host Ddl synthesised D-Ala-D-Ala and the second was characterised originally as a dipeptide ligase with altered specificity and was later demonstrated to catalyse synthesis of the depsipeptide D-Ala-D-lactate (Bugg *et al.*, 1991).

Table 1. Dipeptides and depsipeptides in glycopeptide-resistant bacteria

Phenotype / genus	Di- or depsi-peptide synthesised
VanA, VanB, VanD	D-Ala-D-Lac, D-Ala-D-Ala
VanC, VanE	D-Ala-D-Ser, D-Ala-D-Ala
Glycopeptide producers	D-Ala-D-Lac, [D-Ala-D-Ala]
Lactobacilli (most), Leuconostocs, Pediococci	D-Ala-D-Lac
Vancomycin-dependent enterococci	D-Ala-D-Lac

3.1. THE VANA PHENOTYPE

This initial finding led to the characterisation of a complete operon of plasmid-borne resistance genes in high-level resistant *Enterococcus faecium*. The VanA phenotype is mediated by transposon Tn*1546* or related elements which carries a gene cluster encoding 9 proteins with at least three different types of function, transposition, induction of resistance and the resistance mechanism itself (Arthur *et al.*, 1993); some of the five proteins in the latter group are essential and others accessory for resistance (Arthur *et al.*, 1996b). Synthesis of the enzymes required for the new synthetic pathway is regulated at the transcriptional level by the VanR-VanS two-component regulatory system (Arthur *et al.*, 1992). VanS controls the level of phosphorylation of VanR as a result of its kinase and phosphatase activities (Wright *et al.*, 1993). The phosphorylated VanR binds to two promoters upstream of the *vanR* and *vanH* genes (Arthur *et al.*, 1997). The nature of the signal recognised by VanS has not been fully characterised: it may be a metabolite on the synthetic or degradative pathway of peptidoglycan as peptidoglycan synthesis inhibitors other than glycopeptides (e.g. moenomycin) act as inducers (Allen and Hobbs, 1995; Baptista *et al.*, 1996; Lai and Kirsch, 1996) and the sensor protein in a VanB strain with only 23% identity to VanS does not respond to teicoplanin.

3.1.1. Peptidoglycan Synthesis: a New Pathway

The proteins which catalyse reactions in the new synthetic pathway are VanH (a D-hydroxycarboxyllic acid dehydrogenase (Bugg *et al.*, 1991) and VanA, a D-Ala:D-Ala ligase of altered specificity. These proteins are cytoplasmic and their combined action results in the synthesis of D-Ala-D-lactate. All the enzymes of peptidoglycan synthesis in enterococci that utilise substrates terminating in D-Ala-D-Ala function with substrates terminating with D-Ala-D-lactate. The catalytic mechanism of D-Ala:D-Ala ligases has been studied extensively and inhibitors of the purified enzyme have been identified Ellsworth *et al.*, 1996). A change of a single residue in the active site of the DdlB ligase from *E. coli* from tyrosine to phenylalanine (Y216F) changes the specificity of the enzyme so that it functions as a D-Ala:D-lactate ligase (Park *et al.*, 1996): the catalytic mechanism is believed to be closely analogous to that of the original ligase. Although the mutation of a single amino acid residue in the host Ddl ligase is sufficient to change the substrate specificity so that the enzyme catalyses synthesis of D-Ala-D-lactate in addition to D-Ala-D-Ala, the amino acid sequence of VanA is only 30% identical to that of *E. coli* DdlB: its low sequence identity to other dipeptide ligases, the depsipeptide ligases of lactobacilli, Leuconostocs and the glycopeptide-producing organisms *Amycolatopsis orientalis* and *Streptomyces toyocaensis* (Marshall *et al.*, 1998) suggests that VanA has not evolved recently by direct mutation of an existing ligase. It is important that the activity of VanA as a D-Ala:D-Ala ligase is low so that D-Ala-D-lactate is produced almost exclusively by this enzyme in resistant enterococci. The activity of VanA together with VanH is insufficient for glycopeptide resistance except in the special case of glycopeptide-dependent enterococci. These strains have a defective D-Ala:D-Ala ligase (van Bambeke *et al.*, 1999; Sifaoui and Gutmann, 1997) and rely on the inducible vancomycin-resistance pathway of peptidoglycan synthesis for manufacture of cell wall precursors, hence their dependence on vancomycin for growth. Frameshift and missense mutations have been shown to lead to a defective ddl ligase. The mutations leading to a defective Ddl ligase confer no benefit on the organisms as cessation of vancomycin treatment leads to elimination of the organism: only strains originally containing one of the glycopeptide resistance operons can therefore survive, and then only when vancomycin is present.

3.1.2. Peptidoglycan Synthesis: Control of Normal Host Pathway

Under circumstances in which the host Ddl ligase is active, the degree of resistance of VanA-type strains to glycopeptides is dependent on the ratio of the alternative soluble peptidoglycan precursors UDP-MurNAc-L-Ala-g-D-Glu-L-Lys-D-Ala-D-Ala (pentapeptide) and UDP-MurNAc-L-Ala-g-D-Glu-L-Lys-D-Ala-D-lactate (pentadepsipeptide) (Arthur *et al.*, 1996a). Synthesis of pentapeptide can be almost eliminated by the combined action of two enzymes, a D,D-dipeptidase of strict specificity (VanX) (Reynolds *et al.*, 1994a) and a penicillin-insensitive D,D-carboxypeptidase (VanY) (Wright *et al.*, 1992): the extent of elimination is dependent on gene copy number (see later). VanX hydrolyses D-Ala-D-Ala prior to its addition to tripeptide: if VanX is present in large amounts the whole pathway of peptidoglycan synthesis becomes directed to the exclusive synthesis of pentadepsipeptide. VanX hydrolyses other D,D-dipeptides, though at a reduced rate compared with D-Ala-D-Ala;

it does not function with tripeptides or with D,D-dipeptides substituted at either the N- or C-terminus, and surprisingly it does not hydrolyse the depsipeptide D-Ala-D-lactate (Reynolds *et al.*, 1994a; Wu *et al.*, 1995) even though esters are normally better substrates than peptides for such peptidases. Crystallographic studies and analysis of the active centre of VanX are important for the design of inhibitors of the enzyme but enterococci could overcome the action of an inhibitor, if one were developed, by loss of the host Ddl ligase and mutation of the induction system so that the enzymes essential for vancomycin resistance were synthesised constitutively as has happened in VanD strains (see later). The second control enzyme, VanY has been described as accessory as it is not essential for resistance in VanA-type strains in which the resistance operon is present on a multicopy plasmid (Arthur *et al.*, 1994). Under these circumstances the activity of VanX is normally sufficient to keep the available concentration of D-Ala-D-Ala at such a low level that virtually no pentapeptide is synthesised. VanY catalyses the identical reaction to the low molecular mass D,D-carboxypeptidases that are penicillin-binding proteins (PBPs), namely the removal of D-Ala from compounds terminating in acyl-D-Ala-D-Ala (Arthur *et al.*, 1998), but in VanA and VanB strains it does not contain the motifs characteristic of the penicillin-binding domain of bacterial PBPs. If the activity of VanX is low or the growth medium is supplemented with D-Ala, the presence of VanY is important and its activity results in the conversion of pentapeptide to UDP-MurNAc-L-Ala-g-D-Glu-L-Lys-D-Ala (tetrapeptide) to which glycopeptides do not bind.

The rationale for the resistance mechanism in VanA-type strains is that the replacement of the amide link between the two D-Ala residues at the C-terminus of the peptidoglycan precursors by the ester link in D-Ala-D-lactate results in loss of one of the five hydrogen bonds in the complex between vancomycin and its target. This loss results in at least a 1000-fold lowering of the affinity (Bugg *et al.*, 1991) and raises the MIC value of vancomycin for the organism by a similar factor. D-lactate is one of the few alternative C-terminal residues of peptidoglycan precursors that can be utilised by the penicillin-susceptible transpeptidases (Rasmussen and Strominger, 1978), and consequently stem peptides in the peptidoglycan precursors containing this residue can act as donors in transpeptidation reactions which cross-link the peptidoglycan strands in the wall.

3.2. SIMILARITY AND DIVERSITY OF THE RESISTANCE OPERONS OF VANA-, VANB- AND VAND-TYPE ENTEROCOCCI

The resistance mechanism present in VanA enterococci is essentially similar to that present in VanB- (Evers and Courvalin, 1996) and VanD-type (Périchon *et al.*, 1997) enterococci. Both of these other resistance types have genes encoding a D-lactate dehydrogenase (VanH_B, H_D), an altered ligase (VanB, VanD) which catalyses synthesis of D-Ala-D-lactate, and VanX and VanY-type proteins but the manner of achieving resistance is subtly different between the three types. In VanA strains the location of the genes on multicopy plasmids in clinical isolates leads to high expression of VanX and its activity is sufficient to effectively eliminate D-Ala-D-Ala as it is synthesised before it is added to tripeptide to form the pentapeptide (Reynolds, 1998). In these strains VanY is inessential for resistance and its role may be regarded as accessory: the gene encoding it is located downstream from the essential VanH, A and X genes. The construction of strains containing the regulatory genes and those essential for resistance in the chromosome enabled the effect of gene copy number to be investigated (Arthur *et al.*, 1996a). A single copy of the VanR, S, H, A and X genes resulted in low MIC values of the constructs for vancomycin (MIC 4 - 8 mg/ml), and the two peptidoglycan precursors pentadepsipeptide:pentapeptide were synthesised in a ratio of between 1 - 2. When 2 copies of the genes were present the MIC was raised to 32 mg/ml and the ratio of precursors was 16, whereas with 5 copies the pentadepsipeptide:pentapeptide ratio had risen to 32 (i.e. 97% pentadepsipeptide) and the MIC was 256 mg/ml. The functions of genes encoding glycopeptide resistance are summarised in Table 2.

Table 2. Function of vancomycin resistance genes

Gene	Function
VanR, R_B	Phospho-VanR binds to promoters P_R (for *vanR*, *S*) and P_H (for *VanH*, *A*, *X*)
VanS, S_B	Membrane-bound sensor, detects presence of glycopeptides: phosphorylates VanR and dephosphorylates phospho-VanR
VanA,B,D	D-Ala:D-lactate ligase
VanC,E	D-Ala:D-Ser ligase
VanH, H_B, H_D	D-lactate dehydrogenase
VanX, X_B	D,D-dipeptidase: hydrolyses D-Ala-D-Ala, no activity vs. D-Ala-D-Lac
VanY, *YB*	D,D-carboxypeptidase (penicillin insensitive): hydrolyses D-Ala from acyl-D-Ala-D-Ala, low activity vs. acyl-D-Ala-D-Lac
VanY_D	D,D-carboxypeptidase (penicillin sensitive)
VanXY_C	D,D-dipeptidase/ D,D-carboxypeptidase:
	low activity vs. D-Ala-D-Ser and acyl-D-Ala-D-Ser
VanT_C	Serine racemase (membrane-bound)
VanZ	*Confers low-level teicoplanin resistance*
VanW	*?*

3.2.1. VanB-type resistance

The VanB resistance operon is also located on a transposon (Tn*1547*) but this is present in the chromosome and presumably only single copies of the resistance genes are present (Quintiliani and Courvalin, 1996). The most obvious differences with VanA-type strains are i) induction of resistance is obtained only with vancomycin and not with teicoplanin; ii) the percentage identity of $VanS_B$ with VanS and $VanY_B$ with VanY is low (23% and 30% respectively) and $VanY_B$ is located immediately after $VanS_B$ and is upstream of the $VanH_B$, VanB and $VanX_B$ genes (Evers and Courvalin, 1996). The VanB resistance genes are transcribed as a polycistronic messenger RNA (J. Stigter, unpublished observations) and it is probable that $VanY_B$ is produced in greater amounts than the other resistance proteins. The regulatory system in VanB-type strains appears to be tightly regulated: in the absence of an inducer no pentadepsipeptide is produced and no VanX activity can be detected in extracts. This contrasts with VanA strains in which pentadepsipeptide is synthesised in relatively low amounts and VanX is detectable in extracts prior to induction. Consequently when vancomycin is added to a VanB-type strain prior to the induction of resistance, early events include a massive build-up of pentapeptide in the cytoplasm because lipid molecules that act as transporters of the basic peptidoglycan subunit (a disaccharide-peptide) across the membrane are sequestered by vancomycin on the outside surface, thus preventing their recycling, and so are unavailable for the transfer of further peptidoglycan subunits across the membrane (Reynolds, 1998). Peptidoglycan synthesis is frozen until new C_{55}-undecaprenol carrier lipid molecules are either synthesised or made available by dissociation of the lipid II intermediate-vancomycin complex. In order to prevent precursors containing acyl-D-Ala-D-Ala being added to new lipid carrier molecules, pentapeptide molecules are converted by $VanY_B$ to tetrapeptide which is present in large amounts in the cytoplasm within 30 min of the start of induction. As stated earlier tetrapeptide can be utilised as a precursor of peptidoglycan and the cytoplasmic content decreases gradually as the capacity to synthesise pentadepsipeptide increases during the first two hours of induction of resistance (Reynolds, 1998). In VanB strains the single copy resistance genes result in a slow rate of elimination of D-Ala-D-Ala by VanX so the early presence of $VanY_B$ activity is important in controlling the amount of pentapeptide present.

3.2.2. VanD-type resistance

The glycopeptide resistance operon in the single VanD strain that has been fully characterised is similar in general terms to that in VanA and VanB strains but there are some interesting differences (Casadewall and Courvalin, 1999). The VanD ligase, the $VanH_D$ dehydrogenase and the $VanX_D$ dipeptidase are between 59 - 70% identical to the corresponding proteins in VanA and VanB strains (Table 3).

Table 3. Comparison of VanD-, VanA- and VanB-type proteins

	% Identity with		% Identity of VanA-type
VanD-type	VanA-type	VanB-type	With VanB-type
VanR	58	34	35
VanS	42	19	17
VanY	13	15	30
VanH	59	63	68
VanD	69	69	76
VanX	68	70	75

Resistance is expressed constitutively and this results from mutations in at least two genes. VanD isolates are defective for the host Ddl ligase (Casadewall and Courvalin, 1999) and if this was the only mutation the strain would be vancomycin-dependent. Additionally the two-component regulatory system must be switched on permanently though the mutation(s) - probably in $VanS_D$ - has not been identified because the degree of identity with VanS and $VanS_B$ is relatively low (42% and 19% respectively). Two other differences have been detected in VanD *E. faecium* BM4339. Firstly, although all the amino acids known to be involved in substrate-binding, Zn^{2+}-binding and catalysis in VanX and related proteins (Lessard and Walsh, 1999) are present in the putative $VanX_D$, no VanX-type activity was detected. This is of little consequence as the host Ddl ligase is inactive, but VanD would be able to catalyse synthesis of small amounts of D-Ala-D-Ala if the internal cellular concentration of D-Ala was high. Secondly, no penicillin-insensitive D,D-carboxypeptidase activity is present either in the membrane or cytoplasm, though tetrapeptide peptidoglycan precursors were detected in the cytoplasm after inhibiting peptidoglycan synthesis with ramoplanin (Périchon *et al.*, 1997). This indicates that the strain has the capability to manufacture pentapeptide. However, there is a membrane-bound, penicillin-sensitive D,D-carboxypeptidase present in the membrane in amounts greatly in excess of that in a vancomycin-susceptible *E. faecium*, and the derived amino acid sequence of the 356 amino acid protein, $VanY_D$, contains the motifs characteristic of the penicillin-binding domain of penicillin-binding proteins (PBPs). This is the first example of the direct involvement of a potential PBP in glycopeptide resistance. In common with VanB-type strains the gene encoding $VanY_D$ is upstream of $VanH_D$, VanD and $VanX_D$ (Casadewall and Courvalin, 1999). The characteristics of the VanD phenotype are given in Table 4 and the relative location of genes in vancomycin-resistant enterococci in Table 5.

Table 4. Characteristics of VanD Enterococci

Gene	Function of gene product
vanH$_D$,D	Synthesis of D-Ala-D-Lac
host *ddl*	D-Ala:D-Ala ligase is defective. Similar to vancomycin-dependent strains but vancomycin is not essential for growth
vanX$_D$	D,D-dipeptidase is inactive in strain BM4339: it is not required for resistance as Ddl is inactive
vanY$_D$	D,D-carboxypeptidase: the membrane-bound enzyme is inhibited by penicillin; it contains motifs of a typical low M_r penicillin-binding protein
vanR$_D$	response regulator: switched on permanently
vanS$_D$	senses presence of antibiotic; may contain a mutation as operon is expressed constitutively

Table 5. Comparison of vanA, vanB, vanC and vanD operons

Resistance type	Order of genes									
VanA	R	S			H	A	X		Y	Z
VanB	R_B	S_B	Y_B	W	H_B	B	X_B			
VanD	R_D	S_D	Y_D		H_D	D	X_D			
VanC						C	XY_C	T_C	R_C	S_C

3.3. VANC-TYPE RESISTANCE

As with VanA-, VanB- and VanD-type enterococci, the intrinsically glycopeptide-resistant VanC-type enterococci (*E. gallinarum*, *E. casseliflavus*, *E. flavescens*) have two ligases (Dutka-Malen *et al.*, 1990b, 1992), but differ in that the 'additional' one is a second D,D-dipeptide ligase with altered specificity rather than a depsipeptide ligase. VanC-type strains synthesise UDP-MurNAc-pentapeptide terminating in D-Ser (pentapeptide[Ser]) (Billot-Klein *et al.*, 1994a; Reynolds *et al.*, 1994b) and the purified protein encoded by the *vanC-2* gene of *E. casseliflavus* has been characterised as a D-Ala:D-Ser ligase (Park *et al.*, 1997). A second gene in the resistance operon encodes a membrane-bound serine racemase ($VanT_C$) which provides a supply of D-Ser for the cytoplasm-located D-Ala:D-Ser ligase (Arias *et al.*, 1999). L-Ser has a number of important metabolic roles in cells and it is possible that the membrane-bound domain of $VanT_C$ with its putative 10 transmembrane helices acts as an L-Ser transporter to channel L-Ser directly to the racemase domain of the protein. Modelling studies indicate that the protein may exist as a dimer, as occurs with the alanine racemase from *Bacillus stearothermophilus* (Shaw *et al.*, 1997), with two active sites formed by the interacting soluble domains of the two polypeptide chains (Arias *et al.*, 1999). VanC and $VanT_C$ result in the production of D-Ala-D-Ser in the cytoplasm but the chromosomal Ddl ligase synthesises D-Ala-D-Ala which competes with D-Ala-D-Ser for addition to tripeptide. The third gene essential for VanC-type resistance encodes a soluble protein with both D,D-dipeptidase and D,D-carboxypeptidase activity (Reynolds *et al.*, 1999). It apparently lacks the two amino acids characteristic of specific VanX-type proteins involved in binding the N-terminus of D-Ala-D-Ala, but it contains all the other amino acids important for substrate-binding, Zn^{2+}-binding and catalysis. It differs from VanX in having minimal activity against D-Ala-D-Ser and being able to hydrolyse N-terminal-substituted D-Ala-D-Ala: it therefore has D,D-carboxypeptidase activity in addition to D,D-dipeptidase activity and has consequently been designated $VanXY_C$ (Reynolds *et al.*, 1999). Sequence comparison suggests it has a single active site and that it is related more closely to a VanY-type than to a VanX-type enzyme in spite of its cytoplasmic localisation. The protein lacks a hydrophobic domain close to the N-terminus which anchors VanY and $VanY_B$ to the membrane. The specificity of the enzyme in lacking activity against D-Ala-D-Ser is preserved in relation to its inability to remove D-Ser from pentapeptide[Ser]. In the bacterium the activity of $VanXY_C$ as a D,D-dipeptidase is insufficient to prevent synthesis of pentapeptide[D-Ala] but the D,D-carboxypeptidase activity degrades pentapeptide[Ala] but not pentapeptide[Ser] to tetrapeptide: this does not accumulate in the cytoplasm so it is either used efficiently as a peptidoglycan precursor or is degraded further. *E. gallinarum* BM4174 is constitutive for resistance (Dutka-Malen *et al.*, 1990b) but the VanC operon also contains $vanR_C$ and $vanS_C$ regulatory genes (Table 6) which are located downstream of the 3 essential resistance genes (C.A.Arias, unpublished observations). It will be necessary to sequence large numbers of strains that are inducible and constitutive for resistance to determine the regions of the sensor that are vital for strict regulation of the operon.

Surprisingly, the glycopeptide-susceptible *E. faecalis* JH2-2, the host for recombinant plasmids used to determine the function of the individual genes, synthesised D-Ala-D-Ser, presumably using the single host Ddl ligase, when 50 mM D-Ser was added to the growth medium. However, the absence of $VanXY_C$ resulted in the cosynthesis of pentapeptide [Ala] and pentapeptide[Ser] and virtually no decrease in susceptibility to vancomycin was observed (Reynolds *et al.*, 1999). Although vancomycin forms a complex with acyl-D-Ala-D-Ser, the binding affinity relative to that for acyl-D-Ala-D-Ala is reduced *ca* 7-fold (Billot-Klein *et al.*, 1994b) which results in elevation of the MIC of the organism by the same degree. VanC strains exhibit only low-level vancomycin resistance for this reason and remain susceptible to teicoplanin.

Table 6. Characteristics of VanC enterococci

5 proteins encoded by chromosomal genes	
VanC	D-Ala:D-Ser ligase
$VanXY_C$	D,D-dipeptidase/D,D-carboxypeptidase
$VanT_C$	serine racemase (membrane-bound)
$VanR_C$	response regulator, 50% identity with VanR
$VanS_C$	sensor, 40% identity with VanS

3.3.1. VanE-type Resistance

Recently, a fifth type of vancomycin resistance has been investigated in *E. faecalis* BM4405, designated VanE-type (Fines *et al.*, 1999). A new ligase gene was located and the derived protein had *ca* 50% identity to VanC-1: furthermore the VanE strain synthesised peptidoglycan precursors terminating in D-Ser. Membrane-bound serine racemase activity and cytoplasmic D,D-peptidase and penicillin-insensitive D,D-carboxypeptidase activities have been demonstrated (Fines *et al.*, 1999). These results suggest that the mechanisms of resistance in VanC-type and VanE-type strains are similar.

4. Glycopeptide Resistance in Staphylococci.

In view of the mobility of the Tn*1546* transposon that carries the VanA resistance operon and the large conjugative transposon containing Tn*1547*, harbouring the VanB operon, it is perhaps surprising that high-level inducible glycopeptide resistance has not yet been reported in staphylococci. VanA-type resistance has been transferred to *Staphylococcus aureus* (non MRSA) by mating a vancomycin-resistant *E. faecalis* with a b-lactam-susceptible *S. aureus* on the skin of a mouse (Noble *et al.*, 1992). The resultant recombinant *S. aureus* was inducible resistant to high concentrations of vancomycin and teicoplanin. If such an event occurred naturally with a multiply-resistant *S. aureus* (MRSA) it could render these organisms virtually untreatable with antibiotics currently available. Possible reasons for lack of transfer under hospital conditions (the most high-pressured, selective environment outside the molecular biology laboratory) include i) the different ecological niches occupied by enterococci and staphylococci and ii) the possibility that the mechanisms of b-lactam and glycopeptide resistance in MRSA strains may be incompatible in relation to the biosynthesis of a rigid peptidoglycan. MRSA strains rely on an additional penicillin-binding protein, PBP2' or 2a, to catalyse transpeptidation involved in peptidoglycan cross-linking (Reynolds and Brown, 1985) and this PBP is fastidious in its substrate structural requirements. A number of *fem* (Berger-Bächi *et al.*, 1992) or *aux* (DeLencastre and Tomasz, 1994) mutants, which contain a normal amount of PBP2' but with a variety of defects in peptidoglycan precursor synthesis, are susceptible to b-lactam antibiotics, indicating that PBP2' does not catalyse cross-linking with the altered precursors. It is possible that the substitution of D-lactate for D-Ala is also not tolerated by PBP2'.

Relatively low-level, constitutive resistance to teicoplanin in staphylococci, particularly the coagulase-negative *S. epidermidis* and *S. haemolyticus*, is well documented (see review by Hiramatsu, 1998): an increased amount of PBP2 and the presence of an additional membrane protein (ca 35 kDa) are features common to the majority of these strains. More recently vancomycin-resistant strains of *S. aureus* have been isolated and detailed studies of wall metabolism have been carried out in some of them (e.g. Mu3, Mu50, Hanaki *et al.*, 1998a, 1998b). The degree of resistance is still low (the most resistant Mu50 has an MIC of 8 mg/ml to vancomycin), but the strains exhibit an increased amount of the normal PBP2, a lower degree of cross-linking with a concomitant increase in uncross-linked monomers in the peptidoglycan, an enhanced rate of peptidoglycan synthesis and turnover and a cell wall *ca* 2 times thicker than in a susceptible strain. It has been suggested that as a result of these changes, more vancomycin would be bound to pentapeptides, and therefore sequestered in the outer layers of the wall before any of the antibiotic reached the vital target sites on the outside surface of the cytoplasmic membrane (Hanaki *et al.*, 1998b). If this hypothesis is correct, this type of resistance mechanism is unlikely to result in high-level glycopeptide resistance, unless wall breakdown products released into the growth medium also sequester large amounts of antibiotic: it is therefore difficult to reconcile a limited degree of wall thickening with the large increase in MIC from 2 to 100 mg/ml as reported by Sieradski and Tomasz (1997) for a *S. aureus* isolate trained to grow in increasing concentrations of vancomycin. The details of the resistance mechanism(s) in staphylococci to teicoplanin and vancomycin remain to be fully elucidated, though a recent report that the laboratory-trained glycopeptide-resistant *S. aureus* lacked PBP4 (Sieradzki *et al.*, 1999), which functions as a secondary transpeptidase to increase the degree of cross-linking, is consistent with the suggestion that more pentapeptides would be available in this strain to sequester vancomycin in the outer layers of the wall.

5. The Future

The genes essential for high-level glycopeptide resistance in enterococci are present not only in enterococci but also in glycopeptide-producing organisms (Marshall *et al.*, 1998) and some of them are present in lactobacilli, Leuconostocs and Pediococci (Billot-Klein *et al.*, 1994a). The degree of relatedness of VanH, VanA and VanX varies between 54 - 64% identity with the corresponding genes in glycopeptide-producing organisms suggesting that the genes may have been acquired by enterococci from this source, though regulatory genes are not apparently present. Transfer of resistance to other genera has been demonstrated in the laboratory and the finding of VanA-type resistance in *Streptococcus bovis* (Poyart *et al.*, 1997), *Oerskovia turbata* and *Arcanobacterium haemolyticum* (Power *et al.*, 1995) suggests the genes are mobile under natural conditions. As the pool of glycopeptide-resistance genes is widespread in both hospitals and animals there is concern that the misuse and overuse of glycopeptides will lead to widespread dissemination of resistance amongst Gram-positive pathogenic bacteria. It is therefore important that the use of vancomycin in hospitals in general and intensive care wards in particular, where opportunistic pathogens flourish, is limited to the treatment of essential cases and that it is not used for prophylaxis. The presence of glycopeptide resistance operons in animals almost certainly results from the use of avoparcin, a glycopeptide with an identical mechanism of action to vancomycin and teicoplanin, in animal feed in order to promote growth by the elimination of low grade pathogens from the gut. Large amounts of the antibiotic have been used in the past for this purpose though a number of countries have now banned it for growth promotion purposes.

If in spite of sensible control measures the incidence of vancomycin resistance continues to increase it has been suggested that a super-bug may emerge that is untreatable with conventional antibiotics. To prevent this scenario new compounds are required as antibacterial agents which could be produced either by modification of existing compounds, by synthesis of new compounds using combinatorial chemistry or by screening old libraries of compounds against new targets. One such development involves glycopeptides: it was originally believed that the aglycone portion of the structure determined antimicrobial activity and that the sugar components affected pharmacological properties. It has now been demonstrated that substituents of the sugar moieties and of the peptide backbone increase the apparent affinity of the molecule for molecules terminating in acyl-D-Ala-D-Ala or acyl-D-Ala-D-lactate. This is due to an increased ability of the molecule to dimerise which results in an enhanced affinity of the molecule for cell wall analogues in free solution (Beauregard *et al.*, 1997). Additionally, the presence of a membrane anchor that allows the antibiotic to interact with the cytoplasmic membrane permits a chelate effect enhancement of antibacterial activity. Vancomycin dimerises to only a limited extent (ca 1%) but the compound LY333328 which contains both epivancosamine and chlorobiphenyl substituents dimerises strongly, interacts with glycopeptide-resistant precursors containing acyl-D-Ala-D-lactate (Allen *et al.*, 1997) and inhibits growth of glycopeptide-resistant enterococci at low concentrations (1-2 mg/ml). Recently it has been shown that a molecule lacking the heptapeptide backbone of the glycopeptide is active as a transglycosylase inhibitor (Ge *et al.*, 1999): the concentration required for inhibition *in vitro* is high, *ca* 1000 fold greater than that of the inhibitor moenomycin which inhibits the enzyme directly. It is possible that the action of the molecule, which contains a membrane anchor, could result from a general interference with membrane processes.

An alternative approach to therapy is the identification of new targets. In the pathway of peptidoglycan synthesis the lipid cycle is potentially the Achilles' heel of the process as inhibition of the cycle at any stage will prevent recycling of the undecaprenyl-phosphate carrier molecules and lead to immediate inhibition. One such molecule is ramoplanin which inhibits the addition of GlcNAc from UDP-GlcNAc to lipid I (Somner and Reynolds, 1990).

The identification of low-level glycopeptide resistance in multiply-resistant strains of *S. aureus* (and also in *S. epidermidis* and *S. haemolyticus*) by a mechanism distinct from that encountered in the enterococci emphasises the adaptability of micro-organisms when subjected to selective pressure (Hiramatsu, 1998). If the mechanism proves to be based on sequestration of the antibiotic in the outer layers of the wall due to an enhanced level of groups that can bind glycopeptides, the level of resistance that can be attained may be limited, though the concentration of glycopeptide required to inhibit growth of the organisms could be higher than the concentration easily maintained in the body. The isolation in the laboratory of high-level resistant strains of *S. aureus* trained to grow in the presence of vancomycin (Sieradzki and Tomasz, 1997) or teicoplanin (Sieradzki and Tomasz (1998) is indicative of the situation which clinical microbiologists could encounter in the future.

Acknowledgement

The collaboration of P. Courvalin, M. Arthur and others in the Unité des Agents Antibactériens at the Pasteur Institute has been invaluable in determining the genetics and biochemistry of glycopeptide resistance mechanisms and their contributions are acknowledged as are those of my research students and particularly C. Arias.

References

Allen, N.E., and Hobbs, J.N. (1995) Induction of vancomycin resistance in *Enterococcus faecium* by non-glycopeptide antibiotics, *FEMS Microbiol Lett 132*, 107-114.

Allen, N.E., LeTourneau, D.L., and Hobbs, J.N. (1997) Molecular interactions of a semisynthetic glycopeptide antibiotic with D-alanyl-D-alanine and D-alanyl-D-lactate residues, *Antimicrob Agents Chemother 41*, 66-71.

Arias, C.A., Martin-Martinez, M., Blundell, T.L., Arthur, M., Courvalin, P., and Reynolds, P.E. (1999) Characterisation and modelling of VanT, a novel, membrane-bound, serine racemase from vancomycin-resistant *Enterococcus gallinarum* BM4174, *Mol Microbiol 31*, 1653-1664.

Arthur, M., Molinas, C., and Courvalin, P. (1992) The VanS-VanR two-component regulatory system controls synthesis of depsipeptide peptidoglycan precursors in *Enterococcus faecium* BM4147, *J Bacteriol 174*, 2582-2591.

Arthur, M., Molinas, C., Depardieu, F., and Courvalin, P. (1993) Characterisation of Tn*1546*, a Tn*3*-related transposon conferring glycopeptide resistance by synthesis of depsipeptide peptidoglycan precursors in *Enterococcus faecium* BM4147, *J Bacteriol 175*, 117-127.

Arthur, M., Depardieu, F., Snaith, H.A., Reynolds, P.E., and Courvalin, P. (1994) Contribution of VanY D,D-carboxypeptidase to glycopeptide resistance in *Enterococcus faecalis* by hydrolysis of peptidoglycan precursors, *Antimicrob Agents Chemother 38*, 1899-1903.

Arthur, M., Depardieu, F., Reynolds, P.E., and Courvalin, P. (1996a) Quantitative analysis of the metabolism of soluble peptidoglycan precursors of glycopeptide-resistant enterococci, *Mol Microbiol 21*, 33-44.

Arthur, M., Reynolds, P.E., and Courvalin, P. (1996b) Glycopeptide resistance in enterococci, *Trends in Microbiol 4*, 401-407.

Arthur, M., Depardieu, F., Gerbaud, G., Galimand, M., Leclercq, R., and Courvalin, P. (1997) The VanS sensor negatively controls VanR-mediated transcriptional activation of glycopeptide resistance genes of Tn*1546* and related elements in the absence of induction, *J Bacteriol 179*, 97-106.

Arthur, M., Depardieu, F., Cabanié, L., Reynolds, P.E., and Courvalin, P. (1998) Requirement of the VanY and VanX D,D-peptidases for glycopeptide resistance in enterococci, *Mol Microbiol 30*, 819-830.

Baptista, M., Depardieu, F., Courvalin, P., and Arthur, M. (1996) Specificity of induction of glycopeptide resistance genes in *Enterococcus faecalis*, *Antimicrob Agents Chemother 40*, 2291-2295.

Barna, J.C.J., and Williams, D.H. (1984) The structure and mode of action of glycopeptide antibiotics of the vancomycin group, *Ann Rev Microbiol 38*, 339-357.

Beauregard, D.A., Maguire, A.J., Williams, D.H., and Reynolds, P.E. (1997) Semiquantitation of cooperativity in binding of vancomycin-group antibiotics to vancomycin-susceptible and -resistant organisms, Antimicrob Agents Chemother 41, 2418-2423.

Berger-Bächi, B., Strässle, A., Gustafson, J.E., and Kayser, F.H. (1992) Mapping and characterisation of multiple chromosomal factors involved in methicillin resistance in *Staphylococcus aureus*, *Antimicrob Agents Chemother 36*, 1367-1373.

Billot-Klein, D., Gutmann, L., Sable, S., Guittet, E., and van Heijenoort, J. (1994a) Modification of peptidoglycan precursors is a common feature of the low-level resistant VANB-type Enterococcus D366 and of the naturally glycopeptide-resistant species *Lactobacillus casei, Pediococcus pentosaceus, Leuconostoc mesenteroides* and *Enterococcus gallinarum*, *J Bacteriol 176*, 2398-2405.

Billot-Klein, D., Blanot, D., Gutmann, L., and van Heijenoort, J. (1994b) Association constants for the binding of vancomycin and teicoplanin to *N*-acetyl-D-alanyl-D-alanine and *N*-acetyl-D-alanine-D-serine, *Biochem J 304*, 1021-1022.

Bugg, T.D.H., Wright, G.D., Dutka-Malen, S., Arthur, M., Courvalin, P. and Walsh, C.T. (1991) Molecular basis for vancomycin resistance in *Enterococcus faecium* BM4147: biosynthesis of a depsipeptide peptidoglycan precursor by vancomycin resistance proteins VanH and VanA, *Biochemistry 30*, 10408-10415.

Casadewall, B., and Courvalin, P. (1999) Characterisation of the vanD glycopeptide resistance gene cluster from *Enterococcus faecium* BM4339, *J. Bacteriol 181*, 3644-3648.

DeLencastre, H., and Tomasz, A. (1994) Reassessment of the number of auxiliary genes essential for expression of high-level methicillin resistance in *Staphylococcus aureus*, *Antimicrob Agents Chemother 38*, 2590-2598.

Dutka-Malen, S., Molinas, C., Arthur, M., and Courvalin, P. (1990a) The VANA glycopeptide resistance protein is related to D-alanyl-D-alanine ligase cell wall biosynthesis enzymes, *Mol Gen Genet 224*, 364-372.

Dutka-Malen, S., Leclercq, R., Coutant, V., Duval, J., and Courvalin, P. (1990b) Phenotypic and genotypic heterogeneity of glycopeptide resistance determinants in gram-positive bacteria, *Antimicrob Agents Chemother 34*, 1875-1879.

Dutka-Malen, S., Molinas, C., Arthur, M., and Courvalin, P. (1992) Sequence of the *vanC* gene of *Enterococcus gallinarum* BM4174 encoding a D-alanine:D-alanine ligase-related protein necessary for vancomycin resistance, *Gene 112*, 53-58.

Ellsworth, B.A., Tom, N.J., and Bartlett, P.A. (1996) Synthesis and evaluation of inhibitors of bacterial D-alanine:D-alanine ligases, *Chemistry and Biology 3*, 37-44.

Evers, S., and Courvalin, P. (1996) Regulation of VanB-type vancomycin resistance gene expression by the $VanS_B$-$VanR_B$ two-component regulatory system in *Enterococcus faecalis* V583, *J Bacteriol.178*, 1302-1309.

Fines, M., Perichon, B., Reynolds, P., Sahm, D.F., and Courvalin, P. (1999) VanE, a new type of acquired glycopeptide resistance in *Enterococcus faecalis* BM4405, *Antimicrob Agents Chemother* (In press).

Frère, J-M., and Joris, B. (1985) Penicillin-sensitive enzymes in peptidoglycan biosynthesis, *Crit Rev Microbiol 11*, 299-396.

Ge, M., Chen, Z., Onishi, H.R., Kohler, J., Silver, L.L., Kerns, R., Fukuzawa, S., Thompson, C., and Kahne, D. (1999) Vancomycin derivatives that inhibit peptidoglycan biosynthesis without binding D-Ala-D-Ala, *Science 284*, 507-511.

Hanaki, H., Kuwahara-Arai, K., Boyle-Vavra, S., Daum, R.S., Labischinski, H., and Hiramatsu, K. (1998a) Activated cell-wall synthesis is associated with vancomycin resistance in methicillin-resistant *Staphylococcus aureus* clinical strains, *J Antimicrob Chemother 42*, 199-209.

Hanaki, H., Labischinski, H., Imaba, Y., Kondo, N., Murakami, H., and Hiramatsu, K. (1998b) Increase in glutamine-non-amidated muropeptides in the peptidoglycan of vancomycin-resistant *Staphylococcus aureus* Mu50, *J Antimicrob Chemother 42*, 315-320.

Hiramatsu, K. (1998) Vancomycin resistance in staphylococci, *Drug resistance updates 1*, 135-150.

Jeffs, P.W., and Nisbet, L.J. (1988) Glycopeptide antibiotics: a comprehensive approach to discovery, isolation, and structure determination, in P. Actor, L. Daneo-Moore, M.L. Higgins, M.R.J. Salton and G.D. Shockman (eds), *Antibiotic inhibition of bacterial cell surface assembly and function* American Society for Microbiology, Washington, pp. 509-530.

Lai, M.H., and Kirsch, D.R. (1996) Induction signals for vancomycin resistance encoded by the *vanA* gene cluster in *Enterococcus faecium, Antimicrob Agents Chemother 40*, 1645-1648.

Lessard, I.A.D., and Walsh, C.T. (1999) Mutational analysis of active site residues of the enterococcal D-Ala-D-Ala dipeptidase VanX and comparison with *Escherichia coli* D-Ala:D-Ala ligase and identification of potential active site residues for the D-Ala-D-Ala carboxypeptidase VanY, *Chem Biol 6*, 177-187.

Marshall, C.G., Lessard, I.A.D., Park, I.S., and Wright, G.D. (1998) Glycopeptide antibiotic resistance genes in glycopeptide-producing organisms, *Antimicrob Agents Chemother 42*, 2215-2220.

Noble, W.C., Virani, Z., and Cree R.G.A. (1992) Co-transfer of vancomycin and other resistance genes from *Enterococcus faecalis* NCTC 12201 to *Staphylococcus aureus, FEMS Microbiol Lett 93*, 195-198.

Park, I.S., Lin, C.H., and Walsh, C.T. (1996) Gain of D-alanyl-D-lactate or D-lactyl-D-alanine synthetase activities in three active-site mutants of the *Escherichia coli* D-alanyl-D-alanine ligaseB, *Biochemistry 35*, 10464-10471.

Park, I.S., Chung-Hung, L., and Walsh, C.T. (1997) Bacterial resistance to vancomycin: overproduction, purification and characterisation of VanC-2 from *Enterococcus casseliflavus* as a D-Ala:D-Ser ligase, *Proc Natl Acad Sci USA 94*, 10040-10044.

Périchon, B., Reynolds, P.E., and Courvalin, P. (1997) VanD-type glycopeptide-resistant *Enterococcus faecium* BM4339, *Antimicrob. Agents Chemother. 41*, 2016-2018.

Power, E.G.M., Abdulla, Y.H., Talsania, H.G., Spice, W., Aathithan, S., and French, G.L. (1995) *vanA* genes in vancomycin-resistant clinical isolates of *Oerskovia turbata* and *Arcanobacterium* (*Corynebacterium*) *haemolyticum, J Antimicrob Chemother 36*, 595-606.

Poyart, C., Pierre, C., Quesne, G., Pron, B., Berche, P., and Trieu-Cuot, P. (1997) Emergence of vancomycin resistance in the genus *Streptococcus*: characterization of a vanB transferable determinant in *Streptococcus bovis, Antimicrob Agents Chemother 41*, 24-29.

Quintiliani, Jr, R., and Courvalin, P. (1996) Characterisation of Tn*1547*, a composite transposon flanked by the IS16 and IS256-like elements, that confers vancomycin resistance in *Enterococcus faecalis* BM4281, *Gene 172*, 1-8.

Rasmussen, J.R., and Strominger, J.L. (1978) Utilization of a depsipeptide substrate for trapping acyl-enzyme intermediates of penicillin-sensitive D-alanine carboxypeptidase, *Proc Natl Acad Sci U.S.A. 75*, 84-88.

Reynolds, P.E., and Brown, D.F.J. (1985) Penicillin-binding proteins of b-lactam-resistant strains of *Staphylococcus aureus, FEBS Lett 192*, 28-32.

Reynolds, P.E. (1989) Structure, biochemistry and mechanism of action of glycopeptide antibiotics, *Eur J Clin Microbiol Infect Dis 8*, 943-950.

Reynolds, P.E., Depardieu, F., Dutka-Malen, S., Arthur, M., and Courvalin, P. (1994a) Glycopeptide resistance mediated by enterococcal transposon Tn*1546* requires production of VanX for hydrolysis of D-alanyl-D-alanine, *Mol Microbiol 13*, 1065-1070.

Reynolds, P.E., Snaith, H.A., Maguire, A.J., Dutka-Malen, S., and Courvalin, P. (1994b) Analysis of peptidoglycan precursors in vancomycin-resistant *Enterococcus gallinarum* BM4174, *Biochem J 301*, 5-8.

Reynolds, P.E. (1998) Control of peptidoglycan synthesis in vancomycin-resistant enterococci: D,D-peptidases and D,D-carboxypeptidases, *Cell Mol Life Sci 54*, 325-331.

Reynolds, P.E., Arias, C.A., and Courvalin, P. (1999) Gene $vanXY_C$ encodes D,D-dipeptidase (VanX) and D,D-carboxypeptidase (VanY) activities in vancomycin-resistant *Enterococcus gallinarum* BM4174, *Mol Microbiol* (in press).

Shaw, J.P., Petsko, G.A., and Ringe, D. (1997) Determination of the structure of alanine racemase from *Bacillus stearothermophilus* at 1.9-Å resolution. *Biochemistry 36*, 1329-1342.

Sieradzki, K., and Tomasz, A. (1997) Inhibition of cell wall turnover and autolysis by vancomycin in a highly vancomycin-resistant mutant of *Staphylococcus aureus*, *J Bacteriol 179*, 2557-2566.

Sieradzki, K., and Tomasz, A. (1998) Suppression of glycopeptide resistance in a highly teicoplanin-resistant mutant of *Staphylococcus aureus* by transposon inactivation of genes involved in cell wall synthesis, *Microbial Drug Resistance 4*, 159-168.

Sieradzki, K., Pinho, M.G., and Tomasz, A. (1999) Inactivated pbp4 in highly glycopeptide-resistant laboratory mutants of *Staphylococcus aureus*, *J Biol Chem 274*, 18942-18946.

Sifaoui, F., and Gutmann, L. (1997) Vancomycin dependence in a VanA-producing *Enterococcus avium* strain with a nonsense mutation in the natural D-Ala:D-Ala ligase gene, *Antimicrob Agents Chemother 41*, 1409.

Somner, E.A., and Reynolds, P.E. (1990) Inhibition of peptidoglycan biosynthesis by ramoplanin, *Antimicrob Agents Chemother 34*, 413-419.

van Bambeke, F., Chauvel, M., Reynolds, P.E., Fraimow, H.S., and Courvalin, P. (1999) Vancomycin-dependent *Enterococcus faecalis* clinical isolates and revertant mutants, *Antimicrob Agents Chemother 43*, 41-47.

Wright, G.D., Molinas, C., Arthur, M., Courvalin, P., and Walsh, C.T. (1992) Characterisation of VanY, a D,D-carboxypeptidase from vancomycin-resistant *Enterococcus faecium* BM4147, *Antimicrob Agents Chemother 36*, 1514-1518.

Wright, G.D., Holman, T.R., and Walsh, C.T. (1993) Purification and characterisation of VanR and the cytosolic domain of VanS: a two-component regulatory system required for vancomycin resistance in *Enterococcus faecium* BM4147, *Biochemistry 32*, 5057-5063.

Wu, Z., Wright, G.D., and Walsh, C.T. (1995) Overexpression, purification and characterisation of VanX, a D,D-dipeptidase which is essential for vancomycin resistance in *Enterococcus faecium* BM4147, *Biochemistry 34*, 2455-2463.

β-LACTAMASES, AN OLD BUT EVER RENASCENT PROBLEM

ANDRÉ MATAGNE, MORENO GALLENI, NEZHA LARAKI, GIANFRANCO AMICOSANTE[*], GIANMARIA ROSSOLINI[+] AND JEAN-MARIE FRÈRE

*Centre d'Ingénierie des Protéines et Laboratoire d'Enzymologie, Université de Liège, Institut de Chimie, B6, Sart Tilman, B-4000 Liège, Belgium; *Università degli Studi dell'Aquila, Dipartimento di Scienze e Tecnologie biomediche e di biometria, Via Vetoio, I-67010 Coppito (L'Aquila), Italy; +Università di Siena, Dipartimento di Biologia moleculare, Sezione di Microbiologia, Via Laterina, 8, I-53100 Siena, Italy.*

Abstract

Bacteria utilise several strategies to escape the lethal action of antibiotics. Penicillins and cephalosporins are the most widely used antibacterial agents and the major resistance mechanism involves the predominant participation of β-lactamases, enzymes which catalyse very efficiently the hydrolysis of the amide bond in the β-lactam ring. On the basis of their primary structures, these enzymes have been divided into four distinct classes. Enzymes of class B are metallo-β-lactamases containing one or two zinc ions. They are usually very broad-spectrum enzymes which hydrolyse both classical penams and cephems, but also a wide range of other β-lactam compounds, usually considered as "β-lactamase-stable". In particular, these enzymes confer resistance to imipenem and related carbapenem antibiotics, compounds which escape the activity of most β-lactamases from classes A, C and D.

Classes A, C and D contain active-site serine enzymes. In class A and in a lesser measure, in class D, mutations have allowed enzymatic adaptations resulting in a considerably increased activity *versus* « β-lactamase stable » compounds. The major weapon of class C-producing bacteria has been massive enzyme overproduction by deregulation of the delicate system which controls their biosynthesis. The available results illustrate the diversity and efficiency of bacterial responses to the use and abuse of antibiotics : acquisition of new genes or mutations resulting in modified activity profiles or enzyme overproduction. It is quite clear that the same mechanisms are responsible for resistance to other classes of antibiotics. In this respect, the clinical utilisation of antimicrobial compounds should become more rational and the addition of

A. Van Broekhoven et al. (eds.), Novel Frontiers in the Production of Compounds for Biomedical Use, 117–129.
© 2001 *Kluwer Academic Publishers. Printed in the Netherlands.*

this type of compounds to animal foodstuff should be severely controlled, if not completely banished.

1. Introduction

Recently, the discovery of penicillin has been cited as one of the major scientific achievements of the 20th Century by the American magazine *Time*, a fact which underlines its impact on the general public. Indeed, the introduction of penicillin and related β-lactam antibiotics in the chemotherapeutic arsenal can be considered as a revolution in the fight between mankind and pathogenic bacteria.

However, the emergence of resistant strains has been a recurrent problem from the very beginning of the clinical utilisation of penicillins (but this is also true for the other families of antibiotics). This has resulted in a competition between scientists and bacteria, in which the former tried to find or synthesise new compounds, while the latter continuously improved the mechanisms which allow them to escape the lethal action of antibiotics. In consequence, at the present time, a rather large number of molecules are proposed by the pharmaceutical industry, but none of them can be considered as the « universal drug » which might kill all pathogenic bacteria. By contrast, some bacterial strains have acquired resistance characters which make them resistant to all known antibiotics, a very worrying situation. In this context, it is quite distressing to realise that some major pharmaceutical companies have lost interest in the search for new antibacterial agents.

2. The target of penicillin and other β-lactams

β-Lactams kill bacteria by interfering with the biosynthesis of the cell wall peptidoglycan, a macromolecule which surrounds the bacterial cell and insures its survival by efficiently counteracting its own osmotic pressure. The target enzymes are active-site DD-transpeptidases, also called Penicillin-Binding Proteins or PBPs, which catalyse the formation of the essential peptide cross-links in nascent peptidoglycan. Penicillins acylate the essential serine residue of the DD-transpeptidases (Fig. 1), a reaction which results in the formation of a rather stable acylenzyme which is unable to carry out its physiological function (1,2). In the kinetic scheme depicted by Fig. 1, the value of k_3 is usually so low ($< 10^{-3}s^{-1}$) that it becomes irrelevant and in consequence, the sensitivity of a DD-transpeptidase to a given β-lactam is characterised by the k_2 and K' values (where $K' = (k_{-1}+k_2)/k_{+1}$). Since K' is nearly always larger than the range of attainable antibiotic concentrations, the k_2/K' ratio is often the only significant parameter. It is interesting to note that, even among so-called « sensitive » DD-transpeptidases, the value of k_2/K' can vary very strongly, from 1000 to more than 100,000 $M^{-1}s^{-1}$.

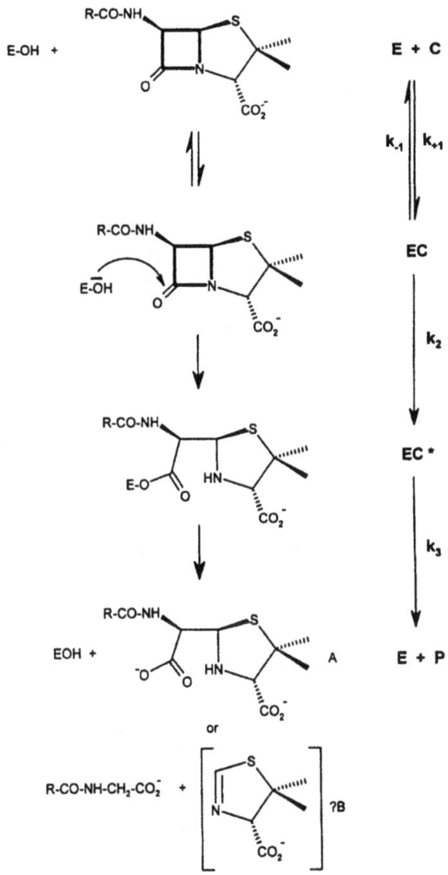

Figure 1. Reaction between a DD-transpeptidase or an active-site serine β-lactamase (E-OH) and a penicillin. The β-lactam ring is in bold type. The deacylation rate is low ($k_3 < 10^{-3}s^{-1}$) with DD-transpeptidases and generally high with β-lactamases (see the text) depending upon the enzyme. The k_3-step can lead to penicilloic acid (A) or fragmentation of the penicilloyl moiety (B). The structure of the molecule between brackets has not been established in an unequivocal manner.

3. Resistance mechanisms

To avoid the lethal effects of β-lactam antibiotics, bacteria utilise one or several of the following mechanisms.

1) Acquisition of a novel penicillin-resistant DD-transpeptidase or the extensive modification of an existing enzyme by genetic recombination of the encoding DNA with

that of a protein of lower affinity. The former mechanism is responsible for the appearance of the infamous methicillin-resistant *Staphylococcus aureus* while the latter is encountered in resistant Pneumococci and *Neisseria gonorrhoeae*. Alternatively, in *Enterococcus hirae*, the synthesis of a pre-existing, minor resistant DD-transpeptidase can be deregulated, and the resistance further increased by point mutations. In all cases, the resistant protein can take over the function of the other, sensitive DD-peptidases after they are inactivated by β-lactams.

2) Synthesis of β-lactamases, enzymes which efficiently hydrolyse the β-lactam amide bond yielding biologically inert products. Most β-lactamases are active-site serine enzymes and act according to the three-step mechanism depicted by Fig. 1 for the DD-peptidases. With the former, however, both the k_2/K' and k_3 values can be extremely high, more than $1 \times 10^6 M^{-1}s^{-1}$ and 1000 s^{-1}, respectively, which led to the proposal by Waley and co-workers (3) that some β-lactamases are « fully efficient enzymes » with their best substrates .

Although major quantitative differences are thus observed, the interactions between β-lactam and β-lactamases or DD-peptidases utilise the same kinetic pathway. Moreover, the elucidation of various 3D-structures has underlined clear similarities in the general fold of the two types of enzymes (4-6) and highlighted the conservation, around the active site, of amino acid side chains exhibiting similar chemical functionalities (Table 1).

*Table 1. The three conserved structural and functional elements of penicillin-recognising enzymes. The active-site serine is indicated by *. The sequences between parentheses occur only in a very limited number of cases.*

	Element 1	Element 2	Element 3
Class A	70 Ser*-Xaa-Xaa-Lys	130 Ser-Asp-Asn (Ser-Asp-Ser) (Ser-Asp-Gly)	234 Lys-Thr-Gly Lys-Ser-Gly Arg-Thr-Gly Arg-Ser-Gly
Class C	64 Ser*-Xaa-Ser-Lys	150 Tyr-Ala-Asn Tyr-Ser-Asn	314 Lys-Thr-Gly
Class D	70 Ser*-Xaa-Xaa-Lys	144 Tyr-Gly-Asn	214 Lys-Thr-Gly
S. R61 DD-peptidase	62 Ser*-Val-Thr-Lys	159 Tyr-Ser-Asn	298 His-Thr-Gly
Other known PBPs	Ser*-Xaa-Xaa-Lys	Ser-Xaa-Asn (Ser-Xaa-Cys) (Tyr-Gly-Asn)	Lys-Thr-Gly Lys-Ser-Gly

A smaller group of β-lactamases are Zn^{++}-containing metallo proteins. Once regarded as biochemical curiosities, they have attracted increasing attention in the recent times, because of their ability to hydrolyse carbapenems. Note that all β-lactamases are not equally active on all β-lactams (see below).
3) In Gram-negative bacteria, decrease of the outer membrane permeability. To reach the DD-transpeptidases, which are bound to the cytoplasmic membrane, the antibiotics must cross the outer membrane *via* the porin channels. Modification of a porin, or its disappearance can decrease the rate of penetration of the antibiotic into the periplasm (7). Bacteria cannot, however, « close » all the porin channels, because they also allow the entry of essential nutrients. The effects of this diffusion barrier are amplified by the presence of a β-lactamase in the periplasm (7,8).
4) In some Gram-negative genera, mainly in *Pseudomonas* spp, active efflux systems can efficiently eject the β-lactam molecules out of the periplasm (9).

4. β-Lactamases

These enzymes constitute the most widespread and ever increasing factor in the bacterial resistance to the lethal action of β-lactam antibiotics. About 300 different enzymes have been described so far (10), and it can be predicted that we have not reached the end of the list. They have been classified on the basis of their functional properties [substrate specificity, sensitivity to inactivators (11)] or of their primary structures (12). The latter classification distinguishes the class B enzymes (Zn^{++}-metallo proteins) from their active-site serine counterparts (classes A, C and D).

Although there is no resemblance in the structures of the Zn^{2+}-containing β-lactamases and the active-site serine enzymes, X-ray diffraction studies have drawn attention to the striking similarities in the organisation of the secondary structure elements of several PBPs and β-lactamases of classes A and C (5,13). Despite the very low degree of sequence similarities between serine DD-transpeptidases and β-lactamases, their similar overall 3D-structures suggest the existence of a close relationship between these two families of penicillin-recognising enzymes, and thus support the hypothesis of a common ancestor protein for DD-transpeptidases and β-lactamases (14,15).

It is interesting to note that the first report (by Abraham and Chain in 1940) of « an enzyme from bacteria able to destroy penicillin » (16) preceded the determination of the exact structure of the molecule. Indeed, as late as 1944, several structures had been proposed for penicillin (Fig. 2) and the β-lactam structure was only firmly established by X-ray diffraction analysis in 1945 (see ref. 17 for a detailed and very entertaining review). On the basis of the few properties described by Abraham and Chain (16), it can be concluded that the first « inactivating enzyme » found in *E. coli* was probably a class C β-lactamase.

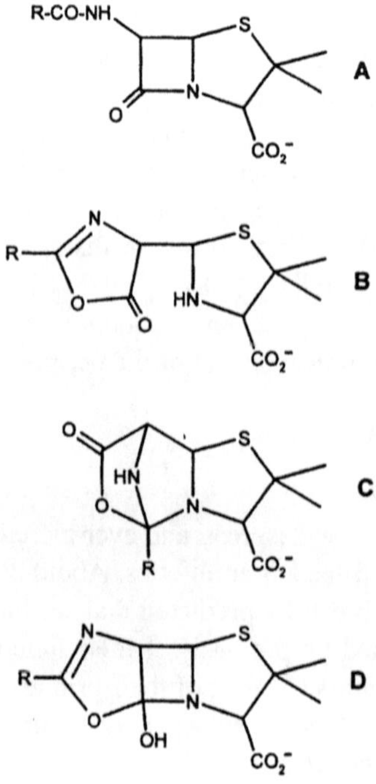

Figure 2. Structures proposed in 1944 for penicillin (17) : A : β-lactam (Abraham and Chain) ; B : thiazoline-oxazolone (Robinson R.) ; C : « tricyclic » (Rohrman and Robinson F.A.) ; D : azlactol (Stodola).

The first clinical problems were due to the appearance of β-lactamase-producing *Staphylococcus aureus* in the late 1940's. The proportion of resistant strains rapidly increased in the London hospitals and the problem was solved by the introduction of methicillin and cephalosporins, which were very poorly hydrolysed by the staphylococcal enzymes. This story set the stage for the latter developments in the field. Indeed, intrinsic (DD-transpeptidase of low sensitivity) or β-lactamase-induced resistance was fought by the introduction of new compounds of increased affinity or which could escape the activity of the known β-lactamases. Bacteria reacted by the production of new, or modified enzymes, which emerged in response to the strong selective pressure created by the intensive utilisation of the new compounds, thus initiating and fuelling the « β-lactamase cycle » (18). In the following paragraphs, we discuss some aspects of this cycle.

5. Carbapenems and carbapenem-hydrolysing β-lactamases

For some time, carbapenems (Fig. 3) appeared to be the « ideal » antibiotics. They exhibited a high to moderate affinity for DD-transpeptidases and were not or poorly hydrolysed by the most prevalent β-lactamases. With class A and class D enzymes, the k_2/K' values were quite low (19) and, although their affinity for the class C enzymes was significantly better (k_2/K' = 30,000 - 150,000 $M^{-1}s^{-1}$), the k_3 values remained rather low (0.002 – 0.02 s^{-1}) (20). They were sensitive to the only known class B metallo-β-lactamase, but this enzyme was produced by the rather innocuous *Bacillus cereus* and was not considered as dangerous. However, more than 15 metallo-β-lactamases produced by pathogenic strains have now been described. Several three-dimensional structures have been solved (21-24). Most of these enzymes are monomers containing 230-250 amino acid residues, but the *Stenotrophomonas maltophilia* L1 enzyme is a tetramer devoid of allosteric properties with most substrates (25). As currently purified, the enzymes contain one or two Zn ions at the active site. The second Zn ion is inhibitory in the *Aeromonas hydrophila* enzyme and appears to be slightly activating in the other cases. All metallo β-lactamases display very similar folds, consisting of an αββα sandwich with two helices on each face. This fold is totally different from those of other metallo-peptidases, active-site serine β-lactamases, soluble DD-peptidases and PBPs (21). A similar fold is however observed in some other metalloproteins such as human glyoxalase II (26). The N-terminal domain of this protein contains two Zn binding sites at the hedge of a four-stranded β-sandwich structure.

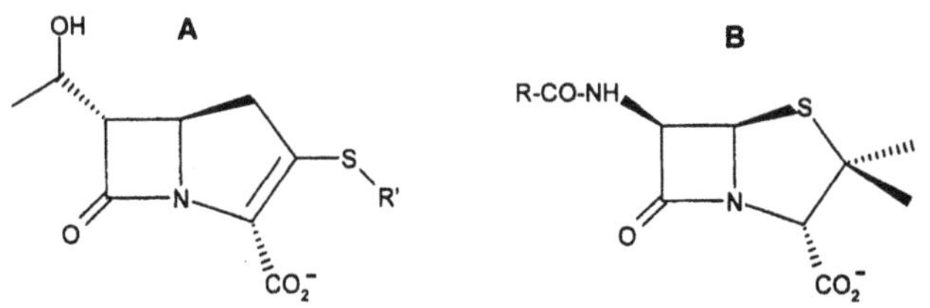

Figure 3. Comparison of the structures of carbapenems (A) and penicillins (B). Note the different stereochemistries at the level of C_6. Imipenem (A) : R = CH_3-CHOH-, R' = -S-CH_2-CH_2-NH-CH=NH. Benzylpenicillin (B) : R = C_6H_5-CH_2-CO-NH-.

A distinct feature of most class B enzymes is that they catalyse the hydrolysis of virtually all β-lactam antibiotics used for therapeutic purposes, with the sole exception of monobactams (25). By contrast, the Zn^{2+}-β-lactamases produced by *Aeromonas hydrophila* appear to be specific carbapenemases and exhibit a very poor activity *versus* penicillins and cephalosporins. The most threatening enzyme is probably the extended-spectrum Imp-1 β-lactamase, whose gene is part of an integron and whose plasmid-mediated synthesis has been observed in several species, in Japan and England, very distant geographical places (27). Although no major outbreak of resistant bacteria has been reported due to dissemination of this enzyme yet, the fact that it is encoded by an integron raises the possibility that this β-lactamase (and related class B enzymes) may become a serious clinical problem (10,28).

A second group of carbapenem-hydrolysing enzymes consists of three class A plasmid-encoded β-lactamases from *Enterobacter cloacae* (NMC-1 and IMI-1) and *Serratia marcescens* (Sme-1) which share about 70 % of sequence identity (29). These are very broad-spectrum enzymes which efficiently hydrolyse both « classical » β-lactams and a wide-range of compounds usually considered as resistant to class A β-lactamases. They exhibit a relatively low (about 50 %) degree of sequence analogy with the TEM-1 β-lactamase (whose carbapenemase activity is very low) but the 3D-structure of NMC-A is not very different from that of the latter protein (30). The main differences are a disulfide bridge just behind the active site of NMC-A and the fact that the side-chain of Asn-132, a highly conserved residue in all active-site serine β-lactamases and PBPs is about 0.1 nm further away from the active-site serine hydroxyl group in NMC-A than in TEM-1 (30). It is interesting to note that site-directed mutagenesis of the Asn-132 residue in the *Streptomyces albus* G β-lactamase resulted in drastic modifications of the specificity profile of this enzyme (31,32). These data underline the importance of the Asn-132 residue in determining the substrate profile of class A β-lactamases, but the phenomenon is still poorly understood, since the short half-lives of the Henri-Michaelis complexes have precluded the determination of their 3D-structures, even with poor or very poor substrates.

6. Hydrolysis of third-generation cephalosporins : the TEM and SHV variants

The plasmid-encoded TEM-1, TEM-2 and SHV-1 class A β-lactamases hydrolyse a large number of classical penicillins and first-generation cephalosporins. Second-, and more significantly, third-generation cephalosporins such as cefotaxime and ceftazidime (Fig. 4) escape their hydrolytic activity and were thus extensively utilised to fight the numerous strains which produce these enzymes. Within a few years however, the appearance of a variety of resistant strains was recorded.

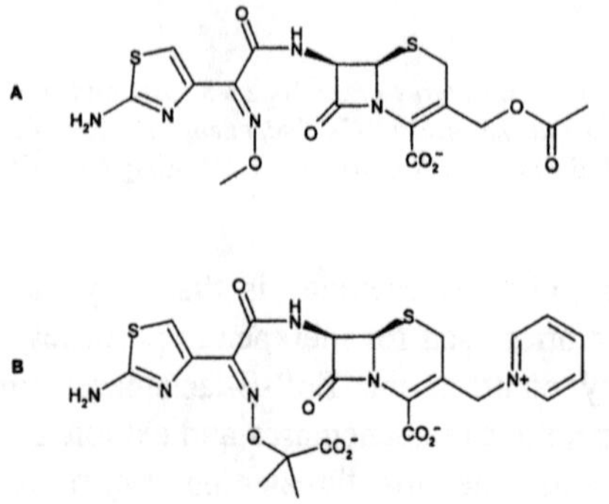

Figure 4. Structures of cefotaxime (A) and ceftazidime (B), third-generation cephalosporins.

In most cases, this could be attributed to the production of new, « extended-spectrum » β-lactamases (ESBLs) with modified substrate profiles. Although some of these were entirely new « exotic » enzymes (e.g. PER-1/2, MEN-1, TOHO-1), the large majority belong to the SHV and TEM families (10,13,33). About 20 SHV and 60 TEM variants have been identified so far (for up-to-date information, see the website http://www.lahey.org/studies/web.htm), which differ from the parent enzymes by a very limited number of amino acid substitutions.

The TEM and SHV families now contain a large number of different proteins which hydrolyse second- and third-generation cephalosporins with various efficiencies (12,13,34). Unfortunately, and despite many efforts, these variants have, up to the present time, escaped X-ray structural analysis and the explanation of their catalytic properties rests on molecular modelling studies based on the known 3D-structures of related enzymes or, in a more reliable way, on that of the TEM-1 parent enzyme (see for instance ref. 35). The following mutations have been identified as crucial in the extension of the TEM enzyme substrate specificity : E104K, R164H or R164S, G238S and E240K. The latter two mutations are also found in some SHV variants, but this list is not exhaustive (see refs. 12, 13 and 34 for more details). Most of the residues involved in the extension of the substrate profile are found in the vicinity of the active-site cleft, but are not directly involved in the catalytic mechanism (13). Another group of extended-spectrum β-lactamases is represented by the class D OXA enzymes (36). In contrast to most class A and to all class C enzymes, class D β-lactamases hydrolyse oxacillin and cloxacillin very efficiently, and hence they are usually considered as « oxacillinases ». One should be aware, however, that in contrast to its TEM and SHV counterparts, the OXA family is rather heterogeneous. Indeed, besides enzymes which differ by a very restricted number of point mutations, class D contains proteins exhibiting strongly diverging sequences. In that respect, the utilisation of the term « OXA family » as a synonym for « class D » is misleading as it tends to obscure the exact relationship between the various enzymes.

In class C, mutations resulting in substrate profile extension have been exceptional (37). The gene encoding these enzymes was first found on the chromosome, but, more recently, many new plasmid-encoded extended spectrum enzymes have been discovered (see 12 for a review).

7. Inhibitor-resistant enzymes

Most class A β-lactamases, including ESBLs are sensitive to inactivators such as clavulanic acid (Fig. 5), sulbactam and tazobactam. Combination of such compounds with classical, β-lactamase-sensitive antibiotics (e.g. amoxycillin) has been quite successful in fighting β-lactamase-producing strains and are currently widely utilised. However, various TEM variants have emerged, exhibiting increased resistance to clavulanic acid. They are sometimes referred to as « IRT's » (for Inhibitor Resistant TEM). The most frequent mutations involve residues Arg-244 and Met-69, which are close to the active site (12).

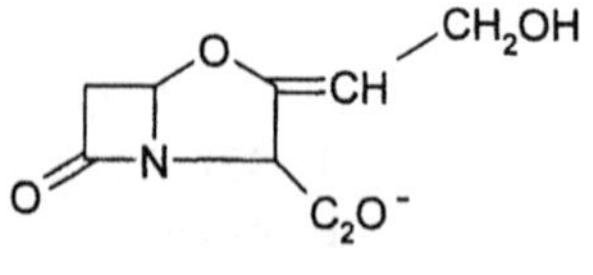

Figure 5. Structure of clavulanate.

8. Overproduction by deregulation of the induction system

As mentioned above, the chromosome-encoded class C β-lactamases have not been modified to acquire new properties in a very significant way. Here, bacteria have used a different weapon : massive overproduction. In most *Enterobacteriaceae* and *Pseudomonas* species, synthesis of the β-lactamase is exquisitely controlled by a complex system which rests on the balance between the concentrations of peptidoglycan precursors and degradation products. The inactivation of the AmpD amidase, an enzyme involved in the recycling of the degradation products, leads to constitutive overproduction of the β-lactamase (38). In some of these deregulated « monsters », easily selected by the antibiotic pressure in hospitals, the periplasmic concentration of β-lactamase has been calculated to be close to 1 mM.

In the absence of an active efflux mechanism and if only one β-lactamase is produced, the MIC of a Gram-negative strain is given by (7,8,39)

$$\mathrm{MIC} = \mathrm{Ipl} + \frac{\mathrm{Ipl}V}{\mathrm{P.A}(K_m + \mathrm{Ipl})} \qquad \text{(eq 1)}$$

where Ipl is the periplasmic concentration of antibiotic necessary to kill the bacterium [it depends on the sensitivity of the essential DD-transpeptidases], V is the maximum rate of the β-lactamase (in nmoles hydrolysed per second and mg of dry weight), P the rate of diffusion of the antibiotic through the outer membrane (in cm s^{-1}), A the surface area of the bacterial cell (132 cm^2 per mg of dry weight for *Enterobacteriaceae*) and K_m is that of the β-lactamase. The Ipl value can be considered as equal to the MIC of an isogenic strain which does not produce any β-lactamase. This equation has been used by several authors (7,40,41) to predict the MIC values of β-lactamase-producing strains within a factor of 2 or 4.

Overproduction of a β-lactamase can increase the MIC value of even poor substrates by one or several orders of magnitude, the more so if the P value is low (8). When the MIC is significantly larger than Ipl and if Ipl is, in turn, larger than K_m (and the K_m values of class C β-lactamases are very often quite low, mainly for poor substrates), eq. 1 simplifies to

$$MIC = \frac{V}{P.A} \qquad \text{(eq 2)}$$

Equation 2 leads to a rather unexpected conclusion : the MIC value becomes essentially independent of Ipl, thus of the sensitivity of the DD-peptidase (8). In consequence, it is not surprising that in Gram-negative strains, β-lactamase overproducing mutants are much more often isolated than strains where the DD-transpeptidase sensitivity is decreased. It is indeed much easier to deregulate β-lactamase production by an inactivating mutation in the AmpD amidase than to modify the DD-transpeptidase so that it retains its enzymatic activity but exhibits a decreased sensitivity to the antibiotics. In this respect, the search for efficient inactivators of class C β-lactamases should be strongly encouraged for solving the clinical problems caused by these overproducing mutants.

In the presence of an efflux system and/or if several different β-lactamases are produced, eq 1 can be modified, yielding eq 3

$$MIC = Ipl + \sum \frac{Ipl.V_i}{P.A(Km_i + Ipl)} + \frac{Ipl.V_e}{P.A(Km_e + Ipl)} \qquad \text{(eq 3)}$$

where V_i, Km_i, V_e, Km_e are those of the individual β-lactamases and of the efflux system, respectively (8). Unfortunately, at the present time, little is known about the kinetic parameters of efflux systems.

9. Conclusion

These examples show how bacteria have used every possible mechanism to avoid the lethal effects of β-lactams. Although the search for new targets is expected to result in the discovery of new antibiotics, it can be predicted that bacteria will find original ways to escape their action. A good example is supplied by the emergence of vancomycin resistance in *Enterococci* (42).

In our opinion, β-lactams, or other compounds interfering with the DD-transpeptidase reaction remain choice compounds to fight a large number of bacterial infections. But the adaptability of bacteria will require continuous research efforts to better understand the mechanism of action and the factors governing the specificity of β-lactamases and of other, still unknown, inactivating enzymes.

Acknowledgements

This work was supported in part by the Belgian programme on Interuniversity Poles of Attraction initiated by the Belgian State, Prime Minister's Office, *Services fédéraux des affaires scientifiques, techniques et culturelles* (PAI n° P4/03) and by the European Union, Training and mobility of Researchers Program (CT98-0232). AM is *Chercheur qualifié* of the *Fonds National de la Recherche Scientifique* (FNRS, Brussels).

References

(1)Frère, J.M., Duez, C., Ghuysen, J.M., and Vandekerckhove, J. (1976) Occurrence of a serine residue in the penicillin-binding site of the exocellular DD-carboxypeptidase-transpeptidase from *Streptomyces* R61. *FEBS Lett. 70*, 257-260.

(2)Frère, J.M. and Joris, B. (1985) Penicillin-sensitive enzymes in peptidoglycan biosynthesis. *CRC Critical Reviews in Microbiology 11*, 299-396.

(3)Christensen, H., Martin, M.T., and Waley, S.G. (1990) β-Lactamases as fully efficient enzymes. Determination of all the rate constants in the acylenzyme mechanism. *Biochem. J. 266*, 853-861.

(4) Kelly, J.A., Dideberg, O., Charlier, P., Wéry, J.P., Libert, M., Moews, P.C., Knox, J.R., Duez, C., Fraipont, C., Joris, B., Dusart, J., Frère, J.M., and Ghuysen, J.M. (1986) On the origin of bacterial resistance to penicillin. Comparison of an active-site serine β-lactamase and an active-site serine D-alanyl-D-alanine cleaving peptidase. *Science 231*, 1429-1431.

(5)Knox, J.R., Moews, P.C., and Frère, J.M. (1996) Molecular evolution of bacterial β-lactam resistance. *Chemistry and Biology 3*, 937-947.

(6)Pares, S., Mouz, N., Petillot, Y., Hakenbeck, R., and Dideberg, O. (1996) X-ray structure of *Streptococcus pneumoniae* PBP2x, a primary penicillin target enzyme. *Nat. Struct. Biol. 3*, 284-289.

(7)Nikaido, H. and Normark, S. (1987) Sensitivity of *Escherichia coli* to various β-lactams is determined by the interplay of outer membrane permeability and degradation by periplasmic β-lactamase : a quantitative predictive treatment. *Mol. Microbiol. 1*, 29-36.

(8)Lakaye, B., Dubus, A., Lepage, S., Groslambert, S., and Frère, J.M. (1999) When drug inactivation renders the target irrelevant to antibiotic resistance : a case story with β-lactams. *Molecular Microbiology 31*, 89-101.

(9)Nikaido, H. (1996) Multidrug efflux pumps of Gram-negative bacteria. *J. Bacteriol. 178*, 5853-5859.

(10)Bush, K. (1999) β-Lactamases of increasing clinical importance. *Curr. Pharm. Dis. 11*, 839-845.

(11)Bush, K., Jacoby, G.A., and Medeiros, A.A. (1995) A functional classification scheme for β-lactamases and its correlation with molecular structures. *Antimicrob. Ag. Chemother. 39*, 1211-1233.

(12)Matagne, A., Dubus, A., Galleni, M., and Frère, J.M. (1999) The β-lactamase cycle : a tale of selective pressure and bacterial ingenuity. Nat. Prod. Rep. **16**, 1-19.

(13)Matagne, A., Lamotte-Brasseur, J., and Frère, J.M. (1998) Catalytic properties of class A β-lactamases : efficiency and diversity. *Biochem. J. 330*, 581-598.

(14)Tipper, D.J., and Strominger, J.L. (1965) Mechanism of action of penicillins : a proposal based on the structural similarity to acyl-D-alanyl-D-alanine. *Proc. Natl. Acad. Sci. USA 54*, 1131-1141.

(15)Ghuysen, J.M. (1991) Serine β-lactamases and penicillin-binding proteins. *Annu. Rev. Microbiol. 45*, 37-67.

(16)Abraham, E.P. and Chain, E. (1940) An enzyme from bacteria able to destroy penicillin. *Nature (London) 146*, 837.

(17)Abraham, E.P. (1987) Sir Robert Robinson and the early history of Penicillin. *Nat. Prod. Rep. 4*, 41-46.

(18)Sykes, R. and Bush, K. (1982) Physiology and biochemistry of β-lactamases, in M. Morin and M. Gorman (eds.), *The Chemistry and Biology of β-Lactam Antibiotics, vol. 3*, Academic Press, New York, pp. 155-207.

(19)Matagne, A., Lamotte-Brasseur, J., and Frère, J.M. (1993) Interactions between active-site serine β-lactamases and so-called β-lactamase stable antibiotics : kinetic and molecular modelling studies. *Eur. J. Biochem. 217*, 61-67.

(20)Galleni, M., Amicosante, G., and Frère, J.M. (1988) A survey of the kinetic parameters of class C β-lactamases. II. Cephalosporins and other β-lactam compounds. *Biochem. J. 255*, 123-129.

(21)Carfi, A., Pares, S., Duée, E., Galleni, M., Duez, C., Frère, J..M., and Dideberg, O. (1995) The 3D-structure of a zinc metallo-β-lactamase from *Bacillus cereus* reveals a new type of protein fold. *EMBO J. 14*, 4914-4921

(22)Concha, N.O., Rasmussen, B.A., Bush, K., and Herzberg, O. (1994) Crystal structure of the wide-spectrum binuclear zinc β-lactamase from *Bacteroides fragilis*. *Structure 4*, 823-836.

(23)Fitzgerald, P.M.D., Wu, J.K., and Toney, J.H. (1998) Unanticipated inhibition of the metallo-β-lactamase from *Bacteroides fragilis* by MES. A crystallographic study at 1.85 Å resolution. *Biochemistry 37*, 6791-6800.

(24)Ullah, J.H., Walsh, T.R., Taylor, I.A., Emery, D.C., Verma, C.S., Gamblin, S.J., and Spencer, J. (1998) The crystal structure of the L1 metallo β-lactamase from *Stenotrophomonas maltophilia* at 1.7 Å resolution. *J. Mol. Biol. 284*, 125-136.

(25)Felici, A., Amicosante, G., Oratore, A., Strom, R., Ledent, Ph., Joris, B., Fanuel, L., and Frère, J.M. (1993). An overview of the kinetic parameters of class B β-lactamases. *Biochem. J. 291*, 151-155.

(26)Cameron, A.D., Ridderström, M., Olin, B., and Mannervik, B. (1999) Crystal structure of human glyoxalase II and its complex with a glutathione thiolester substrate analogue. *Structure 7*, 1067-1078.

(27)Woodford, N., Palepou, M.F., Babini, G.S., Bates, J., and Livermore, D.M. (1998) Carbapenemase-producing *Pseudomonas aeruginosa* in UK. *Lancet 352*, 546-547.

(28)Lauretti, L., Riccio, M.L., Mazzariol, A., Cornaglia, G., Amicosante, G., Fontana, R., and Rossolini, G.M. (1999) Cloning and characterization of blaVIM, a new integron-borne metallo-β-lactamase gene from a *Pseudomonas aeruginosa* clinical isolate. *Antimicrob. Agents Chemother. 43*, 1584-1590.

(29)Rasmussen, B.A., and Bush, K. (1997) Carbapenem-hydrolyzing β-lactamases. *Antimicrob. Agents Chemother. 41*, 223-232.

(30)Swarén, P., Maveyraud, L., Raquet, X., Cabantous, S., Duez, C., Pédelacq, J.D., Mariotte-Boyer, S., Mourey, L., Labia, R., Nicolas-Chanoine, M.H., Nordmann, P., Frère, J.M., and Samama, J.P. (1998) X-ray analysis of the NMC-A β-lactamase at 1.64-Å resolution, a class A carbapenemase with broad substrate specificity *J. Biol. Chem. 273*, 26714-26721.

(31)Jacob, F., Joris, B., Dideberg, O., Dusart, J., Ghuysen, J.M., and Frère, J.M. (1990) Engineering of a novel β-lactamase by a single point mutation. *Protein Engineering 4*, 79-86.

(32)Jacob, F., Joris, B., Lepage, S., Dusart, J., and Frère, J.M. (1990) Role of the conserved residues of the "SDN" loop in a class A β-lactamase studied by site-directed mutagenesis. *Biochem. J. 271*, 399-406.

(33)Medeiros, A.A. (1997) Evolution and dissemination of β-lactamases accelerated by generations of β-lactam antibiotics. *Clin. Infect. Dis. 24* (Suppl. 1), S19-S45.

Tsouvelekis, L.S. and Bonomo, R.A. (1999) SHV-type β-lactamases. *Current Pharmaceutical Design 5*, 847-864.

(35)Raquet, X., Lamotte-Brasseur, J., Fonzé, E., Goussard, S., Courvalin, P., and Frère, J.M. (1994) TEM β-lactamase mutants hydrolysing third-generation cephalosporins. A kinetic and molecular modelling analysis. *J. Mol. Biol. 244*, 625-639.

(36)Naas, T. and Nordmann, P. (1999) OXA-Type β-lactamases. *Current Pharmaceutical Design 5*, 865-879.

(37)Nukaga, M., Haruta, S., Tanimoto, K., Kogure, K., Taniguchi, K., Tamaki, M., and Sawai, T. (1995) Molecular evolution of a class C β-lactamase extending its substrate specificity. *J. Biol. Chem. 270*, 5729-5735.

(38)Jacobs, C., Frère, J.M., and Normark, S. (1997) Cytosolic intermediates for cell wall biosynthesis and degradation control inducible β-lactam resistance in Gram-negative bacteria. *Cell 88*, 823-832.

(39)Zimmermann, E. and Rosselet, A. (1977) The function of the outer membrane of *Escherichia coli* as a permeability barrier to β-lactam antibiotics. *Antimicrob. Ag. Chemother. 12*, 368-372.

(40)Waley, S.G. (1987) An explicit model for bacterial resistance : application to β-lactam antibiotics. *Microbiol. Sci. 4*, 143-146.

(41)Frère, J.M. (1989) Quantitative relationship between sensitivity to β-lactam antibiotics and β-lactamase production in Gram-negative bacteria. I. Steady-state treatment. *Biochem. Pharmacol. 38*, 1415-1426.

(42)Reynolds, P.E. (1998) Control of peptidoglycan synthesis in vancomycin-resistant enterococci : DD-peptidases and DD-carboxypeptidases. *CMLS Cell. Mol. Life Sci. 54*, 325-331.

(29)[illegible], Raleigh, [illegible], Isato, [illegible] and [illegible] (199[illegible]) [illegible]

(38)[illegible] (199[illegible]) [illegible] *Cell* [illegible]

(39)[illegible] (199[illegible]) [illegible]

(44)[illegible] (1994) [illegible] *J. Antimicrob. Chemother.* [illegible]

METABOLIC FLUX ANALYSIS IN STREPTOMYCES COELICOLOR:

Effect of nitrogen source

FERESHTEH NAEIMPOOR AND FERDA MAVITUNA
Department of Chemical Engineering, UMIST,
PO Box 88, M60 1QD, UK E-mail: f.mavituna@umist.ac.uk

Abstract

Metabolic flux analysis was applied to *Streptomyces coelicolor* growth under nitrogen limitation with nitrate or ammonia as the nitrogen source. In all cases, maximum specific growth rates corresponding to the same specific glucose uptake rates were calculated in addition to the fluxes of some 200 metabolic reactions. Use of nitrate resulted in lower specific growth rates compared to ammonia. Excretion of some organic metabolites was observed in both limitation cases.

1. Introduction

1.1 STREPTOMYCETES

Streptomycetes are a group Gram-positive bacteria belonging to the Actinomycetes. They have a high G+C ratio (69-78%) in their DNA. They are found in the soil and they have adapted their metabolism to survive in this environmental niche. They are obligatory aerobic bacteria. They produce a diverse range of metabolic products, some of which have important roles in medicine and veterinary science. Although the most common use of these products is as antibiotics, there are also other applications such as the modulation of the immune system and enzyme inhibitors (Kleinkauf and von Dohren, 1997). Some examples with commercial applications are antibacterial compounds, such as chloramphenicol, streptomycine and tetracycline, antiparasitic drugs such as avermectin, fungicidal agents such as polyoxin, antitumor drugs such as adriamycin, immunosuppressive drugs such as rapamycin, a herbicide (biolaphos), and β-lactamase inhibitor clavulanic acid. There is also a great deal of interest in

A. Van Broekhoven et al. (eds.), Novel Frontiers in the Production of Compounds for Biomedical Use, 131–145.
© 2001 *Kluwer Academic Publishers. Printed in the Netherlands.*

streptomycetes enzymes; for example, streptomycetes xylose isomerase is used a great deal in food industry. There are several excellent literature reviews on the physiology and genetics of streptomycetes (Chater and Hopwood, 1993; Piepersberg, 1993; Chater and Bibb, 1997; Hopwood, 1999; Hodgson, in press).

Soil is a nutrient-limited environment rich in carbon and poor in nitrogen and phosphate. Since plant-derived material is the main nutrient in the soil, streptomycetes have evolved with the capability of secreting a wide range of extracellular enzymes that break down the plant material. Their metabolic pathways for carbohydrate catabolism and the control of primary metabolism reflect the nutrient availability and the variety of carbohydrates in the soil (Hodgson, in press). The carbohydrate catabolic systems are inducible. They exhibit catabolite repression which is not yet well understood and is different from other bacteria. In the amino acid and nucleotide catabolism, half of the pathways are inducible and half are expressed at low levels and are constitutive. Repression of amino acid biosynthesis seems to be absent. An interesting aspect of streptomycete metabolism is that the primary metabolism is relatively poorly controlled whereas the secondary metabolism is tightly controlled (Hodgson, in press).

Streptomycetes are unique amongst bacteria in their complex morphological differentiation resembling that of filamentous fungi and involving at least three distinct cell types (Chater and Losick, 1997). In the vegetative phase, they grow as a network of highly branched substrate mycelium (on solid agar media). As the nutrients become exhausted, the substrate mycelia lyse to provide nutrients for the production of aerial hyphae. These eventually metamorphose into spores. In liquid culture, the cells are exclusively of the substrate mycelium type, eventhough some strpetomycetes can sporulate in liquid culture (Hodgson, 1992). The discoveries of the underlying gene regulatory cascade for their sophisticated developmental cycle have significant implications. For example, antibiotic production in surface-grown cultures of streptomycetes generally coincides with the onset of morphological differentiation (Chater, 1993; Horinouchi and Beppu, 1994), most probably to prevent the competitors from the use of the products of lysis from the substrate mycelia.

After the discovery of genetic conjugation in streptomycetes in the 1950s, research has led to a detailed understanding of streptomycete genetics. The streptomycete chromosomes which are 8 megabases and linear are amongst the largest for bacteria. Many streptomycetes also have very large linear, as well as the covalently closed circular plasmids. A characteristic of streptomycetes is the relative instability of their genomes. Some regions of the chromosome are devoid of genetic markers and appear to be dispensable. Significant deletions of chromosomal DNA, especially at these silent regions, can occur without detectable effect on the physiology of the organism. These extensive deletions are often associated with massive tandem amplification that can account for 10-50% of the total DNA (Flett and Cullum, 1987; Chater *et al.*, 1988; Muth *et al.*, 1997; Volff and Altenbuchner, 1997; Pandza *et al.*,, 1997). Reversible instability, typically the reversion of an antibiotic-sensitive mutant phenotype to the original resistant phenotype, is observed in several streptomycetes. The genes encoding functionally related enzymes of catabolic pathways as well as those of antibiotic biosynthesis are often clustured, and furthermore the resistance and regulatory genes of an antibiotic lie adjacent to the structural genes. The clustering of biosynthetic, resistance and regulatory genes for any antibiotic pathway greatly facilitates the isolation of the genes for the entire pathway. The formation of secondary metabolites, especially those with antimicrobial activities is usually in response to the nutritional limitations and stress conditions that are unfavourable to growth (Demain *et al.*, 1988). So far, there is no simple unifying model for interlinking the various genetic and physiological factors which influence the onset of antibiotic production in streptomycetes.

1.2 STREPTOMYCES COELICOLOR

Williams *et al.* (1983) classified over 400 species belonging to the genus *Streptomyces* into 19 major cluster groups designated with a letter, A to J, and 40 minor cluster designated with an Arabic numeral. The minor clusters were considered as species. *Streptomyces coelicolor* A3(2) is a member of the largest major cluster group A, and the minor cluster group 21.

Streptomyces coelicolor A3(2) produces four known antibiotics; actinorhodin, undecylprodigiosin, methylenomycin and the calcium-dependent antibiotic (CDA). Actinorhodin is red in acid and blue in alkali while undecylprodigiosin is normally red but tends to turn orange at low pH. Actinorhodin is a polyketide, a dimeric molecule derived from one acetyl-CoA precursor and seven malonyl-CoA extender units. Undecylprodigiosin is a tripyrrole derivative, the major component of a mixture of four related prodigionines. Methylenomycin is a cyclic compound with a penta-carbon ring and a carboxyl moiety on one of the branch carbons. CDA is a non-ribosomally synthesized cyclic lipopeptide consisting of 11 unusual amino acid residues with a 2,3-epoxyhexanoyl lipid moiety attached at the N-terminus of the peptide (Chong *et al.*, 1998). Since these are weak antibiotics without commercial value and the pigmentation allows visual observation of the onset of antibiotic production, *S. coelicolor* A3(2) has been used as a model microorganism for detailed and extensive genetic and physiological studies.

Streptomyces coelicolor A3(2) is therefore, genetically the most-studied strain in streptomycetes along with *Streptomyces griseus*. An excellent review of the development of *Streptomyces coelicolor* A3(2) genetics is provided by Hopwood (1999). Redenbach *et al.* (1996) provide a combined genetic and physical map of the 8 Mb linear chromosome of *S. coelicolor* A3(2). SCP1 and SCP2 are the two indigenous plasmids *of S. coelicolor*. SCP1 is a 350 kb plasmid containing the biosynthetic *loci* for methylenomycin. SCP2 is a 30 kb plasmid which is used as a high capacity cloning system. Information on the complete chromosome sequencing project for *Streptomyces coelicolor* A3(2) at Sanger Centre, Cambridge can be found at:

http://www.sanger.ac.uk/Projects/S_coelicolor/.

Although this model microorganism has attracted attention of many geneticist and a lot of work has been done on its physiology and biology, less attention has been paid to the biochemical engineering aspects of its antibiotic production. The regulation of actinorhodin production by carbon, nitrogen and phosphate limitation and media composition (Doul and Vining, 1990; Hobbs *et al.*, 1992; Ozergin-Ulgen and Mavituna,

1993; Melzoch *et al.*, 1997; Elibol and Mavituna, 1998), effect of cell immobilisation (Ozergin-Ulgen and Mavituna, 1994), oxygen limitation (Elibol and Mavituna, 1995; 1996; 1999; Ulgen and Mavituna, 1998) and mode of bioreactor operation (Ates *et al.*, 1997) have been studied in order to understand the environmental factors affecting the antibiotic production in *S. coelicolor*.

1.3 POLYKETIDES, PEPTIDE ANTIBIOTICS AND STREPTOMYCETES

Polyketides are a large family of structurally diverse natural products with a broad range of biological activities such as antibiotic, anticancer, antiparasitic and immunosuppressant properties. Polyketides are produced predominantly by streptomycetes and related filamentous bacteria. The enzymes involved in the biosynthesis of polyketides are called polyketide synthases (PKSs). They are large multifunctional enzymes that are structurally and mechanistically related to fatty acid synthases. PKSs catalyse iterative decarboxylative condensation between coenzymeA thioesters (acetyl, propionyl, malonyl or methylmalonyl). Polyketides can be grouped into two structurally diverse classes; aromatic such as actinorhodin, and complex such as erythromycin. Complex polyketides are made by type I polyketide synthases (PKSs) which consist of multiple sets or modules of enzymes. Structural variation in polyketides arises from the number and type of modules in a pathway, the stereochemical outcome of the condensation, and the extent and stereochemistry of reduction at each cycle the molecule being synthesised goes through the PKS modules. The type II PKSs that synthesise aromatic polyketides typically consist of a single set of enzymes. The modular arrangement of type I PKSs raises the exciting prospect of obtaining novel polyketides simply by genetic manipulation of modules to engineer new hybrid PKSs.

From a biosynthetic point of view, non-ribosomally synthesised peptides are one of the most interesting classes of natural products (Kleinkauf and von Dohren, 1996; Stein and Vater, 1996; Chong *et al.*, 1998). They include antibiotics such as the β-lactams, gramicidin, and tyrocidine, immunosuppressants like cyclosporin and toxins such as enniatin. The biosynthesis of these compounds is directed by peptide synthetases which are large multi-functional enzymes analogous in some way to the type I PKSs. Peptide synthetases and the corresponding biosynthetic genes have been isolated from a variety of fungi and bacteria including Actinomycetes.

It is therefore possible to genetically manipulate the microorganisms to design novel modules of PKSs and peptide synthetases to produce novel compounds. There is even more exciting scope for the synthesis of novel unnatural compounds by the combination of biosynthetic ability of the genetically engineered modules of polyketide synthases (PKSs) and peptide synthetases with combinatorial chemistry (McDaniel *et al.*, 1994; Khosla, 1997; Hopwood, 1997; Stachelhaus *et al.*, 1996). Metabolic engineering, as described briefly below, will be a powerful tool in this quest for novel rationally designed products.

1.4 METABOLIC ENGINEERING

Metabolic Engineering can be broadly defined as the study of microbial, plant and animal metabolism using scientific and engineering tools in order to; a) understand it better under normal and abnormal (mutant, genetically altered, diseased) conditions, b) identify and test metabolic bottlenecks for targeted genetic engineering, c) design and genetically modify metabolic pathways for the increased the formation of desired metabolites and decreased formation of unwanted metabolites in bioprocesses or in cells, tissues and organs. Metabolic engineering is also very effective in understanding the actions of hormones, drugs and chemical agents and thus aids the search for drugs that affect metabolism (host or pathogen) by identifying targets.

There are some excellent books on metabolic engineering which in turn refer to some excellent review articles (Reich and Sel'kov, 1981; Heinrich and Schuster, 1996; Fell, 1997; Stephanopoulos *et al.*, 1998).

Metabolic flux analysis and metabolic control analysis form two very powerful tools of metabolic engineering. The theoretical aspects of metabolic flux analysis involve the construction of a matrix of stoichiometrically balanced biochemical reactions of the metabolic pathways of interest, making a material balance on the cells' metabolism and solving the metabolic reaction network matrix which yields rates (fluxes) of individual metabolic reactions.

Although most of the bioreactions, especially those which are related to cell growth are common in all living organisms, there are certain metabolic pathways, and regulatory functions leading to the production of secondary metabolites specific to each microorganism. In order to increase the product yield one should know how the carbon source is directed to biomass and product formation. Having known the metabolic flux distribution within the cells one can manipulate the genes of the cells or their physicochemical environment in order to achieve higher desired production rates.

2. Metabolic Flux Analysis in *S. coelicolor*

Metabolic flux analysis (MFA) is a method for the determination of metabolic pathway fluxes (specific rates of reactions) through a stoichiometric model of the cellular pathways, using mass balances for intracellular metabolites. It is a powerful tool for the identification of the important bioreactions in the metabolism for further genetic manipulations (Varma & Palsson, 1994, Stephanopoulos *et al.*, 1998) and for selecting the physicochemical conditions for improved bioprocesses.

The cellular reaction rates can be expressed as the specific rate of reaction (fluxes) with the units of m moles (g DW h)$^{-1}$. By performing mass balances for each intracellular metabolite under quasi-steady state assumption the steady state balance equation is written as:

$$S \,.\, v = b$$

where **S** is the stoichiometric matrix, **v** is the vector of fluxes, and **b** is the vector of net specific excretion rates from the cell (in the case of nutrient uptake from the environment, the elements of **b** will be negative). Since the number of reactions is usually greater than the number of metabolites, the number of unknown fluxes will be greater than the number of balance equations leading to an underdetermined system of linear algebraic equations. This situation leads to infinite solutions. By using linear programming, a defined objective function can be optimised and the associated metabolite fluxes can be obtained (Naeimpoor and Mavituna, 2000).

2.1 METABOLITE AND PRODUCT EXCRETIONS

Wild-type cultures tend to direct most of their resources towards biomass production when they are cultivated under proper physicochemical conditions using readily available nutrients (Shapiro, 1989). However, this does not necessarily mean that no product, either desired or undesired, is excreted during their growth. Streptomycetes spp. are well-known for the production of a variety of secondary products of which many are antibiotics (Shapiro, 1989). These species are known to show overflow metabolism under carbon-excess conditions leading to the excretion of some organic compounds the type of which is profoundly dependent on the nature of the growth limitation and the physiology of species (Madden, *et al.*, 1996; Hobbs *et al.*, 1992; Delkleva & Strohl, 1987; Ahmed *et al.*, 1984).

3. Results and Discussion

3.1 MODEL DESCRIPTION

The biochemical reaction network used in this work to simulate *S. coelicolor* metabolism was very similar to that described in Naeimpoor and Mavituna (2000). The network consisted of more than 200 biochemical reactions of the major metabolic pathways of such as glycolysis, pentose phosphate (PP), tricarboxylic acid (TCA), glyoxylate, as well as the anaplerotic reactions, sulphate assimilation, electron transport, folic acid and thioredoxin reactions, the biosynthesis of aromatic, aspartate, glutamate, pyruvate, serine family amino acids and histidine, the biosynthesis of pyrimidine and purine nucleotides, the biosynthesis of macromolecular components of biomass such as RNA, DNA, protein, phospholipids, carbohydrate and actinorhodin production.

The amount of protein, RNA and DNA in the dry biomass of *Streptomyces coelicolor* was taken from Shahab *et al.* (1996). The phospholipid content of the biomass was assumed to be the same as *E. coli* (Ingraham *et al.*, 1983) and the rest of the biomass was assumed to be made up of carbohydrate. The amount of precursors required to produce one gram of protein, RNA, DNA and phospholipids as well as the polymerisation energies required for the biosynthesis of these macromolecules were assumed to be the same as in *E. coli* (Ingraham *et al.*, 1983). The cell composition was assumed to be constant. P/O ratio was assumed to be 1.9 in NADH oxidation reactions and two thirds of this amount for $FADH_2$ oxidation reactions.

The mathematical model, which consisted of a set of linear algebraic equations, was obtained by applying pseudo-steady state mass balance for all metabolites and neglecting the changes in pool metabolites. In the model, the number of biochemical reactions (unknown metabolic fluxes) was higher than the number of metabolites involved in the network (number of mass balance equations). Therefore, this underdetermined set of linear equations could only be solved by optimising a defined objective function (Stephanopoulos *et al.*, 1998). In this work the objective function was defined as the specific growth rate and maximisation of this objective function was the direction of optimisation. The solution then gave not only the maximum value of the objective function but also the values of the unknown metabolic fluxes, including the specific substrate uptake and product formation rates, within the network. It should be noted that the solution presents a theoretical case in which all the calculated fluxes are the optimised values needed to achieve the objective of optimisation. The experimental values for various specific rates are used as constraints in the solution of the model.

3.2 EFFECT OF DIFFERENT NITROGEN SOURCES ON BIOMASS YIELD

Many physico-chemical factors such as temperature, concentration and type of the nutrients, and pH affect the growth rate and biomass yield on a carbon source. In order to investigate the effect of nitrogen source on biomass yield two different nitrogen sources, ammonia and nitrate were examined in this work. By defining the stoichiometric specific growth rate as the objective function to be maximised and using a specific glucose uptake rate of 1 m mol glucose (g DW h)$^{-1}$ in both cases, the model was solved via linear programming in GAMS (General Algebraic Modelling System) environment (Brooke *et al.*, 1992). The calculated values of the maximum specific growth rates and specific uptake rates of oxygen and nitrogen sources resulting from this optimisation are given in Table 1.

Table 1- The maximised specific growth rates and the corresponding rates of specific oxygen and nitrogen source uptake, and carbon dioxide production for nitrate or ammonia as the nitrogen source. The specific glucose uptake rate is 1 m mol(g DW h)$^{-1}$.

Nitrogen Source	**Maximum specific growth rate, μ_{max}, h^{-1}**	**Specific Rates, m mol (g DW h)$^{-1}$**		
		q_{O_2}	q_{N_2} *	q_{CO_2}
Ammonia	0.119	0.902	0.869	1.148
Nitrate	0.092	0.699	0.674	2.237

* Specific uptake rate of nitrogen source, m mol nitrogen (g DW h)$^{-1}$.

When ammonia is the nitrogen source, the results in Table 1 show a higher maximum stoichiometric specific growth rate. Since both specific growth rates were obtained with the same specific glucose uptake rate of 1 m mol glucose (g DW h)$^{-1}$, this means a higher biomass yield on glucose when the nitrogen source is ammonia compared to the case of nitrate. Accordingly, the calculated values of specific ammonia and oxygen uptake rates are higher when ammonia is used as the nitrogen source. Now, the question one may ask is what happens to that part of the glucose taken up but not directed to biomass formation in the case of nitrate. This question can be answered by considering the concept of generalised degree of reduction first introduced by Roels (1983), which plays an important role in the biomass yield on different types of substrates. Since the cells need to reduce nitrate to ammonia first before being able to use it in their metabolic reactions, they need the reducing power in the form of NADPH for nitrate assimilation. Cells obtain this reducing power from the oxidation of glucose which is the electron donor. Consequently, this decreases the amount of glucose which can be directed to biomass formation and hence the maximum value of the specific growth rate shows a decrease compared to growth on ammonia. Furthermore, despite the fact that the value of maximum specific growth rate is 30.4 % lower in the case of nitrate compared to ammonia, the specific carbon dioxide production is much higher when nitrate is used. In the next section, we show the flux of carbon into different metabolic pathways for the two nitrogen sources.

3.3 METABOLITE EXCRETION WITH DIFFERENT NITROGEN SOURCES

Since nitrogen source limitation is one the most important nutrient limitations that can lead to an increase in the formation of secondary metabolites, we investigated its effect on the changes in metabolic flux distribution in *S. coelicolor*. In our computational programme, the excretion of metabolites such as pyruvate, acetate, formate, glycerol, and actinorhodin is allowed if necessary in order to close the material and redox balances. For comparison purposes, the values of the specific glucose uptake rates were again kept constant at 1 m mol (g DW h)$^{-1}$. Since the specific nitrate uptake rate corresponding to the maximum growth rate with this specific glucose uptake rate using nitrate was 0.67 m mol (g DW h)$^{-1}$ from Table 1, any limitation of nitrogen source should occur with uptake rates below this value. Therefore, in order to simulate the limitation of nitrogen source in both cases, the nitrogen source uptake rates were reduced from 0.6 m mol nitrogen source (g DW h)$^{-1}$ down to zero by steps of 0.1 unit. In mathematical terms, such limitation is expressed as an added constraint to the linear system of equations and can be solved via linear programming. Since the specific glucose uptake rate is fixed at 1 m mol (g DW h)$^{-1}$ and the biomass composition is assumed constant not all of the glucose taken up can be directed towards biomass formation. Consequently, a portion of it is diverted to the formation and excretion of some metabolites.

Table 2- The maximum specific growth rate and the percentage of consumed carbon directed to excreted metabolites at different limited specific uptake rates of nitrogen sources, ammonia or nitrate, for specific glucose uptake rate of 1 m mol (g DW h)$^{-1}$.

Specific Nitrogen Source Uptake Rate *	Specific Growth Rate h^{-1}	Percentage of Consumed Carbon Directed to Excreted Metabolites
0	0	100
0.1	0.014	90.69
0.2	0.027	81.38
0.3	0.041	72.07
0.4	0.055	62.75
0.5	0.069	53.45
0.6	0.082	44.13

* m mol nitrogen (g DW h)$^{-1}$.

Table 2 gives the maximised specific growth rates and the corresponding total percentages of the consumed carbon directed to the excreted metabolites at various nitrogen limitations. The results show that the highest total percentage of consumed carbon excretion occurs at zero specific growth rate where the specific nitrogen uptake rate is zero. As the specific uptake rate of the nitrogen source is increased, more carbon is used for biomass formation and hence specific growth rate increases, but the total percentage of the consumed carbon directed to excreted metabolites decreases.

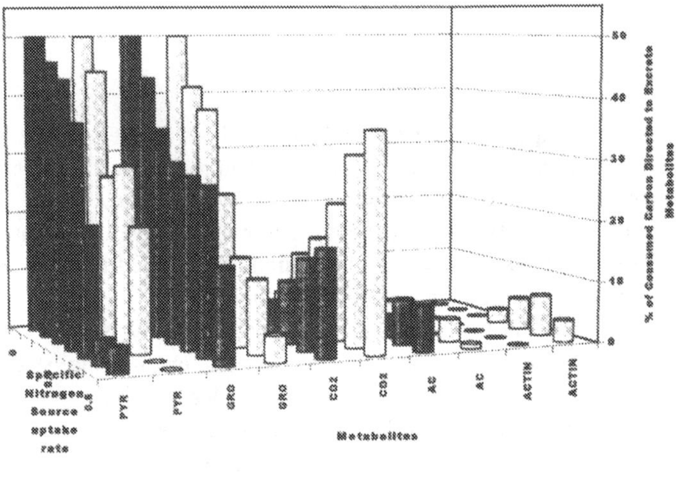

Figure 9- Percentage of consumed carbon that is directed to the excreted metabolites for the two different nitrogen sources. The unit of specific nitrogen source uptake rate is m mole (g DW h)$^{-1}$. Ammonia, dark grey bars; Nitrate, light grey bars.

Figure (1) shows to which excreted metabolite and to what extent the consumed carbon is directed, with either ammonia or nitrate as the nitrogen source. There are noticeable differences between the percentages of the consumed carbon excreted as carbon dioxide at various limitations of nitrate and ammonia. Since the total amount of carbon excretion at the same specific uptake rate of nitrate and ammonia is the same (see Table 2), the higher amount of carbon dioxide produced should be compensated by lower amount of carbon directed to other excreted metabolites in the case of nitrate compared to ammonia. Figure (1) shows that apart from carbon dioxide, pyruvate and glycerol are the other main metabolites excreted in both cases. However, up to 6.6% of carbon uptake is directed to excreted actinorhodin with nitrate limitation while there is no actinorhodin excretion with ammonia limitation.

Figures (2) and (3) present an overview of the important parts of the metabolic flux maps for ammonia and nitrate limitations, respectively, for the same specific glucose and nitrogen source uptake rates in each case (based on 1 m mole (g DW h)$^{-1}$). In the case of ammonia, Figure (2), the carbon dioxide produced in pentose phosphate pathway contributes to nearly 50% of the total specific carbon dioxide production, whereas the contribution of pentose phosphate pathway when nitrate is used, Figure (3), is around 80%. In addition, the amount of specific generation rate of NADPH in pentose phosphate pathway in the case of nitrate is nearly three times greater than that in the case of ammonia which is in accordance with the cell's requirement for the reducing power for nitrate reduction (Stephanopoulous *et al.*, 1998). Most of the energy required for the biosynthesis of biomass seems to be provided by the electron transfer reactions, although glycolysis provides a part of this energy. TCA cycle provides some of the precursors required for the biosynthesis of other metabolites, but according to the detailed results of the model solution it is not following a complete cycle (detail not shown in the maps), and its contribution to production of the reducing power is not noticeable.

4. CONCLUSION

According to the computational metabolic flux analysis applied to *S. coelicolor* here, assimilation of ammonia as nitrogen source results in higher maximum specific growth rates compared to nitrate assimilation. Glycerol, pyruvate, citrate, oxoglutarate, actinorhodin and acetate as well as carbon dioxide are excreted as organic metabolites in all nitrogen limited cases. Actinorhodin excretion at limited nitrogen uptake rates is at the expense of excretion of some organic metabolites.

Metabolic engineering tools such as metabolic flux analysis and metabolic control should indicate genetic, combinatorial and process engineering strategies for the industrial production of exciting novel unnatural products such as polyketides and non-ribosomally synthesised peptides. It has been shown by research and industrial practice that very large increases in process performance can be achieved by genetic modifications of metabolic control systems. Past improvements in the performance of a process by modification of the control structures were mainly based on trial and error methods and on well-understood, relatively simple metabolic pathways. Intuitive and trial and error methods however, become increasingly ineffective as the complexity of metabolic pathways of interest increases. Therefore, it is important to have guidance as to what changes in metabolism and its regulation might be of greatest benefit to achieve the bioprocess targets. Metabolic flux analysis and other tools of metabolic engineering can aid us in the comparison of sets of admissible metabolic routes for the wild type and the genetically altered cells. It can also be used to test the metabolism's response to genetic manipulations especially through the metabolic bypasses.

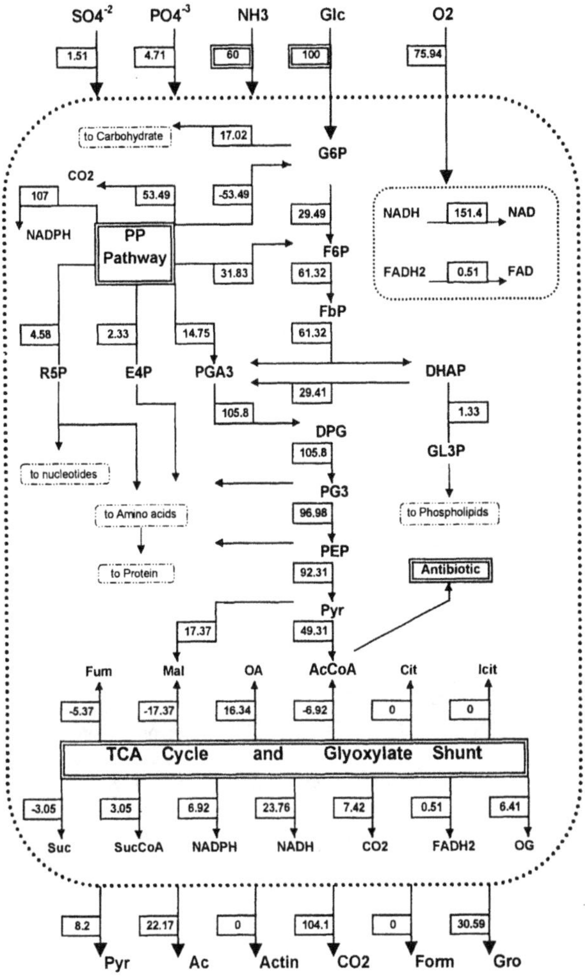

Figure 10- Metabolic flux map for maximised specific growth rate. The values of the metabolic fluxes are normalised by setting the specific glucose uptake rate as 100 m mole (g DW h)$^{-1}$. The ammonia uptake rate is taken 60 m mole (g DW h)$^{-1}$. The boxes with double line borders present the defined fluxes, while the others present the calculated fluxes. Negative values indicate that the reaction proceeds in the opposite direction of the arrow. Not all fluxes are shown.

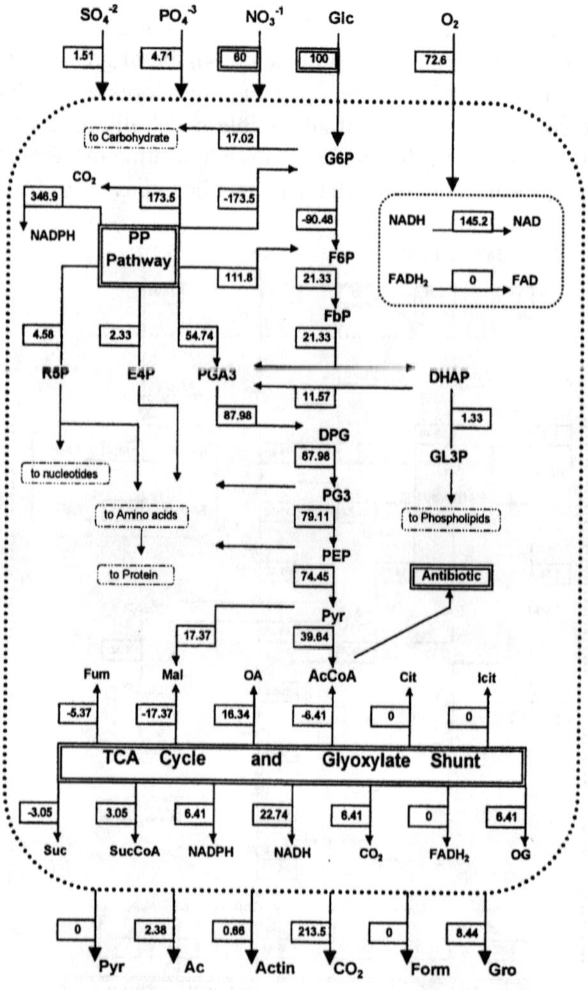

Figure 11- Metabolic flux map for maximised specific growth rate. The values of the metabolic fluxes are normalised by setting the specific glucose uptake rate as 100 m mole (g DW h)$^{-1}$. The nitrate uptake rate is taken 60 m mole (g DW h)$^{-1}$. The boxes with double line borders present the defined fluxes, while the others present the calculated fluxes. Negative values indicate that the reaction proceeds in the opposite direction of the arrow. Not all fluxes are shown.

Acknowledgements

F. Naeimpoor is grateful to the Department of Chemical Engineering, Iran University of Science and Technology (IUST) for providing the opportunity to do this research and the Ministry of Science, Research, and Technology in I.R. of Iran for its financial support.

Nomenclature

AcCoA	Acetyl coenzyme A
Actin	Actinorhodin
Cit	Citrate
CO2	Carbon dioxide
DHAP	Dihydroxyacetone phosphate
DNA	Deoxyribonucleic acid
DPG	1- and 3-diphosphoglycerate
E4P	Erythrose 4-phosphate
F6P	β-D-Fructose 6-phosphate
FAD	Flavine adenine dinucleotide
FADH2	Flavine adenine dinucleotide (reduced)
FbP	β -D-Fructose 1 and 6-bis-phosphate
Form	Formate
Fum	Fumarate
G6P	Glucose 6-phosphate
GL3P	Glycerol 3-phosphate
Glc	Glucose
Gro	Glycerol
Icit	Malate
NAD	Nicotineamide-adeninedinucleotide
NADH	Nicotineamide-adeninedinucleotide (reduced)
NADPH	Nicotineamide-adeninedinucleotide phosphate (reduced)
NH_3	Ammonia
NO_3^{-1}	Nitrate
02	Oxygen
OA	Oxaloacetate
OG	2-Oxoglutarate
PEP	Phosphoenolpyruvate
PG3	3-Phospho-D-glycerate
PGA3	Glyceraldehyde-3-phosphate
PO_4^{-3}	Phosphate
Pyr	Pyruvate
R5P	Ribose 5-phosphate
RNA	Ribonucleotide Acid
Ru5P	Ribulose 5-phosphate
SO_4^{-2}	Sulphate
Suc	Succinate
SucCoA	Succinyl-CoA

References

Ahmed, Z. U., Shapiro, S., and Vining L. C. (1984) Excretion of α-keto acids by strains of *Streptomyces venezuelae, Can. J. Microbiol.* 30, 1014-1021.

Ates, S., Elibol, M., and Mavituna, F. (1997) Characteristics of antibiotic production in by *Streptomyces coelicolor* in batch and fed-batch cultures, *Process Biochem.* 32(4), 273-278.

Bailey, J. E. (1991) Towards a science of metabolic engineering, *Science* 252, 1668-1674.

Brooke, A, Kendrick D., and Meeraus A. (1992) *GAMS: A User's Guide, release 2.25*, The scientific Press.

Chater, K. F. (1993) Genetics of differentiation in *Streptomyces, Ann. Rev. Microbiol.* 47, 685-713.

Chater, K. F. and Bibb, M. J. (1997) Regulation and Bacterial Antibiotic Production, in H. -J. Rehm, G. Reed, A. Puhler and P. J. W. Stadler (series eds), *Biotechnology*, Vol. 7, Products of Secondary Metabolism, H. Kleinkauf and H. von Dohren (Vol. eds.), VCH, Weinheim, Germany, pp. 57- 105.

Chater, K. F. and Hopwood, D. A. (1993) Streptomycetes, in A. L. Sonenshein (Ed.-in-Chief), *Bacillus subtilis and Other Gram-positive Bacteria; Biochemistry, Physiology and Molecular Genetics*, J. A. Hoch and R. Losick (eds.), Am. Soc. Microbiol., Washington DC, Chapter 6, pp. 83-100.

Chater, K. F. and Losick, R. (1997) Mycelial life style of *Streptomyces coelicolor* A3(2) and its relatives, in J. A. Shapiro and M. Dworkin (eds.), *Bacteria as Multicellular Organisms*, Oxford University Press, New York.

Chater, K. F., Henderson, D. J., Bibb, M. J. and Hopwood, D. A. (1988) Genome flux in *Streptomyces coelicolor* and other streptomycetes and its possible relevance to the evolution of mobile resistance determinants, in A. J. Kingsman, K. F. Chater, and S. M. Kingsman (eds.), *Transposition*, Cambridge University Press, pp. 7-42.

Chong, P. P., Podmore, S. M., Kieser, H. M., Redenbach, M., Turgay, K., Marahiel, M., Hopwood, D. A., and Smith, C. P. (1998) Physical identification of a chromosomal locus encoding biosynthetic genes for the lipopeptide calcium-dependent antibiotic (CDA) of *Streptomyces coelicolor* A3(2), *Microbiol.* 144, 193-199.

Dekleva, M. L. and Strohl W. R. (1987) Glucose-stimulated acido-genesis by *Streptomyces peucetius, Can. J. Microbiol.* 33, 1129-1132.

Demain, A. L., Aharonowitz, Y., and Martin, J. F. (1988) Metabolic control of secondary biosynthetic pathways, in L. C. Vining, (ed), *Biochemistry and genetic regulation of commercially important antibiotics,* Addison-Wesley, Reading, Mass. pp. 49-72.

Doull, J.F. and Vining, L.C. (1990) Nutritional control of actinorhodin production by *Streptomyces coelicolor* A3(2), *Appl. Microbiol. Biotechnol.* 32, 449-454.

Elibol, M. and Mavituna, F. (1995) Effect of perfluorodecalin as an oxygen carrier on actinorhodin production by *Streptomyces coelicolor* A3(2), *Appl. Microbiol. Biotechnol.* 43, 206-210.

Elibol, M. and Mavituna, F. (1996) Use of perfluorocarbon for oxygen supply to immobilised *Streptomyces coelicolor* A3(2), *Process Biochem.* 31(5), 507-512.

Elibol, M. and Mavituna, F. (1998) Effect of sucrose on actinorhodin production by *Streptomyces coelicolor* A3(2), *Process Biochem.* 33(2), 307-311.

Elibol, M. and Mavituna, F. (1999) A remedy to oxygen limitation in antibiotic production: addition of perfluorocarbon, *Biochem. Engg. J.* 3, 1-7.

Fell, D. (1997) *Understanding the Control of Metabolism*, Portland Press, London.

Flett, F. and Cullum, J. (1987) DNA deletions in spontaneous chloramphenicol-sensitive mutants of *Streptomyces coelicolor* A3(2) and *Streptomyces lividans* 66, *Mol. Gen. Genet.* 207, 499-502.

Heinrich, R. and Schuster, S. (1996) The regulation of cellular systems, Chapman & Hall, New York.

Hobbs, G., Obanye A. I. C., Petty J., Mason J. C., Barratt E., Gardner D. C. J, Flett F., Smith C. P., Broda P., and Oliver S. G. (1992) An integrated approach to studying regulation of production of the antibiotic methylenomycin by *Streptomyces coelicolor* A3(2), *J. Bacteriol.* 174, 1487-1494.

Hodgson, D. A. (1992) Differentiation in actinomycetes, in S. Mohan, C. Dow, and J. A. Cole (eds), *Prokaryotic Structure and Function,* Cambridge University Press, Cambridge, pp. 407-440.

Hodgson, D. A., (in press) Primary metabolism and its control in Streptomycetes: a most unusual group of bacteria, *Adv. in Microbiol Physiol.* 42.

Hopwood, D. A. (1997) Genetic contributions to understanding polyketide synthases, *Chem. Rev.* 97, 2465-2497.

Hopwood, D. A. (1999) Forty years of genetics with *Streptomyces*: from *in vivo* through *in vitro* to *in silico. Microbiology* 145, 2183-2202.

Horinouchi, S. and Beppu, T. (1994) A-factor as a microbial hormone that controls cellular differentiation and secondary metabolism in *Streptomyces griseus. Mol. Microbiol.* 12, 859-864.

Khosla, C. (1997) Harnessing the biosynthetic potential of modular polyketide synthases, *Chem. Rev.* 97, 2577-2590.

Kleinkauf, H. and von Dohren, H. (1996) A non-ribosomal system of peptide biosynthesis, *Eur. J. Biochem.* 236, 335-351.

Kleinkauf, H. and von Dohren, H. (Volume eds.) (1997) *Products of Secondary Metabolism*, in H. -J. Rehm, G. Reed, A. Puhler and P. J. W. Stadler (series eds), *Biotechnology,* Vol. 7, VCH, Weinheim, Germany.

Madden, T., Ward J. W., and Ison A. P. (1996) Organic acid excretion by *Sreptomyces lividans*TK24 during growth on defined carbon and nitrogen sources, *Microbiology* 142, 3181-3185.

Mandelstam, J., McQuillen K., and Dawes I. (1985) *Biochemistry of Bacterial Growth*, 3rd Ed., Blackwell Scientific Publications.
McDaniel, R., Ebert-Khosla, S., Hopwood, D. A., and Khosla, C. (1994) Engineered biosynthesis of novel polyketides: Influence of a downstream enzyme on the catalytic specificity of a minimal aromatic polyketide synthase, *Proc. Natl. Acad. Sci., Biochemistry* 91, 11542-11546.
Melzoch, K., Teixeira de Mattos, M. J., and Neijssel, O. M. (1997) Production of actinorhodin by *Streptomyces coelicolor* A3(2) grown in chemostat culture, *Biotechnol. and Bioeng.* 54(6), 557-582.
Muth, G., Frese, D., Kleber, A., and Wohlleben, W. (1997) Mutational analysis of the *Streptomyces lividans recA* gene suggests that only mutants with residual activity remain viable, *Mol. Gen. Genet.* 255, 420-428.
Naeimpoor, F. and Mavituna, F. (2000) Metabolic flux analysis in *Streptomyces coelicolor* under various nutrient limitations, *Metabolic Engineering*, 2(2), 140-148.
Nielsen, J. and Villadsen, J. (1994) Bioreaction engineering principles, Plenum Press, New York, London.
Ozergin-Ulgen, K., and Mavituna, F. (1993) Actinorhodin production by *Streptomyces coelicolor* A3(2): kinetic parameters related to growth, substrate uptake and production, *Appl. Microbiol. Biotechnol.* 40, 457-462.
Ozergin-Ulgen, K. and Mavituna, F. (1994) Production of actinorhodin by immobilised and freely suspended *Streptomyces coelicolor* A3(2), *Appl. Microbiol. Biotechnol.* 41, 197-202.
Pandza, K., Pfalzer, G., Cullum, J., and Hranueli, D. (1997) Physical mapping shows that the unstable oxytetracycline gene cluster of *Streptomyces rimosus* lies close to one end of linear chromosome, Microbiology-UK 143, 1493-1501.
Piepersberg, W. (1993) Streptomycetes and Corynebacteria, in H. -J. Rehm, G. Reed, A. Puhler and P. J. W. Stadler (series eds), *Biotechnology*, H. Sahm (vol. ed.) *Vol. 1, Biological Fundamentals*, VCH, Weinheim, Germany, pp 433-468.
Redenbach, M., Kieser, H. M., Denapaite, D., Eichner, A., Cullum, J., Kinashi, H., and Hopwood, D. A. (1996) A set of oredered cosmids and a detailed genetic and physical map for the 8 Mb *Streptomyces coelicolor* A3(2) chromosome, *Mol. Microbiol.* 21, 77-96.
Reich, J. G. and Sel'kov, E. E. (1981) *Energy metabolism of the cell*, Academic Press, London.
Roels, J. A. (1983) *Energetics and kinetics in biotechnology*, Elsevier Biomedical, Amsterdam, Oxford.
Shapiro, S. (1989) Nitrogen assimilation in actinomycetes and the influence of nitrogen nutrition on actinomycete secondary metabolism, in S. Shapiro (ed.), *Regulation of Secondary Metabolism in Actinomycetes*, CRC Press, Boca Raton, pp 135-213.
Stachelhaus, T., Schneider, A., and Marahiel, M. A. (1996) Engineered biosynthesis of peptide antibiotics, Biochem. *Pharmacology* 52, 177-186.
Stein, T. and Vater, J. (1996) Amino acid activation and polymerisation at modular multienzymes in non-ribosomal peptide biosynthesis, *Amino Acids* 10, 201-227.
Stephanopoulos, G. N., Aristidou, A. A., and Nielsen, J. (1998) *Metabolic Engineering: Principles and Methodologies*, Academic Press, San Diego.
Strohl, W. R. and Connors, N. C. (1992) Micro Review: Significance of anthraquinone formation and the cloning of actinorhodin genes in heterogeneous streptomyctes, *Molecular Microbiology* 6(2), 147-152.
Takac, S., Calik, G., Mavituna, F., and Dervakos, G. (1998) Metabolic flux distribution for the optimised production of L-glutamate, *Enzyme and Microbial Technology* 23, 286-300.
Ulgen, K. and Mavituna, F. (1998) Oxygen transfer and uptake in *Streptomyces coelicolor* A3(2) culture in a batch bioreactor, *J. Chem. Technol. Biotechnol.* 73, 243-250.
Varma, A. and Palsson, B. O. (1994) Metabolic flux balancing: basic concepts, scientific and practical use, *Bio/ Technology* 12, 994-998.
Volff, J. N. and Altenbuchner, J. (1998) MicroReview: Genetic instability of the *Streptomyces* chromosome, *Mol. Microbiol.* 27, 239-246.
Williams, S. T., Goodfellow, M., Alderson, G., Wellington, E. M. H., Sneath, P. H. A., and Sackin, M. J. (1983) Numerical classification of *Streptomyces* and related genera, *J. Gen. Microbiol.* 129, 1743-1813.

METABOLIC ENGINEERING OF THE LYSINE PATHWAY FOR β-LACTAM OVERPRODUCTION IN *PENICILLIUM CHRYSOGENUM*

CASQUEIRO, J.[1], BAÑUELOS, O.[2], GUTIÉRREZ, S.[2,3] AND MARTÍN, J.F.[1,2*]

*[1]Institute of Biotechnology (INBIOTEC), Avda. del Real, nº 1, 24006 León, [2]Area of Microbiology, Faculty of Biology, University of León, 24071 Leon, and [3]ESTIA, Campus de Ponferrada, Avda. de Astorga, s/n, 24400 Ponferrada, León, Spain. *Corresponding author. Tel: 34-987-291505. Fax: 34-987-291506. E-mail: degjmm@unileon.es.*

Abstract

α-Aminoadipic acid is a well known precursor in the biosynthesis of penicillin. α-Aminoadipic acid is formed in *Penicillium chrysogenum* as an intermediate in the lysine biosynthetic pathway. The lysine pool has a great influence on penicillin production since lysine is a potent inhibitor of penicillin biosynthesis. The relationship between intracellular α-aminoadipic acid availability and the penicillin production rate of one strain is well established. Metabolic engineering of the lysine pathway has shown that it is possible to channel the metabolic flux of lysine precursors towards penicillin biosynthesis by target inactivation of the α-aminoadipate reductase (the first enzyme of the lysine pathway after the branch point). In addition, *lys*1 (encoding homocitrate synthase) gene amplification and gene overexpression studies showed that there are more than one rate-limiting step in the direct supply of α-aminoadipic acid for penicillin biosynthesis from the first part of the lysine pathway. In summary, gene disruption of diverging branches of a biosynthetic pathway leads to channelling of the intermediates towards the desired end products whereas amplification of a specific gene encoding a putative bottleneck enzyme in a biosynthetic pathway does not necessarily result in overproducing strains since other bottlenecks may still occur after removal of the first one.

1. Lysine biosynthesis: synthesis of α-aminoadipic acid a precursor of β-lactam antibiotics

The essential amino acid, lysine, is the only one in nature that it is synthesised through two completely different pathways (Bhattacharjee 1985). The so-called diaminopimelic acid pathway is used to synthesise lysine in bacteria, some lower fungi and green plants (Bryan 1980); this pathway has been extensively studied in *Escherichia coli* (Patte 1983; Cohen and Saint-Girons 1987) and corynebacteria (Malumbres and Martín 1996).

A. Van Broekhoven et al. (eds.), Novel Frontiers in the Production of Compounds for Biomedical Use, 147–159.
© 2001 *Kluwer Academic Publishers. Printed in the Netherlands.*

However, there is another pathway for the biosynthesis of lysine, which is called the α-aminoadipate pathway (Fig. 1). This pathway is present in some lower fungi (Quitidrials, Blastocladials and Mucorals), higher fungi (Ascomycetes, Yeasts, Basidiomycetes and Deuteromycetes) and Euglenoids (Lejohn 1971; Rothstein and Saffran 1963; Vogel 1960). Recently, this pathway was also found to occur in the thermophilic bacteria *Thermus thermophilus* (Kosuge and Hoshino 1998; Kobashi et al., 1999).

The α-aminoadipate pathway has been studied in detail in *Saccharomyces cerevisiae* and has been taken as a model for other microorganisms using the same lysine biosynthetic pathway. Lysine biosynthesis (Fig. 1) starts with the condensation of α-ketoglutarate and acetyl-coenzymeA to form homocitrate, which is latter converted by isomerisation, oxidative decarboxylation and maintain steps into homoisocitrate, α-ketoadipate and α-aminoadipic acid. In the second part of the lysine biosynthetic pathway, α-aminoadipic acid is converted into α-aminoadipic acid-δ-semialdehyde, saccharopine and finally converted into lysine (reviewed in Bhattacharjee 1985).

L-α-aminoadipic acid is a well known precursor in the biosynthesis of penicillin by *Penicillium chrysogenum*, cephalosporins by *Acremonium chrysogenum* and cephamycins by Actinomycetes (Aharanowitz et al., 1992; Demain 1983; Martin and Liras 1989; Martin et al., 1997).

The interest in the relationship between lysine metabolism and β-lactam antibiotic biosynthesis began with the observation made by Bonner (1947) that 25% of the *P. notatum* lysine auxotrophs were unable to produce penicillin. Demain (1957) discovered ten years later that lysine is a potent inhibitor of penicillin biosynthesis. Later, it was found that α-aminoadipic acid is able to revert lysine inhibition; α-aminoadipic acid is also able to increase penicillin biosynthesis when lysine is not present in culture broths (Somerson et al., 1961). Additional evidences of the α-aminoadipic acid role in the β-lactam antibiotics biosynthesis were obtained by Arnstein et al (1960) when they discovered the formation of the tripeptide, δ-(α-aminoadipyl)-cysteinil-valine (ACV) in *P. chrysogenum* mycelia. Later, Loder and Abraham (1971) isolated the same tripeptide from *A. chrysogenum* mycelia; they studied its structure and found that it has the LLD configuration. The same ACV peptide was found to be present in *Streptomyces clavuligerus* (Abraham 1978). In parallel, Flynn et al (1962) discovered one β-lactam antibiotic, that had α-aminoadipic acid as side chain; this hydrophilic β-lactam named isopenicillin N was isolated from *P. chrysogenum*. Finally, Warren et al (1967) showed with radioactive α-aminoadipic acid, that this compound is one of the precursors of cephalosporins.

Radiochemical studies during the 1940s and 1950s (reviewed in Corbett 1990; Brakhage 1998) revealed that penicillins and cephalosporins are naturally synthesised from the three amino acid precursors: L-α-aminoadipic acid, L-cysteine and L-valine. Although only cysteine and valine are finally incorporated in the β-lactam nucleus, the requirement for L-α-aminoadipic acid has since long been established (Demain 1983; Martín and Aharonowitz, 1983). Cysteine and valine are proteinogenic amino acids, whereas L-α-aminoadipic acid is an intermediate of the lysine biosynthetic pathway in *P. chrysogenum* and *A. chrysogenum*. Thus L-α-aminoadipic acid is a branching point intermediate for the L-lysine and penicillin and cephalosporin biosynthesis (Fig. 1).

2. Relationships between lysine and penicillin biosynthesis

As mentioned above, one of the penicillin biosynthesis precursors, L-α-aminoadipic acid, is an intermediate in the biosynthesis of lysine in *P. chrysogenum*. Although it has been described that lysine catabolism could supply some of the L-α-aminoadipic acid for penicillin biosynthesis in *P. chrysogenum* (Kurzatkowski et al., 1990; Esmahan et al., 1994) and *Aspergillus nidulans*, the lysine biosynthesis route seems to be the main supplier of L-α-aminoadipic acid for penicillin production.

It is well known that lysine inhibits penicillin production in *P. chrysogenum* (Demain 1957; Masurekar and Demain 1972). Since both molecules (penicillin and lysine) are the end products of a common pathway that has the branch point at the L-α-aminoadipic acid, it was proposed that lysine inhibits the homocitrate synthase (the first enzyme of the lysine pathway) leading to a decrease in the L-α-aminoadipic acid availability (Demain and Masurekar 1974; Demain 1974). *In vivo* studies suggested that lysine acts at the homocitrate synthase level during both the growth and the idiophase (Demain and Masurekar 1974). Friedrich and Demain (1977) supported these observation showing that addition of homocitrate to a resting cell system could revert the inhibition of penicillin biosynthesis mediated by lysine. Later on, we showed that lysine acts by inhibiting homocitrate synthase and to a lesser extent by repressing its synthesis (Luengo et al., 1979; 1980). Repression of *lys*1 encoding homocitrate synthase was recently shown to exert a minor effect on regulation of the lysine pathway (Bañuelos et al., 1999a).

In *A. nidulans*, using reporter gene fusion with the *pcb*AB (*acv*A) and *ipn*A promoters, it was described that addition of lysine to culture broths affects the transcription of both genes and could be the reason for decreasing the penicillin biosynthesis in this microorganism (Brakhage and Turner 1992). In *A. chrysogenum*, supplementation of the cultures with low (1 mM) lysine concentrations increases cephalosporin C production. However addition of high concentration of lysine decreases cephalosporin C production. Similarly to what occurs in *P. chrysogenum*, in resting cells systems lysine inhibition of cephalosporin C biosynthesis could be reverted by addition of L-α-aminoadipic acid (Mehta et al., 1979).

The negative effect of lysine on penicillin biosynthesis is reverted by addition of L-α-aminoadipic acid (Somerson et al., 1961), homocitrate and to a lesser extent α-ketoadipate (Friedrich and Demain 1977). These three intermediates of the first part of the lysine pathway have also an increasing effect on penicillin biosynthesis when lysine is not present.

The L-α-aminoadipic acid has a stimulatory effect on penicillin production since addition of this amino acid (Friedrich and Demain 1978; Revilla et al., 1986) or the use of conditions that increases the L-α-aminoadipic acid intracellular pool (Hölinger and Kubicek 1989a) enhances ACV synthesis and penicillin production. Jaklitsch et al (1986) showed a direct correlation between penicillin production and the L-α-aminoadipic pool; this correlation was not observed for valine nor cysteine, the other precursors for penicillin biosynthesis (Revilla et al., 1986).

The L-α-aminoadipate reductase (the first enzyme after the branch point in lysine biosynthesis) is inhibited by lysine *in vivo* and *in vitro* (Affenzeller et al., 1989).

Hölinger and Kubicek (1989b) showed a reduction in the conversion of L-α-aminoadipic acid into L-lysine in the strain P2 as compared with other strains with lower penicillin production ability. It was also shown that the L-α-aminoadipate reductase of *P. chrysogenum* strains with increased penicillin production ability had lower affinity for L-α-aminoadipic acid than the lower penicillin production strains (Lu et al., 1992).

These authors suggested that a decrease in the L-α-aminoadipic acid metabolism could be the mechanism by which high penicillin production strains can keep high intracellular L-α-aminoadipic acid concentrations (Hölinger and Kubicek 1989a, 1989b). The amino acid precursors for penicillin biosynthesis are provided from vacuoles (Hölinger and Kubicek 1989a) and differences in the kinetic exchange of valine have also been observed in strains with different penicillin production ability (Affenzeller and Kubicek 1991). Thus, it is possible that the high L-α-aminoadipic acid concentrations available in penicillin overproducer strains are due to a more efficient precursor transport into vacuoles as compared with transport in low penicillin production wild type strains (Lendenfeld et al., 1993).

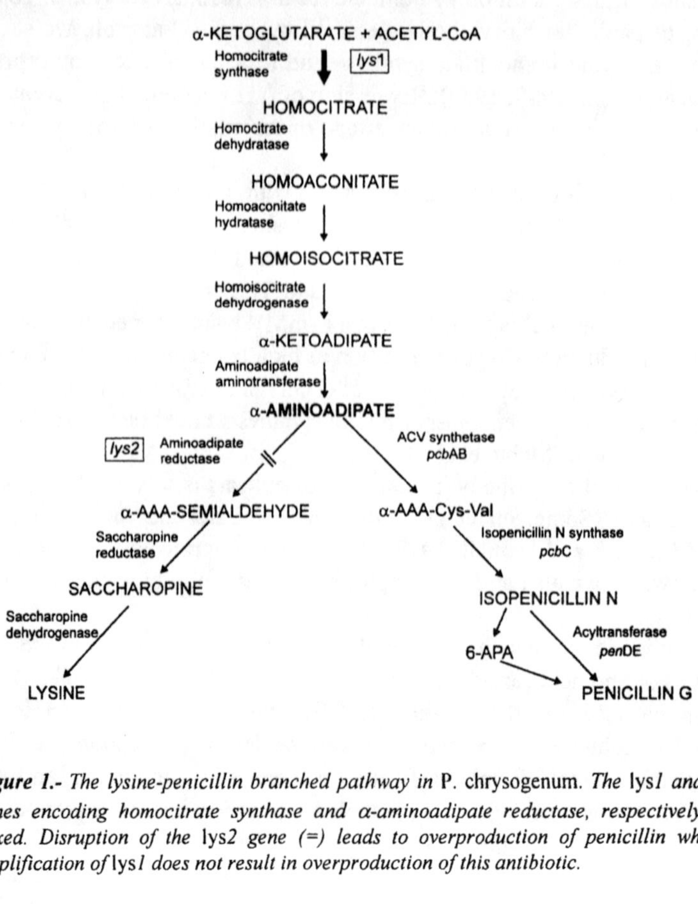

Figure 1.- *The lysine-penicillin branched pathway in* P. chrysogenum. *The* lys1 *and* lys2 *genes encoding homocitrate synthase and α-aminoadipate reductase, respectively, are boxed. Disruption of the* lys2 *gene (=) leads to overproduction of penicillin whereas amplification of* lys1 *does not result in overproduction of this antibiotic.*

3. Metabolic engineering of the lysine pathway in *P. chrysogenum*

As shown above, L-α-aminoadipic acid plays an important role in penicillin biosynthesis. However, there is a scarce knowledge at the molecular level dealing with the α-aminoadipic acid biosynthesis in filamentous fungi. Only three genes of the lysine pathway have been characterised so far: the *lys*F gene which encodes homoaconitase in *A. nidulans* (Weidner et al., 1997), the *lys*2 gene encoding the large subunit of the α-aminoadipate reductase (Casqueiro et al., 1998) and the *lys*1 gene which encodes the homocitrate synthase in *P. chrysogenum* (Bañuelos et al., 1999a).

Based on this background information, there were two targets for metabolic engineering: i) the channelling of the lysine metabolic flux towards penicillin biosynthesis by targeted inactivation of the α-aminoadipate reductase, which is the first enzyme after the branch point for lysine and penicillin biosynthesis, and ii) the increase of the supply of α-aminoadipic acid from the first part of the lysine pathway by modification of homocitrate synthase levels.

3.1.- METABOLIC ENGINEERING AT THE α-AMINOADIPATE REDUCTASE LEVEL: CHANNELLING OF LYSINE METABOLIC FLUX TOWARDS PENICILLIN BIOSYNTHESIS.

In the lysine pathway α-aminoadipic acid is converted into α-aminoadipate-δ-semialdehyde by the α-aminoadipate reductase (Sagisaka and Simura 1960, 1962; Affenzeller et al., 1989) whereas in the penicillin pathway α-aminoadipic acid is condensed with L-cysteine and D-valine to form the tripeptide δ-L-(α-aminoadipyl)-L-cysteinyl-D-valine (ACV) by the ACV synthetase (Fig. 1). In *P. chrysogenum* it should be possible to increase the α-aminoadipic acid intracellular pool available for penicillin biosynthesis by targeted inactivation of the α-aminoadipate reductase (EC. 1.2.1.31) (also known as α-aminoadipate-δ-semialdehyde dehydrogenase), which is the first enzyme of the lysine pathway (Fig. 1) after the branch point. This enzyme catalyses the formation of α-aminoadipate-δ-semialdehyde in a reaction that includes activation of the α-aminoadipic δ-carboxyl group to form α-aminoadipyladenylate, followed by a reduction leading to α-aminoadipate-δ-semialdehyde (Sagisaka and Simura 1960, 1962; Sinha and Bhattacharjee 1971). In *S. cerevisiae* the α-aminoadipate reductase is composed of two subunits; the large one is encoded by the *LYS2* gene (Eibel and Philippsen 1983; Barnes and Thorner 1986; Morris and Jinks-Robertson 1991) and seems to be responsible of the enzymatic activity. The small subunit, is encoded by the *LYS5* gene (Miller and Bhattacharjee 1996) and has recently been shown to correspond to the phosphopantetheinyl transferase of the large subunit (Lambalot et al., 1996; Ehmann et al., 1999).

The *lys*2 gene of *P. chrysogenum* was cloned from a genomic library of this fungus (Casqueiro et al., 1998). It encodes a protein of 1409 amino acids (Mr 154,859) with strong similarity to *S. cerevisiae* (49.9% identity), *Schizossacharomyces pombe* (51.3% identity) and *Candida albicans* (48.12 % identity) α-aminoadipate reductases (Morris and Jinks-Robertson 1991; Miller and Bhattacharjee 1996; Suvarna et al., 1998) and a lesser degree of identity to the amino acid activating domains of the non-ribosomal peptide synthetases (Kleinkauf and von Döhren 1996), including the α-aminoadipate-activating domain of the ACV synthetase of *P. chrysogenum*. The *lys*2 encoded protein is a member of the aminoacylated-forming enzyme family with a reductase domain in its C-terminal region (Casqueiro et al., 1998).

Gene disruption is a difficult task in filamentous fungi (Fincham 1989). There are several species including *P. chrysogenum* where gene disruption is even much more difficult to achieve due to the high frequency of ectopic recombination. The first work dealing with gene disruption in *P. chrysogenum* was the inactivation of the *lys*2 gene (Casqueiro et al., 1999). In this work the *lys*2 gene was inactivated by two different strategies: i) the single integration method (Shortle et al., 1982) or ii) the one step gene replacement method (Rothstein 1983).

In the first strategy, disruption of the *lys*2 gene was obtained by single crossing-over between the endogenous *lys*2 and a fragment of the same gene located in an integrative plasmid. The *lys*2 disrupted mutants were obtained with a 1.6% efficiency (2 inactivated mutants in 127 transformants studied) when the *lys*2 homologous region was 4.9 kb (Fig. 2A) but no disruption of *lys*2 was observed with constructions containing shorter homologous regions. Similarly, *lys*2 disrupted mutants were obtained by double crossing-over (gene replacement) with an efficiency of 0.14% (1 mutant in 728 transformants) by using two *lys*2 homologous regions of 4.3 and 3.0 kb flanking the *pyr*G gene marker (Fig. 2B); no homologous recombination was detected when the selectable marker was flanked by short *lys*2 homologous genomic DNA. By both methods, three mutants were isolated and, as expected, all mutants were lysine auxotrophs unable to grow in Czapek minimal medium supplement with α-aminoadipic acid. Disruption of the *lys*2 gene was confirmed by Southern blot analysis (Fig. 2A and 2B). The *lys*2-disrupted mutants lacked α-aminoadipate reductase and showed specific penicillin yields double than those of the parental non-disrupted strain Wis 54-1255 (Fig. 2C) (Casqueiro et al., 1999). Moreover, the *lys*2 disrupted mutants were studied by HPLC, and showed higher intracellular α-aminoadipic acid pool as compared to the control strain Wis 54-1255 (Bañuelos et al., 1999b). In summary, in the mutants lacking α-aminoadipate reductase the α-aminoadipic acid precursor is channelled towards penicillin biosynthesis.

3.2.- METABOLIC ENGINEERING AT THE HOMOCITRATE SYNTHASE LEVEL

α-Aminoadipic acid is synthesised in *P. chrysogenum* as an intermediate in the lysine biosynthesis. This pathway begins with the condensation of acetyl-coenzymeA and α-ketoglutarate to form homocitrate (Fig. 1). The enzyme that catalyses this reaction, homocitrate synthase (EC. 4.1.3.21) appears to be a rate-limiting step in the supply of α-aminoadipic acid for penicillin biosynthesis. This hypothesis is supported by several experimental data: i) feedback inhibition of homocitrate synthase by lysine decreases the penicillin production rate by reducing the flux of intermediates for the synthesis of α-aminoadipic acid (Demain and Masurekar, 1974; Jaklitsch and Kubicek, 1990; Luengo et al., 1980), ii) homocitrate synthases from high penicillin production strains seem to be less sensible to lysine regulation than enzymes from low penicillin production strains (Luengo et al., 1979, 1980), and iii) high penicillin production strains keep high *lys*1 transcription levels and maintains elevated homocitrate synthase activity during the late idiophase as compared with low production strains (Bañuelos et al., 1999b).

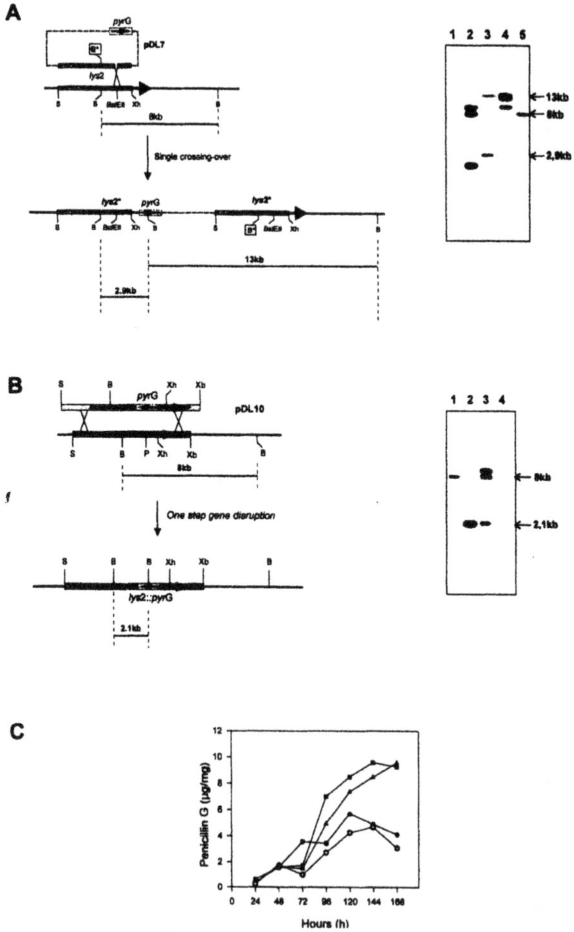

***Figure 2.-** (A) Inactivation of the* lys2 *gene by single crossing-over integration. The Southern blot analysis (on the right) revealed that the parental strain, Wis 54-1255 (Lane 5) hybridises with an 8 kb band while the* lys2 *mutants (lanes 4 and 3) lack this hybridisation band. The mutant in lane 3 has the canonical pattern of single crossing-over while the mutant in lane 4 suffered an additional recombination process. **(B)**. Inactivation of the* lys2 *gene by the one-step gene disruption method. Southern blot analysis revealed that the parental strain Wis 54-1255 (lane 1) hybridised with an 8 kb band while the mutant (lane 2) showed a 2.1 kb hybridisation band. (C) Specific penicillin production of the* lys2 *disrupted mutants in DP medium:* P. chrysogenum *Wis 54-1255 supplemented with 4 mM lysine (○) or without supplementation (•); and the* lys2 *inactivated mutants (▲ and ■)*

The gene encoding the *P. chrysogenum* homocitrate synthase (named *lys*1) has been cloned, sequenced and characterised (Bañuelos et al., 1999a). The *lys*1-encoded protein showed 71.1% identical amino acid residues with the *Yarrowia lipolytica* homocitrate synthase (Pérez-Campo et al., 1997) and 71,7% identity with *S. cerevisiae* homologous enzyme. The *lys*1 gene transcription levels were high in *P. chrysogenum* Wis 54-1255 cultures in defined production medium at 24 and 48 hours coinciding with the rapid growth phase, but clearly decreased during the penicillin production phase and the same profile was followed by the homocitrate synthase activity. In a high penicillin production strain (*P. chrysogenum* AS-P-99) *lys*1 transcription level and homocitrate synthase activity were keep high during the late idiophase (Bañuelos et al., 1999a, 1999b), suggesting that α-aminoadipic acid formation for penicillin biosynthesis may be limiting at the homocitrate synthase level in low penicillin production strains.

To study the possible presence of a rate-limiting step at the beginning of the lysine pathway, the homocitrate synthase levels were modified by increasing the *lys*1 gene dosage and by expression of the *lys*1 gene under the control of strong fungal promoters (Bañuelos et al., 1999b).

Four strategies were designed to overexpress the native *lys*1 gene by combining an early (P*lys*1) or a late (P*pcb*C) promoter (from primary or secondary metabolism) with integrative or autonomous replicating plasmids. The increase in the copy number of the *lys*1 gene (with its native promoter) was found to be the best strategy to increase expression of the *lys*1 gene and the homocitrate synthase activity in *P. chrysogenum* Wis 54-1255. A good correlation between the gene copy number and the homocitrate synthase activity level was evident in two transformants named TAR9 and TI3 (Fig. 3). The TAR9 transformant contains about 30 additional copies of the *lys*1 gene in an autonomous replicating plasmid (Fierro et al., 1996; Bañuelos et al., 1999b) and showed higher *lys*1 transcript levels and homocitrate synthase activity than the parental strain (Fig. 3). Transformant TI3 with only two additional copies of the *lys*1 gene, showed a small increase in the expression of the *lys*1 gene that did not lead to a detectable increase in the homocitrate synthase activity; expression of integrated genes (as in TI3) is highly dependent on the genetic environment at the integration locus. The high homocitrate synthase activity in transformant TAR9 did not result in detectable increases in penicillin biosynthesis. Analysis by HPLC of the α-aminoadipic acid intracellular pool in the TAR9 transformant showed no difference with the parental untransformed strain.

These results suggest that: i) other steps of the common stem of the lysine pathway are saturated and therefore the α-aminoadipic acid pool do not increase despite the increment in homocitrate synthase activity, or ii) *lys*1 overexpression leads to a higher flux of metabolites towards α-aminoadipic acid, but an active α-aminoadipate reductase prevents the accumulation of this intermediate by converting it into lysine (Lu et al., 1992). These two hypothesis were tested by overexpressing the *lys*1 gene in the *lys*2-disrupted strain. In this strain, the α-aminoadipic acid pool is much higher than in the wild type; however amplification of the *lys*1 gene in the *lys*2-disrupted strain do not lead to an increase in the α-aminoadipic acid pool or to an increment in penicillin production, indicating that there are additional rate-limiting steps in the common stem of the lysine pathway between homocitrate and α-aminoadipate.

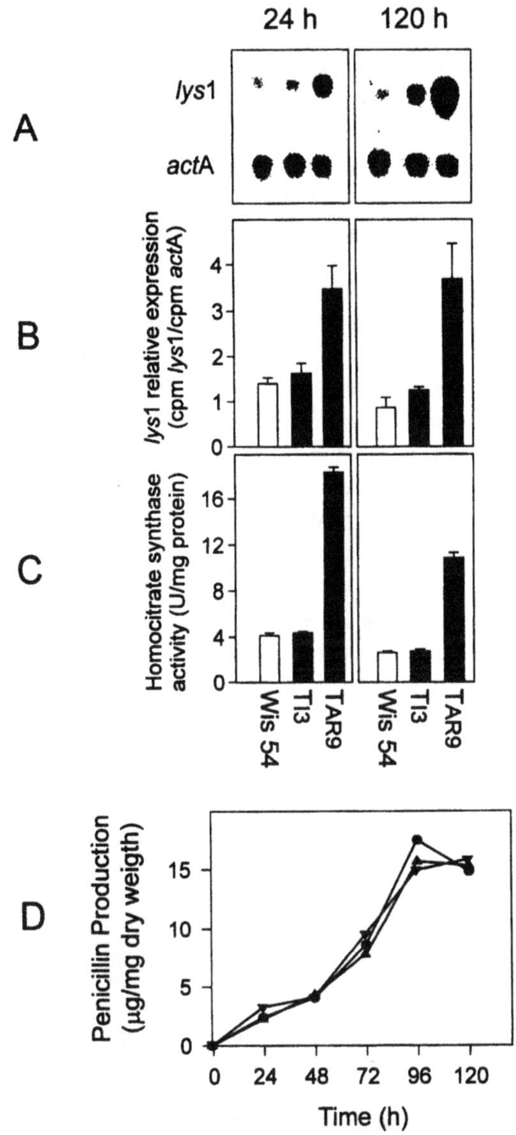

***Figure 3.*-** *Transcript levels and homocitrate synthase activity in the control strain Wis 54-1255, in the TI3 and TAR9 transformants in DP medium. (A). Northern hybridisation of total RNA from 24 and 120 hour cultures. (B). Relative expression of the* lys1 *gene with respect to the* actA *transcript level. (C). Homocitrate synthase activity of Wis 54-1255 and the TI3 and TAR9 transformants. (D). Specific penicillin production of the TI3 (•), TAR9 (◆) transformants and Wis 54-1255 (▼).*

4.- Future perspectives

The direct relationship between the α-aminoadipic acid pool and penicillin production is well established. By metabolic engineering it was possible to provide more α-aminoadipic acid for penicillin biosynthesis by blocking the branching lysine pathway at the α-aminoadipate reductase level. It is clear that inactivation of this enzyme channels the metabolic flux of the common stem of the lysine pathway towards penicillin biosynthesis. However, disruption of the *lys*2 gene introduces a requirement for lysine (*lys*2 mutants are lysine auxotrophs). Directed mutations of the *lys*2 gene –now in progress in our laboratory- should provide improved strains with reduced α-aminoadipate reductase activity but still prototrophic, thus avoiding the lysine requirement for growth. Increasing the direct supply of α-aminoadipate by the lysine pathway needs more studies because there is still an scarce knowledge about the bottlenecks and regulatory mechanisms of the lysine pathway in *P. chrysogenum*.

The future work is clearly oriented in two directions. The first one, is searching for a regulator of lysine pathway in order to modify it for increasing the supply of α-aminoadipic acid. It looks more reasonable to modify one regulator gene that the whole pathway in order to increase the α-aminoadipic acid supply. The second approach will be to explore lysine catabolism. It is known from preliminary studies that lysine when is catabolised in *P. chrysogenum* and *A. nidulans* is converted into α-aminoadipic acid (Esmahan et al., 1994). The real contribution of the lysine catabolism to the α-aminoadipic acid pool available for penicillin biosynthesis is still an open question.

Acknowledgements.

This work was supported by grants of the European Union (BIO4-CT96-0535) and the CICYT (Madrid) (BIO97-0289-C02-02). We thank M. Corrales for excellent technical assistance.

References

Abraham EP (1978). Developments in the chemistry and biochemistry of β-lactam antibiotics. In "Antibiotic and other secondary metabolites". Hütter R, Leisinger T, Nuesch J, Wehrli W (Eds). Academic Press, London, pp. 141

Affenzeller K, Jaklitsch WM, Hönlinger C, Kubicek CP (1989). Lysine biosynthesis in *Penicillium chrysogenum* is regulated by feedback inhibition of α-aminoadipate reductase. FEMS Microbiol. Lett. 58:293-298.

Affenzeller K, Kubicek CP (1991). Evidence for a compartmentation of penicillin biosynthesis in a high and a low-producing strain of *Penicillium chrysogenum*. J. Gen. Microbiol. 137:1653-1660.

Aharonowitz Y, Cohen G, Martín JF (1992). Penicillin and cephalosporin biosynthetic genes: Structure, organization, regulation and evolution. Annu. Rev. Microbiol. 46:461-495.

Arnstein HRV, Artman M, Morris D, Toms EJ (1960). Sulfur-containing amino acids and peptides in the mycelium of *Penicillium chrysogenum*. Biochem. J. 76:353-357.

Bañuelos O, Casqueiro J, Fierro F, Hijarrubia MJ, Gutiérrez S, Martín JF (1999a). Characterization and lysine control of expression of the *lys*1 gene of *Penicillium chrysogenum* encoding homocitrate synthase. Gene 226:51-59.

Bañuelos O, Casqueiro J, Gutiérrez S, Martín JF (1999b). Overexpression of *lys1* gene in *Penicillium chrysogenum*: homocitrate synthase levels α-aminoadipate pool and penicillin production (submitted for publication).

Barnes DA, Thorner J (1986). Use of the *LYS2* gene for gene disruption, gene replacement and promoter analysis in *Saccharomyces cerevisiae*. In "Gene manipulation in fungi". Benett JW, Lasure LL (Eds). Academic Press, Inc. pp. 197-226.

Bhattacharjee JK (1985). α-aminoadipate pathway for the biosynthesis of lysine in lower eukaryotes. In "Critical reviews in microbiology". CRC Press. Boca Raton, FL, vol. 12, pp. 131-151.

Bonner D (1947). Studies on the biosynthesis of penicillin. Arch. Biochem. 13: 1-9

Brakhage AA (1998). Molecular regulation of β-lactam biosynthesis in filamentous fungi. Microbiol. Mol. Biol. Rev. 62:547-585.

Brakhage AA, Turner G (1992). L-lysine repression of penicillin biosynthesis and the expression of penicillin biosynthesis genes *acvA* and *ipnA* in *Aspergillus nidulans*. FEMS Microbiol. Lett. 98:123-128.

Bryan JK (1980). Synthesis of aspartate family and branched-chain amino acids. In "The biochemistry of plants". Miflin, B. J (Ed). Academic Press. New York, vol. 5. pp. 402-440.

Casqueiro J, Gutiérrez S, Bañuelos O, Fierro F, Velasco J, Martín JF (1998). Characterization of the *lys2* gene of *Penicillium chrysogenum* encoding α-aminoadipic acid reductase. Mol. Gen. Genet. 259:549-556.

Casqueiro J, Gutiérrez S, Bañuelos O, Hijarrubia MJ, Martín JF (1999). Gene targeting in *Penicillium chrysogenum:* disruption of the *lys2* gene leads to penicillin overproduction. J. Bacteriol. 181:1181-1188.

Cohen NH, Saint-Girons I (1987). Biosynthesis of threonine, lysine and methionine. In "*Escherichia coli* and *Salmonella typhymurium* cellular and molecular biology". Neihardt, F. C (Ed). ASM. Washington. pp. 429-444.

Corbett K (1990). The history of antibiotic production. The Biochemist 12:8-13.

Demain AL (1957). Inhibition of penicillin formation by lysine. Arch. Biochem. Biophys. 67:244-245.

Demain AL (1974). Biochemistry of penicillin and cephalosporins biosynthesis fermentations. Lloydia 37:147-167.

Demain AL (1983). Biosynthesis of β-lactam antibiotics. In "Antibiotics containing the β-lactam structure". Demain AL, Solomon NA (Eds). Springer-Verlag, Berlin. pp. 189-228.

Demain AL, Masurekar PS (1974). Lysine inhibition of *in vivo* homocitrate synthesis in *Penicillium chrysogenum*. J. Gen. Microbiol. 82:143-151.

Ehmann DE, Gehring AM, Walsh CT (1999). Lysine biosynthesis in *Saccharomyces cerevisiae*: mechanism of alpha-aminoadipate reductase (*LYS2*) involves posttranslational phosphopantetheinylation by Lys5. Biochemistry 38:6171-6177.

Eibel H, Philippsen P (1983). Identification of the cloned *Saccharomyces cerevisiae LYS2* gene by an integrative transformation approach. Mol. Gen. Genet. 191:66-77.

Esmahan C, Alvarez E, Montenegro E, Martín JF (1994). Catabolism of lysine in *Penicillium chrysogenum* leads to formation of 2-aminoadipic acid, a precursor of penicillin biosynthesis. Appl. Environ. Microbiol. 60:1705-1710.

Fierro F, Kosalková K, Gutiérrez S, Martín JF (1996). Autonomously replicating plasmids carrying the *AMA1* region in *Penicillium chrysogenum*. Curr. Genet. 29:482-489.

Fincham JRS (1989). Transformation in fungi. Microbiol. Rev. 53:148-170.

Flynn EH, McCormick MH, Stamper MC, De Valeria H, Godzeski DW (1962). A new natural penicillin from *Penicillium chrysogenum*. J. Am. Chem. Soc. 84:4594-4595.

Friedrich CG, Demain AL (1977). Homocitrate as a crucial site of the lysine effect on the penicillin biosynthesis. J. Antibiot. 9:760-761.

Friedrich CG, Demain AL (1978). Uptake and metabolism of alpha-aminoadipic acid by *Penicillium chrysogenum* Wis 54-1255. Arch. Microbiol. 119:43-47.

Hönlinger C, Kubicek CP (1989a). Metabolism and compartmentation of α-aminoadipic acid in penicillin-producing strains of *Penicillium chrysogenum*. Biochim. Biophys. Acta 993:204-211.

Hönlinger C, Kubicek CP (1989b). Regulation of δ-(L-α-aminoadipyl)-L-cysteinyl-D-valine and isopenicillin N biosynthesis in *Penicillium chrysogenum* by the α-aminoadipate pool size. FEMS Microbiol Lett. 65:71-76.

Jaklitsch WM, Kubicek CP (1990) Homocitrate synthase from *Penicillium chrysogenum:* localization, purification of the cytosolic isoenzyme, and sensitivity to lysine. Biochem. J. 269:247-253.

Jaklitsch WM, Hampel W, Röhr M, Kubicek CP (1986). α-aminoadipate pool concentration and penicillin biosynthesis in strains of *Penicillium chrysogenum*. Can. J. Microbiol. 32:473-480.

Kleinkauf H, von Döheren H (1996). A nonribosomal system of peptide biosynthesis. Eur. J. Biochem. 236:335-351.

Kobashi N, Nishiyama M, Tanokura M (1999). Aspartate kinase-independent lysine synthesis in an extremely termophilic bacterium, *Thermus thermophilus*: lysine is synthetized via α-aminoadipic acid not via diaminopimelic acid. J. Bacteriol. 181:1713-1718.

Kosuge T, Hoshino T (1998). Lysine is synthetized through the alpha-aminoadipate pathway in *Thermus thermophilus*. FEMS Microbiol. Lett. 169:361-367.

Kurzatkowski W, Kurzatkowski JD, Filipek J, Solecka J, Kurylowitz W (1990). Reversion of L-lysine inhibition of penicillin G biosynthesis by 6-oxopiperidine-2-carboxylic acid in *Penicillium chrysogenum* PQ-96. Appl. Microbiol. Biotechnol. 34:397-398.

Lambalot RH, Gehting AM, Flugel RS, Zuber P, LaCelle M, Marahiel MA, Reid R, Khosla C, Walsh CT (1996). A new enzyme superfamily the phosphopantetheinyl transferases. Chemistry and Biology 3:923-936.

Lejohn HB (1971). Enzyme regulation, lysine pathways, and cell structures as indicators of major lines of evolution in fungi. Nature (London) 231:164-168.

Lendenfeld T, Ghali D, Wolschek M, Kubicek-Praz EM, Kubicek CP (1993). Subcellular compartmentation of penicillin biosynthesis in *Penicillium chrysogenum*. J. Biol. Chem. 268:665-671.

Loder PB, Abraham EP (1971). Biosynthesis of peptides containing α-aminoadipic acid and cysteine in extracts of a *Cephalosporium acremonium* sp. Biochem. J. 123:477-482.

Lu Y, Mach RL, Affenzeller K, Kubicek CP (1992). Regulation of α-aminoadipate reductase from *Penicillium chrysogenum* in relation to the flux from α-aminoadipate into penicillin biosynthesis. Can. J. Microbiol. 38:758-763.

Luengo JM, Revilla G, Lopez MJ, Villanueva JR, Martín JF (1980). Inhibition and repression of homocitrate synthase by lysine in *Penicillium chrysogenum*. J. Bacteriol. 144:869-876.

Luengo JM, Revilla G, Villanueva JR, Martín JF (1979). Lysine regulation of penicillin biosynthesis in low-producing and industrial strains of *Penicillium chrysogenum*. J. Gen. Microbiol. 115:207-211.

Malumbres M, Martín JF (1996). Molecular control mechanisms of lysine and threonine biosynthesis in amino acid-producing corynebacteria: Redirecting the carbon flow. FEMS Microbiol. Lett. 143:103-114.

Martín JF, Aharonowitz Y (1983). Regulation of biosynthesis of β-lactam antibiotics. In "Antibiotics containing the β-lactam structure". Demain, A.L., Solomon, N.A. (Ed)., New York: Springer. pp. 189-228.

Martín JF, Gutiérrez S, Demain AL (1997). β-lactams. In "Fungal biotechnology". Anke, T (Ed). Weinheim: Chapman and Hall, pp. 91-127.

Martín JF, Liras P (1989). Enzymes involved in penicillin, cephalosporin and cephamycin biosynthesis. In "Advances in biochemical engineering/biotechnology". Fiechter, A (Ed). Vol. 39. Springer-Verlag. Berlín. pp. 153-187

Masurekar PS, Demain AL (1972). Lysine control of penicillin biosynthesis. Can. J. Microbiol. 18:1045-1048.

Mehta RJ, Speth JL, Nash CH (1979). Lysine stimulation of cephalosporin C synthesis in *Cephalosporium acremonium*. Eur. J. Appl. Microbiol. Biotechnol. 8:177-182.

Miller KG, Bhattacharjee JK (1996). The *LYS5* of *Saccharomyces cerevisiae*. Gene 172:167-168.

Morris ME, Jinks-Robertson S (1991). Nucleotide sequence of the *LYS2* gene of *Saccharomyces cerevisiae*: homology to *Bacillus brevis* tyrocidine synthetase I. Gene 98:141-145.

Patte JC (1983). Diaminopimelate and lysine. In "Amino acids: biosynthesis and genetic regulation". Hermann, K. M. and Somerville, R. L (Eds). Adison-Wesley Reading. Mass. pp. 213.

Pérez-Campo FM, Nicaud JM, Gaillardin C, Domínguez A (1996). Cloning and sequencing of the *LYS1* gene encoding homocitrate synthase in the yeast *Yarrowia lipolytica*. Yeast 12:1459-1469.

Revilla G, Ramos FR, López-Nieto MJ, Alvarez E, Martín JF (1986). Glucose represses formation of δ-(L-α-aminoadipyl)-L-cysteinyl-D-valine and isopenicillin N synthase but not penicillin acyltransferase in *Penicillium chrysogenum*. J. Bacteriol. 168:947-952

Rothstein M, Saffran EM (1963). Lysine biosynthesis in algae. Arch. Biochem. Biophys. 101:373.

Rothstein RJ (1983). One-step gene disruption in yeast. Method. Enzymol. 101:202-211.

Sagisaka S, Shimura K (1960). Mechanism of activation and reduction of α-aminoadipic acid by a yeast enzyme. Nature (London) 188:1189-1190.

Sagisaka S, Shimura K (1962). Studies on lysine biosynthesis: IV. Mechanism of activation and reduction of α-aminoadipic acid. J. Biochem. (Tokyo) 52:155.

Shortle D, Novick P, Botsein D (1982). Lethal disruption of yeast actin gene by integrative DNA transformation. Science 217: 371-373.

Sinha AK, Bhattacharjee JK (1971). Lysine biosynthesis in *Saccharomyces cerevisiae*: conversion of α-aminoadipate into α-aminoadipic-δ-semialdehyde. *Biochem. J.* 125:743-749.

Somerson NL, Demain AL, Nunheimer TD (1961). Reversal of lysine inhibition of penicillin production by α-aminoadipic acid. Arch. Biochem. 93:238-241.

Suvarna K, Seah L, Bhattaharjee V, Bhattacharjee JK (1998). Molecular analysis of the *LYS2* gene of *Candida albicans*: homology to peptide antibiotic synthetases and the regulation of the α-aminoadipate reductase. Curr. Genet. 32:268-275.

Vogel HJ (1960). Two models of lysine synthesis among lower fungi: evolutionary significance. Biochim. Biophys. Acta 41:172.

Warren SC, Newton GGF, Abraham EP (1967). Use of α-aminoadipic acid for the biosynthesis of penicillin N and cephalosporin C in *Cephalosporium* sp. Biochem. J. 103: 891-901.

Weidner G, Stefan B, Brakhage AA (1997). The *Aspergillus nidulans lysF* gene encodes homoaconitase an enzyme involved in the fungus-specific lysine biosynthesis pathway. Mol. Gen. Genet. 355:237-247.

GLYCOSYLATION OF ANTIBIOTICS AND OTHER AGENTS FROM ACTINOMYCETES

WOLFGANG PIEPERSBERG
Chemische Mikrobiologie, FB9 - Chemie, Bergische Universit,, t GH Wuppertal, Gauss-Str. 20, D-42097 Wuppertal; FAX (+49)-202-439-2521; e-mail: piepersb@uni-wuppertal.de

Summary

The diverse glycosidic side chains of many natural products from Actinomycetes, e.g. antibiotics, are relevant for their biological activity. They have been started to be used as targets for biocombinatorial product design together with various aglycones. Examples of pathways for the activation and biosynthesis of NDP-sugar derivatives and the glycosyltransferases transferring them are reported.

1. Introduction

In the past two decades the carbohydrate moieties in extracellular homo- and heteropolymers or in low-molecular weight products, mostly disregarded before as being rather non-specific structural, conditioning or chemically protecting groups in biological material, have become fully accepted for their equivalent use in creating biological specificity. The secondary carbohydrates from Actinomycetes are mainly rare sugar-based or sugar-containing (i.e. sugar derived components are substructures) compounds of very diverse structures and functions, such as (amino)glycosidic or glycosylated polyketidic and peptidic antibiotics, glycosidase inhibitors or compounds with neither activity (Liu and Thorson, 1994; Kirschning et al., 1997; Piepersberg, 1994, 1997; Piepersberg and Distler, 1997). They form a particular interesting group of communication metabolites because of their resemblance to the glycoconjugates of eukaryotes in providing specific surface- or ligand shaping and recognition mediating chemistry in other major groups of organisms, e.g. eukaryotes. Also, cell-wall structuring or cell-surface marking sugar conjugates in other groups of bacteria, such as O-antigen chains (lipopolysaccharides, LPS) or extracellular polysaccharides (EPS) in the gram-negative bacteria might be also regarded as secondary carbohydrates, which are functionally and/or evolutionarily parallel inventions of an even more generally distributed communication metabolism (Piepersberg, 1992). This is also reflected by the

A. Van Broekhoven et al. (eds.), Novel Frontiers in the Production of Compounds for Biomedical Use, 161–168.
© 2001 *Kluwer Academic Publishers. Printed in the Netherlands.*

use of a common, highly versatile and non-taxonomically distributed gene pool for secondary carbohydrate metabolism in all bacteria which becomes now more and more apparent in the fully sequenced bacterial mosaic genomes. This enables closely related bacteria to synthesise different strain-specific extracellular material individualising biovars much below the species limits of taxonomic distance.

2. Sugars and Cyclitols as Building Blocks in Actinomycete Secondary Metabolites: Pathways for Modified Sugars and Cyclitols.

2.1. General Scope of Secondary Carbohydrate Metabolism.

As secondary carbohydrates is here defined as the mostly multi-step biosynthesis of modified and not generally distributed derivatives of central sugar metabolism, i.e. from both the glycolytic and pentosephosphate-linked intermediary metabolism. The products can, therefore, be derived from all intermediates comprising the C5 to C9 sugars, where the C6 (hexose-derived) metabolites are the most predominant. Some examples are compiled in Figure 1. The secondary branches of intermediary metabolism leading to these compounds are always based on precursors activated by either phosphorylation or nucleotidylation (NDP-activated), though being highly diverse otherwise (Piepersberg, 1997). In NDP-activated sugars the nucleotide moiety can vary - in Actinomycetes there is evidence for participation of dTDP-, UDP-, GMP- or CDP-activated sugars (cf. Fig. 1) - in order to sort the routes and end products of secondary modification. Alternatively, the sugar precursors can become cyclitols first after C-C bond-forming ring closure reactions to yield similar building blocks. The condensation reactions of the modified intermediates are also quite diverse. They either require specific glycosyltransferases in case of the NDP-activated precursors or other types of condensing transferases. They form either O-, C- or N-bound glycosides or (pseudo-)oligosaccharides. Also, amide-, imino-, ester or C-C bond formations are observed as condensing reactions. Thus, sugar derived building blocks for actinomycete secondary metabolites are a group of compounds at least as versatile as the usual and unusual amino acids in secondary peptides or the oligomerised acyl residues in polyketides together with whom they frequently occur in heterogeneously built end products.

2.2. 6-DEOXYHEXOSES (6DOH) IN GLYCOSYLATION OF ANTIBIOTICS AND OTHER BIOACTIVE SECONDARY METABOLITES.

The most abundant and diverse group of secondary carbohydrate moieties are the 6-deoxyhexoses (6DOH) (Liu and Thorson 1994; Piepersberg, 1994, 1997; Kirschning et al., 1997). The biosynthetic formation of 6DOH occurs on the NDP-activated hexose precursors dTDP-D-glucose (most 6DOH's in actinomycete secondary metabolites or the ubiquitous L-rhamnose), CDP-D-glucose (3,6-dideoxyhexoses in gram-negative bacteria or possibly N-methyl-L-glucosamine in streptomycin) or GDP-D-mannose in basically two or three enzyme-catalysed steps for the D- and L-6DOHs, respectively (cf. Fig. 1).

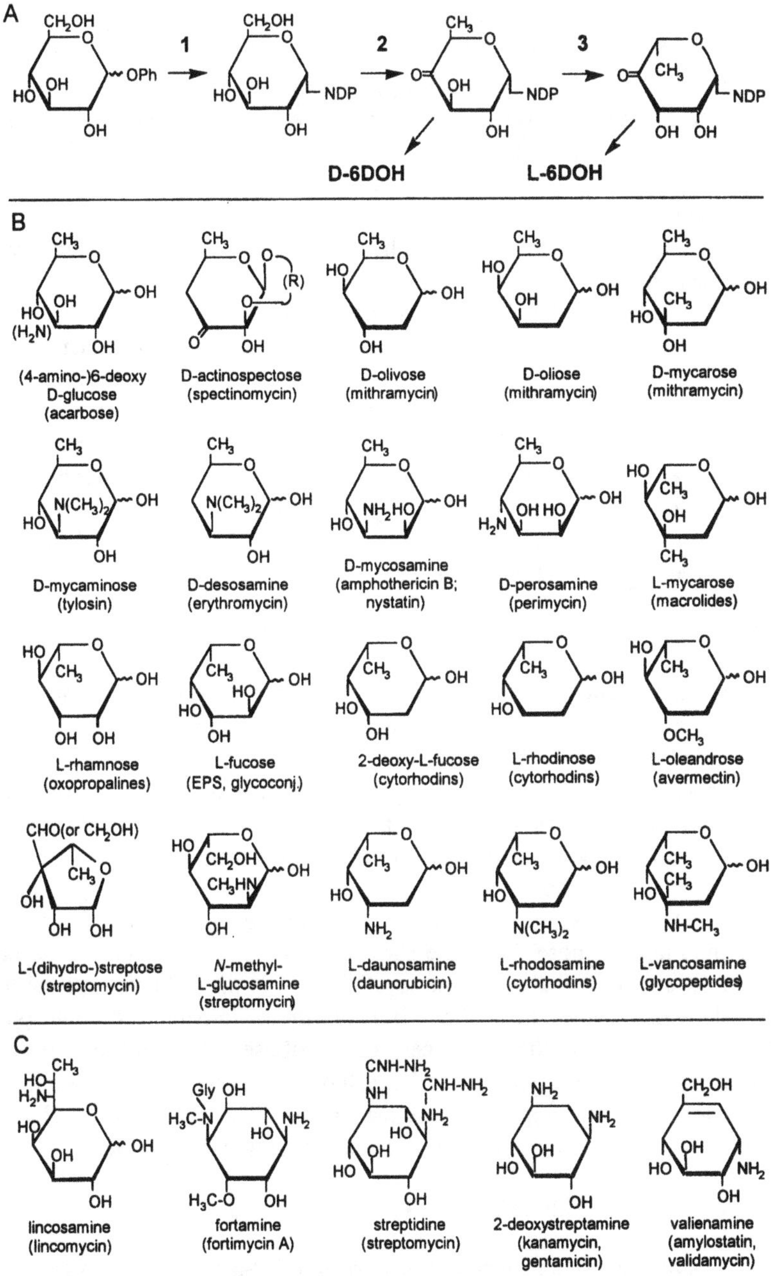

Figure 1. Examples of modified sugars and cyclitols as building blocks of actinomycete secondary carbohydrate anabolism. A. Basic biosynthetic pathway to D- and L-6DOHs; steps 1 to 3 are explained in the text. B. Chemical structures of some relevant D- and L-6DOHs. C. Chemical structures of an octose derivative and some (amino-)cyclitols.

The activating enzymes, NDP-sugar synthases or hexose-1-phosphate nucleotidylyltransferases (pyrophosphorylases; step 1 in Fig. 1A), belong to nucleotide-specific subfamilies of one single enzyme family of moderate sequence conservation. The second step enzymes, NAD+-dependent NDP-hexose 4,6-dehydratases, forming an NDP-4-keto-6-deoxy intermediate (step 2 in Fig. 1A), introduce the 6-deoxy group and are probably the most highly conserved enzyme family in this basic pathway. This conservation allows to use their genes for the use as pathway-specific gene probes to identify gene clusters for secondary metabolites involving 6DOH components with high frequency(Stockmann, 1992).
A third family of enzymes, the NDP-hexose 3,5-epimerases generally convert the NDP-activated intermediates to their L-forms, though it is not yet known, whether they always catalyse the third step (cf. Fig. 1A). These enzymes again are less conserved than the NDP-hexose 4,6-dehydratases and, therefore, less suitable for gene probe design. Subsequently, the further conversion of the keto group in C4-position either involves its stereochemical reduction, complete removal of the oxygen or introduction of other groups, such as an amino group. All other hydroxyls or the originally hydroxylated C-atoms of the sugar might also be modified in many different ways. Some steps of modification even can occur after glycosyltransfer to an aglycone.

2.3. CYCLITOLS.

According to our current knowledge, the cyclitol biosyntheses in Actinomycetes can involve two different mechanisms (Piepersberg, 1997; Piepersberg, 1997). These are exemplified by (1) the NAD+-dependent myo-inositolphosphate synthase, which yields D-myo-inositol-3-phosphate a precursor e.g. of streptidine and fortamine (cf. Fig. 1C), and (2) the dehydroquinate synthase-related, NAD+-dependent and dephosphorylating scyllo-inosose synthase, which is utilised for the first biosynthetic step in the formation of the 2-deoxystreptamine moiety (cf. Fig. 1C) of many aminoglycosides(Yamauchi and Kakinuma, 1993). Recently, we have started to investigate the biosynthesis of the alpha-glucosidase inhibitor acarbose. We have shown that an enzyme of the dehydroquinate synthase family catalyses the biosynthesis of the C7-cyclitol 2-epi-5-epi-valiolone, a precursor of valienamine (cf. Fig. 1C), from sedo-heptulose-7-phosphate in Actinoplanes sp. 50/110 the producer of acarbose. To built-up further the pseudodisaccharidic precursor of acarbose, acarviose (valieniminyl-4,6-dideoxyglucose), the 2-epi-5-epi-valiolone has at least to undergo 2-epimerization and 5,6-dehydratation, besides the formation of the condensing imino bridge.

2.4. GLYCOSYLTRANSFERASES.

Several families of glycosyltransferases transferring NDP-activated sugars to O, C, or N have been identified in all organisms; their evolutionary relatedness is low or absent between those families. These enzymes are either soluble or bound to the cytoplasmic surface of the cell membrane in bacteria; in eukaryotes they might be organelle-specific, especially in the endoplasmic reticulum and the golgi apparatus. Their two substrates, aglycones and the specific NDP-activated sugar as co-substrate, have to be recognised by the glycosyltransferases, which in general requires a high degree of specificity for both reaction partners. In the actinomycete secondary anabolism a particular family of apparently soluble (or membrane-attached?) glycosyltransferases have been identified. The first representatives have been detected among antibiotic resistance evolving macrolide-modifying 2-O-glucosyltransferases, using UDP-D-glucose, enzymes which occur in two forms (Cundliffe, 1992; Vilches et al., 1992). Recent examples from studies on glycopeptides (Solenberg et al., 1997), macrolides (Olano et al., 1998) and aromatic polyketides (Fernandez et al., 1998) suggest that this same family of glycosyltransferases is engaged in their glycosylation. Also there is evidence that part of these enzymes are not too specific for their aglycones and, thus, are possible tools in biocombinatorial use for the production of new glycosylated hybrid products. - Alternative mechanisms for the formation of glycosidic bonds are the transglycosidases, many of which are related to alpha-amylases or other groups of extracellular proteins. These, however, use homooligo- or homopolymeric glycosides for forming a glycosidic bond, are mostly polymerising and do not have a specific requirement for a particular aglycone. One example for use of an alpha-amylase/cyclodextran synthase-related enzyme was also detected to be involved in the metabolism of actinomycete secondary metabolites.

3. Examples.

3.1 THE STREPTOMYCIN PATHWAY.

Streptomycin is built-up from streptidine, L-streptose, and N-methyl-L-glucosamine. The biosynthesis of the aminocyclitol streptidine (for review see Piepersberg, 1997). The activated sugar dTDP-L-dihydrostreptose is postulated to be the precursor of the neutral sugar streptose. We have demonstrated recently that dTDP-L-rhamnose is probably a precursor of L-streptose since it is made by the branching pathway catalysed by the enzymes StrD (dTDP-D-glucose synthase), StrE (dTDP-D-glucose 4,6-dehydratase), StrM (dTDP 4-keto-6-deoxy-D-glucose 3,5-epimerase), and StrL (dTDP-L-rhamnose synthase) (S. Verseck, and W. Piepersberg, unpublished). Also, we have shown that part of the sugar activating and modifying enzymes encoded by genes of the str/sts-gene clusters in the (5'-hydroxy-)streptomycin producers Streptomyces griseus and S. glaucescens differ considerably in demonstrating the existence of a pathway starting from CDP-D-glucose in S. glaucescens, but not in S. griseus (Beyer et al., 1996). The enzyme StrQ from S. glaucescens is a highly specific CDP-D-glucose synthase. This might even indicate the use of different pathways for the same end product, e.g. N-methyl-L-glucosamine.

3.2 MACROLIDE SUGARS.

The macrolides are mostly modified by one to three sugars. Characteristic are the neutral L-6DOH's mycarose or oleandrose and the aminated D-6DOH's mycaminose or desosamine (cf. Fig. 1), which are postulated to be synthesised on partially common routes, respectively. The full set of genes/enzymes involved in the biosynthesis of mycarose and desosamine have been reported so far only from the erythromycin producer Saccharopolyspora erythrea (Summers et al., 1997; Gaisser et al., 1998; Salah-Bey et al., 1998). We have started to analyse the enzymes involved in the L-mycarose pathway from Sa. erythrea by expression in heterologous hosts and by identification of their enzymology (P. Weingarten, M.-C. Raynal, P. Leadley, L. Elling, and W. Piepersberg, unpublished). Also, the genes/enzymes for D-mycaminose biosynthesis from S. mycarofaciens are under similar investigation (L. Cong and W. Piepersberg, unpublished).

3.3 LINCOSAMINE.

Genetic and metabolite-labelling studies of the production of lincomycin A indicated an unusual sugar synthesis for lincosamine, the octose underlying the methylthiolincosaminide sugar moiety this antibiotic (cf. Fig. 1). Our recent analysis of the postulated enzymes involved in this pathway by mutagenesis, enzyme expression, and enzyme studies is based on the lmb-genes putatively involved (Peschke et al., 1995). It allows to postulate a sequence of reactions involving a pathway starting from ribose-5-phosphate and either sedoheptulose-7-phosphate or another C3-donor condensed to an octulose-8-phosphate via a (trans-)aldolase (LmbR; A. Arnold and W. Piepersberg, unpublished). The further steps then could be isomerisation (LmbN), dephosphorylation at C8 (LmbK), phosphorylation at C1 (LmbP), NDP-octose synthesis (LmbO), NDP-octose 4-epimerization (LmbM), NDP-octose 6,8-dehydratation (LmbL/Z), NDP-6-keto-8-deoxyoctose aminotransfer (LmbS) to yield the NDP-lincosamine intermediate which further could be thiomethylated at C1 by an enzyme related to methyltransferases, putatively LmbW. The thus formed methylthiolincosaminide is condensed with the second subunit of lincomycin A, propylproline, in a still unknown mechanism of amide bond formation which seems to involve aminoacyl-adenylate activation of the amino acid moiety.

4. Conclusion

The diverse glycosidic side chains of many natural products from Actinomycetes are relevant for the biological activity of these metabolites and, therefore, interesting targets for use in biocombinatorial product design in combination with variable aglycones. The biochemical tools required are the genes and enzymes for the nucleotidediphosphate (NDP)-activation and biosynthesis of the sugar derivatives, being highly modified in most cases, and the glycosyltransferases transferring them to the acceptor groups in the aglycones. Currently, the still lacking basic knowledge on these aspects is accumulating rapidly and several systems, e.g. macrolides, aromatic polyketides with antitumor activity, and glycopeptides, are worked out as model systems for hybrid glycosylation. Others, e.g. aminoglycosides, lincosamides, amylostatins, but also the pathways for lipo- and other exopolysaccharides in gram-negative bacteria, could yield the biosynthetic potential for a largely extended diversity of secondary carbohydrate building blocks. Several strategies could be envisaged in order to employ their potential: (1) Cloning and expression of biosynthetic and (glycosyl-)transferase genes into new hosts producing a suitable aglycone would be feasible (combinatorial biochemistry; pathway engineering); (2) Alternatively, the opposite direction could be followed in recombining the genetic material for the aglycone pathway into a "glycosylation host" or feeding it with the aglycone directly (biotransformation); (3) In vitro synthesis of NDP-activated sugar derivatives and their transfer to the aglycone by cell-free and/or otherwise engineered glycosyltransferase systems (enzyme technology). In any case, an extended collection of natural and engineered glycosyltransferases will be needed in order to produce hybrid glycosylated products, since the specificity for both a given donor-substrate (NDP-sugar) and the accompanying acceptor-substrate (aglycone) seems to be high in most cases.

References

Beyer, S., Distler, J., Piepersberg, W. (1996) New operons in the str genecluster for the biosynthesis of 5_-hydroxystreptomycin in Streptomyces glaucescens GLA.O (ETH 22794) and evidence for pathway-specific regulation by StrR. Mol. Gen. Genet. 250: 775-784.

Cundliffe, E. (1992) Glycosylation of macrolide antibiotics in extracts of Streptomyces lividans. J. Biol. Chem. 36: 348-352.

Fernandez, E., Weissbach, U., Sanchez Reillo, C., Brana, A. F., Mendez, C., Rohr, J., Salas, J. A. (1998) Identification of two genes from Streptomyces argillaceus encoding glycosyltransferases involved in transfer of a disaccharide during biosynthesis of the antitumor drug mithramycin. J. Bacteriol. 180: 4929-4937.

Gaisser, S., Bohm, G. A., Staver, M. J., Wendt-Pienkowski, E., Hutchinson, C. R., Katz, L. (1998) Analysis of eryBI, eryBIII and eryBVII from the erythromycin biosynthetic gene cluster in Saccharopolyspora erythraea. Mol. Gen. Genet. 258: 78-88.

Liu, H.-W., Thorson, J.S. (1994) Pathways and mechanisms in the biogenesis of novel deoxysugars by bacteria. Ann. Rev. Microbiol. 48: 223-256.

Olano, C., Rodriguez, A. M., Michel, J. M., Mendez, C., Raynal, M. C., Salas, J. A. (1998) Analysis of a Streptomyces antibioticus chromosomal region involved in oleandomycin biosynthesis, which encodes two glycosyltransferases responsible for glycosylation of the macrolactone ring. Mol. Gen. Genet. 259: 299-308.

Piepersberg, W. (1997) Molecular biology, biochemistry, and fermentation of aminoglycoside antibiotics (Ch. 4). In: Biotechnology of Industrial Antibiotics (2nd Ed.) (Strohl WR, Ed). New York: Marcel-Dekker Inc., pp. 81-163.

Piepersberg, W., Distler, J. (1997) Aminoglycosides and sugar components in other secondary metabolites (Ch. 10). In: Biotechnology (2nd Ed.), Vol. 7: Products of Secondary Metabolism (Rehm H-J, Reed G, P□hler A, Stadler P, Eds). Weinheim: VCH-Verlagsgesellschaft, pp. 397-488.

Pissowotzki, K., Mansouri, K., Piepersberg, W. (1991) Genetics of streptomycin production in Streptomyces griseus. Molecular structure and putative function of genes strELMB2N. Mol. Gen. Genet. 231: 113-23.

Peschke, U., Schmidt, H., Zhang, H. Z., Piepersberg, W. (1995) Molecular characterization of the lincomycin-production gene cluster of Streptomyces lincolnensis 78-11. Mol. Microbiol. 16: 1137-1156.

Salah-Bey, K., M. Doumith, Michel, J. M., Haydock, S., Cortes, J., Leadlay, P. F., Raynal, M. C. (1998) Targeted gene inactivation for the elucidation of deoxysugar biosynthesis in the erythromycin producer Saccharopolyspora erythraea. Mol. Gen. Genet. 257: 542-53.

Solenberg, P. J., Matsushima, P., Stack, D. R., Wilkie, S. C., Thompson, R. C., Baltz, R. H. (1997). Production of hybrid glycopeptide antibiotics in vitro and in Streptomyces toyocaensis. Chem. Biol. 4: 195-202.

Stratmann, A., Mahmud, T., Lee, S., Distler, J., Floss, H.G., Piepersberg, W. (1999) The AcbC protein from Actinoplanes sp. is a C7-cyclitol synthase related to 3-dehydroquinate synthases and is involved in the biosynthesis of the alpha-glucosidase inhibitor acarbose. J. Biol. Chem. 274: in press.

Summers, R. G., Donadio, S., Staver, M. J., Wendt-Pienkowski, E., Hutchinson, C. R., Katz, L. (1997). Sequencing and mutagenesis of genes from the erythromycin biosynthetic gene cluster of Saccharopolyspora erythraea that are involved in L-mycarose and D-desosamine production. Microbiol. 143: 3251-3262.

Vilches, C., Hernandez, C., Mendez, C., Salas, J. A. (1992) Role of glycosylation and deglycosylation in biosynthesis of and resistance to oleandomycin in the producer organism, Streptomyces antibioticus. J. Bacteriol. 174: 161-165.

Yamauchi, N, Kakinuma, K. Biochemical studies on 2-deoxy-scyllo-inosose an early intermediate in the biosynthesis of 2-deoxystreptamine. VI. A clue to the similarity of 2-deoxy-scyllo-inosose synthase to dehydroquinate synthase. J Antibiot 1993; 46:1916-8.

ENZYMATIC SYNTHESIS OF AMOXICILLIN

Process integration using multiphase systems

A.C. SPIESS AND V. KASCHE
Technical University Hamburg-Harburg, Biotechnology II, Denickestr. 15, 21071 Hamburg. Mail: a.spiess@tu-harburg.de, Fax.: ++49 40 428 78 2127, Tel.: ++49 40 428 78 3018

1. Introduction

In the past five years more attention is being paid to enzyme catalysed conversions with suspended substrates and precipitating products (Erbeldinger *et al.* 1998 a). The main motivation is the high reactant concentration required to achieve a suitable productivity and a simple down stream processing for the industrial application of biocatalysis. Since many natural products exhibit only low to moderate solubility organic solvents were used to enhance the substrate supply. However, safety, toxicity (especially in case of food and pharmaceutical industry), and environment considerations as well as the poor stability of biocatalysts in presence of organic solvents favour the use of aqueous suspensions.

Present and potential industrial products from suspension bio-conversions cover steroid and non-steroid anti-inflammatory drugs (Maxon *et al.* 1966; Miller 1985, Furui *et al.* 1996), antibiotics (Stambolieva *et al.* 1992), peptides as pharmaceuticals or food additives (Eichhorn *et al.* 1997 a), amino acid derivatives (Takahashi 1986; Lee *et al.* 1999), organic acids (Kitahara *et al.* 1955, van der Werf 1995), and chiral synthons (Sih *et al.* 1988) as intermediates for the fine chemicals industry, surfactants (Cao *et al.* 1997), and emulsifiers (Fregapane *et al.* 1991, Bornscheuer 1995).

Halling *et al.* (1995) investigated the thermodynamics of solid-to-solid conversion, and explained the high yields obtained with a switch-like behaviour of the enzymatic conversion, which proceeds until complete dissolution of the solid substrate or product. This is common to equilibrium controlled syntheses, independent on the aqueous, organic or eutectic type of the liquid phase (Kuhl *et al*; 1992, Kuhl *et al.* 1995; Gill, Vulfson 1994). Often, mass transfer limitations were observed (Wolff *et al.* 1997) and appropriate measures, such as intensive mixing or sonication and addition of detergents or seed crystals, were suggested (Eichhorn *et al.* 1997 b, Erbeldinger *et al.* 1998 b).

A. Van Broekhoven et al. (eds.), Novel Frontiers in the Production of Compounds for Biomedical Use, 169–192.
© 2001 *Kluwer Academic Publishers. Printed in the Netherlands.*

Several peptides and amoxicillin could not (or only with low yields) be synthesised with the reverse hydrolysis reaction (Eichhorn *et al.* 1997 a; Diender *et al.* 1998). Using activated acyl donors, higher yields than the corresponding equilibrium conversions may be obtained. The feasibility of these kinetically controlled bioconversions with suspended substrates was shown by applying high nucleophile to substrate ratio in the liquid phase at equimolar substrate supply in the suspension gaining high yields and high rates (Cerovsky 1992; Kasche, Galunsky 1995; Eichhorn *et al.* 1997 b).

The aim of the present work is to study the influence of suspended substrate (HPGA) on the rate, yield and selectivity of kinetically controlled synthesis and to propose simple process engineering options (catalyst selection and process integration) for the improvement of above parameters. The model reaction is the penicillin amidase (PA from *E. coli*, E.C. 3.5.1.11) catalysed synthesis of amoxicillin (Amox) from p-hydroxy-R-phenylglycine amide (HPGA) and 6-aminopenicillanic acid (6-APA) (Figure 1).

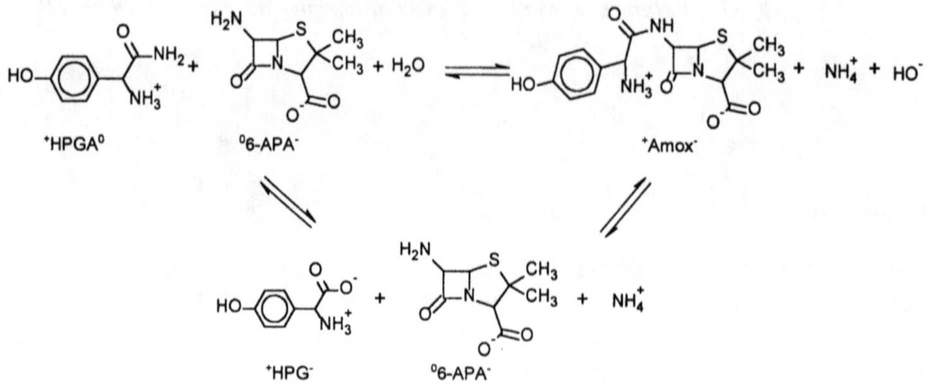

Figure 1. Synthesis of amoxicillin from HPGA and 6-APA catalysed by E. coli *PA. The indicated charges correspond to neutral pH values. HPG: p-hydroxy-R-phenylglycine.*

2. Theory

For the theoretical understanding of suspension systems, two process steps are considered: The dissolution of the solid substrate and the enzyme catalysed conversion of the dissolved substrate into product (Wolff *et al.* 1997; Michielsen *et al.* 1998; Lee *et al.* 1999). In the following, the two processes are discussed and the bottlenecks in the complete process are identified.

2.1. SOLID PHASE AND DISSOLUTION

The aqueous solubility c^* describes the phase equilibrium between the pure solute and water. It is determined by the molecular interactions of the component with water. For dissociating species, the solubility is pH dependent. The neutral (or amphoteric) form exhibits the lowest solubility. The solubility increases with both more acidic and more alkaline pH. This may be calculated by (Tomlinson, Regosz 1985):

$$c^* = c^{*,\circ} \cdot \left(\frac{c_{H^+}}{K_1} + 1 + \frac{K_2}{c_{H^+}} \right) \tag{1}$$

where $c^{*,\circ}$ is the intrinsic solubility, and K_1 and K_2 denote the dissociation constants of two different dissociating groups. The amount of acid or base added at a constant pH value is stoichiometric with the dissolution of a dissociating solute.

In ideal i.e. dilute solutions VAN'T HOFFS law expresses the temperature dependence of the solubility (Grant, Higuchi 1990):

$$\ln \frac{c^*}{c^*_{ref}} = -\frac{\Delta H^s}{R} \cdot \left(\frac{1}{T} - \frac{1}{T_{ref}} \right) \tag{2}$$

where ΔH^s denotes the heat of dissolution and the index reference stands for the reference state (25 °C and 1.013 MPa).
Two consecutive processes determine the rate of dissolution v_{diss}: The dissolution reaction at the particle surface and the diffusional transport through the laminar boundary layer. In 90 % of all cases, the transport is rate limiting (Grant, Higuchi 1990). Then the dissolution rate is expressed as:

$$v_{diss} = k_L \cdot a \cdot \left(c^* - c^b \right) \tag{3}$$

where c^b is the concentration of the solute A in the bulk liquid B and a the specific surface area of the particles, i.e. the area of the particle surface per liquid volume V.

The mass transfer coefficient k_L is dependent on the average particle size d_P, the viscosity of the medium ν, the diffusion coefficient D_{AB} and the mixing conditions. For very small particles, Harriott (1962) applied the semiempirical FRÖSSLING equation to estimate the minimum value of k_L:

$$k_L \geq 2 \cdot \frac{D_{AB}}{d_P} + 0.62 \cdot \frac{D_{AB}{}^{2/3} \cdot u_T{}^{1/2}}{d_P{}^{1/2} \cdot \nu^{1/6}} \tag{4 a}$$

The terminal slip velocity relative to the fluid u_T may be approximated by the slip velocity of particle and fluid under gravity $u_{T,S}$ using the STOKES law and a correction factor, which is dependent on the corresponding particle REYNOLDS number ($\mathbf{Re}_P = u_{T,S} \cdot d_P / \nu$) (Perry *et al.* 1997):

$$u_{T,S} = \frac{d_P^2 \cdot (\rho_B - \rho_A) \cdot g}{18 \cdot \eta_B} \tag{4 b}$$

The true k_L value is 1.5 to 8-fold the value obtained by Equation 2.4 a. The minimum SHERWOOD number (**Sh** = $k_L \cdot d_P$ / D_{AB}) is 2. Since the particle size varies during dissolution, the k_L value is time dependent (Wolff *et al.* 1997).

The diffusion coefficient D_{AB} for the diffusion of solute A in solvent B at infinite dilution may be estimated with ± 20 % error using the WILKE-CHANG correlation, which is an extension of the STOKES EINSTEIN theory to smaller molecules (Perry *et al.* 1997):

$$D_{AB} = \frac{7.5 \cdot 10^{-8} \cdot (\Phi_B M_B)^{1/2} \cdot T}{\eta_B \cdot V_A^{0.6}} \tag{5}$$

where Φ_B is the hydrogen association parameter (water: 2.26), M_B the molar mass, η_B the dynamic viscosity of the solvent water (B), and V_A the molar volume of the solute A.

2.2. SUSPENSION TO SUSPENSION CONVERSION

Opposite to the case of cellulase, which adsorbs onto the solid surface (Walker, Wilson 1991), penicillin amidase similar to protease catalyses the reaction exclusively in the aqueous phase (Erbeldinger *et al.* 1998 b). Thus the concentration of substrate A relevant for the conversion rate is constant and equal to the solubility c_A^* or a constant steady state concentration c_A^{ss} (Wolff *et al.* 1997).

If only one substrate is present as a solid phase, the rate of the enzyme catalysed conversion may be estimated using MICHAELIS – MENTEN kinetics (1). If $K_m >> c_A^*$ the MICHAELIS – MENTEN kinetic expression will reduce to the first order term with respect to c_A^* (Wolff *et al.* 1997).

$$v = \frac{k_{cat}}{K_m} \cdot c_{E,0} \cdot c_A^* \tag{6 a}$$

An increase in solubility will also increase the rate of substrate conversion. If $K_m << c_A^*$ the reaction proceeds with the maximal enzyme activity present in the reaction system (Kasche, Galunsky 1995), independent of the solubility.

$$v = k_{cat} \cdot c_{E,0} = V_{max} \tag{6 b}$$

For immobilised enzymes, the apparent kinetic constants k_{cat}^{app} and K_m^{app} should be used instead of the intrinsic k_{cat} and K_m values to account for the mass transfer resistance into the carrier and for conformational changes due to the immobilisation.

Opposite to equilibrium controlled reactions, where the yield is dependent on the thermodynamic equilibrium and not on enzyme characteristics (Halling *et al.* 1995), the yield in kinetically controlled reactions is dependent both on the selectivity v_T/v_H of the enzyme for the nucleophile NH, expressed by the transferase to hydrolase ratio $(k_T/k_H)^{app}$ or the partition constant p, and on the specificity k_{cat}/K_m of the enzyme for activated substrate AB or product AN (Kasche *et al.* 1984, Könnecke *et al.* 1984).

$$\frac{v_T}{v_H} = \frac{k_T}{k_H} \cdot \frac{c_{NH}}{c_W} = \frac{c_{NH}}{p} \tag{7}$$

For homogenous reaction mixtures, Kasche and Michaelis (1990) developed a short cut formula for the estimation of the maximal concentration of the condensation product AN in a kinetically controlled reaction, valid for the case of considerable product hydrolysis:

$$c_{AN,max} = \frac{v_T}{v_T + v_H} \cdot \frac{\left(\frac{k_{cat}}{K_m}\right)_{AB,NH}}{\left(\frac{k_{cat}}{K_m}\right)_{AN,NH}} \cdot c_{AB,max} \tag{8}$$

For suspension systems with a high excess of AB, the level of AB in the reaction remains constant, such that $c_{AB,max}$ may be substituted by $c_{AB,0}$ in (8). Under the assumptions

- that the variation of enzyme selectivity during the reaction is negligible and equals the initial enzyme selectivity $(v_T/v_H)_0$, and
- that the specificity constants k_{cat}/K_m are independent on the binding of NH for the hydrolysis of both the substrate AB and the product AN,

$c_{AN,max}$ may be estimated with good agreement to experimental values (Mincheva *et al.* 1996).

2.3. LIMITING REGIMES

Under steady state conditions in the dissolution – reaction system, the rates of dissolution and enzymatic conversion equal each other. The substrate concentration under these conditions takes a value between zero and the substrate solubility dependent on whether the system is dissolution or reaction limited. For the case of a irreversible reaction with $K_m >> c_A^*$ Wolff *et al.* (1997) derived the following correlation:

$$c_A^{ss} = \frac{c_A^*}{\frac{k_{cat} \cdot c_{E,0}}{K_m \cdot k_L \cdot a} + 1} \tag{9 a}$$

Alternatively, for $K_m << c_A^*$

$$c_A^{ss} = c_A^* - \frac{k_{cat} \cdot c_{E,0}}{k_L \cdot a} \tag{9 b}$$

Whether the reaction is mass transfer or reaction limited can be decided on base of the DAMKÖHLER number **Da,** which is the ratio of maximal enzyme conversion rate and maximal mass transfer rate.

$$\mathbf{Da} = \frac{k_{cat} \cdot c_E}{k_L \cdot a \cdot c_A^*} \tag{10}$$

For **Da** > 10 a pure mass transfer limitation may be assumed; for **Da** < 0.1, reaction limitation.

3. Materials and Methods

3.1ENZYMES AND REAGENTS.

Soluble penicillin amidase from *E. coli* (PA, E.C. 3.5.1.11) was purchased from Sigma, Deisenhofen, DE and kindly donated by Sclavo, Milano, IT. Immobilised PA preparation (Eupergit PcA 600) was donated by Röhm, Darmstadt, DE and Gist brocades, Delft, NL (Assemblase). Enzyme carriers were supplied by Röhm, Darmstadt, DE (Eupergit C, Eupergit C250L, PBA – Eupergit) or were purchased: Trisoperl (Controlled pore glass) from Schuller, Steinach, DE and CNBr - Sepharose 4B from Pharmacia, Uppsala, SE. para-Hydroxy-D-phenylglycine amide (HPGA) and para-hydroxy-D-phenylglycine (HPG) were donated by DSM, Geleen, NL. Amoxicillin and 6-aminopenicllanic acid (6-APA) were purchased from Fluka, Buchs, CH. Fluoresceinisothiocyanat (FITC) and tetramethylrhodaminisothiocyanat (TRITC) were purchased from Molecular Probes Inc., Eugene, USA. Nitrophenylaceticbenzoic acid (NiPAB) was donated by GBF, Braunschweig, DE. Phenylmethylsulphonylfluoride (PMSF) was purchased from Sigma, Deisenhofen, DE. All other chemicals were of reagent grade. Ion exchange and bipolar membranes from Tokuyama Soda Inc. were kindly donated by Dr. Schmoldt, Stantech, Hamburg, DE.

Immobilisation of enzymes. Immobilisation on Eupergit C, C250L and PBA Eupergit as well as CNBr - Sepharose was conducted according to manufacturer's procedures. The immobilisation onto Schuller CPG was conducted according to Janasek and Spohn (1998).

3.2 ENZYME ACTIVITY ASSAY.

Suitably diluted enzyme solution (25 μL) was added to a cuvet containing 950 μL sodium potassium phosphate buffer (PBS) at pH 7.5, I = 0.2 M and 25 °C. Reaction starts upon addition of 25 μL 5 mM NiPAB solution. The absorbance change at 380 nm was recorded with a spectrophotometer (Kontron Uvikon 860). One absorbance unit min^{-1} corresponds to 0.33 mg PA/ml solution or 14 U_{PenG}/ml enzyme solution (3,51 U_{NiPAB}/ml). Using immobilised enzyme 5 to 25 mg wet carrier were weighed into a stirring cuvet and suspended in 975 µL PBS with Hellma Cuv-o-stir Model 333. The rest of the procedure was conducted as described above (Galunsky *et al.* 1994).

3.3 HPLC ANALYSIS.

Samples of 25 µL were injected to a 2 ml/min flow of 95 % 67 mM KH_2PO_4 pH 6.0 and 5 % MeOH (Pump Kontron 322 System). The HPLC separation was conducted by a Merck LiChroSpher RP8 125/5 kept at 56 °C. HPG, HPGA, 6-APA and Amox eluted had retention times of 0.8, 0.9, 1.2 and 1.8 min, respectively. The peak areas were detected at 225 nm and quantified using calibration standards (Kontron HPLC 322 Detector and Kontron software).

3.3 SOLUBILITY MEASUREMENTS.

Water was added to a given amount of solid reactant. The pH - value was adjusted by addition of 0.1 M HCl or NaOH and the sample was thermostated in a water bath (5 to 35 °C) under constant stirring. After at least 4 hours samples of the supernatant were filtered through a 0.2 µm syringe filter (Nalgene Cat. 176, Nylon), diluted 1:100 in eluent and analysed by HPLC (Kontron HPLC 360 Autosampler). This was repeated until the concentration of the supernatant remained constant.

3.4 DETERMINATION OF K_{CAT} AND K_M.

Varying concentrations of amoxicillin (0.2 to 20 mM) in PBS 0.2 M, pH 7.5 were hydrolysed by soluble penicillin amidase in a 1 ml stirred tank reactor with thermostatic jacket. The initial rate of product formation (4 to 6 samples in less than 60 min) was observed by HPLC. The parameters were determined by non-linear regression analysis using GraFit 3.0 (Erithacus Software Ltd.). The enzyme concentration was determined by active site titration using PMSF as penicillin amidase inhibitor (Svedas *et al.* 1977) and subsequent activity assay.

3.5 SYNTHESIS OF AMOXICILLIN IN HOMOGENEOUS REACTION.

Substrate solution (4 ml HPGA and 6-APA both in the concentration range 10 to 100 mM in water or 50 to 100 mM PBS, pH 5.5 to 7.5) were added to a thermostated (5 to 35 °C) and magnetically stirred glass vessel. The reaction was started by addition of soluble PA (0.02 to 0.5 mg/ml). pH is kept constant by suitable buffering or titration with 0.1 M HCl or H_3PO_4 (Metrohm pH stat or Radiometer phm 290 and abu 901). The reaction progress was observed with an online HPLC where the samples are drawn, diluted and injected automatically.

3.6 SYNTHESIS OF AMOXICILLIN IN HETEROGENEOUS REACTION.

A given amount of solid reactants (30 to 1200 mM 6-APA and HPGA) is suspended in 10 ml buffer or water in a thermostated (5 to 25 °C) glass stirred tank reactor (In case of soluble enzyme: Magnetic stirrer Omnilab RW 3001 K, in case of immobilised enzyme: Janke-Kunkel RW 10 R with propeller stirrer). The pH was controlled using 0.1 M HCl or 1 M H_3PO_4. The reaction is started by addition of enzyme. The reaction progress was observed by HPLC. At regular time intervals (15 to 120 min) samples were withdrawn. For total concentration samples are taken with a widened pipette and the solid reactants are dissolved by addition of an appropriate amount of MeOH. The soluble enzyme was inhibited by addition of a 5 to 10 fold excess of PMSF. Immobilised enzyme was removed by filtration through a 0.2 µm syringe filter. The supernatant concentration samples were filtered through a 0.2 µm syringe filter (Nalgene Cat. 176 Nylon). Soluble enzyme was then inhibited using PMSF. The samples were diluted and analysed using the autosampler (Kontron HPLC 360).

3.7 SUSPENSION PICTURES.

Diluted (1:5 with water or PBS) reaction solution was observed with a Leica Diavert Microscope under 10-fold magnification. Images of the solution and of a length scale were taken with a digital camera (Hitachi KP-M1K/K and software from Scanalytics, MA) and saved as *.tif format.

3.8 CONJUGATION WITH FLUORESCENT DYES.

Each 3 µmol FITC and TRITC were dissolved in 150 µL of DMSO, and diluted 1:100 with Carbonate buffer, pH 9.0, I = 0.1 M (according to manufacturer). To e.g. 2 ml carrier suspension 1 ml dye mixture was added to yield 67 µM dye concentration. The mixture was incubated for 1 hour at room temperature. Excess dye was removed by washing with storage buffer. The samples were protected against light exposure.

3.9 CLSM (CONFOCAL LASER SCANNING MICROSCOPY) IMAGE ACQUISITION.

The CLSM system used was the confocal microscope DM IRBE with acquisition and evaluation software TCS NT, both from Leica, Heidelberg, DE. The excitation light source was a Kr/Ar – laser. The object was observed with 10-fold magnification. The emitted light was filtered through a FITC / TRITC filter set. After adjusting the reactor position, both the photo-multiplier voltage and the initial and final position of the z-drive were adjusted under continuous (x/y-) scan. Image stacks were taken for calibration, bleaching and reaction experiments. Each slice was scanned 4 times and averaged. The stack size was 8 slices of 512 x 512 pixel, corresponding to 1000 µm x 1000 µm real size.

3.10 REACTOR CYCLE:

The calibration buffers and reaction solutions were pumped (Watson-Marlow 101 U) at a constant flow rate of ~20 ml/min through a modified chromatography column packed with immobilised enzyme. This reactor was a DURAN® glass cylinder (d_{inner} = 11.1 mm, d_{outer} = 11.3 mm), lengthways face ground and polished to roughly cover glass thickness (250 - 350 µm) and closed with chromatographic fittings. The reactor was taped onto the microscope table.

3.11 IMAGE PROCESSING:

The quantitative evaluation was conducted with the physiology module of the TCS NT software. ROIs (Regions of interest) of constant size were defined, both for several particles and as background (either uncoloured particle or void volume). The background signal was subtracted automatically from each channel (FITC, TRITC), and the ratio of FITC to TRITC channel was calculated for each slice. Post-processing of the data was conducted with Microsoft Excel 97.

3.12 PH CALIBRATION AND PH MEASUREMENT USING CLSM.

Buffer solutions (PBS pH 6 to 8 in 0.4 unit steps, I = 50 mM + 1 M KCl to reduce double layer influences at the particle surface) and reaction solutions (PenG hydrolysis: 100 mM PenG in PBS I = 50 mM, pH 8.0; Cex-synthesis: each 200 mM (20 mM) 7-ADCA and PGA in water or PBS I = 20 mM, pH 7; Amox-synthesis: each 20 mM HPGA and 6-APA in water or PBS, I = 20 mM, pH 6.5), respectively, were pumped subsequently through the fixed bed reactor. Images were taken 1 to 3 minutes after solution change. The FITC/TRITC ratio was correlated to the pH using the dissociation equation. The pK value was assumed to be constant and known (6.5) throughout the data set.

3.13 BIPOLAR MEMBRANE MODULE. CONSTRUCTION AND OPERATION.

The bipolar membrane module consists of six 5 mm Plexiglas frames with each two inlet and outlet tubes serving as compartments and two Plexiglas end sheets supporting the electrodes (anode titanium with RuO modified surface, cathode V2A stainless steel). The compartments are separated by the ion exchange and bipolar membranes (~ 17.4 cm^2) sealed with silicon sheets and clamped by a metal end sheet. Four pumps (Watson Marlowe 502 S, Watson Marlowe 501 U, ismatec MV JPN 4 and B. Braun FE 411 or Watson Marlowe 101 U) were used to circulate the electrode compartments, the rinsing compartments and the anode and cathode compartments and their respective supplies through silicon tubes. The power supply was ES 030-5 from Delta Elektronika, SE.

3.14 MODULE CHARACTERISATION.

Varying the medium (aqueous salt solution with different buffer amount), the flow rates and the applied current, the pH progress in the anode and cathode supply was observed both in batch and continuous mode, until constant pH values were found or steady state was achieved. At regular time intervals, the current and the pH value in the anode and cathode compartment were measured. The current and voltage measurements were conducted with a Fluke 77 Multimeter. pH measurements were conducted with Schott pH Meter CG 822 using a Festo electrode.

4. Results and Discussion

4.1. SOLUBILITY AND RATE OF DISSOLUTION

For all components involved in the kinetically controlled synthesis of amoxicillin the solubility in water was determined as a function of pH and temperature to define the concentration limits for the homogeneous and heterogeneous reactions (Figure 2). The solubility of HPG in water at 25°C was constant around 105 mM in the pH range investigated, indicating no charge variation in the molecule under those conditions. This corresponds well to the minimal solubility (115 mM) and the pK values (2.2 and 9.2) reported by Diender *et al.* (1998). From the solubility curve of HPGA, only one dissociating group could be found. Its pK value lies in the range of 7.4, i.e. the dissociation of the second amino group in the molecule is much easier than in HPG. The minimal solubility amounts to roughly 2.4 mM (1). For 6-APA and Amox only few data points were taken to assess the method by the available literature data. From the data of Diender *et al.* (1998) an intrinsic solubility of 10.5 mM for 6-APA (pK_1 = 2.5, pK_2 = 4.9) and of 7.7 mM for Amox in water (pK_1 = 2.9, pK_2 = 7.4) could be determined using (1). A good agreement was observed (Figure 2).

The solubility of all components is only slightly dependent on temperature. The heat of dissolution in the range from 5 to 35 °C was determined to 9.8 and 14.3 kJ/mol for HPG and HPGA, respectively (2). This differs by less than a factor 1.5 from the heat of dissolution of Amox (12.5 kJ/mol) and 6-APA (9 kJ/mol) (data from Tomlinson, Regosz 1985).

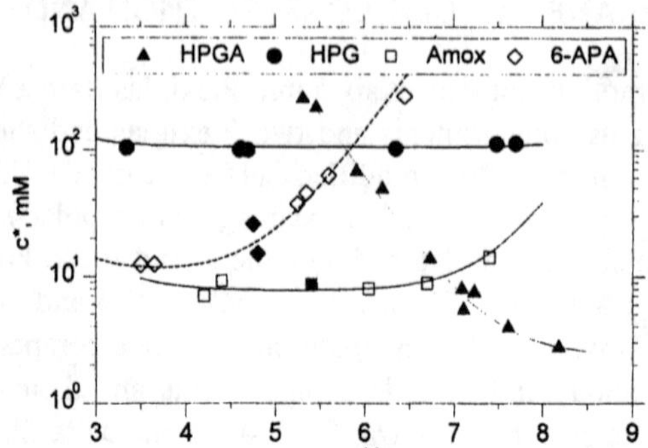

Figure 2. Solubility of HPG, HPGA, 6-APA and Amox as single solute in water as a function of pH. Closed symbols refer to data measured at 25 °C in water. Open symbols are values reported by Diender et al. (1998). The lines are fits according to (1).

To describe the dynamics of the dissolution process, information on the available surface area and thus the particle size is needed. Classic techniques as the Malvern Particle Sizer and sedimentation analysis fail for soluble species. Thus, the initial particle size of the HPGA suspension was estimated from microscopic images. Figure 3 A shows the needle like crystals. Their width is maximal 10 μm, their length maximal 100 μm. The diameter of the surface equal sphere is estimated to ~ 24 μm.

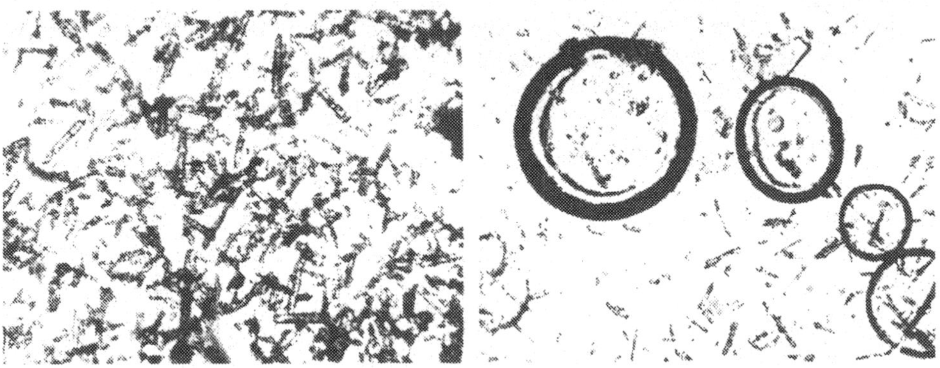

*Figure 3. Microscopic image of 1200 mM HPGA suspension in 400 mM dissolved 6-APA before the reaction start (**A**) and after ~ 9 hours (**B**). The suspension was diluted 1:5 with water. The circles in **B** are the immobilised enzyme particles (Eupergit PcA 600). Aqueous solution was adjusted to pH 6.5 at 5 °C.*

For a lower estimate of the mass transfer rate, typical conditions of pH 6, 25 °C and 400 mM HPGA are assumed. From (1) a solubility of 60.3 mM is calculated. The suspended substrate crystals have a mass concentration of 56.5 g/L with a volume specific surface of 17550 m²/m³. With (4 b) one estimates a terminal flow velocity $u_{T,S}$ of $7.18 \cdot 10^{-6}$ m/s using the dynamic viscosity of water of 0.014 Pa·s.. $\mathbf{Re_p}$ then amounts to $\sim 10^{-5}$. This is so small, that $u_T \sim u_{T,S}$. Thus a k_L value of $4.5 \cdot 10^{-5}$ m/s is obtained (4 a). The diffusion coefficient is calculated according to WILKE-CHANG (5) to $5.22 \cdot 10^{-10}$ m²/s using the molar volume of HPGA of 125.9 ± 3 cm³/mol. The total initial mass transfer then amounts to 47.6 mM/s (3).

4.2. KINETIC PARAMETERS OF AMOXICILLIN SYNTHESIS AND HYDROLYSIS

The k_{cat} and K_m values for amoxicillin hydrolysis by PA from *E. coli* were determined in order to evaluate the potential enzyme activity and the specificity of synthesis and hydrolysis of the condensation product. The measured k_{cat} amounts to two thirds of the one of PA from *X. citri* (Kato *et al.* 1980). The turnover numbers of amoxicillin, ampicillin (Ampi) and NiPAB hydrolysis are of the same order of magnitude (Table 1). They are roughly half of the value of the specific penicillin amidase substrate penicillin G. Larger differences are found for the K_m value. The lowest K_m values are found for PenG and NiPAB. Amox has a 5-fold lower K_m than HPGA, the substrate. I.e. the free enzyme preferably binds amoxicillin.

The specificity constants k_{cat}/K_m show that PA is more specific for the hydrolysis of the condensation product Amox than for of the activated substrate HPGA. The ratio of specificity constants of 2 indicates that the antibiotic product underlies severe hydrolysis at higher product concentrations.

Under the conditions in Table 1 (pH 7.5, $T = 25$ °C) the solubility of amoxicillin is evaluated to 17.4 mM and the one of HPGA to 4.8 mM. Thus, for the activity estimation, (6 b) applies for amoxicillin hydrolysis ($v = 23\ s^{-1} \cdot e_0$) and (6 a) for HPGA hydrolysis ($v = 19.5\ s^{-1} \cdot e_0$). Typical enzyme concentrations are in the range of 1 µM. Thus ~1.2 mM/min is the maximal substrate conversion.

Table 1. Kinetic rate constants for the hydrolysis of several substrates by penicillin amidase from E. coli *at 25 °C, pH 7.5 in 0.2 M PBS.* [1]*: this work.* [2]*: RIEKS (1997).*

Component	k_{cat}, s^{-1}	K_m, mM	k_{cat}/K_m, $s^{-1}\,M^{-1}$
Amox[1]	23	2.8	8,200
HPGA[2]	65	16	4,100
Ampi[2]	28	9	3,100
PenG[2]	50	0.01	5,000,000
NiPAB[2]	26	0.016	1,625,000

4.3. EVALUATION OF AMOXICILLIN SYNTHESIS PROGRESS: PH, T, *I*-DEPENDENCE

The effect of reaction conditions (pH, *T*, and *I*), suspended substrates, and enzyme on the synthesis of amoxicillin was determined by evaluation of the reaction progress. Figure 4 shows a typical progress curve of a heterogeneous reaction. Both Amox and HPG were formed from the reaction start on without any lag time. Amoxicillin concentration travelled through the kinetically controlled maximum, whereas HPG concentration increased steadily. HPGA was completely consumed during the reaction. 6-APA concentration reached its minimum value when amoxicillin hydrolysis prevailed over its synthesis.

The measured dissolved concentrations of the substrate HPGA and the product Amox correspond to roughly double the calculated solubility of the components in water at pH 6.5 and 25 °C (21.5 mM and 8.67 mM). This may be caused by co-dissolution of the components. For a short period of time the dissolved amoxicillin concentration exceeded the expected solubility. Actually the hydration of the formed amoxicillin leads to amoxicillin trihydrate with a lower solubility (Grant, Higuchi 1990), which was determined above (Figure 2)

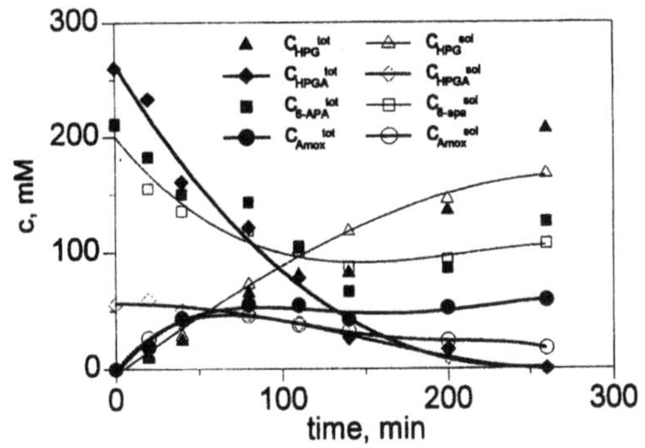

Figure 4. Progress curve of amoxicillin synthesis from each 200 mM HPGA and 6-APA catalysed by immobilised E. coli *PA (Eupergit PcA 600) at 25 °C and pH 6.5.*

In the subsequent text, the progress curves are described by the following parameters:

- The selectivity v_T/v_H or is the ratio of the initial rate of Amox formation over the initial rate of HPG formation. Since it is dependent on the concentration of the nucleophile 6-APA, sometimes the transferase to hydrolase ratio k_T/k_H is used instead.
- The enzyme activity at the specific reaction conditions is expressed in terms of HPGA hydrolysis and equals the sum of v_T and v_H. For standardisation it is divided by the NiPAB activity at 25 °C, pH 7.5 in 0.2 M PBS and denoted as $v_{0,ref}$.
- The product concentration an_{max} is the Amox concentration at the kinetically controlled maximum. The yield is the ratio of an_{max} and the initial 6-APA concentration.
- The space-time-yield STY or productivity of the reactor equals the maximal product concentration divided by the standardised time required to reach the kinetically controlled maximum.

In amoxicillin synthesis with free enzyme, the selectivity of PA decreased with increasing temperature (from 5 °C to 35 °C). The opposite was observed with respect to the activity and productivity, which were enhanced with temperature. From a VAN'T HOFF plot of standardised initial activity, the activation energy could be determined to 42.2 kJ/mol. Both effects are opposite, so that the maximal product concentration was independent of temperature (all data between 2.5 and 3 mM under the conditions in Figure 5). For amoxicillin synthesis with PA from *X. citri*, an optimal synthesis temperature of 20 °C with respect to the yield was found (Kato *et al.* 1980). For ampicillin synthesis with PA from *E. coli* the same activity and selectivity dependence as above was observed. Moreover, the yield was increasing with decreasing temperature (until -5 °C). The stability of the catalyst was enhanced more than the activity was reduced, thus improving the catalysts overall productivity (Baldaro 1991).

Similar to the temperature dependence, it was observed that PA selectivity decreases with increasing pH in the range of 6.5 to 8.5 and that the activity increased with pH. Marconi *et al.* (1975) showed the same for a pH range from 6 to 7. Again, for *X. citri* enzyme, a clear optimal pH of 6.5 with respect to the yield was determined (Kato *et al.* 1980). Ospina *et al.* (1996) studied the pH influence on ampicillin synthesis using PA from *E. coli* and showed a maximal selectivity at pH 6, decreasing both at lower and higher pH values, with a minimum at pH 7.5 and an increasing activity in the pH range from 5.5 to 7.2. The pH dependence of both selectivity and activity can be explained by the dissociation of functional groups involved in substrate binding and conversion in the active site (Kasche, Michaelis 1990). As only the non-protonated amino group of the nucleophile 6-APA binds into the active site, the conversion is reduced at pH values lower than 5.5.

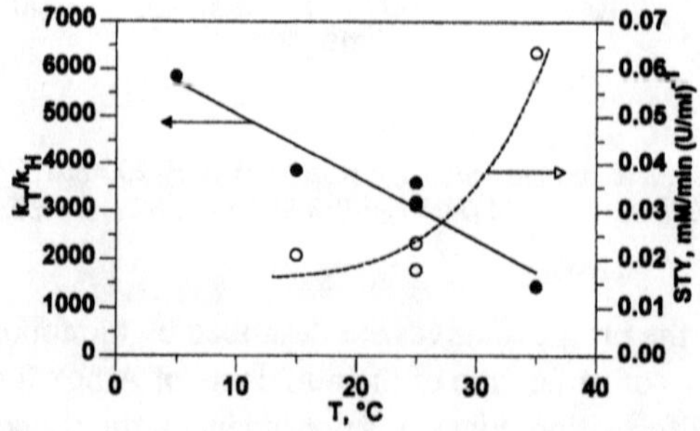

Figure 5. Temperature dependence of PA selectivity, expressed in k_T/k_H, *and of the STY. All reactions at pH 6.5 with 10 mM HPGA and 50 mM 6-APA using soluble PA from* E. coli. *One data point for space-time yield is lacking because the maximum was not properly reached.*

With immobilised PA the picture became more complex. The initial activity and the initial productivity increased as expected both with pH and temperature like for free enzyme. However, the selectivity showed both different pH dependence at low temperature (5 °C) and different temperature dependence at low pH (5.5). The performance of the maximal product concentration and the space time yield corresponded to the selectivity. This may be a hint that the enzyme activity was already strongly suppressed at pH 5.5. Then the increased selectivity could not counteract the reduced activity. The second explanation is that 6-APA was partly undissolved at the low pH experiments and that the reduced 6-APA concentration in the liquid phase influenced the selectivity (7).

Neither selectivity nor activity of amoxicillin synthesis were significantly dependent on ionic strength or buffer concentration. No enhancement of synthetic activity with reduced ionic strength could be observed in opposite to findings for amoxicillin synthesis with *X. citri* (Kato *et al.* 1980) and other β-lactam antibiotic syntheses (Kasche *et al.* 1984). This may be due to the fact, that the contribution of the substrates (20 - 400 mM) to the ionic strength was higher than that of the buffer (I = 0 - 200 mM) and that latter concentration was below measurable influence as in the cephalexin synthesis reported by Knauseder and Palma (1984).

4.4. EFFECT OF CONCENTRATION VARIATION AND PURE SUBSTRATE PHASE

The nucleophile variation showed the expected linear increase of the selectivity v_T/v_H with 6-APA concentration (Figure 6). No effect of the nucleophile on the transferase to hydrolase ratio k_T/k_H was found. The errors in k_T/k_H estimation increase with decreasing nucleophile concentration (Kasche *et al.* 1984).

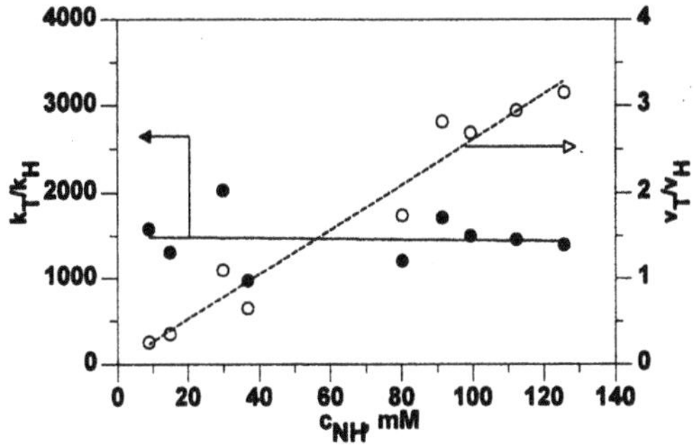

Figure 6. Nucleophile specificity of free PA from E. coli *(Sigma) shown by the selectivity* v_T/v_H *and the transferase to hydrolase ratio* k_T/k_H. *All experiments were done at 25 °C, pH 7.5, 10 mM HPGA.*

With the same data a decrease in enzyme activity (HPGA consumption rate) with increasing nucleophile was observed. This indicates nucleophile inhibition of the conversion. The inhibition constant was determined to $K_i = 5.8$ mM from a LINEWEAVER BURK plot, which corresponds well to the data of Svedas (1980 b). Both maximal product concentration and productivity increased with the nucleophile concentration. Improved maximal concentration and reduced yield with increasing 6-APA concentration was common to all antibiotics and enzymes investigated (Marconi *et al.* 1975, Kato *et al.* 1980, Knauseder, Palma 1984).

Increasing the HPGA concentration means exceeding the solubility limit, which is calculated to be 21.5 mM at pH 6.5 and 25 °C. The initial rate of HPGA hydrolysis increased with the concentration (Figure 7). An apparent K_m of HPGA could be determined, which amounted to 29 mM with a $V_{max,ref}$ of 0.21. This corresponds to the K_m extrapolated by Rieks (1997) from homogeneous reaction. The selectivity decreased with increasing HPGA concentration, although it should be influenced solely be the conditions in the liquid phase. At the high concentrations applied, substrate inhibition might contribute. The maximal product concentration increased with HPGA supply, but less than proportionally, so that the percentage yield with respect to HPGA was reduced. This might be due to the selectivity reduction caused by the 6-APA consumption in the later phase of the reaction. The maximal product concentration was found at roughly equimolar substrate supply.

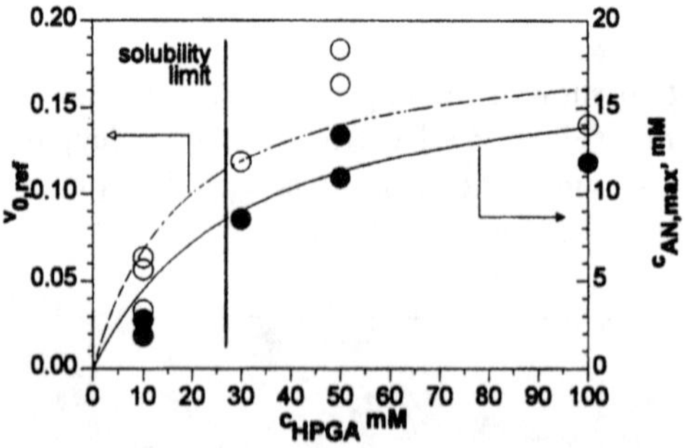

Figure 7. Dependence of initial rate of hydrolysis of HPGA and of maximal amoxicillin concentration on HPGA supply. Points on the left of the solubility limit are data from homogeneous reactions, on the right are from reactions with suspended HPGA. Lines are MICHAELIS-MENTEN plots. Other conditions are 25 °C, pH 6.5 and 50 mM 6-APA and soluble PA from E. coli.

Increasing both HPGA and 6-APA concentration simultaneously, the above effect was overcome. The final amoxicillin concentration increased strongly and the relative yield remained at 25 % with respect to the activated substrate. A considerable decrease in initial activity (rate of HPGA hydrolysis) was observed with increasing substrate concentration. This may be due to 6-APA inhibition as indicated before. The selectivity v_T/v_H increased while k_T/k_H decreased with increasing content of both substrates. For the reactions with immobilised enzyme, where the HPGA excess was considerable, the results were less pronounced. Again, for low pH and temperature the rules with respect to selectivity were inverted. Generally, the selectivity and activity in suspension systems was correlated more to the dissolved substrate concentration than to the total concentration.

4.5. COMPARISON OF SOLUBLE AND IMMOBILISED ENZYME

The effect of suspended substrate on free enzyme was studied by determination of the residual NiPAB activity in the supernatant. Only 25 % of the expected activity could be found. Converting the measured initial activity of HPGA consumption $v_{0,ref}$ to the theoretical NiPAB activity using the respective k_{cat} values, only 11 % of the expected activity could be found. This may be due to adsorption of the soluble enzyme onto the hydrophobic surface of the suspended HPGA particles. This problem may be easily overcome by immobilisation of the enzyme.

However, immobilisation leads to a considerable loss in selectivity: At pH 6.5, 25 °C and each 200 mM substrates, free enzyme had a selectivity of 4.2, immobilised enzyme only one of 1.2. This is caused partly by the coupled transport and reaction in the enzyme carrier, where the nucleophile depletes with increasing depth in the particle. Since the selectivity is linear with nucleophile concentration, the net particle selectivity decreases as well.

Since the benefit in activity exceeds the loss in selectivity, both the product concentration and the productivity of amoxicillin formation are much improved for immobilised enzyme. Thus, no alternative to the use of immobilised enzyme is given.

4.6. ACTIVITY AND SELECTIVITY OF IMMOBILISED ENZYMES

Table 2. Activity and selectivity of immobilised and free enzymes at 25 °C. NiPAB activity was detected in PBS pH 7.5, I = 0.2 M. HPGA activity and selectivity at pH 6.5, with each 50 mM HPGA and 6-APA.

Enzyme	NiPAB activity U/g wet weight	HPGA activity U/g wet weight	Selectivity v_T/v_H
Assemblase®	0.6	16.7	1.1
Eupergit PcA C250L	9.4	15.7	1.5
Eupergit PcA 600	8.1	18.5	1
Sclavo PA beads	4.5	25.7	1.1
Eupergit PBA	8	n.d.	n.d.
Trisoperl CPG	8-10	n.d.	n.d.
Sepharose	3.7	n.d.	n.d.
free PA from *E. coli*	63.2 U/ml	10.1 U/ml	~ 3

In order to choose a suitable enzyme for reactions in suspension, a range of industrially applied immobilised PA preparations were tested for amoxicillin synthesis selectivity (Table 2, first four lines). Obviously, the immobilised enzymes showed a much reduced selectivity of 1 to 1.5 compared to the free enzyme (~ 3) under the same conditions.

Although in all cases *E. coli* PA was immobilised, the enzymes exhibited different selectivity and different activities with respect to hydrolysis and synthesis. These parameters did not correlate. NiPAB activity is smaller than HPGA activity since the effectiveness factor is smaller for NiPAB than for HPGA (Kasche 1986).

The NiPAB activity of some prepared immobilisates is given for comparison (Table 2). The immobilisation yield calculated from the mass balances of the free enzyme in the incubation solution and the washing buffers amounted from 70 to 100 % for the different carriers. The best result was obtained for Eupergit C and the worst for Eupergit PBA. The CPG binding yield was very high (> 90 %), except for the largest particles with small pores, where diffusion hindrance of the enzyme might play a role. The immobilisation yields, expressed as ratio of the theoretical activity to the activity measured with the immobilised enzyme, amounts to only 10 to 20 %. This is again a hint for severe transport limitation (or bad mixing due to small density differences between carrier and fluid). The transport limitation, however, does explain the reduced activity, but not the reduced selectivity.

4.7. PH PROFILES IN IMMOBILISED BIOCATALYSTS UNDER REACTION

A pH value in the carriers different to that in bulk solution might explain the reduced overall selectivity. To study this hypothesis, the pH value under reaction is determined using confocal laser scanning microscopy (Spiess and Kasche, *subm.*). Therefore, a fluorescent pH indicator (FITC) is bound together with the reference substance TRITC covalently to the immobilised enzyme and the ratio of the fluorescence signals under flow of reaction solution compared with that under flow of calibration buffer solutions.

The shape of the profiles was as expected. For PenG hydrolysis the pH decreased from the particle surface to the core; for Assemblase® even below the pH range (< pH 5). For the synthesis reaction, the pH value increased from the surface to the core; for Eupergit PcA 600 a constant pH plateau was reached, for other carriers the pH increased monotonously. The pH profiles measured in CPG were unexpected, since all pH values were lower than the bulk pH. This is a hint for a reduced synthetic activity; eventually the pure hydrolysis of PGA to PG led to an acidification of the particle (Figure 8). The pH level and the shape of the pH profile obviously depend on the carrier, which may explain the different influence of immobilisation matrices on the selectivity of the biocatalyst.

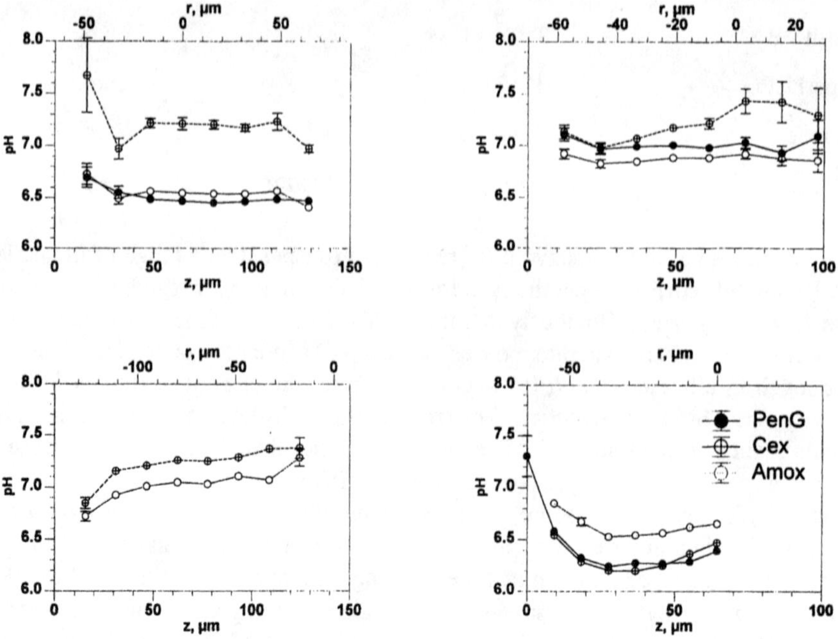

Figure 8. pH profiles in immobilised enzymes during the reaction. The reactions are PenG: Hydrolysis of 100 mM PenG in PBS, bulk pH 8, I = 50 mM, Amox: Synthesis of Amox from each 20 mM HPGA and 6-APA in water, bulk pH 6.5, Cex: Synthesis of Cex from each 200 mM PGA and 7-ADCA in water, bulk pH 7.0. ***A*** *Eupergit PcA 600, d_P = 175 µm.* ***B*** *Eupergit C - PA positively charged, d_P = 135 µm.* ***C*** *Assemblase®, d_P = 285 µm.* ***D*** *Trisoperl CPG (d_{Pore} = 30 nm), d_P = 188 µm.*

4.8. MASS TRANSFER LIMIT AND YIELD PREDICTION

In order to evaluate the reaction dissolution process, the process limitation must be identified. The observed dissolved concentration of the substrate HPGA was higher than or close to the calculated solubility in water. This corresponds both to (9 a) and (9 b), since the ratio of conversion to mass transfer is dwarfed by the $k_L a$ contribution and thus $c_A^{ss} \sim c_A^*$. Similarly, the ratio of the maximal observed HPGA conversion rate of 1.3 mM/min and the calculated mass transfer of ~ 48 mM/s results in a DAMKÖHLER number **Da** of ~ 0.0005 (10), indicating a purely reaction limited conversion. This relation holds for the whole observed parameter space. The calculations are supported by the observed immediate increase in product concentration upon reaction start. Dissolution limited behaviour in suspension conversion was observed only for 10 to 100 fold larger particles (Wolff *et al.* 1997, Lee *et al.* 1999).

The measured maximal product concentration was compared to the calculated one (8) to assess its use in suspension reactions. Assuming that the ratio of specificity constants does not vary with pH and temperature, the values at pH 7.5 and 25 °C from Table 1 can be applied. The measured initial selectivity was assumed to be constant during the reaction. Generally, the use of total initial HPGA concentration gave an upper estimate of maximal product concentration, whose goodness decreased with both increasing HPGA and 6-APA concentration. For homogenous reaction, the HPGA concentration at the maximum should be used, eventually producing more accurate results. Using the dissolved initial concentration for the suspension reactions underestimated the yield, since it does not account for the permanent dissolution of substrate.

4.9. PROPOSAL FOR INTEGRATED REACTION SEPARATION PROCESS

The analysis of (8) under different reaction conditions revealed that the use of suspended substrates could raise the amoxicillin yield only up to ~ 25 % with respect to 6-APA and lower with respect to HPGA. Higher yields can only be achieved with an integrated process. Such a process taking advantage of the pH dependence of the solubility of the respective components (Figure 2) is sketched in Figure 9. Under the high pH value of e.g. 7.5 in the reactor chamber, HPGA is mostly suspended and a high excess of 6-APA is given in the liquid phase. The filtered supernatant is passed through the ED chamber and acidified until precipitation of Amox, e.g. at pH 6.5. Latter pH must be high enough that 6-APA does not precipitate yet. The filtrate is passed again through the ED module and the pH lifted to the original reaction pH. In principle the pH shift could also be accomplished be acid and base addition, however to the price of dilution and effective salt addition. Electrodialysis with bipolar membranes is the more economic and ecologic alternative.

The pH progress over time in such an electrodialysis module with bipolar membranes was observed for 350 ml 0.5 M NaCl solution circulating in the compartments adjacent to the BPM. The pH progress was reproducible and nearly independent on the initial pH values in the supply tanks. The pH progress curves apparently tend towards a constant pH value. High current yields of up to 100 % were obtained. The reactants of amoxicillin synthesis are buffering substances, which associate with protons upon acidification. They require a much higher acid amount to reach a lower pH value during titration than a salt solution. This was shown for solutions with increasing buffer concentration, which are pumped through the module and where the steady state pH values were observed (Table 3). Indeed, the observed pH shift reduced considerably with increasing buffer concentration.

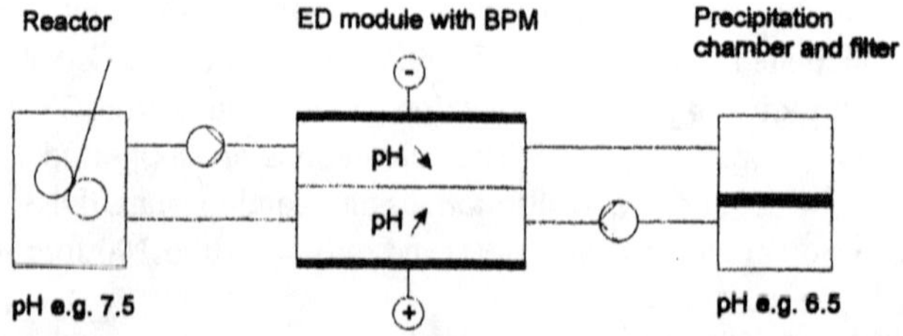

Figure 9. Schematic set-up of integrated reaction - separation process. A bypass stream is pumped from the reactor through the electrodialysis module where the pH is reduced. This leads to precipitation of the product Amox which is retained by the filter. The fluid is pumped back to the reactor and readjusted in pH.

Table 3. Effect of the buffer concentration on the steady state pH difference in a bipolar membrane module. Membrane (Tokuyama Soda) area ~ 17 cm², supply volume each 350 ml, volumetric flow ~ 100 ml/min, constant current ~ 100 mA.

Medium	$pH_{Reaction}$	$pH_{Precipitation}$
0.5 M NaCl	9,4	3,9
2 mM PBS + 0.5 M NaCl	8,75	6,45
20 mM PBS + 0.5 M NaCl	7,2	7,0
50 mM PBS + 0.5 M NaCl	7,05	6,95

To define suitable pH values for reaction and product precipitation, the definition of the kinetically controlled maximum of Amoxicillin formation is analysed (8). The intrinsic Amox solubility of 7.7 mM in water has to be exceeded to precipitate Amox. In a suspension system, the HPGA concentration at the kinetically controlled maximum corresponds to the solubility of HPGA at the reaction pH value. The pH dependent kinetic parameters, the selectivity parameter $K_{i,nh}$ and the specificity constants k_{cat}/K_m of HPGA and Amox, are determined by the catalyst. Although these kinetic parameters are not known in detail for Amox synthesis, the magnitude and pH dependence of their product $1/K_{i,nh} \cdot (k_{cat}/K_m)_{HPGA} \cdot (k_{cat}/K_m)_{Amox}^{-1}$ may be estimated, assuming $hpga_{max} \sim 1/3\ hpga_0$ (Marconi et al. 1975) and $apa_{max} \sim apa_0$. At 25 °C values of ~ 0.01 mM^{-1} for free enzyme and ~ 0.005 mM^{-1} for immobilised enzyme (Eupergit PcA 600) were found for this lumped parameter independent on pH. Thus, the 6-APA concentration required to exceed the solubility limit of Amox may be calculated as a function of pH. This curve is shown in Figure 10 together with the 6-APA concentration required to exceed the Amox solubility at the respective pH value.

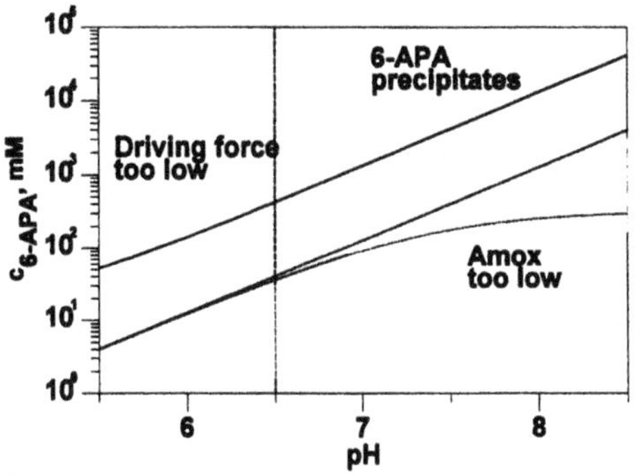

Figure 10. Process window (grey shaded) of 6-APA concentration over pH value. The left line indicates the solubility of 6-APA in water at 25 °C. The right lines indicate the minimum 6-APA concentration to reach the minimal Amox solubility (7.7 mM, lower line) and the Amox solubility at the respective pH value (upper line). The vertical distance between those lines shows the maximal concentration difference for the precipitation.

These curves form the lower limit of nucleophile concentration at the given reaction pH value. For the precipitation pH value, the 6-APA solubility determines its the maximal concentration, since only Amox shall be precipitated. Thus, a process window (Woodley, Titchener-Hooker 1996) with respect to the pH value of reaction and precipitation and the nucleophile concentration is defined. Below a reaction pH of 6.5, the driving force for the precipitation, i.e. the concentration difference between saturation at reaction pH value and solubility at precipitation pH value, becomes too small. The activity optimum of PA lies around the isoelectric point of 6.7 - 7. With increasing pH value the selectivity decreases and thus the amount of hydrolysed activated substrate increases.

5. Conclusion and prospects

The kinetically controlled synthesis of amoxicillin from HPGA and 6-APA catalysed by *E. coli* penicillin amidase both in homogeneous and heterogeneous (with dissolved and suspended substrates) reaction systems was described systematically. pH, temperature, buffer strength and substrate concentrations as well as the catalyst type (free or immobilised enzyme) were varied and the progress curves of the synthesis were evaluated with respect to activity, selectivity and maximal product concentration.

The amoxicillin synthesis was dependent on pH and temperature as described for other β-lactam antibiotic syntheses. No buffer influence could be observed. An increase in substrate concentration led to an enhanced productivity of the reactor. It was shown that in the simultaneous dissolution and reaction of suspended HPGA the enzyme catalysed conversion was process limiting. Precipitated 6-APA limited the nucleophile concentration in the liquid phase and thus reduced the selectivity and the product concentration. Hence, optimum reaction conditions can be found where the HPGA total concentration is high and beyond solubility limit, allowing a high molar ratio of 6-APA and HPGA in the liquid phase. The solubility limit of 6-APA should not be exceeded.

The activity of soluble enzyme was reduced in comparison with its theoretical value due to adsorption. Thus, immobilised enzyme should be used. However, the selectivity of immobilised enzyme is generally lower than the selectivity of soluble enzyme due to mass transfer limitations. The second factor is an indirect effect of diffusion limitation: Due to the 6-APA consumption in the transfer reaction, the pH rises with increasing conversion. This led to a pH value in the carrier, which were up to 0.8 pH units higher than in the bulk solution. Since the selectivity is pH dependent and decreases with increasing pH, this effect amplifies the pure diffusion limitation.

The integration of the production with product separation and substrate recycling is required to increase the yields. The feasibility of process integration of the kinetically controlled synthesis of amoxicillin from HPGA and 6-APA in suspension systems with the selective precipitation of Amox using the pH shift generated by the BPM module in continuous mode was shown. A process window for the 6-APA concentration and the pH value could be sketched, which defined the suitable operating conditions for reaction and precipitation. Theoretically, the Amox yield could then be increased considerably with respect to both the activated substrate and the nucleophile.

References

Baldaro, E.M.: Effect of temperature on enzymatic synthesis of cephalosporins. pp. 237-240 in: Pandit, U.K.; Alderweireldt, F.C. (eds.): Bioorganic Chemistry in Healthcare and Technology (*NATO ASI Series A* **207**). New York: Plenum Press, 1991.

Bornscheuer, U.: Lipase-catalyzed syntheses of monoacylglycerols *Enzyme and Microbial Technology* **17**(7): 578-586, 1995.

Cao, L.; Fischer, A.; Bornscheuer, U.T.; Schmid, R.D.: Lipase-catalyzed solid phase synthesis of sugar fatty acid esters. *Biocatalysis and Biotransformation* **14**: 269-281, 1997.

Cerovsky, V.: Protease-catalysed peptide synthesis in solvent-free system *Biotechnology Techniques* **6**(2): 155-160, 1992.

Diender, M.B.; Straathof, A.J.J.; van der Wielen, L.A.M.; Ras, C.; Heijnen, J.J.: Feasibility of the thermodynamically controlled synthesis of amoxicillin. *Journal of Molecular Catalysis / B.* **5**(1-4): 259-254, 1998.

Eichhorn, U.; Bommarius, A.S.; Drauz, K.; Jakubke, H.-D.: Synthesis of Dipeptides by Suspension-to-suspension Conversion via Thermolysin Catalysis: From Analytical to Preparative Scale. *Journal of Peptide Science* **3**: 245-251, 1997.

Eichhorn, U.; Beck-Piotraschke, K.; Schaaf, R.; Jakubke, H.-D.: Solid-phase Acyl Donor as a Substrate Pool in Kinetically Controlled Protease-catalysed Peptide Synthesis. *Journal of Peptide Science* **3**: 261-266, 1997.

Erbeldinger, M.; Ni, X.; Halling, P.J.: Effect of Water and Enzyme Concentration on Thermolysin-Catalyzed Solid-to-Solid Peptide Synthesis. *Biotechnology and Bioengineering* **59**(1): 68-72, 1998.

Fregapane, G.; Sarney, D.B.; Vulfson, E.N.: Enzymic solvent-free sythesis of sugar acetal fatty acid esters. *Enzyme and Microbial Technology* **13**(10): 796-800, 1991.

Furui, M.; Furutani, T.; Shibatani, T.; Nakamoto, Y.; Mori, T.: A Membrane Bioreactor Combined with Crystallizer for Production of Optically Active (2R, 3S)-3-(4-Methoxyphenyl)-Glycidic Acid Methyl Ester. *Journal of Fermentation and Bioengineering* **81**(1): 21-25, 1996.

Galunsky, B.; Schlothauer, R.-C.; Böckle, B.; Kasche, V.: Direct Spectrophotometric Measurement of Enzyme Activity in Heterogeneous Systems with Insoluble Substrate or Immobilized Enzyme. *Analytical Biochemistry* **221**: 213-214, 1994.

Gill, I.; Vulfson, E.: Enzymic catalysis in heterogeneous eutectic mixtures of substrates. *Trends in Biotechnology* **12**: 118-122, 1994.

Grant, D.J.W.; Higuchi, T.: Solubility behavior of organic compounds (*Techniques of chemistry* **Volume XXI**). New York: John Wiley & Sons, 1990.

Halling, P.J.; Eichhorn, U.; Kuhl, P.; Jakubke, H.-D.: Thermodynamics of solid-to-solid conversion and application to enzymic peptide synthesis. *Enzyme and Microbial Technology* **17**: 601-606, 1995.

Harriott, P.: Mass Transfer to Particles: Part I. Suspended in Agitated Tanks *AIChE Journal* **8**: 93-102, 1962.

Janasek, D.; Spohn, U.: Poröses Glas als Enzymträger. *BIOforum* **21**: 108-109, 1998.

Kasche, V.: Mechanism and yields in enzyme catalysed equilibrium and kinetically controlled synthesis of beta-lactam antibiotics, peptides and other condensation products. *Enzyme and Microbial Technology* **8**(1): 4-16, 1986.

Kasche, V.; Galunsky, B.: Enzyme Catalyzed Biotransformations in Aqueous Two-Phase Systems with Precipitated Substrate and/or Product. *Biotechnology and Bioengineering* **45**: 261-267, 1995.

Kasche, V.; Haufler, U.; Zöllner, R.: Kinetic Studies on the Mechanism of the Penicillin Amidase-Catalysed Synthesis of Ampicillin and Benzylpenicillin *Hoppe-Seyler's Zeitung für Physiologische Chemie* **365**: 1435-1443, 1984.

Kasche, V.; Michaelis, G.: Peptide Bond Synthesis by Enzyme-Catalyzed Acyl-Transfer. pp. 81-100 in: Kleinkauf, H.; von Döhren, H. (eds.): Biochemistry of Peptide Antibiotics Recent Advances in the Biotechnology of beta-Lactams and Microbial Bioactive Peptides. Berlin: Walter de Gruyter, 1990.

Kato, K.; Kawahari, K.; Takahashi, T.; Igarasi, S.: Enzymatic Synthesis of Amoxicillin by the Cell-Bound α-Amino Acid Ester Hydrolase of *Xanthomonas citri*. *Agricultural and Biological Chemistry* **44**(4): 821-825, 1980.

Kitahara, K.; Fukui, S.; Misawa, M.: Preparation of L-malate from fumarate by a new process "enzymatic transcrystallization". *Journal of General Applied Microbiology* **6**(2): 108-116, 1955.

Knauseder, F.; Palma, N.: Enzymatic Synthesis of cephalexin by immobilized penicillin acylase from *E. coli*. pp. 431-438 in: *European Congress of Biotechnology*, 1984.

Könnecke, A.; Schellenberger, V.; Hofmann, H.-J.; Jakubke, H.-D.: Die Partitionskonstante als Effizienzparameter von Nucleophilen bei enzymkatalysierten kinetisch kontrollierten Peptidsynthesen. *Pharmazie* **39**(11): 785-786, 1984.

Kuhl, P.; Eichhorn, U.; Jakubke, H.-D.: Thermolysin- and Chymotrypsin-catalysed peptide synthesis in the presence of salt hydrates. pp. 513-519 in: Tramper, J.; Vermue, M.H., Beeftink, H.H.; Stockar, U. von (eds..): Biocatalysis in Non-Conventional Media Proceedings of an International Symposium Noordwijkerhout, 26-29 April 1992 (*Progress in Biotechnology* **8**). Amsterdam: Elsevier, 1992.

Kuhl, P.; Eichhorn, U.; Jakubke, H.-D.: Enzymic Peptide Synthesis in Microaqueous, Solvent-Free Systems. *Biotechnology and Bioengineering* **45**: 276-278, 1995.

Lee, D.-C.; Park, J.-H.; Kim, G.-J.; Kim, H.-S.: Modeling, Simulation, and Kinetic Analysis of a Heterogeneous Reaction System for the Enzymatic Conversion of Poorly Soluble Substrate. *Biotechnology and Bioengineering* **64**(3): 272-283, 1999.

Marconi, W.; Bartoli, F.; Cecere, F.; Galli, G.; Morisi, F.: Synthesis of Penicillins and Cephalosporins by Penicillin Acylase Entrapped in Fibres. *Agricultural and Biological Chemistry* **39**: 277-279, 1975.

Maxon, W.D.; Chen, J.W.; Hanson, F.R.: Simulation of a steroid bioconversion with a mathematical model. *Industrial Engineering and Chemistry Process Design and Development* **5**(3): 285-289, 1966.

Michielsen, M.J.F.; Reijenga, K.A.; Wijffels, R.H.; Tramper, H.; Beeftink, R.H.: Dissolution kinetics of Ca-maleate crystals: Evaluation for Biotransformation Reactor Design. *Journal of Chemical Technology and Biotechnology* **73**(1): 13-22, 1998.

Miller, T.L.: Steroid Fermentations. pp. 297-318 in: Moo-Young, M. (ed.): Comprehensive biotechnology. Oxford: Pergamon Press, 1985.

Mincheva, Z.; Stambolieva, N.; Petrova, K.; Galunsky, B.: Penicillin amidase-catalysed preparative synthesis of cephem-7-(2-benzoxazolon-3-yl-acetamido) - desacetoxycephalosporanic acid using a non-specific polyethyleneglycol-modified acyl donor. *Biotechnology techniques* **10**(10): 727-730, 1996.

Ospina, S.; Barzana, E.; Ramirez, O.T.; Lopez-Munguia, A.: Effect of pH in the synthesis of ampicillin by penicillin acylase. *Enzyme and Microbial Technology* **19**: 462-469, 1996.

Perry, R.H.; Green, D.W.; Maloney, J.O. (eds.): Perry's Chemical Engineers' Handbook, 7th edition. New York: McGraw-Hill, 1997.

Rieks, A.: Stereospezifität und chirale Erkennung von Enzymen: Optimierung der Enantioselektivität von Penicillinamidase durch Variation der Reaktionsparameter und der Primärstruktur. TU Hamburg-Harburg, Diss., 1997.

Spieß, A.C.; Kasche, V.: Direct Measurement of pH Profiles in Immobilised Enzymes During Kinetically Controlled Synthesis Using CLSM.. *Biotechnology Progress,* submitted.

Stambolieva, N.; Mincheva, Z.; Galunsky, B.; Kalcheva, V.: Penicillin amidase-catalysed transfer of low specific acyl moiety. Synthesis of 7-benzoxazolonylacetamidodesacetoxycephalosporanic acid. *Enzyme and Microbial Technology* **10**: 496-500, 1992.

Svedas, V.K.; Margolin, A.L.; Sherestyuk, C.F.; Klesov, A.A.; Berezin, I.V.: *Bioorganicheskay Khimia* **3**: 543-546, 1977.

Svedas, V.K.; Margolin, A.L.; Borisov, I.L.; Berezin, I.V.: Kinetics of the enzymatic synthesis of benzylpenicillin. *Enzyme and Microbial Technology* **2**: 313-317, 1980.

Takahashi, S.: Microbial Production of D-p-Hydroxyphenylglycine. pp. 269-279 in: Aida, K.; Chibata, I.; Nakayama, K.; Takinami, K.; Yamada, H. (eds.): Biotechnology of Amino Acid Production. (*Progress in industrial microbiology* **Vol. 24**). Amsterdam: Elsevier, 1986.

Tomlinson, E.; Regosz, A.: Beta - lactam - antibiotics. *IUPAC Solubility data series* **16/17**, 1985.

van der Werf, M.J.; Hartmans, S.; van den Tweel, W.J.J.: Effect of maleate counterion on malease activity: Production of D-malate in a crystal-liquid two-phase system. *Enzyme and Microbial Technology* **17**(5): 430-436, 1995.

Walker, L.P.; Wilson, D.B.: Enzymatic Hydrolysis of Cellulose: An Overview. *Bioresource Technology* **36**: 3-14, 1991.

Wolff, A.; Zhu, L.; Kielland, V.; Straathof, A.J.J.; Jongejan, J.A.; Heijnen, J.J.: Simple Dissolution-Reaction Model for Enzymatic Conversion of Suspension of Solid Substrate. *Biotechnology and Bioengineering* **56**(4): 433-440 ,1997.

Woodley, J.M.; Titchener-Hooker, N.J.: The use of windows of operation as a bioprocess design tool. *Bioprocess Engineering* **14**: 263-268 (1996).

Part 3

PRODUCTION OF THERAPEUTIC ANTIBODIES

NEW RECOMBINANT BI- AND TRISPECIFIC ANTIBODY DERIVATIVES.

NICO MERTENS*, REINILDE SCHOONJANS, AN WILLEMS, STEVE SCHOONOOGHE, JANNICK LEOEN AND JOHAN GROOTEN

*Molecular Immunology Unit, Department of Molecular Biology, Flanders Interuniversity Institute of Biotechnology (VIB), Ghent University, K.L.Ledeganckstraat 35, B-9000 Gent, Belgium, *Tel.: ++-32-9-264 5134, Fax : ++-32-9-264 5348, e-mail : Nico.Mertens@DMB.RUG.AC.BE*

Summary

Bispecific antibodies (BsAb) are promising therapeutic tools in tomorrow's medicine. When constructing BsAbs, the final molecular size should be large enough to avoid rapid renal clearing, but small enough to allow efficient tissue distribution. In order to produce such intermediate sized BsAb, a good heterodimerisation technique will improve existing production methods. When considering recombinant expression of BsAbs, the heterodimerisation motif can be incorporated into the molecule. Recombinant BsAb can e.g. be made by fusing single chain variable fragments (scFv) to a heterodimerisation domain. We compared the efficiency of the isolated CL and CH1 constant domains with complete Fab chains to drive heterodimerisation of BsAbs in mammalian cells. We found that the isolated CL:CH1 domain interaction was inefficient for secretion of heterodimers. However, when the complete Fab chains were used, secretion of a heterodimerised bispecific antibody was successful. By C-terminal fusion of scFv molecules to the Fd- and the L-chains efficient heterodimerisation in mammalian cells was obtained and a novel intermediate sized, disulfide stabilised BsAb could be efficiently produced. Since the Fab chain encodes a binding specificity on its own, bispecific (BsAb) or trispecific (TsAb) antibodies can be made. This gave rise to disulphide stabilised Fab-scFv BsAb (Bibody) or Fab-(scFv)2 TsAb (Tribody) of intermediate molecular size. Heterodimerisation of the L and Fd-containing fusion proteins was very efficient, and up to 90% of all secreted antibody fragments was in the desired heterodimerised format. All building blocks remained functional in the fusion product, and the bispecific character of the molecules as well as the functionality was demonstrated.

A. Van Broekhoven et al. (eds.), Novel Frontiers in the Production of Compounds for Biomedical Use, 195–208.
© 2001 *Kluwer Academic Publishers. Printed in the Netherlands.*

Due to the high heterodimerisation efficiency, the ease of purification of the desired product from by-products and the lack of post-production processing, this method for producing bi- or trispecific antibodies in mammalian cells could become a method of choice for the production of intermediate sized trispecific antibodies, BsAb with monovalent or bivalent binding for one antigen, or immunoconjugates.

1.Introduction

Bispecific antibodies (BsAb) are versatile tools in the development of new, experimental therapies of various diseases. Typically, one part of the BsAb specifically recognises a target molecule or cell (e.g. a cancer cell), while the other part may be directed towards an enzyme, a toxin, a virus or an effector cell (e.g. a cytotoxic T cell).

The problem of producing functional BsAb in sufficient quantity and with high purity is still a hurdle to be taken when considering clinical applications. Production of bispecific antibodies is usually achieved by either chemical crosslinking of the monovalent halves of antibodies (Brennan et al., 1985), complete monoclonal antibodies ($Bs(IgG)_2$) (Segal et al., 1986) or Fab' fragments ($Bs(Fab')_2$). Alternatively, monovalent bispecific IgG molecules could be obtained by the hybrid hybridoma technique (BsIgG) (Milstein and Cuello, 1983). These techniques however require extensive post-production steps, not at least to purify the bispecific moiety from all by-products formed.

Recombinant DNA methodology and antibody engineering techniques have already contributed in facilitated production in mammalian or bacterial expression systems. One approach uses protein engineering of the interface of the immunoglobulin CH3 domains in order to obtain bispecific monoclonal antibodies ("knobs-into-holes" mutagenesis, Ridgway et al., 1996).

Smaller antibody fragments based on the variable regions of the antibodies were constructed that could be efficiently produced in bacterial production systems (Holliger and Winter, 1993; Horn et al., 1996; Zhu et al., 1996). These smaller antibody derivatives could also be dimerised to bispecific molecules ($Bs(scFv)_2$) by chemical cross-linking (Adams, et al., 1993; Kipriyanov et al., 1994, Luo et al., 1997) or chemical coupling to heterodimerising molecules (Chaudri et al., 1999). However, a more efficient preferential heterodimerisation at the time of production will improve the course of downstream processing. Therefore, bispecificity was engineered by direct genetic fusion ($Bssc(Fv)_2$) (Mack et al., 1995), or linked with a helical dimerisation domain ("mini-antibodies") (Pack and Pluckthun, 1992). By incorporating the helical heterodimerisation domain in the linker region connecting two scFv molecules in a $Bs(scFv)_2$, a dimeric bispecific "DiBi-miniantibody" could be produced (Muller et al., 1998a). These heterogeneous heterodimerisation extensions, however, also possibly increase the immunogenicity of the recombinant antibody. It was found that scFv molecules had a tendency to form dimers (Whitlow et al., 1994), and that by shortening the polypeptide linker a stable dimer of the scFv could be obtained ("diabody", Holliger et al., 1993). By permutating the variable domains of two different antibodies (e.g. V_HA-V_LB and V_HB-V_LA), a bispecific diabody can be formed. To increase the formation of the heterodimer (since also the non-functional homodimers are formed), the engineering of complementary "knobs-into-holes" mutations (Zhu et al., 1997) or the formation of single chain diabodies (scDb, (Alt et al., 1999; Kipriyanov et al., 1999) has been proposed.

Table 1:Different expression modes to obtain bispecific antibodies. (a) Molecular weight (MW, in kDa) is roughly estimated. (b) Valency. (c) Key references are: 1: Brennan et al., 1985, 2: Milstein and Cuello, 1983, 3: Ridgway et al., 1996, 4: Karpovsky et al., 1984, 5: Glennie et al., 1987, 6: Kostelny et al., 1992, 7: Kipriyanov et al., 1994, 8: Pack and Pluckthun, 1992, 9: Mack et al., 1995, 10: Hayden et al., 1994, 11: Muller et al., 1998a, 12: Holliger et al., 1993, 13: Zhu et al., 1997, 14: Kipriyanov et al., 1999, 15: Alt et al., 1999, 16: Hu et al., 1996, McGregor et al., 1994, 17: Muller et al., 1998b, 18: Coloma and Morrison, 1997. 19: Schoonjans et al., submitted.

Product	Method	MW [a)]	Val[(b)]	Specific hetero-dimerisation	Fc-tail	Ref [(c)]
BsIg *(hybrid biMab)*	*Chemical reassociation of monovalent L:H fragments*	150	1x1	No	Yes	1
	Hybrid hybridoma	150	1x1	No	Yes	2
	Co-expression of complementary (knobs-into-holes) H-chains	150	1x1	Yes	Yes	3
Bs(IgG)2	*Chemical crosslinking of 2 moAb (heteroconjugate)*	300	2x2	No	Yes	4
Bs(Fab')2	*Chemical crosslinking of 2 Fab'*	100	1x1	No	No	5
	Heterodimerisation via helical domains	100	1x1	Yes	No	6
Bs(scFv)2	*Chemical coupling of 2 scFv*	50	1x1	No	No	7
(mini-Abs)	*Coupling via fusion of a small heterodimerisation domain to svFv*	60	1x1	Yes	No	8
Bssc(Fv)2	*Genetic coupling of 2 scFv*	50	1x1	Yes	No	9
Bs(scFv-Fc)	*Genetic coupling of scFv coupling to Fc*	100	1x1	No	Yes	10
Bs(scFv)2)2 *(BiDi-body)*	*Heterodimerisation of sc(Fv)2*	100	2x2	Yes	No	11
Diabody	*Stable association of VH(a)-VL(b):VH(b)-VL(a)*	50	1x1	No Yes (engineered) Yes	No No No	12 13 14

Table 1:Different expression modes to obtain bispecific antibodies. (a) Molecular weight (MW, in kDa) is roughly estimated. (b) Valency. (c) Key references are: 1: Brennan et al., 1985, 2: Milstein and Cuello, 1983, 3: Ridgway et al., 1996, 4: Karpovsky et al., 1984, 5: Glennie et al., 1987, 6: Kostelny et al., 1992, 7: Kipriyanov et al., 1994, 8: Pack and Pluckthun, 1992, 9: Mack et al., 1995, 10: Hayden et al., 1994, 11: Muller et al., 1998a, 12: Holliger et al., 1993, 13: Zhu et al., 1997, 14: Kipriyanov et al., 1999, 15: Alt et al., 1999, 16: Hu et al., 1996, McGregor et al., 1994, 17: Muller et al., 1998b, 18: Coloma and Morrison, 1997. 19: Schoonjans et al., submitted.

Product	Method	MW [a)]	Val[(b)]	Specific hetero-dimerisation	Fc-tail	Ref [(c)]
				(scDiabody)		
Dimerised **sc-Diabody**	*Single chain diabodies coupled to Ig homodimerisation domain (Fc or CH3)*	125 - 150	2x2	Yes	Yes/ No	15
Bs(scFv-C) (*Minibody*)	*Heterdimerisation via genetic coupling to constant Ig-domain*	75	1x1	No (CL, CH3) Yes (CL:CH1)	No No	16 17
Bs(sIg-scFv)	*Genetic coupling of scFv to CH3 of IgG*	200	2x2	Yes	Yes	18
Bs(Fab-scFv)2	*Genetic coupling of scFv after hinge (Fab')2*	150	2x2	Yes	No	18
Bs(Fab-scFv) (*Bibody*)	*Genetic coupling of scFv to Fab-chain (no hinge)*	75	1x1	Yes	No	19
Bs(Fab-(scFv)2) (*Tribody*)	*Genetic coupling of 2 scFv's to Fab-chain (no hinge)*	100	1x1x1	Yes	No	19

The downsizing of the bispecific antibody derivatives ensured a better tissue penetration as compared with full-sized antibodies, but it was found that due to their small size the bispecific scFv molecules are generally cleared too rapidly from the body to allow efficient accumulation at the tumour site (Venkatachalam and Rennke, 1978, Milenic et al., 1991). It was also shown that an intermediate sized molecule compromises between better tissue penetration while avoiding rapid body clearance in the kidney (Hu et al., 1996). To this end, both the scFv molecules were extended with a non-immunogenic constant immunoglobulin domain, such as CH3 (Hu et al., 1996) or CL (McGregor et al., 1994), to produce "minibodies". This model was further extended by using the heterotypic interaction of the CL and CH1 immunoglobulin domains to drive heterodimerisation of scFv molecules (Muller et al., 1998b). The latter approach offers several advantages. The heterodimerisation domain is non-immunogenic, circumvents

the need for protein engineering to achieve complementarity, and is stabilised by a natural disulphide bridge. However, CL domains are also known to homodimerise, and only 64% heterodimerisation was reported after expression in Escherichia coli. Production in mammalian cells of CL-CH1 heterodimers should improve on the efficiency and specificity of heterodimerisation, since this is enhanced by the chaperoning of the heavy chain-binding protein (BiP) to newly formed CH1 domain the endoplasmic reticulum (Hendershot et al., 1987). BiP is then displaced after expression of the light chain (L), by proper CL-CH1 interaction (Hendershot, 1990).
However, we found that in mammalian cells, the expression of the CL and CH1 domains is not sufficient to allow secretion, and only upon extension of these domains with variable domains (thus forming a complete Fab molecule), a product is formed. We thus propose a model for the production of bispecific antibodies based on the fusion of an scFv to the C-terminus of the Fab heavy chain. This results in a bispecific molecule of intermediate size (~80 kDa), which is formed by preferential heterodimerisation of the L chain with the Fd-scFv fusion. By also extending the L-chain with an scFv (or another type of molecule), trifunctional molecules (~110 kDa) could be created. These molecules are of intermediate size, can be expected to be made with a low immunogenicity, and are immediately produced as stable heterodimers. We believe this model to be a good candidate for production of bispecific antibodies and derivatives in mammalian cells, either in vitro or in vivo.

2.Material and Methods

2.1 CELL LINES

HEK293T, a human embryonic kidney cell line transfected with SV40 large T-antigen (SV40T tsA1609) (DuBridge et al., 1987) was used for transient eukaryotic expression. SP2/0-Ag14 are non-Ig secreting myeloma cells. TE2 cells are murine, CD3 positive Th1-type T-cells (Grooten and Fiers, 1989). MO4I4 cells are MO4 mouse fibrosarcoma cells transfected with the hPLAP gene (Hendrix et al., 1991). BCL1^{vitro} cells are myeloma cells expressing the BCL1 IgM ideotype antigen and adapted for in vitro passage. All cells were cultured as described; all medium components were from GibcoBRL (UK).

2.2 PLASMIDS AND GENE ASSEMBLY

Restriction enzymes and DNA modifying enzymes and polymerases were used as recommended by the manufacturers. DNA amplification was performed with Vent-DNA polymerase (New England Biolabs, MA). E6, B1 and 2c11 denote the genes or gene fragments of an αhPLAP, an αBCL1 and an αmCD3 (De Jonge et al., 1995) monoclonal antibody respectively. Expression plasmids were constructed in pCAGGS (Niwa et al., 1991). The cloning of the light of the parental αhPLAP moAb E6 (IgG2b/κ) in the vector pSV51E6L has been described previously (De Sutter et al., 1992). The E6Fd fragment encodes VH, CH1 and the first five amino acids (not containing cysteines) of the upper hinge region (EPSGP). Fusion with CH1 or CL was carried out before the "elbow" region of the Fab chains (EMKRAD and SAAKTT in the L and Fd chains respectively). Gene assembly was carried out by introduction of suitable restriction sites using modifying PCR primers. All PCR-derived fragments were sequence verified after cloning.

2.3 PRODUCTION AND PURIFICATION OF RECOMBINANT ANTIBODY FRAGMENTS

For transient expression, HEK293T cells were transfected according to the $Ca_3(PO_4)_2$ precipitation method. The cells were covered with supplemented DMEM containing 5 mg/l bovine insulin, 5 mg/l transferrin and 5 µg/l selenium (ITS) replacing the FCS. Medium was harvested every 48h after transfection.

For stable expression lines, SP2/0-Ag14 cells were electroporated, cultured in selective medium, subcloned and screened for production. The secreted Fab-scFv-$(His)_6$ protein was purified under native conditions from the culture supernatant using immobilised metal affinity chromatography (IMAC). Finally, the protein was dialysed to PBS and stored at 4°C. Gel filtration was performed on a calibrated Sephacryl S300 column (Amersham Pharmacia Biotech).

2.4 T-CELL PROLIFERATION ASSAY

For the MO4I4 fibrosarcoma cells or the BCL1^{vitro} lymphoma cells spleen derived T-cells from syngenic C3H/HeOUico or BALB/c mice receptively were used. All mice were purchased from the Charles River group (Sulzfeld, DE). MO4I4 and BCL1^{vitro} tumour cells were pre-treated with 50 µg/ml mitomycin C at 37°C in the dark for 12h and 1.5h respectively. After removal of the mitomycin C, 5 x 10^5 treated cells were co-cultured with 1 x 10^5 splenic T-cells in a round bottom well in the presence of the indicated concentration of the bispecific Fab-scFv BsAb. After 48 h, the T-cells were pulsed with 0,5 µCi of tritium-thymidine ([^{3}H]TdR, 1mCi/ml, Amersham Pharmacia Biotech). 18 h later the cells were disrupted by freeze-thawing, the DNA was spotted on a filter and washed and the incorporated radioactivity was measured by scintillation counting (Top-Count; Packard, CO, USA).

3. Results and discussion.

3.1 HETERODIMERIZATION BY CL-CH1 INTERACTION IN EUKARYOTIC CELLS DEPENDS ON EXTENSION WITH VL AND VH DOMAINS.

Fab fragments can be efficiently produced in mammalian cells by co-expression of genes encoding the heavy chain Fab' fragment (Fd) and the light chain (L). Light chains have been shown to assemble on their own (Bence-Jones proteins, (Roussel et al., 1999), but the Fd chain is withheld in the endoplasmatic reticulum (ER) until the L chain is able to replace the CH1-associated heavy-chain binding chaperone BiP (Hendershot, 1990) (Figure 1A). This method of quality control by mammalian cells maximises the yield of

correct heterodimeric proteins. The production of bispecific minibodies based on CH1:CL heterodimerisation thus could benefit from this quality control mechanism to obtain a maximum amount of heterodimerised molecules. Since the CH1-domain containing molecules are not expected to be produced unless associated with a CL-domain containing molecule, the only molecules expected to be produced are the CL:CL homodimers and, more preferential, the CH1:CL heterodimers.

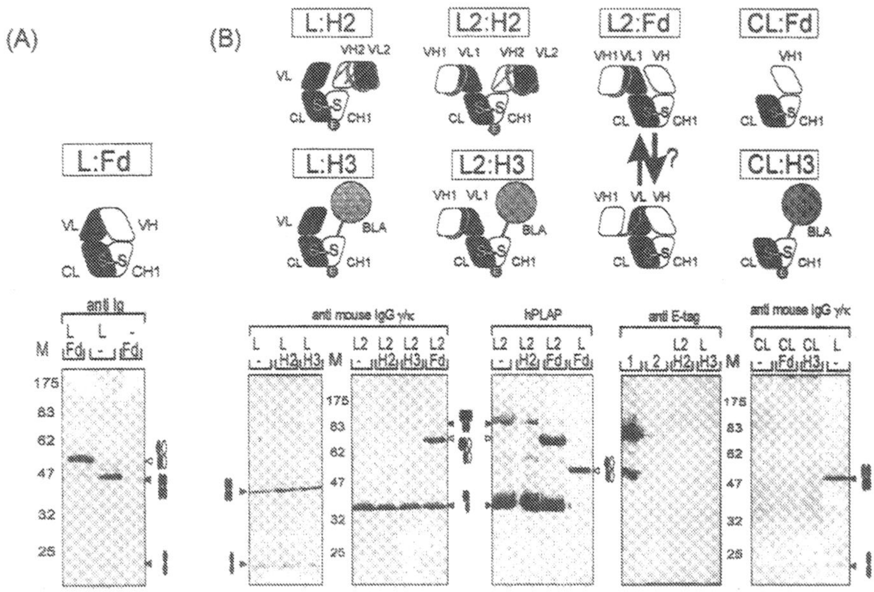

Figure 1: A) *Fab domains and L-chains can be successfully expressed in mammalian cells. Although the L chain on its own is produced (mainly as L:L dimers), the Fd chain however is never produced on its own, but only when associated with an L-chain.* ***B)*** *"Minibodies" made up by fusing scFv molecules to either the CL and CH1 domains are not expressed in mammalian cells. CL:CH1 domains were not able to be produced when their natural extension (VL or VH resp.) was replaced with either a scFv (L2:H2) or a different molecule, such as β-lactamase (L2:H3). Co-expression of such a scFv-C fusion with the natural complementary chain of the Fab (L and Fd), did not result in a detectable product when the scFv-CH1 (or β-lactamase-CH1) fusion was co-expressed with the L-chain. In the reverse case (scFv-CL:Fd) a heterodimer was observed, but can be explained by an alternative domain pairing, reconstituting a Fab chain. VH1 and VL1 (constituting the scFv fused to the CL domain) bind to hPLAP and can be revealed from a Western Blot. To show the lack of expression of the heterodimer, a blot developed with anti E-tag is also shown (CH1-domain contains an E-tag). Finally, also unfused CL-domains are not capable of releasing CH1-domains from the cells.*

Minibodies were constructed using the CL and CH1 domains to promote heterodimerisation of two different scFv molecules fused to their N-terminus. However, after cotransfection of expression plasmids for the scFv(αCD3)-CL (Figure 1B, L2) and the scFv(αhPLAP)-CH1 (Figure 1B, H2) fusion proteins, only the scFv(αCD3)-CL molecule could be detected (Figure 1B, L2:H2). This means the CH1-containing fusion could not be released from the ER by the CL fusion protein.

To eliminate possible sterical hindrance caused by the fusion of the scFv molecules to the CL and CH1 domains, a derivative was made with a longer flexible linker separating the fusion partner from CH1. For simplifying the analysis of proteins produced we used a β-lactamase molecule (bla) as a fusion partner (Figure 1B, H3), which allows for detection of heterodimers simply on the basis of molecular weight. This ß-lactamase molecule is efficiently secreted in mammalian cells (De Sutter et al., 1992). When co-expressing the bla-CH1 fusion with the scFv(αCD3)-CL fusion, again only CL-containing products could be found in the medium (Figuer 1B, L2:H3). This was also true when the bla-CH1-E fusion was co-expressed with a native L chain (Figure 1B, L:H3). However, in the reversed situation when a scFv(αCD3)-CL (L2) fusion was co-expressed with a native Fd chain, a scFv(αCD3)-CL:Fd heterodimer could be formed (Figure 1B, L2:Fd). Apparently, only structures with the possibility of forming a Fab chain give rise to heterodimers upon expression in mammalian cells.

To confirm this, we constructed expression vectors for secretion of the CL domain alone, and combined this with an expression vector for the ß-lactamase-CH1 fusion, which was further modified with a C-terminal E-tag sequence to allow a highly sensitive detection of the product. In this way, the difference in molecular weight should allow for discrimination between the between the homodimers and the heterodimer. However, co-expression of the CL with the bla-linker-CH1-Etag fusion protein (Figure 1B, CL:H3) did not reveal any detectable product in the medium. To asses whether either the VH or the VL domain would be sufficient to allow for progression through the ER, the CL domain was co-expressed with the complete Fd chain (Figure 1B, CL:Fd) and the CH1 domain (in the bla-CH1-E fusion product form) was co-expressed with the native L chain (Figure 1B, L:H3). Also here, no products containing the CH1 domain could be detected.

The failure of the CL and CH1 domains to drive heterodimerisation in mammalian cells when either the VL or the VH domain is removed from the Fab-chains (CL:Fd and CH1:L) suggests that interaction of the complete Fab chains is necessary to obtain heterodimers. Probably, the quality control in the ER is such that both the CL:CH1 and the VL:VH interaction is necessary to obtain heterodimers that can progress through the secretion pathway.

4.2 FAB-SCFV FUSION MOLECULE AS A MODEL SYSTEM FOR INTERMEDIATE SIZED BSAB PRODUCTION

We wanted to investigate whether the Fab-context could be used as a heterodimeric scaffold to build bispecific antibodies. To allow for maximal range of antigen reach, we elongated the Fd(αhPLAP) chain at its C-terminus with a flexible peptide linker and the sequence coding for the scFv(αCD3). The glycine-serine rich peptide linker $(G_4S)_3$ was used to connect both subunits. When transiently co-expressed in HEK293T cells, the plasmid-sets gave rise to the expected (αhPLAP x αCD3) Fab-scFv molecule. As seen on the Western blot developed with anti-mouse IgG, the selective formation of heterodimers was very efficient and there were only a few minor degradation products and some free light chains (Figure 2). The Fab(E6)-scFv(2c11) BsAb was found to heterodimerise very efficiently: up to 90% of all secreted immunoglobulin was in the heterodimeric format. Typically, 1 mg of pure bispecific Fab-scFv could be obtained from 1,5 l culture medium by transient expression. Stable transformed SP2/0 cell lines produced up to ten fold as much. A final gel filtration step on a calibrated gel filtration column showed that the Fab-scFv BsAb was soluble as a monomer (data not shown). The conservation of functional binding specificity of the constituting components of the Fab(E6)-scFv(2c11) BsAb was confirmed with flow-cytometric analysis of binding to the respective target cells. A cellular ELISA confirmed binding of hPLAP while bound to T-cells. The Fab-scFv molecule showed to be able to crosslink two different cells by induction of T-cell proliferation in the presence of antigen expressing target cells (Figure 2).

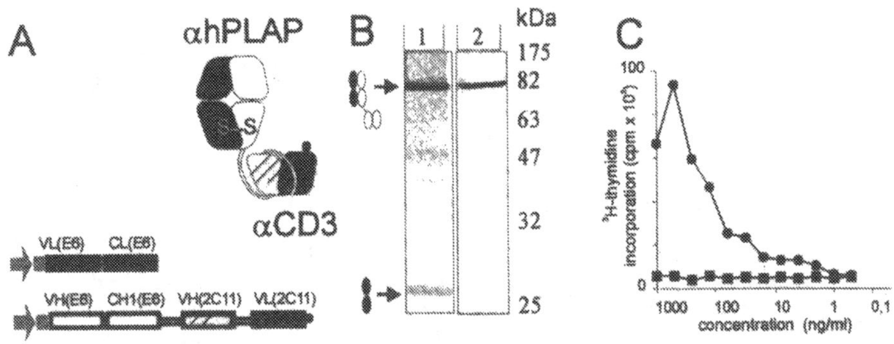

Figure 2: Functional expression of a Fab-scFv fusion as a BsAb. (A) Schematic representation of the gene fusion and of the expected protein formed. (B) Western Blot analysis of culture supernatant by 1) anti mouse IgG or 2) by incubation with hPLAP. (C) T-proliferation in the presence of hPLAP+ MO4I4 cells and a specific (circle) or a non-specific (square) BsAb in the Fab-scFv format.

4.3 FD:L MEDIATED HETERODIMERIZATION OF TWO SCFV MOLECULES LEADS TO EFFICIENT EXPRESSION OF TRISPECIFIC MOLECULES

Analogous to the L:Fd-scFv heterodimerisation, we were also able to obtain efficient L-scFv:Fd heterodimerisation. Next, both the L-scFv and the Fd-scFv fusion molecules were shown to be able to heterodimerise as the major immunoglobulin derived product formed in the culture supernatant. As a model system we used three different specificities: two tumour markers (αhPLAP and αBCL1) and an αCD3 specificity to capture T-cells. To connect the scFv to the L chain we examined two different peptide linkers of 6 and 12 amino acids long. Both linkers gave efficient heterodimerisation of the L:Fd heterodimerised products. ELISA and flow cytometry showed functional binding of all three specificities to their respective target cells. T-cell proliferation could be induced with the (αhPLAP x αBCL1 x αCD3) trispecific antibody in the presence of BCL1-lymfoma cells or in the presence of the $hPLAP^+$ MO4I4 cells, showing that the molecule was able to crosslink two different cells along the (αhPLAP x αCD3) axis and the (αBCL1 x αCD3) axis. Both the 6 and the 12 amino acid linker performed equally well in this assay (Figure 3).

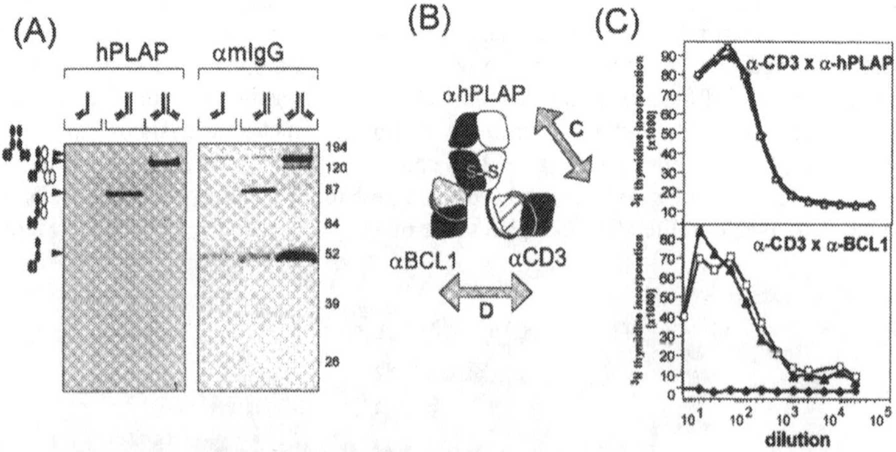

Figure 4: Expression and functionality of trispecific antibodies. (A) Expression detected with hPLAP or with anti mouse IgG serum. (B) Schematic representation of the expected protein to be formed. Arrows C and D depict the axes along which bispecificity is shown by crosslinking two cells in a T-cell proliferation assay (C) T-cell proliferation assay with hPLAP⁺ fibrosarcoma cells or BCL1 as targets. Open and filled symbols depict a variant with a shorter or longer linker connecting the αBCL1 scFv to the L chain. Filled diamonds in the lower graph shows the result with an irrelevant Bibody

In a similar way it was possible to express trifunctional bispecific molecules, with a bivalent binding towards one target cell. Also trivalent binding antibody derivatives could be produced. Upon replacement of one of the scFv molecules with another type of molecule, such as IL2 to enhance immunostimulation, also a bispecific binding immunoconjugate could be produced, showing the versatility of the BsAb design (data not shown).

4.4 INFLUENCE OF LINKER LENGTH AND COMPOSITION ON PRODUCTION AND HETERODIMERIZATION

Expression levels were initially compared using transient expression in HEK293T cells. The promoter used is a hybrid actin / β-globulin promoter enriched with CMV-derived enhancer sequences (Niwa et al., 1991). This expression system usually produces between 0.1 and 1 mg/l, depending on the variable regions used and the format of the molecule (Bibody or Tribody). Other promoters (CMV and EF) were tested and performed less well. For stable production cell lines, SP2/0 myeloma cells were chosen. Transformed cell lines were selected by co-integration of an antibiotic resistance marker (neomycin or zeocin). Often, cell lines could be picked that produced up to five times better than what predicted with the transient expression system. Although the recombinant antibodies were engineered with a HIS-tag for immobilised metal affinity chromatography (IMAC), due to some medium components a first capture step was needed before the IMAC to obtain pure protein. Ion exchange chromatography works fine, but has to be optimised for each individual protein. Hydrophobic interaction chromatography was a more reproducible capture step for all antibodies tested.

In order to test the influence of the linker length between the Fab-chains and the scFv molecules, different linkers were synthesised: H1 (5 aa) and H2 (20 aa) between the H-chain and the scFv1, and L4, L5, L6, L7 and L8 between the L-chain and the scFv1 (6, 12, 6, 15 and 4 aa resp.). L6 and L7 contain 50% proline in their sequence.

Combinations between these linkers showed that most linker combinations could be successfully produced. A slightly better expression level was noted when using longer linkers (Figure 5). This could be expected if the scFv molecules are crowded at the C-terminus of the Fab-fragment. However, combinations using linkers of only a few amino acids still produced functional protein with efficient heterodimerisation.

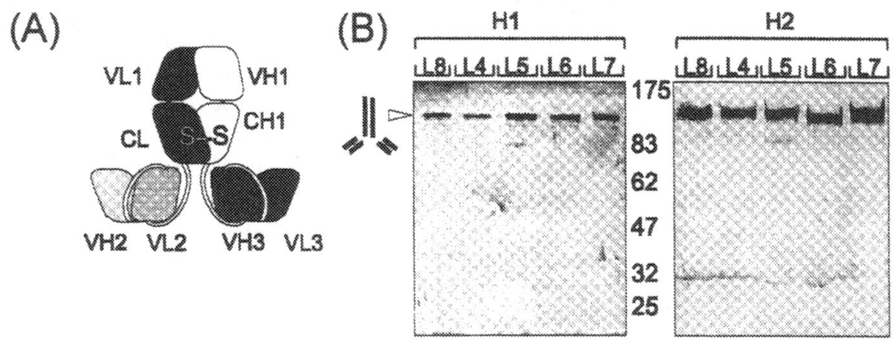

Figure 5: Influence of linker length on expression and heterodimerisation. A series of two linkers attaching the scFv to the Fd chain (H1,5 aa and H2, 20aa) is combined with a series of five different linkers attaching the second scFv to the L-chain (L4-8, 6, 12, 6, 15 and 4 aa resp.). A) Schematic representation of a Tribody. B) An equivalent of 1 ml of culture supernatant of HEK293T cells transfected with the appropriated expression plasmids was resolved by non-reducing SDS-PAGE, blotted and revealed with anti mouse IgG serum.

4. Discussion

A new model for bispecific antibodies is discussed. This model makes use of recombinant production in mammalian cells to produce Fd-scFv and L-scFv fusion chains. Due to the chaperone system found in the ER, a very specific heterodimerisation of both scFv molecules was observed. The heterodimerising scaffold (Fd and L chains) thereby forms a Fab chain that constitutes a third functional binding site of the molecule. The resulting *Bibody* (Fab-scFv) or *Tribody* (Fab-$(scFv)_2$) could be efficiently produced, had a low tendency to aggregate, and was fully functional in crosslinking two different cells. The design of the molecule makes it easy to produce e.g. a monovalent BsAb, a BsAb that bind one determinant bivalently, thus increasing the avidity, a trispecific antibody, or an immunoconjugate. The molecule was chosen not to include an Fc tail, in order to avoid the chance on aspecific crosslinking of two immune effector cells by FcR interaction (Link et al., 1998). Also, the model allows the arm of the antibody directed towards the effector cell to be kept monovalent, to lessen the chance of aggregating the activating receptor on the immune effector cell (e.g. CD3) when not bound to the target cell.

The heterodimerisation observed was very specific, and occurs during the production phase. The only contaminating immunoglobulin molecules produced are light-chain derivatives. Fd-chain derivatives were never observed on secreted without being associated with a light chain derived molecule. A purification method with an affinity step towards either the Fd-scFv chain or for the antigen binding of the Fab eliminates these L-chain derivatives while delivering pure protein with a minimum of purification steps.

In order to maximise the heterodimerisation, mammalian cells are probably the best host, and production in mammalian cells can be more costly than e.g. production in bacteria. However, the direct formation of correctly heterodimerised, functional molecules is harder to obtain in lower organisms. Several cell lines, such as myeloma cells, are adaptable to growth in serum-free medium, seriously reducing the cost of media consumption. Also, in the advent of a more general production in transgenic mammals, *in vivo* production can be considered.

We believe this model to be a serious candidate for developing new antibody-based biologicals.

References

Adams, G.P., McCartney, J.E., Tai, M.S., Oppermann, H., Huston, J.S., Stafford, W.F.d., Bookman, M.A., Fand, I., Houston, L.L. and Weiner, L.M.: *Highly specific in vivo tumour targeting by monovalent and divalent forms of 741F8 anti-c-erbB-2 single-chain Fv.* Cancer Res 53 (1993) 4026-34.

Alt, M., Muller, R. and Kontermann, R.E.: Novel tetravalent and bispecific IgG-like antibody molecules combining single-chain diabodies with the immunoglobulin gamma1 Fc or CH3 region. FEBS Lett 454 (1999) 90-4.

Brennan, M., Davison, P.F. and Paulus, H.: Preparation of bispecific antibodies by chemical recombination of monoclonal immunoglobulin G1 fragments. Science 229 (1985) 81-83.

Chaudri, Z.N., Bartlet-Jones, M., Panayotou, G., Klonisch, T., Roitt, I.M., Lund, T. and Delves, P.J.: Dual specificity antibodies using a double-stranded oligonucleotide bridge. FEBS Lett 450 (1999) 23-6.

Coloma, M.J. and Morrison, S.L.: Design and production of novel tetravalent bispecific antibodies. Nat.Biotechnol. 15 (1997) 159-163.

De Jonge, J., Brissinck, J., Heirman, C., Demanet, C., Leo, O., Moser, M. and Thielemans, K.: Production and characterization of bispecific single-chain antibody fragments. Mol.Immunol. 32 (1995) 1405-1412.

De Sutter, K., Feys, V., Van de Voorde, A. and Fiers, W.: Production of functionally active murine and murine::human chimeric F(ab')2 fragments in COS-1 cells. Gene 113 (1992) 223-230.

DuBridge, R.B., Tang, P., Hsia, H.C., Leong, P.M., Miller, J.H. and Calos, M.P.: Analysis of mutation in human cells by using an Epstein-Barr virus shuttle system. Mol Cell Biol 7 (1987) 379-87.

Glennie, M.J., McBride, H.M., Worth, A.T. and Stevenson, G.T.: Preparation and performance of bispecific F(ab' gamma)2 antibody containing thioether-linked Fab' gamma fragments. J.Immunol. 139 (1987) 2367-2375.

Grooten, J. and Fiers, W.: Acquisition by the murine host of responsiveness toward various neoplastic cell lines, but not toward self, through adoptive transfer of a helper T-lymphocyte clone with antiself specificity. Cancer Res 49 (1989) 3872-8.

Hayden, M.S., Linsley, P.S., Gayle, M.A., Bajorath, J., Brady, W.A., Norris, N.A., Fell, H.P., Ledbetter, J.A. and Gilliland, L.K.: Single-chain mono- and bispecific antibody derivatives with novel biological properties and antitumour activity from a COS cell transient expression system. Ther.Immunol. 1 (1994) 3-15.

Hendershot, L., Bole, D., Kohler, G. and Kearney, J.F.: Assembly and secretion of heavy chains that do not associate posttranslationally with immunoglobulin heavy chain-binding protein. J Cell Biol 104 (1987) 761-7.

Hendershot, L.M.: Immunoglobulin heavy chain and binding protein complexes are dissociated in vivo by light chain addition. J Cell Biol 111 (1990) 829-37.

Hendrix, P.G., Dauwe, S.E., Van de Voorde, A., Nouwen, E.J., Hoylaerts, M.F. and De Broe, M.E.: Radiolocalisation and imaging of stably HPLAP-transfected MO4 tumours with monoclonal antibodies and fragments. Br.J.Cancer 64 (1991) 1060-1068.

Holliger, P. and Winter, G.: Engineering bispecific antibodies. Curr.Opin.Biotechnol. 4 (1993) 446-449.

Holliger, P., Prospero, T. and Winter, G.: "Diabodies": small bivalent and bispecific antibody fragments. Proc.Natl.Acad.Sci.U.S.A. 90 (1993) 6444-6448.

Horn, U., Strittmatter, W., Krebber, A., Knupfer, U., Kujau, M., Wenderoth, R., Muller, K., Matzku, S., Pluckthun, A. and Riesenberg, D.: High volumetric yields of functional dimeric miniantibodies in Escherichia coli, using an optimized expression vector and high- cell-density fermentation under non-limited growth conditions. Appl.Microbiol.Biotechnol. 46 (1996) 524-532.

Hu, S., Shively, L., Raubitschek, A., Sherman, M., Williams, L.E., Wong, J.Y., Shively, J.E. and Wu, A.M.: Minibody: A novel engineered anti-carcinoembryonic antigen antibody fragment (single-chain Fv-CH3) which exhibits rapid, high-level targeting of xenografts. Cancer Res 56 (1996) 3055-61.

Karpovsky, B., Titus, J.A., Stephany, D.A. and Segal, D.M.: Production of target-specific effector cells using hetero-cross-linked aggregates containing anti-target cell and anti-Fc gamma receptor antibodies. J Exp Med 160 (1984) 1686-701.

Kipriyanov, S.M., Dubel, S., Breitling, F., Kontermann, R.E. and Little, M.: Recombinant single-chain Fv fragments carrying C-terminal cysteine residues: production of bivalent and biotinylated miniantibodies. Mol.Immunol. 31 (1994) 1047-1058.

Kipriyanov, S.M., Moldenhauer, G., Schuhmacher, J., Cochlovius, B., Von der Lieth, C.W., Matys, E.R. and Little, M.: Bispecific tandem diabody for tumor therapy with improved antigen binding and pharmacokinetics. J Mol Biol 293 (1999) 41-56.

Kostelny, S.A., Cole, M.S. and Tso, J.Y.: Formation of a bispecific antibody by the use of leucine zippers. J.Immunol. 148 (1992) 1547-1553.

Link, B.K., Kostelny, S.A., Cole, M.S., Fusselman, W.P., Tso, J.Y. and Weiner, G.J.: Anti-CD3-based bispecific antibody designed for therapy of human B-cell malignancy can induce T-cell activation by antigen-dependent and antigen-independent mechanisms. Int J Cancer 77 (1998) 251-6.

Luo, D., Geng, M., Noujaim, A.A. and Madiyalakan, R.: An engineered bivalent single-chain antibody fragment that increases antigen binding activity. J Biochem (Tokyo) 121 (1997) 831-4.

Mack, M., Riethmuller, G. and Kufer, P.: A small bispecific antibody construct expressed as a functional single-chain molecule with high tumor cell cytotoxicity. Proc.Natl.Acad.Sci.U.S.A. 92 (1995) 7021-7025.

McGregor, D.P., Molloy, P.E., Cunningham, C. and Harris, W.J.: Spontaneous assembly of bivalent single chain antibody fragments in Escherichia coli. Mol Immunol 31 (1994) 219-26.

Milenic, D.E., Yokota, T., Filpula, D.R., Finkelman, M.A., Dodd, S.W., Wood, J.F., Whitlow, M., Snoy, P. and Schlom, J.: Construction, binding properties, metabolism, and tumor targeting of a single-chain Fv derived from the pancarcinoma monoclonal antibody CC49. Cancer Res 51 (1991) 6363-71.

Milstein, C. and Cuello, A.C.: Hybrid hybridomas and their use in immunohistochemistry. Nature 305 (1983) 537-540.

Muller, K.M., Arndt, K.M. and Pluckthun, A.: A dimeric bispecific miniantibody combines two specificities with avidity. FEBS Lett 432 (1998a) 45-9.

Muller, K.M., Arndt, K.M., Strittmatter, W. and Pluckthun, A.: The first constant domain (C(H)1 and C(L)) of an antibody used as heterodimerization domain for bispecific miniantibodies. FEBS Lett 422 (1998b) 259-64.

Niwa, H., Yamamura, K. and Miyazaki, J.: Efficient selection for high-expression transfectants with a novel eukaryotic vector. Gene 108 (1991) 193-9.

Pack, P. and Pluckthun, A.: Miniantibodies: use of amphipathic helices to produce functional, flexibly linked dimeric FV fragments with high avidity in Escherichia coli. Biochemistry 31 (1992) 1579-1584.

Ridgway, J.B., Presta, L.G. and Carter, P.: 'Knobs-into-holes' engineering of antibody CH3 domains for heavy chain heterodimerization. Protein Eng. 9 (1996) 617-621.

Roussel, A., Spinelli, S., Deret, S., Navaza, J., Aucouturier, P. and Cambillau, C.: The structure of an entire noncovalent immunoglobulin kappa light-chain dimer (Bence-Jones protein) reveals a weak and unusual constant domains association. Eur J Biochem 260 (1999) 192-9.

Segal, D.M., Perez, P., Karpovsky, B. and Titus, J.A.: Targeting of cytotoxic cells with cross-linked antibody heteroaggregates. Mol Immunol 23 (1986) 1211-4.

Venkatachalam, M.A. and Rennke, H.G.: The structural and molecular basis of glomerular filtration. Circ Res 43 (1978) 337-47.

Whitlow, M., Filpula, D., Rollence, M.L., Feng, S.L. and Wood, J.F.: Multivalent Fvs: characterization of single-chain Fv oligomers and preparation of a bispecific Fv. Protein Eng 7 (1994) 1017-26.

Zhu, Z., Presta, L.G., Zapata, G. and Carter, P.: Remodeling domain interfaces to enhance heterodimer formation. Protein Sci 6 (1997) 781-8.

Zhu, Z., Presta, L.G., Zapata, G. and Carter, P.: Remodeling domain interfaces to enhance heterodimer formation. Protein Sci 6 (1997) 781-8.

Zhu, Z., Zapata, G., Shalaby, R., Snedecor, B., Chen, H. and Carter, P.: High level secretion of a humanized bispecific diabody from Escherichia coli. Biotechnology (N Y) 14 (1996) 192-6.

ADVANTAGES OF SINGLE-DOMAIN ANTIGEN-BINDING FRAGMENTS DERIVED FROM FUNCTIONAL CAMEL HEAVY-CHAIN ANTIBODIES.

Camel Single-domain Antibodies

MUYLDERMANS SERGE, CONRATH KATJA, VU KHOA BANG, SERRAO TERESA, BUSCH MAGNUS, BACKMANN NATASHA, SILENCE KAREN, LAUWEREYS MARC, DESMYTER ALINE.
Ultrastructuur, Vlaams Interuniversitair Instituut voor Biotechnologie, Vrije Universiteit Brussel, Paardenstraat 65, B-1640 Sint Genesius Rode, Belgium.

1. Introduction

For most biotechnological applications, for diagnosis or medicine we would like to have access to molecular recognition units with a minimal size [1]. For *in vivo* use, the smaller formats are expected to possess a better distribution and faster clearance than larger molecules [2]. In addition, to be valuable these small binders should be well expressed, very stable and very soluble. An affinity in the nM range is envisaged since this is sufficient for most applications, and it might be an advantage if the binder could target all possible regions, including cavities on the antigen surface. A low complexity of the binding site is an asset as it simplifies the design of peptide drugs or peptide mimetics even in the absence of structural knowledge [3]. A straightforward method to clone, select, express and purify such optimal binders is also essential.

A variety of methods were proposed to obtain molecular recognition units [4]. First a library is made of a scaffold protein where juxtaposed amino acids are randomised. This library is then searched for the presence of binders by panning [5]. The fragments containing the antigen-binding site of antibodies (Fab or Fv) can also be used as a scaffold [6]. They are the natural solution to generate binders and so they have the advantages that all kinds of target molecules can be antigenic, whether it is as small as a hapten or as large as a virus or any other cell. However, antigen-binding sites of classical antibodies have disadvantages. They are large since at least a pair of domains (VH & VL) is needed to reconstitute the original antigen-binding site [7]. They are unstable due to dissociation of the domains. The linker instability or the aggregation of a covalently linked scFv (single chain Fv) [8] and a low expression yield are additional major hurdles [9].

A. Van Broekhoven et al. (eds.), Novel Frontiers in the Production of Compounds for Biomedical Use, 209–216.
© 2001 *Kluwer Academic Publishers. Printed in the Netherlands.*

In this review we show that the variable domain of the camel heavy-chain antibodies (referred to as VHH) possesses the antigen binding characteristics of antibodies, and that the recombinant single domain meets most of the desired properties of small recognition units as described above.

2. Functional heavy-chain antibodies in sera of camelids

The humoral immune response of camels, dromedaries and llama's is unique because it contains a significant proportion of antibodies composed of heavy chains only without light chains [10]. These immunoglobulins were shown to belong to the IgG isotype. In contrast to similar antibodies devoid of light chains that occasionally occur in human serum as a result of a pathological disorder [11], the camel heavy-chain antibodies were shown to be functional in antigen binding [10,12].

In conventional antibodies containing two heavy - and two light chains it is known that the antigen binding activity resides in the paired N-terminal, variable domains of both the heavy and light chains, the VH and VL respectively [7]. In analogy we surmised that the antigen-binding activity of the heavy-chain camel antibodies should encompass exclusively the variable domain of the heavy chain (VHH).

3. Single antigen-binding domain of camel heavy-chain antibodies

It was expected that the VHH, the single antigen-binding domain of a camel heavy-chain antibody would contain a number of adaptations to cope with the absence of the light chain partner. Especially the removal of the variable N-terminal light chain domain from the antigen-binding site should have serious consequences both in antigen binding and at the surface of the domain that normally interacts with the VL. The first amino acid sequences that became available illustrated how amino acid substitutions (V37F, G44E, L45R and W47G) clustered in the 'VL-side' made this region much more hydrophilic [13,14]. The structure determination of a camel and a llama VHH [15,16] further illustrated how side-chain movements enhanced the hydrophilicity in this region and help to render the isolated domain more soluble.

The sequence analysis of the VHH also indicated that the third antigen-binding loop (or CDR3) of the camel derived heavy-chain antibodies is unusual long [13,14]. This antigen-binding loop is the most variable of all antigen-binding loops both in length and in sequence [17]. It is also the most crucial in antigen binding because it occupies a central location in the antigen-binding site [7,18], and the loop apparently adopts an unlimited number of structures [19]. The long loop in the camel VHH often contains a cysteine. The presence of an additional cysteine residue in the CDR1 (or at position 45) [13,14] suggests that a disulphide bond stabilises and constrains the conformational flexibility of the antigen-binding loops. That the cystine bond is indeed formed was proven by the crystal structure [16] and by a mutagenesis study on camelised human VH where two cysteines introduced on similar loop locations were shown to form a cystine that stabilised the isolated domain [20].

4. Cloning and selecting the camel variable domains of heavy chain antibodies

Immunisation of a dromedary (or camel or llama) - like any other animal - generates a specific immune response for the immunogen. However here it was shown that also the heavy-chain antibodies circulating in the bloodstream were specific for the given antigen [21].

The gene fragments encoding the VHHs can be specifically amplified by PCR on a cDNA template derived from the peripheral blood lymphocytes of the immunised dromedary. All the VHHs of heavy-chain antibodies are members of a single subgroup, subgroup III [13,14]. Therefore only one set of primers is required to amplify the entire VHH repertoire within the blood sample [22]. The repertoire of the VHH gene fragments is then ligated into an expression vector behind a periplasmic secretion signal and in front of gene III of M13 [23]. The expression vector also contains an F1 origin of replication so that the vector genome can be packaged as single stranded DNA into a M13 virion. Expression of the cloned VHH leads to a fusion protein that is incorporated at the tip of this M13 virion. Consequently there is a link between the VHH protein expressed on the tip of the phage and its gene inside the viral particle [6]. Selection of the phage carrying a VHH that recognises the antigen also selects the corresponding gene.

With this technique at hand we succeeded in the identification of several single-domain binders for hen egg-white lysozyme, β-lactamase, α-amylase, RNase A, carbonic anhydrase [21,22]. Even antibodies against haptens could be selected and isolated.

Our method to select specific antigen binders from a cloned antibody repertoire of an immunised dromedary is adapted from similar techniques used to retrieve specific binders from mouse or human svFv (or a Fab) libraries [6]. However, our strategy has the advantage that the VHHs are matured *in vivo* and the antigen-binding site is comprised entirely within this single-domain that is cloned as an entity. In contrast, for the scFv repertoire cloning each gene fragment of the VH-VL pair is first amplified separately and later reassembled. This makes that every functional VH is scrambled with all possible VL partners and *vice versa.* This scrambling means that these scFv libraries need to be 10^4-10^6 times larger than the VHH library which makes both, the generation of the library and the selection of binders more cumbersome.

An alternative approach exists in the generation of extremely large naïve or synthetic libraries (naïve = not matured *in vivo*; in a synthetic library the CDR3 of VH and eventual VL are randomised by a synthetic mutagenic oligonucleotide) from which high affinity binders against a wide variety of antigens were successfully retrieved [24-26]. Evidence was given by Davies and Riechmann [27] that similar single pot libraries of a camelised human VH with a synthetic CDR3 could yield proper binders to haptens. The subsequent randomisation of the CDR1 and CDR2 loops further improved their affinity [28].

5. Characteristics of the single domain antibody fragments

5.1. EXPRESSION YIELD

The production of mg and even kg quantities of the reporter molecule is a necessity for most applications. The expression of the recombinant antibody fragments in *E. coli* is convenient and economic [29]. However, the recombinant expression of a Fv or Fab fragment derived from a monoclonal antibody leads often to poor recoveries [9]. A particular Fv selected from a Fv library in *E. coli* gives possibly better yields. We find that the camel derived VHH fragments produced in *E. coli* can be produced at high levels and we routinely obtain 1-10 mg of recombinant protein per litre in culture flasks. The production of llama VHH in yeast might even give higher yields and 10 g per litre culture were obtained in fermentors. In addition the expression of the camel VHH in plants cloned in proper cloning vehicles seems quite possible and might open new perspectives to protect plants from plant pathogens [30,31].

Also the purification of VHHs is nowadays rather straightforward. The expressed VHH containing a histidine$_6$-tail is directed to the periplasm by the pel B leader signal. The periplasmic extract is already relatively pure, and IMAC and gelfiltration can further purify the VHH to homogeneity. Following this purification scheme we obtain on average 0.5 to 5 mgr VHH per litre culture. On a molar basis this is a very high yield since the molecular weight of the single domain is only 15,000.

An affinity chromatography on a Protein A column is theoretically also possible since the camel VHHs belong to the family III. VHs of family III are known to adsorb on protein A [32]. This is a convenient method to purify the camelised human VH on a small scale [20].

5.2. SOLUBILITY

The isolated single domain VH of a mouse or human antibody is sticky, or insoluble due to the exposure to the solvent of the hydrophobic region that is normally covered by the VL domain [33]. However, the amino acid substitutions observed in this region of a camel VHH constitute a sufficient remedy for the insolubility problems. This was experimentally proven by mutating a human VH in its VL interface region to mimic the amino acids found in camel VHH. This camelisation procedure rendered the human VH more soluble so that its structure could be determined by NMR [34].

5.3. STABILITY

The non-covalently linked VH-VL pairs are particularly sensitive for dissociation and concomitant loss of the antigen-binding activity, whereas the single-domain nature of the camel VHH is more stable. Indeed, the residual functional activity between 80-100 % measured for a purified, dromedary VHH upon prolonged incubation at 37°C is a clear indication for a long shelf-life [22]. In these experiments we noticed that the single-domain binder with the lowest stability lacked a disulphide bond between its antigen binding loops, suggesting that the presence of the inter-loop disulphide bond stabilises the fragment. On the same line Davies and Riechmann [20] observed that the thermal stability of a camelised human VH could also be increased (Tm from 60° to almost 80°) by the introduction of a cystine bond between the CDR1 and CDR3.

5.4. SPECIFICITY AND AFFINITY

The VHs derived from conventional mouse immunoglobulins have the disadvantage that they are sticky, thus probably not very specific. In contrast the VHHs selected after panning from a VHH library of an immunised dromedary turn out to be specific for their antigen [22]. For example, a hen egg-white lysozyme binder failed to recognise the human lysozyme, but could still recognise the turkey lysozyme that has only 7 amino acid differences with hen lysozyme. No cross-reactivity was ever observed for any other non-related protein.

The affinity measured by ELISA or by biosensor resulted in affinities in the range of 1-100 nM for most of the binders [12,21]. This is similar to what is found for conventional Fv fragments.

Although a good immune response in the heavy-chain immunoglobulins was generated against haptens such as phenyloxazolone, we could only retrieve VHH binders with lousy affinities.

5.5. ENZYME INHIBITION

Remarkably, some of the camel VHHs with specificity for enzymes (α-amylase, HEL and carbonic anhydrase) turned out to be specific inhibitors [21,35]. It was proven that for these enzymes, at least, the inhibition was only present in the heavy-chain antibodies and not in the conventional immunoglobulins of the dromedary [21].

The binder of hepatitis C protease, selected from the synthetic camelised human VH library confirms the possibility to find enzyme inhibitors in single domain libraries [36]. It further indicates that the enzyme inhibiting capacity of the VHH derived from heavy-chain antibodies is not due to the specific *in vivo* selection and maturation of the heavy-chain antibody, but results rather from an inherent structural property of the single-domain.

The three-dimensional structure determination of the camel VHH with specificity for lysozyme revealed how the third antigen-binding loop, the long CDR3 protrudes from the remaining antigen-binding site and binds deeply into the active site of the enzyme [16,35]. The presence of a long CDR3 in most of the VHHs indicates that this might be a common mechanism for camel VHHs to produce enzyme inhibitors. However, the presence of enzyme inhibitors with much shorter CDR3 loops (see for example cab-AMD9 in [21]) indicates that other mechanism might occur as well. Probably it is the small antigen-binding site (one domain instead of two domains in Fv) that creates a greater accessibility of the CDR loops due to absence of interference from the VL domain.

5.6. MULTIVALENT CONSTRUCTS

The construction of bispecific (or even multispecific) molecules allows the simultaneous binding of different epitopes that are then kept in each others proximity. In principle this can be achieved by the scFv [37], however, here the presence of a linker in the first scFv and one in the second scFv makes that other complexes can also be generated so that the yield of functional molecules is reduced. A similar drawback is also found for diabodies [38] although additional tailoring might increase the functional yield. With the camel VHHs it is in essence simple to make bispecific constructs. Cloning one VHH after the other eventually spaced by a linker such as the hinge of an antibody should yield functional bispecific constructs.

5.7. INTRABODIES

The intracellular expression of antigen binding fragments (intrabodies) was shown to interfere with the proper functioning of the antigens. For example antibodies against the coat protein of artichoke-mottled-crinkled-virus could reduce the production of infective virus particles in tobacco plants [30]. Also the spreading of HIV could be combated with intrabodies against the HIV proteins Tat, Rev, IN, NC, etc [39]. The method is expected to be far more effective if several viral proteins could be targeted in parallel by intrabodies. Therefore multiple intrabodies should be co-expressed. Unfortunately, this will lead to the production of mixed constructs where the VH of one antibody will interact with the VL of another. These mixed complexes will fail to recognise their targets. The use of single-domain antibodies as intrabodies avoids this shortcoming. It should be no problem to generate multicistronic constructs of multiple camel VHH genes each separated by an IRES (Internal Ribosomal Entry Site) and to co-express these as intrabodies. Their single domain nature will avoid their aggregation and each of them could interfere with the proper function of the target.

Acknowledgements

Much of this work would not have been possible without the initial support of Vlaams Actieprogramma Biotechnologie, the IWT, and VIB.

References

1. Winter, G. and Milstein, C. (1991) Man-made antibodies. *Nature 349*, 293-299.
2. Yokota T., Milenic, D.E., Whitlow, M. and Schlom, J. (1992) Rapid tumor penetration of a singel-chain Fv and comparison with other immunoglobulin forms. *Cancer Res. 52*, 3402-3408.
3. Laune, D., Molina, F., Ferrieres, G., Mani, J.-C., Cohen, P., Simon, D., Bernardi, T., Piechaczyk, M., Pau, B. and Granier, C. (1997) Systematic exploration of the antigen binding activity of synthetic peptides isolated from the variable regions of immunoglobulins. *J.Biol.Chem. 272*, 30937-30944.
4. Nygren, P.-A. and Uhlén, M. (1997) Scaffolds for engineering novel binding sites in proteins. *Curr.Opin.Struc.Biol. 7*, 463-469.
5. Scott, J.K. and Smith, G.P. (1990) Searching for peptide ligands with an epitope library. *Science 249*, 386-390.

6. Winter, G., Griffiths, A.D., Hawkins, R.E. and Hoogenboom, H.R. (1994) Making antibodies by phage display technology. *Annu.Rev.Immunol. 12*, 433-455.

7. Padlan, E.A. (1994) Anatomy of the antibody molecule. *Molecular Immunology 31*, 169-217.

8. Whitlow, M., Bell, B.A., Feng, S.-L., Filpula, D., Hardman, K.D., Hubert, S.L., Rollence, M.L., Wood, J.F., Schott, M.E., Milenic, D.E., Yokota, T. and Schlom, J. (1993) An improved linker for scFv with reduced aggregation and enhanced proteolytic stability. *Protein Engineering 6* , 989-995.

9. Skerra, A. (1993) Bacterial expression of immunoglobulin fragments. *Curr.Op.Immunol. 5*, 256-262.

10. Hamers-Casterman, C., Atarhouch, T., Muyldermans, S., Robinson, G., Hamers, C., Bajyana Songa, E., Bendahman, N. and Hamers, R. (1993) Naturally occurring antibodies devoid of light chains. *Nature 363*, 446-448.

11. Seligmann, M., Mihaesco, E., Preud'homme, J.L., Danon, F. and Brouet, J.C. (1979) Heavy chain diseases: current findings and concepts. *Immunol.Rev. 48*, 145-167.

12. Muyldermans, S. and Lauwereys, M. (1999) Unique single-domain antigen binding fragments derived from naturally occurring camel heavy-chain antibodies. *J.Mol.Recognit. 12*, 1-10.

13. Muyldermans, S., Atarhouch, T., Saldanha, J., Barbosa, J.A.R.G. and Hamers, R. (1994) Sequence and structure of VH domain from naturally occurring camel heavy chain immunoglobulins lacking light chains. *Protein Engineering 7*, 1129-1135.

14. Vu, K.B., Ghahroudi, M.A., Wyns, L. and Muyldermans, S. (1997) Comparison of llama VH sequences from conventional and heavy chain antibodies. *Mol.Immunol. 34*, 1121-1131.

15. Spinelli, S., Frenken, L., Bourgeois, D., de Ron, L., Bos, W., Verrips, T., Anguille, C., Cambillau, C. and Tegoni, M. (1996) The crystal structure of a llama heavy chain variable domain. *Nature Structural Biology 3*, 752-757.

16. Desmyter, A., Transue, T.R., Arbabi Ghahroudi, M., Dao-Thi, M.-H., Poortmans, F., Hamers, R., Muyldermans, S. and Wyns, L. (1996) Crystal structure of a camel single-domain VH antibody fragment in complex with lysozyme. *Nature Structural Biology 3*, 803-811.

17. Wu, T.T., Johnson, G. and Kabat, E.A. (1993) Length distribution of CDR H3 in antibodies. *Proteins: Structure, Function, and Genetics 16*, 1-7.

18. Tomlinson, I.M., Walter, G., Jones, P.T., Dear, P.H., Sonnhammer, E.L.L. and Winter, G. (1996) The imprint of somatic hypermutation on the repertoire of human germline V genes. *J.Mol.Biol. 256*, 813-817.

19. Al-Lazikani, B., Lesk, A.M. and Chothia, C. (1997) Standard conformations for the canonical structures of immunoglobulins. *J.Mol.Biol. 273*, 927-948.

20. Davies, J. and Riechmann, L. (1996) Single antibody domains as small recognition units: design and in vitro antigen selection of camelised, human VH domains with improved protein stability. *Prot.Enging. 9*, 531-537.

21. Lauwereys, M., Ghahroudi, M.A., Desmyter, A., Kinne, J., Hölzer, W., De Genst, E., Wyns, L. and Muyldermans, S. (1998) Potent enzyme inhibitors derived from dromedary heavy-chain antibodies. *EMBO J. 17*, 3512-3120.

22. Ghahroudi, M.A., Desmyter, A., Wyns, L., Hamers, R. and Muyldermans, S. (1997) Selection and identification of single domain antibody fragments from camel heavy-chain antibodies. *FEBS Letters 414*, 521-526.

23. Hoogenboom, H.R., Griffiths, A., , D., Johnson, K.S., Chiswell, D.J., Hudson, P. and Winter, G. (1991) Multi-subunit proteins on the surface of filamentous phage: methodologies for displaying antibody (Fab) heavy and light chains. *Nucl.Acids Res. 19*, 4133-4137.

24. Griffiths, A.D., Williams, S.C., Hartley, O., Tomlinson, I.M., Waterhouse, P., Crosby, W.L., Kontermann, R.E., Jones, P.T., Low, N.M., Allison, T.J., Prosdero, T.D., Hoogenboom, H.R., Nissim, A., Cox, J.P.L., Harrison, J.L., Zaccolo, M., Gherardi, E. and Winter, G. (1994) Isolation of high affinity human antibodies directly from large synthetic repertoires. *EMBO J. 13*, 3245-3260.

25. Nissim, A., Hoogenboom, H.R., Tomlinson, I.M., Flynn, G., Midgley, C., Lane, D. and Winter, G. (1994) Antibody fragments from a 'single pot' phage display library as immunochemical reagents. *EMBO J. 13*, 692-698.

26. Vaughan, T.J., Williams, A.J., Pritchard, K., Osbourn, J.K., Pope, A.R., Earnshaw, J.C., McCafferty, J., Hodits, R.A. and Johnson, K.S. (1996) Human antibodies with sub-nanomolar affinities isolated from a large non-immunized phage display library. *Nature Biotechnology 14*, 309-314.

27. Davies, J. and Riechmann, L. (1995) Antibody VH domains as small recognition units. *Bio/Technology 13*, 475-479.

28. Davies, J. and Riechmann, L. (1996) Affinity improvement of single antibody VH domains. Residues in all three hypervariable regions affect antigen binding. *Immunotechnology 2*, 169-179.
29. Skerra, A. and Plückthun, A. (1988) Assembly of functional immunoglobulin Fv fragment in *Escherichia coli. Science 240*, 1038-1041.
30. Tavladoraki, P., Benvenuto, E., Trinca, S., De Martinis D., Cattaneo A. and Galeffi, P. (1993) Transgenic plants expressing a functional single-chain Fv antibody are specifically protected from virus attack. *Nature 366*, 469-472.
31. Ma, J.K.-C. and Hein, M.B. (1995) Immunotherapeutic potential of antibodies produced in plants. *Tibtech 13*, 522-527.
32. Potter, K.N., Li, Y. and Capra, J.D. (1996) Staphylococcal Protein A simultaneously interacts with framework region 1, complementarity determining region 2 and framework region 3 on human VH3 encoded Igs. *J.Immunol. 157*, 2982-2988.
33. Ward, E.S., Güssow, D., Griffiths, A.D., Jones, P.T. and Winter, G. (1989) Binding activities of a repertoire of single immunoglobulin variable domains secreted from *E. coli. Nature 341*, 544-546.
34. Davies, J. and Riechmann, L. (1994) Camelising human antibody fragments: NMR studies on VH domains. *FEBS Letters 339*, 285-290.
35. Transue, T.R., De Genst, E., Ghahroudi, M.A., Wyns, L. and Muyldermans, S. (1998) Camel single domain antibody inhibits enzyme by mimicking carbohydrate substrate. *Proteins: Structure, Function, and Genetics 32*, 515-522.
36. Martin, F., Volpari, C., Steinkühler, C., Dimasi, N., Brunetti, M., Biasiol, G.A.S., Cortese, R., De Fransesco, R. and Sollazzo, M. (1997) Affinity selection of a camelised VH domain antibody inhibitor of hepatitis C virus NS3 protease. *Prot.Engineering 10*, 607-614.
37. Hudson, P. (1998) Recombinant antibody fragments. *Curr.Op.Biotech. 9*, 395-402.
38. Holliger, P., Prospero, T. and Winter, G. (1993) Diabodies: small bivalent and bispecific antibody fragments. *Proc.Natl.Acad.Sci.USA 90*, 6444-6448.
39. Rondon, I.J. and Marasco, W.A. (1997) Intracellular antibodies (intrabodies) for gene therapy of infectious diseases. *Annu.Rev.Microbiol. 51*, 257-283

Part 4

HETEROLOGOUS PROTEIN PRODUCTION: NEW PRODUCTION STRATEGIES

FURIN AS A TOOL FOR THE ENDOPROTEOLYTIC MATURATION OF SUSCEPTIBLE RECOMBINANT BIOPHARMACEUTICALS

M. HIMMELSPACH, B. PLAIMAUER, F. DORNER AND U. SCHLOKAT[*]
Biomedical Research Center, Hyland Immuno Division, Baxter Healthcare, Uferstrasse 15, 2304 Orth/Donau, Austria
[*]Correspondence: fax +43-1-20100-4000, e-mail schloku@baxter.com

1. Introduction

The ability to produce proteins recombinantly in eukaryotic expression systems has resulted in the development of pharmaceutical proteins, used as therapeutic agents, e.g., for the treatment of patients suffering from haemophilia, heart attack or anaemia. While high expression levels sufficient for industrial purposes are fairly routinely achieved by amplification of the heterologous coding sequence in these systems, post-translational modifications, often a prerequisite for the biological activity of the recombinant protein, frequently become incomplete at overexpression.

Examples for post-translational modifications include glycosylation, phosphorylation, acetylation, sulfation, β-hydroxylation, γ-carboxylation and disulphide-bond formation. In addition, endoproteolytic processing of precursor molecules (pro-proteins) is an essential step in the biosynthesis of many biologically active proteins and peptides. In this case the protein is synthesised as an inactive precursor which, beside the removal of the signal peptide upon translocation into the endoplasmic reticulum, undergoes additional proteolytic cleavage(s) before secretion. Such endoproteolytic processing events typically remove N-terminal propeptides and/or convert a single chain precursor into mature heterodimeric forms or peptides by cleavage C-terminally to basic amino acid motifs. This endoproteolytic maturation process often becomes severely affected at overexpression.

The enzymes responsible for the conversion of precursor molecules into their mature biologically active forms are members of the pro-protein convertases family. Furin, the first mammalian member of this family, is ubiquitously expressed in all cell types examined and was found to process the precursor molecules of a wide variety of physiologically important proteins, such as hormones, growth factors, receptors, plasma proteins, viral envelope proteins and bacterial toxins (table 1).

This report will review the different strategies employing furin and derivatives as used for the maturation of susceptible recombinant proteins of pharmaceutical interest,

A. Van Broekhoven et al. (eds.), Novel Frontiers in the Production of Compounds for Biomedical Use, 219–248.
© 2001 *Kluwer Academic Publishers. Printed in the Netherlands.*

i.e. the plasma factors protein C (PC), factor IX (fIX), factor X (fX) and von Willebrand factor (vWF), all of which require post-translational cleavage events for their conversion into biologically active biopharmaceuticals. Different strategies chosen to ensure complete precursor processing of proteins, expressed either in mammalian cell lines or in transgenic animals, will be described.

2. Sorting and processing of secretory proteins

2.1. THE CONSTITUTIVE AND REGULATED SECRETORY PATHWAYS

Newly synthesised secretory proteins are routed from the rough endoplasmic reticulum (RER) through the Golgi complex (GC) to the trans-Golgi network (TGN). The proteins destined for regulated secretion are segregated from those undergoing constitutive secretion by two sorting events, either in the TGN or the immature secretory granules (ISG), or by a combination of both depending on the cell types (fig. 1; see Urbé et al., 1997, for review). In the TGN, proteins bound for the regulated pathway are actively sorted to the partially clathrin-coated ISGs, while constitutive secretory proteins, transported in secretory vesicles (VS), are continuously delivered therefrom to the plasma membrane. A second sorting event, at regulated secretion, ensures that proteins having entered the ISGs, but not destined to the mature secretory granules (MSG), are removed in clathrin coated vesicles budding from the ISGs. At the same time, regulated secretory proteins retained start to condense, forming the dense core typical to MSGs. Release of proteins from MSGs into the extracellular space is induced by specific stimuli. The proteins retrieved from the ISGs may be targeted to the endosomal pathway, recycled to the TGN, or secreted via a constitutive-like pathway. Beside the segregation of regulated from constitutive secretory proteins, the TGN also represents a sorting compartment for proteins bound for the endosomal/lysosomal pathway. The mechanism by which selective sorting of proteins to the regulated pathway is governed remains elusive; however, different models have been proposed. As a particular feature of proteins destined to the regulated secretory pathway, they form pH-, calcium-, and concentration-dependent homo- and hetero-aggregates from which other proteins are excluded. The association of these aggregates with the luminal membrane has been suggested to auto-induce the formation of secretory granules (Bauerfeind et al., 1993). An alternate mechanism for selective targeting of secretory proteins to their correct destination suggests the direct or indirect interaction of the cargo proteins or aggregates with specific 'sorting receptors', implying the recognition of particular sequences or structural elements in the secretory proteins by the sorting machinery. N-terminal disulphide-linked loop structures, as described for chromogranin B (Chanat et al., 1993) and pro-opiomelanocortin (POMC; Cool et al., 1995), may represent such a molecular targeting signal for the regulated pathway. More recently, carboxypeptidase E, which is involved in the proteolytic maturation of proteins in secretory granules, has been suggested to act as a sorting receptor for POMC, pro-insulin and pro-enkephalin (Cool et al., 1997; Normant and Loh, 1998).

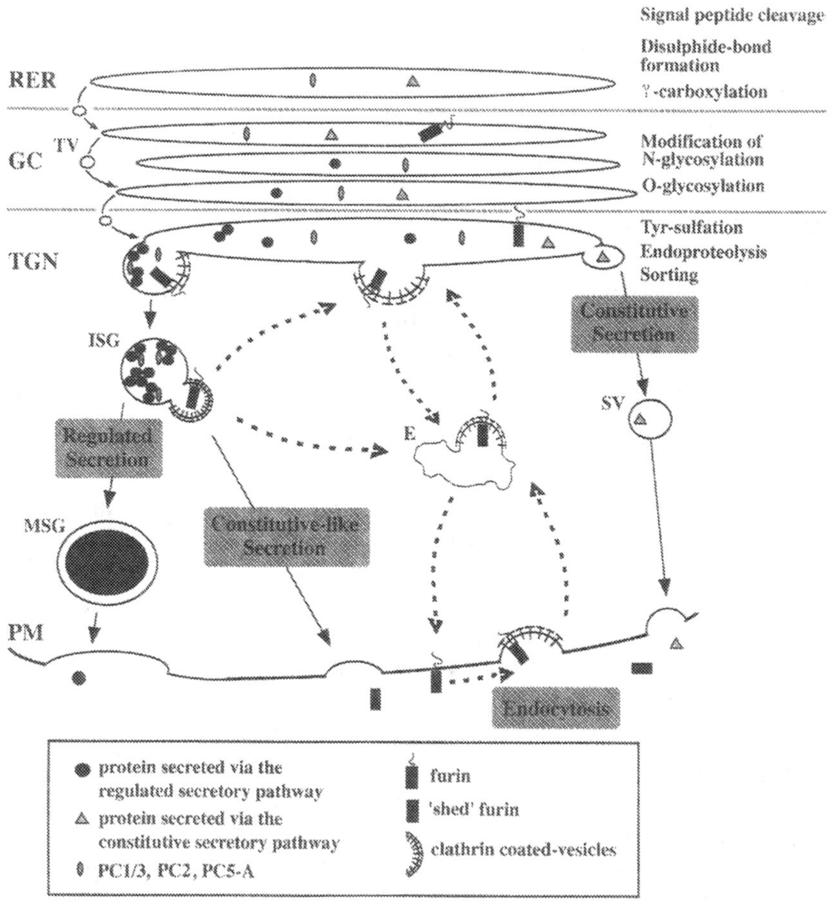

Fig. 1. *Post-translational modification and precursor cleavage of proteins during their transit through the secretory pathway.*

Secretory proteins are transported by transport vesicles (TV) from the rough endoplasmic reticulum (RER) through the Golgi complex (GC) to the trans Golgi network (TGN). In the TGN, proteins are segregated to their respective secretory pathways. In the constitutive secretory pathway, proteins are transport in secretory vesicles (SV) from which they are continuously released to the plasma membrane (PM). Proteins to be secreted in a regulated manner are stored in a condensed form in mature secretory granules (MSG) until they are released upon appropriate stimuli at the PM. During their transit through the different compartments, proteins acquire specific post-translational modifications as indicated. Endoproteolysis of precursor proteins destined for the regulated secretory pathway (black circles) mainly occurs in the immature secretory granules (ISG) by endoproteases such as PC1/3 and PC2 (grey ovals). Furin (black rectangles), a membrane bound endoprotease, is involved in the cleavage of constitutively secreted pro-proteins (grey triangles). Beside its localisation to the TGN, furin cycles between TGN, PM and endosomes (E) also, as indicated by broken arrows. At the PM, furin mediates cleavage of certain bacterial toxins and viral proteins. During its transit through the endosomal system, furin may contribute to the proteolytic cleavage of endocytosed molecules or endosomal precursor proteins. Furin having entered ISGs is actively removed therefrom through clathrin coated vesicles and is recycled back to the TGN either directly or via the endosomal system. Entry of furin into ISGs may be required for the proteolytic maturation of PC2 or for limited cleavage of certain pro-hormones.

Furthermore, evidence that N-linked carbohydrate moieties mediate Golgi to cell membrane transport as well as apical targeting of membrane-bound and secretory proteins suggests that lectins may act as sorting receptors for proteins leaving the TGN (Scheiffele et al., 1995; Gut et al., 1998).

2.2. ENDOPROTEOLYTIC PROCESSING OF PRECURSOR PROTEINS

During their transit through the secretory pathways, proteins may undergo multiple post-translational modifications. Some modifications such as disulphide-bond formation, γ-carboxylation, β-hydroxylation or N-glycosylation occur in the rough endoplasmic reticulum, while others, e. g. O-glycosylation and sulfation, are specific to the different Golgi compartments. In addition to the removal of the signal peptide in the lumen of the rough endoplasmic reticulum (for review see Walter and Johnson, 1994), further endoproteolysis is often required in order to render a secretory protein mature and biologically active (for review see Halban and Irminger, 1994). Cleavage of regulated secretory proteins may occur in the TGN (Davidson et al., 1988; Xu and Shields, 1994) but predominantly takes place within newly formed secretory granules (Orci et al., 1985). Proteolytic processing of constitutive secretory proteins is believed to occur either in the TGN or in secretory vesicles (Nagahama et al., 1991).

Generally, cleavage is performed C-terminal to specific sites consisting of single, paired or multiple basic amino acid motifs. Proteins secreted via the regulated pathway are frequently cleaved after the paired dibasic amino acids R/K-K/R and, occasionally, at mono-arginyl sites (Nakayama et al., 1992). The cleavage site of proteins released from the constitutive pathway requires a more complex basic motif following the rule: an arginine immediately preceding the cleavage site, i.e. at amino acid position -1, and commonly, two additional basic residues at positions -2, -4 or -6 (Watanabe et al., 1993).

3. The pro-protein convertases

3.1. IDENTIFICATION OF EUKARYOTIC PRO-PROTEIN CONVERTASES

The eukaryotic enzymes responsible for the cleavage of pro-proteins at paired or multiple basic amino acid motifs where identified to be calcium-dependent serine proteinases with structural homology to bacterial subtilisins. These proteases are collectively termed pro-protein convertases, pro-hormone convertases (PCs) or subtilisin/kexin-like pro-protein convertases (SPCs). The *fur* gene encoding the first mammalian pro-protein convertase identified, furin, had been discovered in 1986 as an open reading frame located in the fes/fps proto-oncogene upstream region (Roebroek et al., 1986). The identification of the biological role of furin in the endoproteolytic processing of precursor molecules was significantly accelerated by the discovery of the homologous *kex* gene in yeast. Kexin, the translational product of this gene, was found to mediate the cleavage of the precursors of yeast mating factor type α and of the yeast killer toxin (Fuller et al., 1988), as well as of pro-albumin (Bathurst et al., 1987) and pro-opiomelanocortin (POMC; Thomas et al., 1988) at dibasic motifs. Subsequently, additional homologous endoproteases were identified in higher eukaryotes. Presently, the family of mammalian PCs comprises 7 members: furin which is also called PACE (paired basic amino acid cleavage enzyme), PC1/3, PC2, PC4, PACE4, PC5/6 and, finally, PC7 which is also known as LPC, PC8 or SPC7 (Nakayama, 1997; Gensberg et al., 1998; Creemers et al., 1998; Seidah et al., 1998). In addition, isoforms of PACE4 (Zhong et al., 1996; Tsuji et al., 1997; Mori et al., 1997), PC4 (Seidah et al., 1992) and PC5/6 (Nakagawa et al., 1993; De Bie et al., 1996), produced by alternative splicing, have been identified.

3.2. TISSUE DISTRIBUTION, SUBLOCALISATION AND FUNCTION

Each member of the convertase family exhibits a unique tissue distribution; different cell types were found to express individual combinations of these enzymes (Seidah et al., 1994).

PC4 is limited to testicular germ cells (Seidah et al., 1992; 1994). PC1/3 and PC2 are restricted to endocrine and neuroendocrine tissues. PACE4 and PC5 are expressed in several tissues, while PC7 exhibits an even more common tissue distribution. Furin, finally, is expressed ubiquitously.

The cellular sublocalisation of the endoproteases is an important determinant for their physiological function. PC1/3, PC2, and PC5A, an isoform of PC5/6, are involved in the processing of pro-hormones and neuropeptide precursors which are secreted in a regulated manner. PC1/3 and PC2 were shown to cleave neuroendocrine-specific pro-insulin (Smeekens et al., 1992), POMC (Benjannet et al., 1991), pro-glucagon (Rouillé et al., 1994), pro-somatostatin (Xu and Shields, 1994), pro-neurotensin/neuromedin N (proNT/NN; Rovère et al., 1996) and pro-melanin concentrating hormone (MCH; Viale et al., 1999) C-terminally to KR or RR motifs (see Rouillé et al., 1995, for review). PC5A is also involved in the processing of proNT/NN and MCH (Barbero et al., 1998; Viale et al., 1999). Thus, PC1/3, PC2 and PC5A need to be sorted to compartments of the regulated secretory pathway, i.e. secretory granules, where maturation of these precursors occurs (Smeekens and Steiner, 1990; Smeekens et al., 1991; Malide et al., 1995; Tanaka et al., 1996; De Bie et al., 1996; Barbero et al., 1998).

In contrast, furin, PC5B, and PC7, all of which harbour trans-membrane domains, are localised predominantly to the TGN in endocrine and non-endocrine cells, the cellular compartment where cleavage of those protein precursors occurs which are destined for the constitutive pathway (Molloy et al., 1994; Schäfer et al., 1995; De Bie et al., 1996; Seidah et al., 1996; Munzer et al., 1997; Van de Loo et al., 1997). Furin has been shown to cycle between TGN, endosomes and cell surface, thereby playing an important role in the proteolytic processing of both, those precursor molecules transiting through the constitutive secretory pathway as well as of those entering the endocytic pathway. PC7 is involved in the processing of constitutively secreted HIV gp160 (Decroly et al., 1997).

4. The endoprotease furin

4.1. STRUCTURAL ORGANISATION

The cloning of the human *fur* cDNA/gene revealed a primary translation product of 794 amino acids (Roebroek et al., 1986; Wise et al., 1990; Van den Ouweland et al., 1989; Barr et al., 1991). It is structurally composed of a signal peptide, a propeptide, a catalytic moiety, a so-called P-domain, all of which are highly conserved when compared to the other pro-protein convertases, and a COOH-terminus which includes a cysteine-rich region, a trans-membrane domain and a cytoplasmic tail (fig. 2).

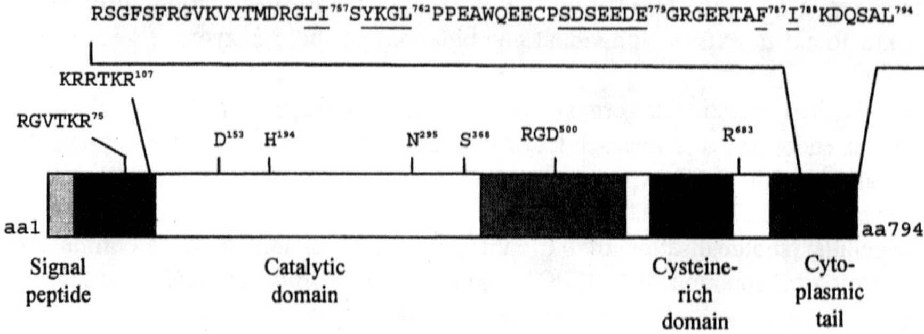

Fig. 2. *Schematic representation of the structural domains of human furin. In addition, individual amino acids, functionally important for maturation, catalytic activity, shedding and trafficking, are given in the one-letter-code and numbered relative to the translational initiator amino acid methionine at position 1. The five targeting motifs of the cytoplasmic tail involved in intracellular trafficking and localisation of the endoprotease are underlined.*

The N-terminal signal peptide indicates that furin is targeted to the secretory pathway. It has been suggested that the adjacent propeptide is crucial to proper folding of the protease during its synthesis in the ER and must be removed in order to render furin functionally active (Rehemtulla et al., 1992). Precursor cleavage occurs in the ER by an autocatalytic intramolecular process after the sequence KRRTKR^{107} (Leduc et al., 1992; Creemers et al., 1993). This cleavage is required for the translocation of furin to the TGN and is essential to, but not sufficient for full activation (Molloy et al., 1994; Vey et al., 1994; Creemers et al., 1995; Anderson et al., 1997). Upon cleavage, the propeptide remains associated with the protease in the neutral environment of the ER where it autoinhibits the enzymatic function (Anderson et al., 1997). Ultimate activation of furin was demonstrated to be a pH and calcium-dependent process, requiring a TGN characteristic environment (pH 6, calcium in the millimolar range), which leads to an additional cleavage within the propeptide C-terminal to the sequence $RGVTKR^{75}$ and its subsequent release. The propeptide thus seems to regulate furin activity in different cellular compartments.

The catalytic domain is located at the N-terminus of the mature protein and consists of approximately 300 amino acids. It contains the conserved catalytic triad D^{153}, H^{194} and S^{368} and the oxyanion hole N^{295} residue. The catalytic triad is essential for autocatalytic maturation and substrate processing. The N^{295} residue is involved in the latter also (Creemers et al., 1993).

The homo-B domain, also called P or middle domain and about 150 amino acids in length, is located adjacent to the catalytic domain. Partial deletion of (Hatsuzawa et al., 1992; Creemers et al., 1993), or mutation within (Takahashi et al., 1995A; Spence et al., 1995) the homo-B domain of furin retained the pro-protease unprocessed in the ER, thus suggesting a role of the homo-B domain in the folding of furin into a conformation appropriate for efficient autocatalytic propeptide removal. This domain also contains the sequence RGD, which is known to mediate integrin binding. For PC1/3, the integrity of its RRGDL motif has been found to be critical for correct sorting of the protease to the regulated secretory pathway, where it is subsequently converted into a more active form by a C-terminal autocatalytic truncation (Lusson et al., 1997). Recently, evidence has been gathered for a role of the homo-B domain in calcium and pH dependence as well as substrate specificity of the individual convertases (Zhou et al., 1998).

The C-terminus of furin contains a cysteine-rich region followed by a trans-membrane domain and, finally, a cytoplasmic tail. To date, the role of the cysteine-rich region remains unclear. The trans-membrane domain anchors furin in the membranes of the TGN/endosomal system. The cytoplasmic tail contains targeting signals essential for compartmentalisation and intracellular trafficking. At present and mostly in analogy with bovine furin, five distinct functional motifs have been identified, i.e. the tyrosine based $YKGL^{762}$ motif, the acidic cluster $SDSEEDE^{779}$, the leucine-isoleucine sequence LI^{757}, and the interwoven mono phenylalanine-based F^{787} and phenylalanine-isoleucine FI^{788} motifs. Tyrosine based motifs conforming the YXXØ consensus sequence (where Ø is an amino acid with a bulky hydrophobic side chain, i.e. L, I, F, M or V) were shown to be responsible for internalisation of receptors from the cell surface, for targeting trans-membrane proteins to the TGN as well as to endosomal/lysosomal compartments, and for basolateral sorting (Marks et al., 1997). This motif has also been shown to interact with the multisubunit clathrin-associated complexes AP1 and AP2 by binding to their 'medium' subunits (Ohno et al., 1996). AP1 and AP2 complexes mediate the interaction of clathrin either with the plasma membrane or the TGN, respectively (Kirchhausen, 1993). Similar trafficking events have been shown to be mediated by di-leucine or leucine-isoleucine motifs (Hunziker and Fumey, 1994; Sandoval and Bakke, 1994). The acidic cluster contains two serines, both of which represent a phosphorylation site for the casein kinase II phosphorylase (Jones et al., 1995; Schäfer et al., 1995; Takahashi et al., 1995; Voorhees et al., 1995).

4.2. SUBCELLULAR LOCALISATION AND TRAFFICKING

While furin is predominantly localised to the TGN, it also cycles between this compartment, the cell surface and endosomes (fig. 1, for review see Molloy et al., 1999). Deletion of the cytoplasmic tail results in mislocalisation of the protease (Bosshart et al., 1994; Molloy et al., 1994; Schäfer et al., 1995; Jones et al., 1995; Voorhees et al., 1995; Dittié et al., 1997). Extensive mutational analysis combined with immunofluorescence studies allowed the assignment of individual functions to the different targeting motifs in the tail and the elucidation of their role in the sorting events in intracellular trafficking.

The acidic cluster is essential for the localisation of furin to the TGN (Jones et al., 1995; Schäfer et al., 1995; Takahashi et al., 1995; Voorhees et al., 1995). Furin exits the TGN to post-TGN vesicular compartments yet to be identified (Wan et al. 1998, Teuchert et al., 1999). The export of furin from the TGN occurs via clathrin-coated vesicles and is mediated by the interaction of the YKGL sequence with the μ1 subunit of the TGN specific AP1 adapter complex (Teuchert et al., 1999). The LI and F motifs participate, to a lesser extent, also. The acidic cluster modulates, in a phosphorylation-dependent manner, the affinity of this interaction (Teuchert et al., 1999). From these endosomal compartments, the endoprotease is retrieved to the TGN via clathrin-coated vesicles containing the AP-1 complex (Wan et al., 1998). The sorting of furin to clathrin-coated vesicles budding from endosomes is mediated by the interaction of the phosphorylated acidic cluster with a cytosolic sorting protein termed PACS-1 (phosphofurin acidic cluster sorting protein). This novel sorting protein serves as the link between molecules such as furin and the mannose-6 phosphate receptor (Johnson and Kornfeld, 1992), and the clathrin associated AP-1 adapter complex (Wan et al., 1998). Similarly, in cells exhibiting the regulated secretory pathway, a phosphorylation-dependent interaction of the acidic cluster with the AP-1 adapter complex is essential for the removal of furin from the ISGs and, hence, for its retrieval to the TGN (Dittie et al., 1997). The involvement of PACS-1 in this sorting event remains to be determined. A parallel cycling loop, where furin is trafficking between the plasma membrane and early endosomes and where the same targeting signals are involved has been proposed recently (Molloy et al., 1998, 1999). Furin present at the cell surface can either be tethered by interaction of the acidic cluster with the cytoskeletal protein ABP-280 (actin binding protein; Liu et al., 1997) or targeted to the endocytic pathway. The internalisation of furin at the plasma membrane occurs via a clathrin-dependent pathway and is mediated by the tyrosine-based motif (Takahashi et al., 1995, Schäfer et al., 1995, Molloy et al., 1998). The LI and F motifs represent two additional yet independent endocytosis signals (Stroh et al., 1999). Once internalised, the endoprotease is routed to early endosomal compartments. The localisation of furin to early endosomes and its recycling to the plasma membrane requires the phosphorylation of the acidic cluster and is PACS-1 dependent (Jones et al., 1995; Molloy et al., 1998). Retrieval of furin from the early endosomes back to the TGN requires the dephosphorylation of the acidic motif. Inhibition of the cellular phosphatases by tautomycin keeps furin cycling between early endosomes and the plasma membrane (Jones et al., 1995; Molloy et al., 1998). The phosphatase responsible for the dephosphorylation of the serine residues from the acidic cluster and, therefore, regulating the retrieval of furin to the TGN has been identified as a protein phosphatase 2A (PP2A) isoform, containing B family regulatory subunits (Molloy et al., 1998). Involvement of this phosphatase in reverse sorting of furin, i.e. from the TGN to the plasma membrane, has also been postulated, however remains to be determined (Molloy et al., 1998). Similarly, it is to be clarified how early endosomes can be distinguished from the post-TGN endosomal compartment where furin is exported to, or which it is retrieved from.

The EEDE sequence of the acidic cluster, in concert with the FI motif, was found to be an essential determinant for the basolateral sorting of furin in polarised epithelial cells (Simmen et al., 1999). Thus, the C-terminal domain is essential for the localisation of furin to specific compartments, where the endoprotease may exert its individual function.

For functional activity, the C-terminus is dispensable. Deletion mutants of furin, lacking the trans-membrane domain and the cytoplasmic tail, were found to exhibit significant enzymatic activity (Hatsuzawa et al., 1992A; Rehemtulla et al., 1992; Molloy et al., 1992; Oda et al., 1992; Takahashi et al., 1994; Bravo et al., 1994; Preininger et al., 1999). Overexpression of full length furin results in the secretion of a naturally truncated form termed 'shed' furin, lacking the trans-membrane and cytosolic domains (Rehemtulla et al., 1992; Vidricaire et al., 1993; Vey et al., 1994; Schäfer et al., 1995). The conversion of furin into this soluble secreted form was shown to be an intracellular process which requires Ca^{2+} ions and an acidic environment (Vey et al. 1994). Recently, R^{683} located in between the cysteine-rich and the trans-membrane domains has been shown to be crucial to shedding, and the internal cleavage site has been mapped (Plaimauer et al., submitted). The enzymes involved in this processing event remain to be determined. Even though an autocatalytic cleavage has been proposed (Rehemtulla et al., 1992), the site identified does not seem to be related to the furin consensus sequence. However, given the occasionally relaxed specificity of furin (Himmelspach et al., in press), the involvement of furin can not entirely be ruled out. The use of specific protease inhibitors may be helpful for the elucidation of this mechanism. Since shedding commonly is observed upon furin overexpression, it remains to be determined whether it represents an emergency system to the host cell, shielding the latter from potentially hazardous effects of excess protease, or whether it is, rather, part of a regulatory phenomenon, modulating intracellular furin concentration/activity by secretion. The latter option has gained considerable support by the recent isolation, from the Golgi fraction of bovine kidney cells, of a truncated endogenous furin molecule (Vey et al., 1994).

4.3. SUBSTRATE SPECIFICITY

The substrate specificity of furin has been extensively analysed by coexpression experiments (Hosaka et al., 1991; Watanabe et al., 1993; Takahashi et al., 1994) and by in vitro studies (Molloy et al., 1992; Hatsuzawa et al., 1992; Oda et al., 1992), the latter employing purified furin or derivatives thereof. Although furin preferentially cleaves C-terminal to the consensus sequence R-X-R/K-R, it is also able to process cleavage sites containing an R at position -1 and two additional amino acid residues, either R or K, at positions -2, -4 or -6. The amino acid at position +1 may modulate the cleavage efficiency; amino acids with a hydrophobic aliphatic side chain such as V, L or I significantly impair cleavage. In a naturally occurring albumin mutant, pro-albumin Bleinheim, which harbours a V rather than the wildtype D residue at position +1, cleavage is entirely abolished (Oda et al., 1992; Brennan and Nakayama, 1994). The substrate specificity of furin has been proposed to be governed by negatively charged amino acid residues located in the vicinity of the substrate binding region (Siezen et al., 1994). Changing the negative charges by mutation altered the substrate specificity (Creemers et al., 1993; Roebroek et al., 1994).

TABLE 1. Protein precursors shown to be susceptible to the endoprotease furin. Listed are the precursors of proteins which have been found to be cleaved by the endoprotease furin either in coexpression experiments or in vitro using partially purified components. The amino acid residues directly preceding the cleavage site are indicated (in the one letter code). K and R residues, which are thought to represent the major determinants of furin specificity, are shown in bold type.

	-6	-5	-4	-3	-2	-1
Furin consensus sequence	**R**	X	**R**	X	**R/K**	**R**
Growth factors and hormones						
• mouse pro-β-nerve growth factor	T	H	**R**	S	**K**	**R**
• porcine pro-brain-derived neutrophic factor	S	M	**R**	V	**R**	**R**
• human pro-neutrophin-3	T	S	**R**	**R**	**K**	**R**
• human pro-transforming growth factor β-1	S	S	**R**	H	**R**	**R**
• rat pro-Müllerian inhibiting substance	**R**	G	**R**	A	G	**R**
• human pro-insulin-like growth factor 1	P	A	**K**	S	A	**R**
• human pro-endothelin-1	L	**R**	**R**	S	**K**	**R**
• human pro-parathyroid hormone related-peptide	L	**R**	**R**	L	**K**	**R**
• human pro-parathyroid hormone	**K**	S	V	**K**	**K**	**R**
Receptors						
• human insulin pro-receptor	P	S	**R**	**K**	**R**	**R**
• human hepatocyte growth factor pro-receptor	E	**K**	**R**	**K**	**K**	**R**
• human pro-LRP	S	N	**R**	H	**R**	**R**
• human integrin α3-chain	P	Q	**R**	**R**	**R**	**R**
• human integrin α6-chain	N	S	**R**	**K**	**K**	**R**
Plasma proteins						
• human pro-albumin	**R**	G	V	F	**R**	**R**

TABLE 1. Protein precursors shown to be susceptible to the endoprotease furin. Listed are the precursors of proteins which have been found to be cleaved by the endoprotease furin either in coexpression experiments or in vitro using partially purified components. The amino acid residues directly preceding the cleavage site are indicated (in the one letter code). K and R residues, which are thought to represent the major determinants of furin specificity, are shown in bold type.

	-6	-5	-4	-3	-2	-1
Furin consensus sequence	**R**	X	**R**	X	**R/K**	**R**
• human pro-factor IX	L	N	**R**	P	**K**	**R**
• human pro-factor X (single chain)	L	E	**R**	**R**	**K**	**R**
• human pro-factor X (propeptide; cleavage in vitro only)	L	A	**R**	V	T	**R**
• human factor X single chain precursor	L	E	**R**	**R**	**K**	**R**
• human pro-protein C	L	**R**	I	**R**	**K**	**R**
• human protein C single chain precursor	**R**	S	H	L	**K**	**R**
• human pro-von Willebrand factor	S	H	**R**	S	**K**	**R**
• rat complement pro-C3	A	A	**R**	**R**	**R**	**R**
Matrix metalloproteinases						
• human stromelysin-3	**R**	N	**R**	Q	**K**	**R**
• human MT-MMP1	N	V	**R**	**R**	**K**	**R**
• mouse MDC 15	A	H	**R**	L	**K**	**R**
Endoproteases						
• human pro-furin (propeptide C-terminus, cleaved in neutr. environ.)	**K**	**R**	**R**	T	**K**	**R**
• human pro-furin (within propeptide, cleavage in acidic environ.)	**R**	G	V	T	**K**	**R**
• rat endopeptidase 3.4.24.18.	P	S	**R**	P	**K**	**R**
Viral envelope glycoproteins						
• HIV gp 160	V	Q	**R**	E	**K**	**R**
• human CMV glycoprotein B	H	N	**R**	T	**K**	**R**
• mouse mammary tumour virus-7 superantigen	E	N	**R**	**K**	**R**	**R**
• avian influenza virus A hemagglutinin	**K**	**K**	**R**	E	**K**	**R**
• measles virus F0	S	**R**	**R**	H	**K**	**R**
• newcastle disease virus F0	G	**R**	**R**	Q	**R**	**R**
• sindbis virus gp E2	S	G	**R**	S	**K**	**R**
• human parainfluenza virus type 3 F0	D	P	**R**	T	**K**	**R**
• ebola virus glycoprotein	G	**R**	**R**	T	**R**	**R**
Bacterial exotoxins						

TABLE 1. Protein precursors shown to be susceptible to the endoprotease furin. Listed are the precursors of proteins which have been found to be cleaved by the endoprotease furin either in coexpression experiments or in vitro using partially purified components. The amino acid residues directly preceding the cleavage site are indicated (in the one letter code). K and R residues, which are thought to represent the major determinants of furin specificity, are shown in bold type.

	-6	-5	-4	-3	-2	-1
Furin consensus sequence	**R**	X	**R**	X	**R/K**	**R**
• anthrax toxin protective antigen	N	S	**R**	**K**	**K**	**R**
• diphteria toxin	G	N	**R**	V	**R**	**R**
• pseudomonas exotoxin A	**R**	H	**R**	Q	P	**R**
• bacterium aeromonas hydrophila aerolysin	**K**	V	**R**	**R**	A	**R**
• shiga toxin	A	S	**R**	V	A	**R**
Other						
• mouse pro-7B2	Q	**R**	**R**	**K**	**R**	**R**
• human salivary prolin-rich protein 1M	**K**	S	**R**	S	P	**R**
• profibrillin-1	**R**	G	**R**	**K**	**R**	**R**

At present, furin has been experimentally shown to process the precursors of a wide variety of proteins which transit the secretory pathway (table 1). These include peptide hormones, growth factors, receptors, coagulation factors, complement proteins and matrix metalloproteinases, as well as several viral envelope proteins and bacterial toxins. The determination of these precursors as furin substrates was primarily based on coexpression and in vitro cleavage experiments. Although suggestive, these approaches are not necessarily conclusive in regard to an actual physiological role of furin in these maturation processes. Expression of the target protein precursors in particular cell types either harbouring distinct combinations of PCs or, alternately, lacking endogenous furin, as well as antisense RNA approaches may support or exclude the actual involvement of endogenous furin. Proteolytic maturation of human pro-albumin (Mori et al., 1999), human pro-factor X (Himmelspach et al., in press), and HIV envelope protein gp 160 (Decroly et al., 1997) was thus shown to be mediated not only by furin, but by other endoproteases as well. Apart from the precursors listed in table 1, several other pro-proteins harbour cleavage sites also matching the furin consensus sequence; however, actual physiological furin involvement in those cases requires experimental confirmation (see also Molloy et al., 1999).

Quite a few of the proteins requiring furin mediated cleavage for maturation represent attractive candidates as potential therapeutic agents, e.g. those proteins which play pivotal roles in blood coagulation and fibrinolysis.

5. Improved biotechnological processes by the use of furin

5.1. DEVELOPMENT OF RECOMBINANT COAGULATION FACTORS

Haemostasis is highly complex and stringently controlled physiologically. Genetic and acquired malfunction or absence of individual proteins involved generally results in severe clinical symptoms of bleeding and thrombosis, respectively. Patients suffering from the corresponding diseases have conventionally been treated with components purified from human plasma. The fear of transmitting blood borne adventitious agents, such as immunodeficiency or hepatitis viruses, resulted in the industrial development of recombinant plasma proteins, i.e. recombinant factors VIII (Recombinate, a trademark of Baxter Healthcare Corp.; Kogenate™; Refacto™), IX (BeneFix™), and activated VII (Novoseven™).

The complex nature of these proteins requires eukaryotic cells as hosts for production, since only the presence of intact and complete post-translational modifications ensures a high degree of functionality and efficacy of the desired recombinant protein. The most prominent modifications include vitamin K-dependent γ-carboxylation of individual glutamic acid residues, essential for interaction of the corresponding factors with phospholipid surfaces, complex N- and O-glycosylation potentially regulating secretion, half-life and biological activity, tyrosine sulfation involved in protein-protein interaction, phosphorylation, β-hydroxylation and proteolytic processing (see Kaufmann, 1998, for review).

Eukaryotic cell lines, e.g. CHO, BHK and HEK 293, are known to support these modifications. The production of large amounts of recombinant protein in mammalian tissue culture systems is routinely achieved by the amplification of the target protein gene using an amplification marker (Schlokat et al., 1997). For the production of therapeutic proteins in the milk of transgenic animals, an approach chosen when even higher yields are desired, the gene of interest is fused to regulatory sequences, which target its expression to the mammary gland, and multiple copies are introduced into fertilised eggs by microinjection (Lubon et al., 1996). Significant yields of the recombinant protein in the milk of the resulting animals are presumably accomplished by the combination of both the densely packed mammary tissue and a comparably high individual expression in the cells of the mammary gland.

Unfortunately, in either system, the cellular machinery involved in post-translational modifications often becomes rapidly insufficient when a foreign protein is overexpressed, resulting in the secretion of only partially active molecules which must subsequently be removed by additional purification steps. For example, undercarboxylation, reduced multimerisation, and incomplete precursor cleavage have been reported at high yield expression.

Incomplete propeptide removal and/or limited conversion of single chain precursors to mature proteins generally abolishes the function of the protein. The necessity of propeptide removal became evident early in 1986 by the characterisation of fIX mutant molecules from haemophilic patients. Amino acid alterations at positions -1 and -4 preceding the cleavage site had resulted in the secretion of propeptide containing non-functional fIX molecules into plasma (Bentley et al., 1986; Diuguid et al., 1986; Ware et al., 1989). The importance of basic amino acid residues for successful propeptide removal at these positions, and at -2, was subsequently confirmed by mutagenesis of the corresponding positions in PC, fIX, and vWF (Foster et al., 1987; Galeffi and Brownlee, 1987; Rehemtulla and Kaufman, 1992A). Apart from propeptide removal, PC and fX require additional proteolysis of the polypeptide chain precursor for their maturation into biologically active heterodimers.

For large scale recombinant production of these proteins, a variety of strategies employing furin and derivatives thereof in coexpression or proteinchemical downstream processing procedures have demonstrated its potential to ensure complete proteolytic cleavage, as discussed below.

5.2. VON WILLEBRAND FACTOR PROPEPTIDE REMOVAL BY FULL LENGTH FURIN

vWF is a multimeric plasma glycoprotein that mediates platelet adhesion to the subendothelium at the site of vascular injury, promotes platelet aggregation and stabilises factor VIII in circulation. Recombinant vWF (rvWF) is currently being developed as a potential therapeutic agent for the treatment of patients suffering from von Willebrand disease (Lethagen, 1995). vWF is synthesised in megakaryocytes and endothelial cells as a large molecule containing a signal peptide 22 amino acids in length, a propeptide harbouring 741 amino acids, and the mature monomer consisting of 2050 amino acids (fig. 3).

After removal of the signal peptide, coinciding with translocation to the endoplasmic reticulum, the precursor molecules undergo dimerisation through intermolecular disulphide linkage between the C-termini of two monomers. Subsequently, the dimers multimerise in order to form high molecular weight molecules up to 20000 kDa in size. These multimers are released constitutively as well as stored in specialised secretory granules called Weibel-Palade bodies (in endothelial cells) and α-granules (in platelets). The release from these storage organelles is triggered by appropriate stimuli. It is thought that the vWF precursors are cleaved and mature vWF multimers and free propeptides are released (see Furlan, 1996, for review). Recent evidence from animal experiments, upon administration of rvWF precursor molecules, suggests the presence of a mechanism for extracellular propeptide removal also (Turecek et al., in press). Despite the necessity of the propeptide for vWF multimerisation (Leyte et al., 1991; Wise et al., 1991), which in turn is believed to be essential for functional activity, its subsequent removal is mandatory to enable vWF to interact with factor VIII.

Human vWF was expressed in a variety of heterologous cells such as monkey kidney derived COS-1 and CV1 cells, baby hamster kidney cells BHK, human and mouse fibroblasts, as well as Chinese hamster ovary cells CHO (Bonthron et al., 1986; Verweij et al., 1987; Meulien et al., 1992). Analysis of secreted rvWF molecules derived from permanent CHO-rvWF cell clones demonstrated that even cells devoid of the regulated secretory pathway are capable of removing the propeptide and assembling high molecular weight vWF multimers. Depending on the expression level achieved, significant amounts of the recombinant molecules were, however, found to still contain propeptide (fig. 4).

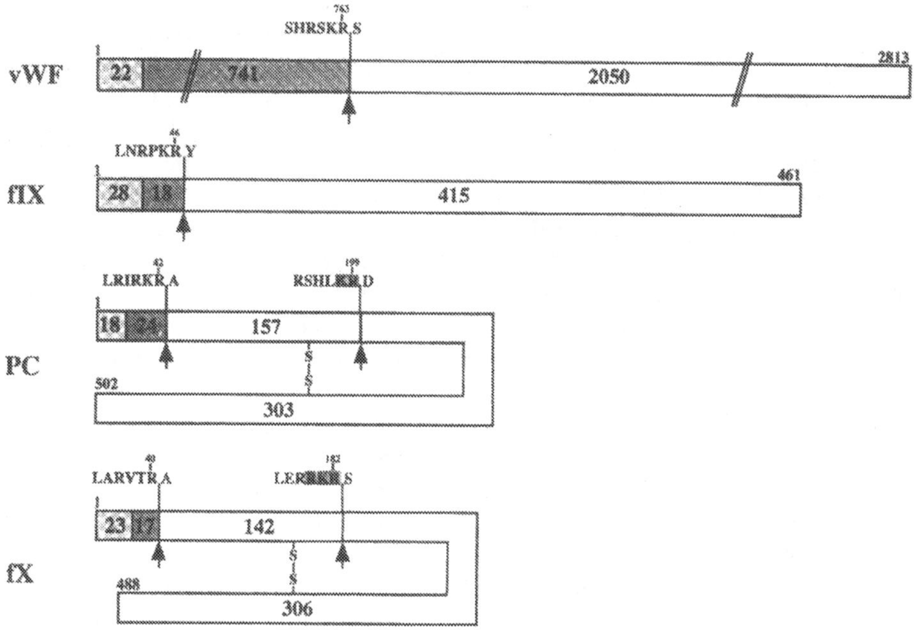

Fig. 3. *Endoproteolytic cleavage of von Willebrand factor, factor IX, protein C, and factor X precursor molecules. These plasma proteins are synthesised as single chain precursor molecules consisting of a signal peptide, a propeptide and the mature molecule, depicted in light grey, dark grey and white, respectively. The sizes of the respective protein moiety are indicated in amino acids. Within the precursor molecules, endoproteolytic maturation sites (other than those involved in signal peptide removal) susceptible to furin are marked by arrows. The amino acid sequences around the cleavage sites are indicated in the one letter code. The position of the amino acid, C-terminally of which cleavage occurs, is given relative to the translational initiation codon methionine. The dipeptide KR^{199} and tripeptide RKR^{182} (shaded in grey) in protein C and factor X, respectively, are removed, after endoproteolysis, by exoproteolytic trimming.*

In conditioned media from cell clones mediating low expression from 50 to 200ng rvWF/10^6 cells x day, only mature rvWF molecules were detectable. In supernatants from amplified CHO/rvWF cells with 100 fold increased expression levels (10µg/10^6 cells x day) approximately 50% of the secreted rvWF molecules were found to still contain the propeptide moiety (Preininger et al., 1999). Propeptide removal thus had become severely insufficient upon overexpression. rvWF multimer analysis with intermediate yield CHO-rvWF cell derived conditioned medium demonstrated that mature monomers and precursor molecules formed high molecular weight heteromultimers (Fischer et al., 1994). The rvWF precursor is the first substrate shown to be cleaved by furin by transient coexpression from furin and vWF cDNAs in COS-1 cells (Wise et al., 1990; Van de Ven et al., 1990). In these cells, processing of rvWF has been postulated to occur exclusively intracellularly when expressed with recombinant furin (Rehemtulla et al., 1992; Rehemtulla and Kaufman, 1992A).

In order to ensure complete rvWF propeptide removal in stably transfected higher yield CHO-rvWF cells also, the cells were additionally transfected with the entire furin cDNA (Fischer et al., 1995; Schlokat et al., 1996). In 24 hours conditioned medium of the resulting CHO-rvWF/rfurin cells, producing intermediate rvWF levels of about 2µg/10^6 cells x day, only propeptide free rvWF molecules were detectable. Coexpression of rvWF and rfurin in BHK cells (Lankhof et al., 1996) yielded similar results, demonstrating that propeptide removal of rvWF could be improved by the expression of full length furin in stably transfected cells. rvWF multimers derived from such a CHO-rvWF/rfurin clone, when produced at industrial scale in a high cell density perfusion bio-reactor, were found to be composed exclusively of entirely propeptide free rvWF monomers and to exhibit an extraordinary structural integrity, significantly surpassing the one shown by plasma derived vWF (Fischer et al., 1995; Fischer et al., 1997).

Interestingly, when the supernatant of the CHO-rvWF/rfurin cells was collected more frequently, i.e. every 8 hours, significant amounts of rvWF precursor molecules were again detectable, and the ratio between precursor and mature rvWF molecules was found to closely resemble that found in the conditioned medium of CHO cells solely expressing rvWF (fig. 5).

This finding suggests that, despite overexpression, intracellular recombinant full length furin does not significantly contribute to rvWF precursor processing, but, rather, that rvWF propeptide removal occurs largely, or even exclusively, in the conditioned medium. Subsequently, 'shed' rfurin was found to be present in the cell culture supernatant. It is this form of furin that accomplishes processing of the rvWF precursor molecules extracellularly, as demonstrated experimentally by mixing conditioned media containing 'shed' furin and incompletely processed rvWF (fig. 6; Schlokat et al., 1996).

These results contrast with earlier reports that rvWF precursor processing in vitro could not be achieved, employing an experimentally truncated furin derivative called PACE SOL (Rehemtulla et al., 1992; Rehemtulla and Kaufman, 1992A). PACE SOL retains the 715 N-terminal amino acids but lacks the trans-membrane and cytosolic domains. However, the different host cells used, the size of PACE SOL (being slightly larger than 'shed' furin) or other factors may explain these discrepancies. Neither 'shed' furin nor rvWF precursors, when derived from CHO-cells, require a cell-membrane association for processing (Schlokat et al., 1996).

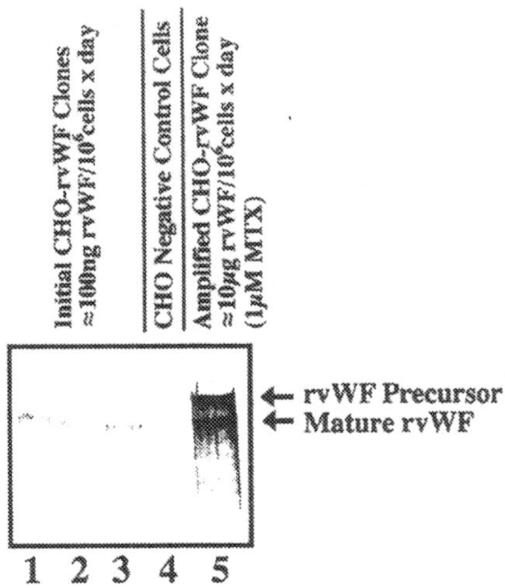

Fig. 4. *Insufficient rvWF precursor processing upon overexpression. rvWF molecules in conditioned media derived from three low and one high yield CHO cell clones were visualised by Western blotting under reducing and denaturing conditions. While endogenous furin levels suffice for complete propeptide removal at low rvWF expression, only a fraction of rvWF precursors are processed at high yield. Lanes 1 to 3, low yield CHO-rvWF clones; lane 4, unmanipulated CHO cells; lane 5, amplified CHO-rvWF cell clone. From Preininger et al. (1999), with kind permission from Kluwer Academic Publishers.*

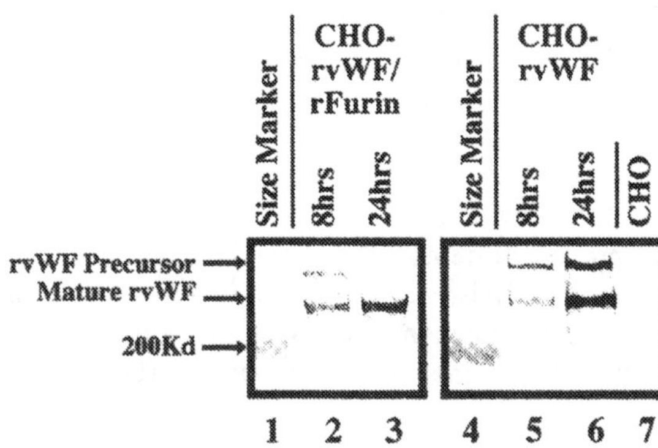

Fig. 5. *Proteolytic processing of rvWF precursor molecules alone and upon coexpression with full length rfurin. rvWF from a stable intermediate yield CHO-rvWF cell clone (lanes 5 and 6) and from a successor CHO-rvWF/rfurin cell clone, additionally transfected in order to express full length rfurin (lanes 2 and 3), were visualised by Western blotting. The conditioned media were collected after 8 (lanes 2 and 5) or 24 hours (lanes 3 and 6). Lanes 1 and 4, 200 kilodalton size marker; lane 7, 24 hours conditioned medium derived from unmanipulated CHO cells. From Preininger et al. (1999), with kind permission from Kluwer Academic Publishers.*

5.3. PRODUCTION OF RECOMBINANT FACTOR IX USING A TRUNCATED SOLUBLE FURIN DERIVATIVE

fIX is a vitamin K-dependent plasma glycoprotein. Its functional deficiency results in haemophilia B. fIX is synthesised as a precursor polypeptide containing a signal peptide, 28 amino acids in length, an 18 amino acids propeptide and a mature protein of 415 amino acids. The propeptide directs the post-translational vitamin K-dependent

γ-carboxylation to the 12 glutamic acid residues of the N-terminus of the mature protein. This modification is required for the molecule's calcium-dependent interaction with phospholipid surfaces (Vermeer, 1990). Removal of the propeptide is crucial to fIX function; fIX is able to acquire its correct conformational structure only after propeptide removal. This process is performed by cleavage after the amino acid sequence LNRPKR[46], which represents the C-terminus of the propeptide (fig. 3).

Since 1985, several different cell lines expressing recombinant human fIX have been established in a variety of attempts to produce biologically active recombinant fIX (rfIX),. In all the systems used (Anson et al., 1985; De la Salle et al., 1985; Busby et al., 1985; Kaufman et al., 1986) only partially active rfIX was obtained, mainly due to insufficient γ-carboxylation and incomplete propeptide removal. Transient transfection of furin, but not PACE4, into CHO-rfIX cells improved the degree of rfIX processing, thereby increasing the specific activity of the secreted fIX 2-3 fold (Wasley et al., 1993). Subsequently, CHO-rfIX cell clones were established, which permanently coexpressed full length furin or PACE SOL. Cell clones yielding approximately 4.2µg and 30µg completely processed rfIX/ml were achieved by coexpression of fIX with full length PACE and PACE SOL, respectively (Wasley et al., 1993). This demonstrated that full length furin as well as its truncated form could mediate complete rfIX processing in stably transfected cells. Processing of the rfIX precursors by the truncated, soluble form of furin was reported to occur intracellularly as well as extracellularly, contrasting the situation reported for rvWF propeptide removal.

Commercially available rfIX is derived from stably transfected CHO cells co-expressing PACE SOL. Current findings indicate that this expression system is highly reliable with regard to propeptide removal. Extracellular precursor molecules, having escaped intracellular endoproteolysis, were shown to be efficiently cleaved later during fermentation while in the culture medium, as well as in the downstream manufacturing process during cell removal and medium concentration (Hamilton and Charlebois, 1997).

5.4. PROCESSING OF RECOMBINANT FACTOR X PRECURSORS USING FURIN DERIVATIVES IN VITRO

fX is a vitamin K-dependent glycoprotein which also plays a key role in haemostasis. fX is present in plasma as a heterodimer composed of a light and a heavy chain linked by a disulphide-bond. During coagulation, it becomes activated to fXa either by the intrinsic fVIIIa/fIXa 'tenase' complex or by the extrinsic fVIIa/tissue factor complex. In combination with factor Va and phospholipids, fXa is the physiological activator of prothrombin. fX deficiency results in an enhanced bleeding tendency (see Watzke and High, 1995, for review). fX is synthesised as a precursor molecule consisting of a signal peptide of 23 amino acids in length, a propeptide of 17 amino acids and a 139 amino acids light chain linked to a 306 amino acids heavy chain by the tripeptide RKR[182]. After signal peptide cleavage, fX precursor molecules require two additional endoproteolytic processing steps to reach their biologically active form. Cleavage after the amino acid sequence RVTR[40] releases the propeptide, while proteolysis after the amino acid sequence RRKR[182] generates the heterodimer consisting of a 142 amino acids light chain and the heavy chain (fig. 3). Subsequent exoproteolytic trimming of the connecting tripeptide yields the mature light chain C-terminus.

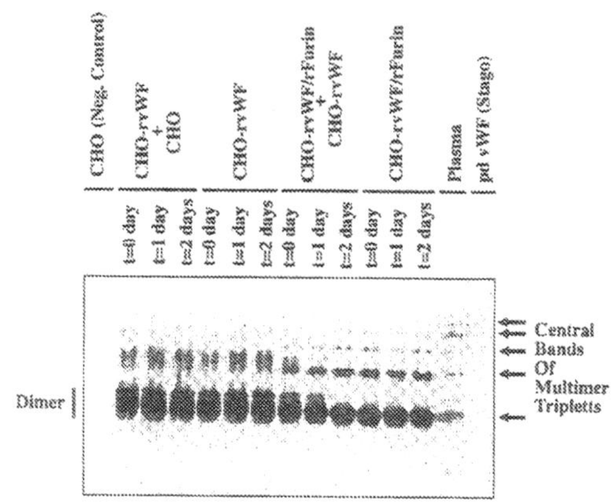

Fig. 6. 'Shed' rfurin mediated in vitro cleavage of an incompletely processed rvWF multimer preparation. High resolution multimer analysis was performed using aliquots of conditioned medium from CHO, CHO-rvWF and CHO-rvWF/rfurin cells, incubated alone or after mixing, as indicated. Propeptide containing heterodimers, depending on the individual combination of covalently linked propeptide free and propeptide containing monomers, are trailing in the immediate vicinity but behind their corresponding fully cleaved, entirely propeptide free multimeric counterparts which migrate the fastest. Plasma and plasma derived vWF, as well as CHO cell derived supernatant, were used as controls. The primary rvWF dimer bands, as well as the central (entirely propeptide free) bands of plasma derived vWF dimer, tetramer, hexamer, octamer and decamer triplets are indicated by arrows on the right. Incubation time (t) is given in days. From Schlokat et al. (1996), with kind permission from Portland Press, Ltd.

Even at modest expression levels of 1-2µg rfX/10^6 cells x day in stably transfected HEK 293 cells, a fraction of secreted rfX molecules was shown to already retain covalently linked propeptide (Rudolph et al., 1997). In stable CHO cells at similar expression levels, significant amounts of rfX were found to be secreted even as single chain molecules (Wolf et al., 1991). The difference of rfX processing status observed in the two cell lines is likely to reflect different expression levels of endogenous endoproteases involved in this process. By replacement of the T residue by a K at position -2, preceding the cleavage site in the rfX propeptide, a cleavage sequence, more closely matching the furin consensus motif, was created (Rudolph et al., 1997). This modification resulted in complete propeptide removal in HEK 293 cells expressing 1.5µg rfX/10^6 cells x day. Upon amplification, CHO cells were recently established that stably express up to 78µg rfX/10^6 cells x day (120µg/ml x day). At such high expression levels, only 50% of the molecules were present as heterodimers. The remaining molecules were found to represent the single chain precursor. N-terminal amino acid sequence analysis revealed that propeptide removal had also become incomplete. When expressed in furin-deficient cells rfX single chain precursor cleavage was entirely abolished, whereas propeptide removal was still accomplished. This finding indicated that an endoprotease different from furin mediates rfX propeptide removal, while furin is responsible for single chain precursor cleavage (Himmelspach et al., in press).

In order to achieve complete cleavage of the precursors even at this enormous yield, a proteinchemical downstream in vitro processing procedure had to be established. Different C-terminally truncated, secreted furin derivatives were constructed and stably expressed for this purpose. rfurinΔCys-4xGly-10xHis was a particularly suitable derivative, which lacks the cysteine-rich, transmembrane and cytosolic domains, but retains the 577 N-terminal amino acids, followed by a short G residue spacer, to ensure steric flexibility, and by a 10 H residue affinity tag, to facilitate purification. The molecules were purified almost to homogeneity by a one step procedure from a Ni^{2+}-NTA resin by an imidazole gradient (Preininger et al., 1999). rfurinΔCys-4xGly-10xHis was found to efficiently perform both rfX cleavage reactions, i.e. single chain precursor processing as well as propeptide removal, in vitro. The protease cleavage site $RVTR^{40}$ in the propeptide does not represent a typical furin site. Since furin was nonetheless able to successfully mediate propeptide removal extracellularly, it exhibits relaxed specificity and may use related, but atypical, sites in vitro.

5.5. USE OF FURIN IN TRANSGENIC ANIMALS

PC also is a vitamin K-dependent plasma glycoprotein, which acts, in concert with its cofactor protein S, as a regulator of haemostasis by inactivating coagulation factors Va and VIIIa. After signal peptide removal, two additional endoproteolytic cleavages C-terminal to the amino acid sequences $IRRK^{42}$ and $HLKR^{199}$ must be performed for maturation of PC. The first reaction releases the 24 amino acids propeptide. Similar to fX, the second cleavage is required for the conversion of the single chain polypeptide into a heterodimer consisting of a 157 amino acids light chain and a 303 amino acids heavy chain, both of which are linked by a disulphide-bond. The C-terminal dipeptide KR^{199} is subsequently removed from the light chain by exoproteolytic trimming (fig. 3). Recombinant PC (rPC) represents a potential therapeutic agent for the treatment of patients suffering from congenital protein C deficiency, disseminated intravascular coagulation and other thrombotic complications.

Human PC was the first vitamin K-dependent protein successfully produced at high yield in the milk of genetically engineered pigs (Velander et al., 1992; Lee et al., 1995) and mice (Velander et al., 1992A; Drews et al., 1995). Microinjection of a PC encoding DNA sequence, under the control of the whey acidic protein promoter, allowed the specific expression of the foreign protein in the mammary glands of lactating animals. Maximal secretion rates of 1mg rPC/ml in swine and 1.65mg rPC/ml in mice were achieved. Similar to mammalian cell systems, limitations of the endogenous processing machinery also occured in the mammary gland tissue of these transgenic animals, resulting in the secretion of 20-30% propeptide-containing and 40-60% single chain rPC forms in milk.

In order to achieve complete processing, bigenic mice were subsequently established by co-injection of the PC gene and full length furin cDNA (Drews et al., 1995). rPC molecules present in the milk of these animals were correctly and almost completely converted into their mature forms. Further analysis revealed that 'shed' furin was present in the milk and responsible for mediating single chain precursor processing and propeptide removal extracellularly, as demonstrated by mixing the milk from transgenic mice expressing only rPC with milk derived from the bigenic PC/furin animals. Thus, maturation of rPC precursors which had escaped intracellular endoproteolysis proceeded in the milk (Paleyanda et al., 1997).

6. Perspectives

The discovery of the mammalian family of subtilisin-like pro-protein convertases has led to a better understanding of the processes involved in the maturation of precursor proteins. Recombinant forms of these proteases were found to improve maturation of recombinant target proteins requiring endoproteolytic cleavages. Coexpression of rvWF with full length rfurin and of rfIX with an experimentally truncated form of furin, which is rapidly released into the conditioned medium, both successfully mediated complete precursor processing in tissue culture. Furin has also proven to be a useful tool for the improvement of production of pro-proteins in insect cells (Laprise et al., 1998). rPC processing in the milk of transgenic animals was similarly accomplished by full length furin coexpression. It is predominantly extracellular 'shed' rfurin, upon full length rfurin overexpression, that seems to be responsible for the completion of target protein precursor cleavage. As a rule of thumb, up to 200ng target protein precursor/10^6 cells x day can be cleaved by endogenous furin present in CHO cells. Processing of target proteins at expression levels up to 2µg and 20µg/10^6 cells x day requires coexpression of full length rfurin and rfurin lacking the transmembrane domain, respectively (Preininger et al., 1999).

It may be possible to modulate processing efficiency, to some extent, by mutational optimisation of the cleavage site. In addition to R residues at amino acid positions -1 and -4 (preceding the cleavage site), basic amino acids at positions -2 and-6 were demonstrated to be beneficial for cleavage. Using prorenin cleavage site mutants, sensitivity to cleavage by rfurin was found to decrease in the following order: R-X-R-X-K-R (best cleavage) > X-X-R-X-K-R > R-X-R-X-X-R > R-X-X-X-K-R > X-X-R-X-X-R (worst cleavage; Takahashi et al., 1994). The nature of the amino acid residues at positions +1 and -3 may also have some modulatory influence. Amino acids with a hydrophobic aliphatic side chain such as V, L and I at position +1 strongly impair furin cleavage; similarly, acidic residues at position -3 are unsuitable (Jean et al., 1995). In contrast, substitution of Y by A at position +1 resulted in enhanced propeptide removal from fIX precursors by endogenous furin (Meulien et al., 1990). Upon the introduction of amino acid alterations at the benefit of process development or yield, recombinant therapeutic proteins may however elicit some undesired consequences, e.g. immune response. Optimisation of cleavage efficiency or modulation of cleavage specificity by mutating relevant amino acids in the target protein propeptide, which will be absent from the therapeutic protein preparation upon complete cleavage, may be a better alternative (Rudolph et al., 1997).

In cases of recombinant protein expression exceeding yields of 20µg/10^6 cells x day, complete processing can be achieved by longer exposure of the target protein precursor to rfurin (when coexpressed) in the conditioned medium. Where prolonged exposure cannot be tolerated, due to lability of the target protein, an in vitro cleavage process needs to be established. Truncated and affinity epitope tagged rfurin derivatives were expressed and could be purified almost to homogeneity by a one step purification procedure. Purified protein precursor molecules were shown to be rapidly and specifically cleaved by purified rfurin in vitro. More basic or acidic conditions during in vitro cleavage may potentially modulate the susceptibility of individual sites. After completion of cleavage, rfurin could easily be recovered via its affinity tag and used anew. More economical would be the immobilisation of rfurin on a column matrix. Passing the insufficiently processed target protein preparation or tissue culture supernatant through the column would allow complete cleavage. In fact, cleavage of a fluorogenic peptide substrate has been successfully demonstrated employing a resin bound rfurin derivative (unpublished).

Large amounts of rfurin derivatives may be required for industrial scale-up. Unfortunately, recombinantly expressed rfurin cannot be recovered in a functional form from prokaryotes. By engineering the catalytic domain of the prokaryotic convertase subtilisin BPN' an enzyme, called furilisin, with a substrate cleavage specificity resembling the one exhibited by furin was generated (Ballinger et al., 1996). Recombinantly engineered bacterial endoproteases may thus eventually prove to be very useful tools.

Significant overexpression of full length rfurin, compared to endogenous furin, did not improve the degree of rvWF propeptide removal intracellularly. Attempts to further increase full length rfurin expression by amplification also failed, possibly because of potential toxicity to the host cell, and/or uncontrolled intracellular dissemination and mislocalisation (Ayoubi et al., 1996). The inability of intracellular full length rfurin molecules to contribute to rvWF precursor processing, beyond the endogenous processing level, suggests the need for cofactors which may be crucial to rvWF precursor processing but which are present in only limiting amounts. Further support for the presence of intracellular cofactors stems from the relaxed specificity of rfurin in vitro observed at rfX maturation. Cofactors may direct the substrate molecules into the vicinity of furin and/or aid the target molecules to acquire a favourable conformation which renders the molecules more accessible to furin and more susceptible to cleavage. Smaller furin molecules, such as 'shed' rfurin or experimentally truncated forms, may not need cofactors as the result of reduced steric constraints due to their smaller size. Cis-acting elements, e.g. γ-carboxylated E residues in vitamin K- dependent proteins, may influence cleavage efficiency favourably (Himmelspach et al., in press). Also, it remains possible that pro-proteins, found to be cleaved by furin in coexpression experiments or in vitro studies, are actually processed by proteases different from but related to furin under physiological conditions (i.e. when intracellular). The maturation of a given pro-protein by endogenous PACE4, PC5B or PC7, which exhibit a cleavage specificity and a tissue distribution similar to that of furin, cannot be excluded on the basis of these experimental approaches. For a definitive answer, expression of the target protein precursor must be performed in furin-deficient cell lines, e.g. human colon carcinoma LoVo, CHO-RPE.40 or CHO-FD11 (Takahashi et al., 1995A; Spence et al., 1995; Gordon et al., 1997). These cell lines may also be of considerable interest for the expression of proteins which commonly are toxic to the host cell; these proteins could be expressed as functionally inactive precursors first and, via native or experimentally introduced furin sites, 'matured' in vitro lateron. By the use of rfurin, activation bears the additional advantage to be accomplished solely recombinantly, as opposed to the currently more popular use of plasma derived factors IIa or Xa.

Recombinant proteins of pharmaceutical interest may exhibit enormous complexity, requiring a wide variety of post-translational modifications frequently crucial to the proper function of the protein. While amplification of heterologous genetic information in tissue culture cells, the most suitable hosts for complex protein production, has become quite a routine process, post-translational modifications often become incomplete. Thus, recombinant cell clones, ultimately used for production, favour optimal functional activity and complete post-translational modifications, at the expense of the yield potentially achievable. Employing recombinant furin, endoproteolytic maturation of pharmaproteins is the first post-translational modification whose insufficiency could be successfully overcome at high yield and industrial scale expression.

Acknowledgements

We thank A. Preininger, G. Mohr, A. Kyd-Rebenburg, V. Stichler, E. Muhr and W. Wernhart for their contributions to rvWF and rfX precursor processing by furin. Also, we are grateful to M. O'Rourke for critical reading of the manuscript.

References

Anderson, E.D., VanSlyke, J.K., Thulin, C.D., Jean, F., and Thomas, G. (1997) Activation of the furin endoprotease is a multiple-step process: requirements for acidification and internal propeptide cleavage. *EMBO J.* **16,** 1508 - 1518.

Anson, D.S., Austen, D.E.G., and Brownlee, G.G. (1985) Expression of active human clotting factor IX from recombinant DNA clones in mammalian cells. *Nature* **315,** 683 - 685.

Ayoubi, T.A.Y., Meulemans, S.M.P., Roebroek, A.J.M., and Van de Ven, W.J.M. (1996) Production of recombinant proteins in chinese hamster ovary cells overexpressing the subtilisin-like proprotein converting enzyme furin. *Mol. Biol. Rep.* **23,** 87 - 95.

Ballinger, M.D., Tom, J., and Wells, J.A. (1996) Furilisin: a variant of subtilisin BPN' engineered for cleaving tribasic substrates. *Biochemistry* **35,** 13579 - 13585.

Barbero, P., Rovève, C., De Bie, I., Seidah, N.G., Beaudet, A., and Kitabgi, P. (1998) PC5-A-mediated processing of pro-neurotensin in early compartments of the regulated secretory pathway of PC5-transfected PC12 cells. *J. Biol. Chem.* **273,** 25339-25346.

Barr, P.J., Mason, O.B., Landsberg, K.E., Wong, P.A., Kiefer, M.C., and Brake, A.J. (1991) cDNA and gene structure for a human subtilisin-like protease with cleavage specificity for paired basic amino acid residues. *DNA and Cell Biol.* **10,** 319-328.

Bathurst, I.C., Brennan, S.O., Carrell, R.W., Cousens, L.S., Brake, A.J., and Barr, P.J. (1987) Yeast KEX2 protease has the properties of a human proalbumin converting enzyme. *Science* **235,** 348-350.

Bauerfeind, R., and Huttner, W.B. (1993) Biogenesis of constitutive secretory vesicles, secretory granules and synaptic vesicles. *Curr. Opin. Cell Biol.* **5,** 628 - 635.

Benjannet, S., Rondeau, N., Day, R., Chrétien, M., and Seidah, N.G. (1991) PC1 and PC2 are proprotein convertases capable of cleaving proopiomelanocortin at distinct pairs of basic residues. *Proc. Natl. Acad. Sci. USA* **88,** 3564 - 3568.

Bentley, A.K., Rees, D.J., Rizza, C., and Brownlee, G.G. (1986) Defective propeptide processing of blood clotting factor IX caused by mutation of arginine to glutamine at position -4. *Cell* **45,** 343 - 348.

Bonthron, D.T., Handin, R.I., Kaufman, R.J., Wasley, L.C., Orr, E.C., Mitsock, L.M., Ewenstein, B., Loscalzo, J., Ginsburg, D., and Orkin, S.H. (1986) Structure of pre-pro-von Willebrand factor and its expression in heterologous cells. *Nature* **324,** 270 - 273.

Bosshart, H., Humphrey, J., Deignan, E., Davidson, J., Drazba, J., Yuan, L., Oorschot, V., Peters, P.J., and Bonifacino, J.S. (1994) The cytoplasmic domain mediates localization of furin to the trans-Golgi network. En route to the endosomal/lysosomal system. *J. Cell Biol.* **126,** 1157 - 1172.

Bravo, D.A., Gleason, J.B., Sanchez, R.I., Roth, R.A., and Fuller, R.S. (1994) Accurate and efficient cleavage of the human insulin proreceptor by the human proprotein-processing protease furin. *J. Biol. Chem.* **269,** 25830 - 25837.

Brennan, S.O., and Nakayama, K. (1994) Cleavage of proalbumin peptides by furin reveals unexpected restrictions at the P_2 and P'_1 sites. *FEBS Lett.* **347,** 80 - 84

Busby, S., Kumar, A., Joseph, M., Halfpap, L., Insley, M., Berkner, K., Kurachi, K., and Woodbury, R. (1985) Expression of active human factor IX in transfected cells. *Nature* **316,** 271 - 273.

Chanat, E., Weiss, U., Huttner, W.B., and Tooze, S.A. (1993) Reduction of the disulphide bond of chromogranin B (secretogranin I) in the trans-Golgi network causes its missorting to the constitutive secretory pathways. *EMBO J.* **12,** 2159 - 2168.

Cool, D.R., Fenger, M., Snell, C.R., and Loh, Y.P. (1995) Identification of the sorting signal motif within pro-opiomelanocortin for the regulated secretory pathway. *J. Biol. Chem.* **270,** 8723 - 8729.

Cool, D.R., Normant, E., Shen, F., Chen, H.C., Pannell, L., Zhang, Y., and Loh, Y.P. (1997) Carboxypeptidase E is a regulated secretory pathway sorting receptor: genetic obliteration leads to endocrine disorders in Cpe(fat) mice. *Cell* **88,** 73 - 83.

Creemers, J.W.M., Siezen, R.J., Roebroek, A.J.M., Ayoubi, T.A.Y., Huylebroeck, D., and Van de Ven, W.J.M. (1993) Modulation of furin-mediated proprotein processing activity by site-directed mutagenesis. *J. Biol. Chem.* **268,** 21826 - 21834.

Creemers, J.W.M., Vey, M., Schäfer, W., Ayoubi, T.A.Y., Roebroek, A.J.M., Klenk, H.-D., Garten, W., and Van de Ven, W.J.M. (1995) Endoproteolytic cleavage of its propeptide is a prerequisite for efficient transport of furin out of the endoplasmic reticulum. *J. Biol. Chem.* **270,** 2695 - 2702.

Creemers, J.W.M., Jackson, R.S., and Hutton, J.C. (1998) Molecular and cellular regulation of prohormone processing. *Sem. Cell. and Dev. Biol.* **9,** 3 - 10.

Davidson, H.W., Rhodes, C.J., and Hutton, J.C. (1988) Intraorganellar calcium and pH control proinsulin cleavage in the pancreatic beta cell via two distinct site-specific endopeptidases. *Nature* **333,** 93-96.

De Bie, I., Marcinkiewicz, M., Malide, D., Lazure, C., Nakayama, K., Bendayan, M., and Seidah, N.G. (1996) The isoforms of proprotein convertase PC5 are sorted to different subcellular compartments. *J. Cell Biol.* **135,** 1261 - 1275.

Decroly, E., Benjannet, S., Savaria, D., and Seidah, N.G. (1997) Comparative functional role of PC7 and furin in the procesing of the HIV envelope glycoprotein gp160. *FEBS Lett.* **405,** 68-72.

De la Salle, H., Altenburger, W., Elkaim, R., Dott, K., Dieterlé, A., Drillien, R., Cazenave, J.-P., Tolstoshev, P., and Lecocq, J.-P. (1985) Active γ-carboxylated human factor IX expressed using recombinant DNA techniques. *Nature* **316,** 268 - 270.

Dittie, A.S., Thomas, L., Thomas, G., and Tooze, S.A. (1997) Interaction of furin in immature secretory granules from neuroendocrine cells with the AP-1 adaptor complex is modulated by casein kinase II phosphorylation. *EMBO J.* **16,** 4859 - 4870.

Diuguid, D.L., Rabiet, M.J., Furie, B.C., Liebman, H.A., and Furie, B. (1986) Molecular basis of haemophilia B: a defective enzyme due to an unprocessed propeptide is caused by a point mutation in the factor IX precursor. *Proc. Natl. Acad. Sci. USA* **83,** 5803 - 5807.

Drews, R., Paleyanda, R.K., Lee, T.K., Chang, R.R., Rehemtulla, A., Kaufman, R.J., Drohan, W.N., and Lubon, H. (1995) Proteolytic maturation of protein C upon engineering the mouse mammary gland to express furin. *Proc. Natl. Acad. Sci. USA* **92,** 10462 - 10466.

Fischer, B., Mitterer, A., Schlokat, U., DenBouwmeester, R., and Dorner, F. (1994) Structural analysis of recombinant von Willebrand factor: identification of hetero- and homo-dimers. *FEBS Lett.* **351,** 345 - 348.

Fischer, B.E., Schlokat, U., Mitterer, A., Reiter, M., Mundt, W., Turecek, P.L., Schwarz, H.P., and Dorner, F. (1995) Structural analysis of recombinant von Willebrand factor produced at industrial scale fermentation of transformed CHO cells co-expressing recombinant furin. *FEBS Lett.* **375,** 259 - 262.

Fischer, B.E., Schlokat, U., Reiter, M., Mundt, W., and Dorner, F. (1997) Biochemical and functional characterization of recombinant von Willebrand Factor produced at large scale. *Cell. Mol. Life Sci.* **53,** 943 - 950.

Foster, D.C., Rudinski, M.S., Schach, B.G., Berkner, K.L, Kumar, A.A., Hagen, F.S., Sprecher, C.A., Insley, M.Y., and Davie, E.W. (1987) Propeptide of human protein C is necessary for gamma-carboxylation. *Biochem.* **26,** 7003 - 7011.

Fuller, R.S., Sterne, R.E., and Thorner, J. (1988) Enzymes required for yeast prohormone processing. *Annu. Rev. Physiol.* **50,** 345-362.

Furlan, M. (1996) Von Willebrand factor: molecular size and functional activity. *Ann. Hematol.* **72,** 341 - 348.

Galeffi, P., and Brownlee, G.G. (1987) The propeptide region of clotting factor IX is a signal for a vitamin K dependent carboxylase: evidence from protein engeneering of amino acid -4. *Nucl. Acids Res.* **15,** 9505 - 9513.

Gensberg, K., Jan, S., and Matthews, G.M. (1998) Subtilisin-related serine proteases in the mammalian constitutive secretory pathway. *Sem. Cell and Dev. Biol.* **9,** 11 - 17.

Gordon, V.M., Rehemtulla, A., and Leppla, S.H. (1997) A role for PACE4 in the proteolytic activation of anthrax toxin protective antigen. *Infect. Immun.* **65,** 3370 - 3375.

Gut, A., Kappeler, F., Hyka, N., Balda, M.S., Hauri, H.P., and Matter, K. (1998) Carbohydrate-mediated Golgi to cell surface transport and apical targeting of membrane proteins. *EMBO J.* **17,** 1919 - 1929.

Halban, P.A., and Irminger, J-C. (1994) Sorting and processing of secretory proteins. *Biochem. J.* **299,** 1-18.

Hamilton, M.A., and Charlebois, T.S. (1997) Extracellular propeptide processing of recombinant human factor IX by a secreted form of the endoprotease, PACE, in M.J.T. Carrondo, B. Griffiths and J.L.P. Moreira (eds.), *Animal Cell Technology*, Kluwer Academic Publishers, pp. 495 - 501.

Hatsuzawa, K., Nagahama, M., Takahashi, S., Takada, K., Murakami, K., and Nakayama, K. (1992) Purification and characterization of furin, a kex2-like processing endoprotease, produced in chinese hamster ovary cells. *J. Biol. Chem.* **267,** 16094 - 16099.

Hatsuzawa, K., Murakami, K., and Nakayama, K. (1992A) Molecular and enzymatic properties of furin, a Kex2-like endoprotease involved in precursor cleavage at Arg-X-Lys/Arg-Arg sites. *J. Biochem.* **111,** 296 - 301.

Himmelspach, M., Pfleiderer, M., Fischer, B., Plaimauer, B., Antoine, G., Falkner, F.G., Dorner, F., and Schlokat, U. Recombinant human factor X: high yield expression and the role of furin in proteolytic maturation in vivo and in vitro. *Thromb. Res.*, in press.

Hosaka, M., Nagahama, M., Kim, W.S., Watanabe, T., Hatsuzawa, K., Ikemizu, J., Murakami, K., and Nakayama, K. (1991) Arg-X-Lys/Arg-Arg motif as a signal for precursor cleavage catalyzed by furin within the constitutive secretory pathway. *J. Biol. Chem.* **266,** 12127 - 12130.

Hunziker, W., and Fumey, C. (1994) A di-leucine motif mediates endocytosis and basolateral sorting of macrophage IgG Fc receptors in MDCK cells. *EMBO J.* **13,** 2963-2969.

Jean, F., Boudreault, A., Basak, A., Seidah, N.G., and Lazure, C. (1995) Fluorescent peptidyl substrates as an aid in studying the substrate specificity of human prohormone convertase PC1 and human furin and designing a potent irreversible inhibitor. *J. Biol. Chem.* **270,** 19225 - 19231.

Johnson, K.F., and Kornfeld, S. (1992) A His-Leu-Leu sequence near the carboxyl terminus of the cytoplasmic domain of the cation-dependent mannose 6-phosphate receptor is necessary for the lysosomal enzyme sorting function. *J. Biol. Chem.* **267,** 17110-17115.

Jones, B.G., Thomas, L., Molloy, S.S., Thulin, C.D., Fry, M.D., Walsh, K.A., and Thomas, G. (1995) Intracellular trafficking of furin is modulated by the phosphorylation state of a casein kinase II site in its cytoplasmic tail. *EMBO J.* **14,** 5869-5883.

Kaufman, R.J. (1998) Post-translational modifications required for coagulation factor secretion and function. *Thromb. Haemost.* **79,** 1068-1079.

Kaufman, R.J., Wasley, L.C., Furie, B.C., Furie, B., and Shoemaker, C.B. (1986) Expression, purification, and characterization of recombinant γ-carboxylated factor IX synthesized in chinese hamster ovary cells. *J. Biol. Chem.* **261,** 9622 - 9628.

Kirchhausen, T. (1993) Coated pits and coated vesicles: sorting it all out. *Curr. Opin. Struct. Biol.* **3,** 182-188.

Lankhof, H., Van Hoeij, M., Schiphorst, M.E., Bracke, M., Wu, Y.-P., Ijsseldijk, M.J.W., Vink, T., De Groot, P.G., and Sixma, J.J. (1996) A3 domain is essential for interaction of von Willebrand factor with collagen type III. *Thromb. Haemost.* **75,** 950 - 958.

Laprise, M.-H., Grondin, F., and Dubois, C.M. (1998) Enhanced TGFβ1 maturation in high five cells coinfected with recombinant baculovirus encoding the convertase furin/PACE: improved technology for the production of recombinant proproteins in insect cells. *Biotechnol. Bioeng.* **58,** 85-91.

Leduc, R., Molloy, S.S., Thorne, B.A., and Thomas, G. (1992) Activation of human furin precursor processing endoprotease occurs by an intramolecular autoproteolytic cleavage. *J. Biol. Chem.* **267,** 14304 - 14308.

Lee, T.K., Drohan, W.N., and Lubon, H. (1995) Proteolytic processing of human protein C in swine mammary gland. *J. Biochem.* **118,** 81 - 87.

Lethagen, S.R. (1995) Pathogenesis, clinical picture and treatment of von Willebrand's disease. *Annals of Medicine* **27,** 641 - 651.

Leyte, A., Voorberg, J., Van Schijndel, H.B., Duim, B., Pannekoek, H., and Van Mourik, J.A. (1991) The pro-polypeptide of von Willebrand factor is required for the formation of a functional factor VIII-binding site on mature von Willebrand factor. *Biochem. J.* **274,** 257 - 261.

Liu, G., Thomas, L., Warren, R.A., Enns, C.A., Cunningham, C.C., Hartwig, J.H., and Thomas, G. (1997) Cytoskeletal protein ABP-280 directs the intracellular trafficking of furin and modulates proprotein processing in the endocytic pathway. *J. Cell Biol.* **139,** 1719 - 1733.

Lubon, H., Paleyanda, R.K., Velander, W.H., and Drohan, W.N. (1996) Blood proteins from transgenic animal bioreactors. *Transfus. Med. Rev.* **10,** 131 - 143.

Lusson, J., Benjannet, S., Hamelin, J., Savaria, D., Chrétien, M., and Seidah, N.G. (1997) The integrity of the RRGDL sequence of the proprotein convertase PC1 is critical for its zymogen and C-terminal processing and for its cellular trafficking. *Biochem. J.* **326,** 737 - 744.

Malide, D., Seidah, N.G., Chrétien, M., and Bendayan, M. (1995) Electron microscopic immunocytochemical evidence for the involvement of the convertases PC1 and PC2 in the processing of proinsulin in pancreatic beta-cells. *J. Histochem. Cytochem.* **43,** 11 - 19.

Marks, M.S., Woodruff, L., Ohno, H., and Bonifacino, J.S. (1996) Protein targeting by tyrosine- and di-leucine-based signals: evidence for distinct saturable components. *J. Cell Biol.* **135,** 341 - 354.

Marks, M.S., Ohno, H., Kirchhausen, T., and Bonifacino, J.S. (1997) Protein sorting by tyrosine-based signals: adapting to the Ys and wherefores. *Trends Cell Biol.* **7,** 124-128.

Meulien, P., Balland, A., Lepage, P., Mischler, F., Dott, K., Hauss, C., Grandgeorge, M., and Lecocq, J.P. (1990) Increased biological activity of a recombinant factor IX variant carrying alanine at position +1. *Prot. Engineering* **3,** 629 - 633.

Meulien, P., Nishino, M., Mazurier, C., Dott, K., Pietu, G., Jorieux, S., Pavirani, A., Girma, J.P., Oufkir, D., Courtney, M., and Meyer, D. (1992) Processing and characterization of recombinant von Willebrand factor expressed in different cell types using a vaccinia virus vector. *Thromb. Haemost.* **67,** 154 - 160.

Molloy, S.S., Bresnahan, P.A., Leppla, S.H., Klimpel, K.R., and Thomas, G. (1992) Human furin is a calcium-dependent serine endoprotease that recognizes the sequence Arg-X-X-Arg and efficiently cleaves anthrax toxin protective antigen. *J. Biol. Chem.* **267,** 16396 - 16402.

Molloy, S.S., Thomas, L., VanSlyke, J.K., Stenberg, P.E., and Thomas, G. (1994) Intracellular trafficking and activation of the furin proprotein convertase: localization to the TGN and recycling from the cell surface. *EMBO J.* **13,** 18 - 33.

Molloy, S.S., Thomas, L., Kamibayashi, C., Mumby, M.C., and Thomas, G. (1998) Regulation of endosome sorting by a specific PP2A isoform. *J. Cell Biol.* **142,** 1399-1411.

Molloy, S.S., Anderson, E.D., Jean, F., and Thomas, G. (1999) Bi-cycling the furin pathway: from TGN localization to pathogen activation and embryogenesis. *Tremds Cell Biol.* **9,** 28-35.

Mori, K., Kii, S., Tsuji, A., Nagahama, M., Imamaki, A., Hayashi, K., Akamatsu, T., Nagamune, H., and Matsuda, Y. (1997) A novel human PACE4 isoform PACE4E is an active processing protease containing a hydrophobic cluster at the carboxy terminus. *J. Biochem. (Tokyo)* **121,** 941 - 948.

Mori, K., Imamaki, A., Nagata, K., Yonetomi, Y., Kiyokage-Yoshimoto, R., Martin, T.J., Gillespie, M.T., Nagahama, M., Tsuji, A., and Matsuda, Y. (1999) Subtilisin-like proprotein convertases, PACE4 and PC8, as well as furin, are endogenous proalbumin convertases in HepG2 cells. *J. Biochem.* **125,** 627-633.

Munzer, J.S., Basak, A., Zhong, M., Mamarbachi, A., Hamelin, J., Savaria, D., Lazure, C., Benjannet, S., Chrétien, M., and Seidah, N.G. (1997) In vitro characterization of the novel proprotein convertase PC7. *J. Biol. Chem.* **272,** 19672 - 19681.

Nagahama, M., Ikemizu, J., Misumi, Y., Ikehara, Y., Murakami, K., and Nakayama, K. (1991) Evidence that differenciates between precursor cleavages at dibasic and Arg-X-Lys/Arg-Arg sites. *J. Biochem.* **110,** 806 - 811.

Nakagawa, T., Murakami, K., and Nakayama, K. (1993) Identification of an isoform with an extremely large Cys-rich region of PC6, a Kex2-like processing endoprotease. *FEBS Lett.* **327,** 165 - 171.

Nakayama, K. (1997) Furin: a mammalian subtilisin/Kex2p-like endoprotease involved in processing of a wide variety of precursor proteins. *Biochem. J.* **327,** 625 - 635.

Nakayama, K., Watanabe, T., Nakagawa, T., Kim, W.-S., Nagahama, M., Hosaka, M., Hatsuzawa, K., Kondoh-Hashiba, K., and Murakami, K. (1992) Consensus sequence for precursor processing at mono-arginyl sites. *J. Biol. Chem.* **267,** 16335- 16340.

Normant, E., and Loh, Y.P. (1998) Depletion of carboxypeptidase E, a regulated secretory pathway sorting receptor, causes misrouting and constitutive secretion of proinsulin and proenkephalin, but not chromogranin A. *Endocrinology* **139,** 2137 - 2145.

Oda, K., Misumi, Y., Ikehara, Y., Brennan, S.O., Hatzuzawa, K., and Nakayama, K. (1992) Proteolytic cleavages of proalbumin and complement pro-C3 in vitro by a truncated soluble form of furin, a mammalian homologue of the yeast Kex2 protease. *Biochem. Biophys. Res. Comm.* **189,** 1353 - 1361.

Ohno, H., Fournier, M.C., Poy, G., and Bonifacino, J.S. (1996) Structural determinants of interaction of tyrosine-based sorting signals with the adaptor medium chains. *J. Biol. Chem.* **271,** 29009 - 29015.

Orci, L., Ravazzola, M., Amherdt, M., Madsen, O., Vassalli, J.D., and Perrelet, A. (1985) Direct identification of prohormone conversion site in insulin-secreting cells. *Cell* **42,** 671 - 681.

Paleyanda, R.K., Drews, R., Lee, T.K., and Lubon, H. (1997) Secretion of human furin into mouse milk. *J. Biol. Chem.* **272,** 15270 - 15274.

Plaimauer B., Mohr, G., Wernhart, W., Himmelspach, M., Dorner, F., and Schlokat, U. 'Shed' Furin: Identification of the internal cleavage site. Submitted.

Preininger, A., Schlokat, U., Mohr, G., Himmelspach, M., Stichler, V., Kyd-Rebenburg, A., Plaimauer, B., Turecek, P.L., Schwarz, H.P., Wernhart, W., Fischer, B.E., and Dorner, F. (1999) Strategies for recombinant furin employment in a biotechnological process: complete target protein precursor cleavage. *Cytotechnology* **30,** 1-15.

Rehemtulla, A., Dorner, A.J., and Kaufman, R.J. (1992) Regulation of PACE propeptide-processing activity: requirement for a post-endoplasmic reticulum compartment and autoproteolytic activation. *Proc. Natl. Acad. Sci. USA* **89,** 8235 - 8239.

Rehemtulla, A., and Kaufman, R.J. (1992A) Preferred sequence requirements for cleavage of pro-von Willebrand factor by propeptide-processing enzymes. *Blood* **79,** 2349 - 2355.

Roebroek, A.J.M., Schalken, J.A., Bussemakers, M.J.G., van Heerikhuizen, H., Onnekink, C., Debruyne, F.M.J., Bloemers, H.P.J., and Van de Ven, W.J.M. (1986) Characterization of human c-fes/fps reveals a new transcription unit (fur) in the immediately upstream region of the proto-oncogene. *Mol. Biol. Rep.* **11,** 117 - 125.

Roebroek, A.J.M., Creemers, J.W.M., Ayoubi, T.A.Y., and Van de Ven, W.J.M. (1994) Furin-mediated proprotein processing activity: involvement of negatively charged amino acid residues in the substrate binding region. *Biochimie* **76,** 210 - 216.

Rouillé, Y., Westermark, G., Martin, S.K., and Steiner, D.F. (1994) Proglucagon is processed to glucagon by prohormone convertase PC2 in αTC1-6 cells. *Proc. Natl. Acad. Sci. USA* **91,** 3242 - 3246.

Rouillé, Y., Duguay, S. J., Lund, K., Furuta, M., Gong, Q., Lipkind, G., Oliva Jr., A.A., Chan, S.J., and Steiner, D.F. (1995) Proteolytic processing mechanisms in the biosynthesis of neuroendocrine peptides: the subtilisin-like proprotein convertases. *Front. Neuroendocrinol.* **16,** 322 - 361.

Rovère, C., Barbero, P., and Kitabgi, P. (1996) Evidence that PC2 is the endogenous pro-neurotensin convertase in rMTC 6-23 cells and that PC1- and PC2-transfected PC12 cells differentially process pro-neurotensin. *J. Biol. Chem.* **271,** 11368-11375.

Rudolph, A.E., Mullane, M.P., Porche-Sorbet, R., and Miletich, J.P. (1997) Expression, purification and characterization of recombinant human factor X. *Prot. Express. Purif.* **10,** 373 - 378.

Sandoval, I.V., and Bakke, O. (1994) Targeting of membrane proteins to endosomes and lysosomes. *Trends Cell Biol.* **4,** 292-297.

Schäfer, W., Stroh, A., Berghofer, S., Seiler, J., Vey, M., Kruse, M.L., Kern, H.F., Klenk, H.D., and Garten, W. (1995) Two independent targeting signals in the cytoplasmic domain determine trans-Golgi network localization and endosomal trafficking of the proprotein convertase furin. *EMBO J.* **14,** 2424-2435.

Scheiffele, P., Peranen, J., and Simons, K. (1995) N-glycans as apical sorting signals in epithelial cells. *Nature* **378,** 96 - 98.

Schlokat, U., Fischer, B.E., Herlitschka, S., Antoine, G., Preininger, A., Mohr, G., Himmelspach, M., Kistner, O., Falkner, F.G., and Dorner, F. (1996) Production of highly homogeneous and structurally intact recombinant von Willebrand Factor multimers by furin-mediated propeptide removal in vitro. *Biotechnol. Appl. Biochem.* **24,** 257 - 267.

Schlokat, U., Himmelspach, M., Falkner, F.G., and Dorner, F. (1997) Permanent gene expression in mammalian cells: gene transfer and selection, in H. Hauser and R. Wagner (eds.), *Mammalian Cell Biotechnology in Protein Production*, Walter de Gruyter, Berlin, New York, pp. 33 - 52.

Seidah, N.G., Day, R., Hamelin, J., Gaspar, A., Collard, M.W., and Chrétien, M. (1992) Testicular expression of PC4 in the rat: Molecular diversity of a novel germ cell-specific Kex2/subtilisin-like proprotein convertase. *Mol. Endocrinol.* **6,** 1559 - 1570.

Seidah, N.G., Chrétien, M., and Day, R. (1994) The family of subtilisin/kexin like pro-protein and pro-hormone convertases: divergent or shared functions. *Biochimie* **76,** 197 - 209.

Seidah, N.G., Hamelin, J., Mamarbachi, M., Dong, W., Tadros, H., Mbikay, M., Chrétien, M., and Day, R. (1996) cDNA structure, tissue distribution, and chromosomal localization of rat PC7, a novel mammalian proprotein convertase closest to yeast kexin-like proteinases. *Proc. Natl. Acad. Sci. USA* **93,** 3388 - 3393.

Seidah, N.G., Day, R., Marcinkiewicz, M., and Chrétien, M. (1998) Precursor convertases: An evolutionary ancient, cell-specific, combinatorial mechanism yielding diverse bioactive peptides and proteins. *Ann NY Acad. Sci.* **839,** 9 - 24.

Siezen, R.J., Creemers, J.W.M., and Van de Ven, W.J.M. (1994) Homology modelling of the catalytic domain of human furin. A model for the eukaryotic subtilisin-like proprotein convertases. *Eur. J. Biochem.* **222,** 255 - 266.

Simmen, T., Nobile, M., Bonifacino, J.S., and Hunziker, W. (1999) Basolateral sorting of furin in MDCK cells requires a phenylalanine-isoleucine motif together with an acidic amino acid cluster. *Mol. Cell. Biol.* **19,** 3136-3144.

Smeekens, S.P., and Steiner, D.F. (1990) Identification of a human insulinoma cDNA encoding a novel mammalian protein structurally related to the yeast dibasic processing protease Kex2. *J. Biol. Chem.* **265,** 2997 - 3000.

Smeekens, S.P., Avruch, A.S., LaMendola, J., Chan, S.J., and Steiner, D.F. (1991) Identification of a cDNA encoding a second putative prohormone convertase related to PC2 in AtT20 cells and islets of Langerhans. *Proc. Natl. Acad. Sci. USA* **88,** 340 - 344.

Smeekens S.P., Montag, A.G., Thomas, G., Albiges-Rizo, C., Carroll, R., Benig, M., Phillips, L.A., Martin, S., Ohagi, S., Gardner, P. et al. (1992) Proinsulin processing by the subtilisin-related proprotein convertases furin, PC2, and PC3. *Proc. Natl. Acad. Sci. USA* **89,** 8822 - 8826.

Spence, M.J., Sucic, J.F., Foley, B.T., and Moehring, T.J. (1995) Analysis of mutations in alleles of the fur gene from an endoprotease-deficient chinese hamster ovary cell strain. *Somat. Cell Mol. Gen.* **21,** 1-18.

Stroh, A., Schäfer, W., Berghöfer, S., Eickmann, M., Teuchert, M., Bürger, I., Klenk, H.-D., and Garten, W. (1999) A mono phenylalanine-based motif (F_{790}) and a leucine-dependent motif (LI_{760}) mediate internalization of furin. *Eur. J. Cell Biol.* **78,** 151-160.

Tanaka, S., Kurabuchi, S., Mochida, H., Kato, T., Takahashi, S., Watanabe, T., and Nakayama, K. (1996) Immunocytochemical localization of prohormone convertases PC1/PC3 and PC2 in rat pancreatic islets. *Arch. Histol. Cytol.* **59,** 261 - 271.

Takahashi, S., Hatsuzawa, K., Watanabe, T., Murakami, K., and Nakayama, K. (1994) Sequence requirements for endoproteolytic processing of precursor proteins by furin: transfection and in vitro experiments. *J. Biochem.* **116,** 47 - 52.

Takahashi, S., Nakagawa, T., Banno, T., Watanabe, T., Murakami, K., and Nakayama, K. (1995) Localization of furin to the trans-Golgi network and recycling from the cell surface involves Ser and Tyr residues within the cytoplasmic domain. *J. Biol. Chem.* **270,** 28397 - 28401.

Takahashi, S., Nakagawa, T., Kasai, K., Banno, T., Duguay, S.J., Van de Ven, W.J.M., Murakami, K., and Nakayama, K. (1995A) A second mutant allele of furin in the processing-incompetent cell line, LoVo. *J. Biol. Chem.* **270,** 26565 - 26569.

Teuchert, M., Schäfer, W., Berghöfer, S., Hoflack, B., Klenl, H.-D., and Garten, W. (1999) Sorting of furin at the trans-Golgi network. *J. Biol. Chem.* **274,** 8199-8207.

Thomas, G., Thorne, B.A., Thomas, L., Allen, R.G., Hruby, D.E., Fuller, R., and Thorner, J. (1988) Yeast KEX2 endopeptidease correctly cleaves a neuroendocrine prohormone in mammalian cells. *Science* **241,** 226-230.

Tsuji, A., Hine, C., Tamai, Y., Yonemoto, K., Mori, K., Yoshida, S., Bando, M., Sakai, E., Mori, K., Akamatsu, T., and Matsuda, Y. (1997) Genomic organization and alternative splicing of human PACE4 (SPC4), kexin-like processing endoprotease. *J. Biochem.* **122,** 438 - 452.

Turecek, P.L., Pichler, L., Auer, W., Eder, G., Varadi, K., Mitterer, A., Mundt, W., Schlokat, U., Dorner, F., Drouet, L., Roussi, J., van Mourik, J.A., and Schwarz, H.P. Evidence for extracellular processing of pro-von Willebrand Factor after infusion in animals with and without severe von Willebrand disease. *Blood,* in press.

Urbé, S., Tooze, S.A., and Barr, F.A. (1997) Formation of secretory vesicles in the biosynthetic pathway. *Biochem. Biophys. Acta* **1358,** 6-22.

Van de Loo, J.W., Creemers, J.W., Bright, N.A., Young, B.D., Roebroek, A.J., and Van de Ven, W.J. (1997) Biosynthesis, distinct post-translational modifications, and functional characterization of lymphoma proprotein convertase. *J. Biol. Chem.* **272,** 27116 - 27123.

Van den Ouweland, A.M.W., Van Groningen, J.J.M., Roebroek, A.J.M., Onnekink, C., and Van de Ven, W.J.M. (1989) Nucleotide sequence analysis of the human *fur* gene. *Nucleic Acids Res.* **17,** 7101 - 7102.

Van de Ven, W.J.M., Voorberg, J., Fontijn, R., Pannekoek, H., Van den Ouweland, A.M.W., Van Duijnhoven, H.L.P., Roebroek, A.J.M., and Siezen, R.J. (1990) Furin is a subtilisin-like pro-protein processing enzyme in higher eukaryotes. *Mol. Biol. Rep.* **14,** 265 - 275.

Velander, W.H., Johnson, J.L., Page, R.L., Russell, C.G., Subramanian, A., Wilkins, T.D., Gwazdauskas, F.C., Pittius, C., and Drohan, W.N. (1992) High-level expression of a heterologous protein in the milk of transgenic swine using the cDNA encoding human protein C. *Proc. Natl. Acad. Sci. USA* **89,** 12003 - 12007.

Velander, W.H., Page, R.L., Morcol, T., Russell, C.G., Canseco, R., Young, J.M., Drohan, W.N., Gwazdauskas, F.C., Wilkins, T.D., and Johnson, J.L. (1992A) Production of biologically active protein C in the milk of transgenic mice. *Ann. NY Acad. Sci.* **665,** 391 - 403.

Vermeer, C. (1990) γ-Carboxyglutamate-containing proteins and the vitamin K-dependent carboxylase. *Biochem. J.* **266,** 625 - 636.

Verweij, C.L., Hart, M., and Pannekoek, H. (1987) Expression of variant von Willebrand factor (vWF) cDNA in heterologous cells: requirement of the pro-polypeptide in vWF multimer formation. *EMBO J.* **6,** 2885 - 2890.

Vey, M., Schäfer, W., Berghöfer, S., Klenk, H.D., and Garten, W. (1994) Maturation of the trans-Golgi network protease furin: compartmentalization of propeptide removal, substrate cleavage, and COOH-terminal truncation. *J. Cell Biol.* **127,** 1829 - 1842.

Viale, A., Ortola, C., Hervieu, G., Furuta, M., Barbero, P., Steiner, D.F., Seidah, N.G., and Nahon, J-L. (1999) Cellular localisation and role of prohormone convertases in the processing of pro-melanin concentrating hormone in mammals. *J. Biol. Chem.* **274,** 6536-6545.

Voorhees, P., Deignan, E., van Donselaar, E., Humphrey, J., Marks, M.S., Peters, P.J., and Bonifacino, J.S. (1995) An acidic sequence within the cytoplasmic domain of furin functions as a determinant of trans-Golgi network localization and internalization from the cell surface. *EMBO J.* **14,** 4961 - 4975.

Vidricaire, G., Denault, J.B., and Leduc, R. (1993) Characterization of a secreted form of human furin endoprotease. *Biochem. Biophys. Res. Commun.* **195,** 1011 - 1018.

Walter, P., and Johnson, A.E. (1994) Signal sequence recognition and protein targeting to the endoplasmic reticulum membrane. *Annu. Rev. Cell Biol.* **10,** 87-119.

Wan, L., Molloy, S.S., Thomas, L., Liu, G., Xiang, Y., Rybak, S.L., and Thomas, G. (1998) PACS-1 defines a novel gene family of cytosolic sorting proteins required for trans-Golgi network localization. *Cell* **94,** 205-216.

Ware, J., Diuguid, D.L., Liebman, H.A., Rabiet, M.-J., Kasper, C.K., Furie, B.C., Furie, B., and Stafford, D.W. (1989). Factor IX San Dimas. *J. Biol. Chem.* **264,** 11401 - 11406.

Wasley, L.C., Rehemtulla, A., Bristol, J.A., and Kaufman, R.J. (1993) PACE/Furin can process the vitamin K-dependent pro-factor IX precursor within the secretory pathway. *J. Biol. Chem.* **268,** 8458 - 8465.

Watanabe, T., Murakami, K., and Nakayama, K. (1993) Positional and additive effects of basic amino acids on processing of precursor proteins within the constitutive secretory pathway. *FEBS Lett.* **320,** 215 - 218.

Watzke, H.H., and High, K.A. (1995) Factor X, in K.A. High and H.R. Roberts (eds.), *Molecular Basis of Thrombosis and Haemostasis,* Marcel Dekker, Inc., pp 239 - 255.

Wise, R.J., Barr, P.J., Wong, P.A., Kiefer, M.C., Brake, A.J., and Kaufman, R.J. (1990) Expression of a human proprotein processing enzyme: correct cleavage of the von Willebrand factor precursor at a paired basic amino acid site. *Proc. Natl. Acad. Sci. USA* **87,** 9378 - 9382.

Wise, R.J., Dorner, A.J., Krane, M., Pittman, D.D., and Kaufman, R.J. (1991) The role of von Willebrand factor multimers and propeptide cleavage in binding and stabilization of factor VIII. *J. Biol. Chem.* **266,** 21948 - 21955.

Wolf, D.L., Sinha, U., Hancock, T.E., Lin, P.-H., Messier, T.L., Esmon, C.T., and Church, W.R. (1991) Design of constructs for the expression of biologically active recombinant human factors X and Xa. *J. Biol. Chem.* **266,** 13726 - 13730.

Xu, H., and Shields, D. (1994) Prosomatostatin processing in permeabilized cells. *J. Biol. Chem.* **269,** 22875 - 22881.

Zhong, M., Benjannet, S., Lazure, C., Munzer, S., and Seidah, N.G. (1996) Functional analysis of human PACE4-A and PACE4-C isoforms: identification of a new PACE4-CS isoform. *FEBS Lett* **396,** 31 - 36.

Zhou, A., Martin, S., Lipkind, G., LaMendola, J., and Steiner, D.F. (1998) Regulatory roles of the P domain of the subtilisin-like prohormone convertases. *J. Biol. Chem.* **273,** 11107 - 11114.

DEVELOPMENT OF BIOPROCESSES FOR THE GENERATION OF ANTI-INFLAMMATORY, ANTI-VIRAL AND ANTI-LEUKAEMIC AGENTS

MAHMOUD MAHMOUDIAN*
Glaxo Wellcome Research and Development, Medicines Research Centre,
Gunnels Wood Road, Stevenage, Herts, SG1 2NY, UK.
E-mail: *mm6382@glaxowellcome.co.uk*
Telephone: ***+44 (0) 438 76 3759***
Facsimile: ***+44 (0) 438 76 3624***

Abstract

This chapter is based on a lecture presented at the 9th European Congress on Biotechnology held in Belgium, Brussels (11-15th July 1999) and gives an overview of several bioprocesses that have recently been developed at Glaxo Wellcome. Three bioprocesses, to produce key chiral intermediates for the synthesis of drug candidates, are described. 1) The nucleoside oxidase from *Stenotrophomonas maltophilia* was used to generate 5'-carboxylic acid derivatives of nucleoside analogues as key intermediates in the synthesis of a novel group of compounds with broad anti-inflammatory properties. The synthetic utility of the enzyme was exploited to produce 5'-carboxylates of several purine nucleoside analogues, including the carbocyclic nucleosides aristeromycin and neplanocin A on a preparative scale. The enzyme was found to have surprisingly wide substrate specificity toward unnatural nucleosides especially in the base moiety. 2) A serine-type protease from an alkalophilic *Bacillus sp.* was used to resolve racemic *N*-substituted lactams as key intermediates in the synthesis of abacavir (Ziagen™), a selective and potent reverse transcriptase inhibitor which was recently approved by the FDA for the treatment of human immunodeficiency virus (HIV) and hepatitis B virus (HBV) infections in adults and children. A simple and efficient process, using commercially available hydrolytic enzymes, was developed to produce enantiomerically pure *N*-substituted γ-lactams. 3) A co-immobilised uridine phosphorylase and purine nucleoside phosphorylase preparation from recombinant *E. coli* strains was used in the production of the anti-leukaemic agent 506U78. Fermentation and bioconversion conditions were optimised and scaled up with up to 200g/L of substrate input and the co-immobilised enzymes could be re-used several times in bioreactors. In a parallel process an immobilised preparation of *Candida antarctica* lipase was used to produce esters of 506U78 which were shown to have a better water solubility and bioavailability. Vinyl

A. Van Broekhoven et al. (eds.), Novel Frontiers in the Production of Compounds for Biomedical Use, 249–265.
© 2001 *Kluwer Academic Publishers. Printed in the Netherlands.*

acetate was used as an acyl donor and reactions were carried out in anhydrous 1,4-dioxane with up to 100g/L of substrate input.

1. Introduction

The use of enzymes and micro-organisms is widely spread in many academic laboratories for the preparation of enantiomerically pure intermediates. A number of pharmaceutical companies (notably Merck, Bristol-Myers Squibb and Schering-Plough) have set up biotransformation groups to complement their in-house chemical research programmes.

We, at Glaxo Wellcome, have also been active in this area for a number of years, routinely utilising enzymes and recombinant micro-organisms in most aspects of our drug discovery programmes; for example to i) generate complex templates, which would otherwise be difficult to synthesise chemically, for use in combinatorial libraries, ii) produce natural products and secondary metabolites *via* fermentation processes, iii) functionalise novel molecules for SAR (structure activity relationship) to improve activity, potency and /or water solubility, iv) use microbial models to generate metabolites that can be used as standards in mammalian metabolism (impurity profiling), and v) produce gram to tonne quantities of key enantiomerically pure intermediates for the synthesis of development compounds. Biocatalytic approach is often considered at the very inception of a chemical route for example, to by-pass an environmentally "unfriendly" step (e.g. use of transitional metal oxidants such as potassium permanganate) or even replacing several chemical stages with a single enzymatic step.

This chapter is based on a lecture presented at the *9th European Congress on Biotechnology* held in Belgium, Brussels (11-15th July 1999). Biocatalytic production of several key chiral intermediates in the synthesis of anti-inflammatory, anti-viral and anti-leukaemic agents is reviewed. These include i) a nucleoside oxidase from *Stenotrophomonas maltophilia* to oxidise unnatural purine nucleosides, ii) a serine-type protease from an alkalophilic *Bacillus sp.* to resolve racemic *N*-substituted lactams, and iii) a co-immobilised uridine phosphorylase and purine nucleoside phosphorylase preparation from recombinant *E. coli* strains that catalyse base transfer reactions.

These case studies highlight the close integration of biocatalytic approaches within exploratory, discovery and chemical development programmes at Glaxo Wellcome. The potential benefits of biotransformations will be fully realised as more companies adopt a similar strategy. Biocatalysis will undoubtedly continue to play a pivotal role in both academic research and industrial laboratories in the new millennium.

2. Process development for the generation of nucleoside 5'-carboxylic acids

The nucleoside analogue **[2]**, (1-[2-chloro-6-[(2,2-diphenylethyl)amino]-9*H*-purin-9-yl]-1-deoxy-β-D-ribofuranuronic acid) is a key intermediate in the synthesis of a novel group of compounds with broad anti-inflammatory properties (Gregson *et al.*, 1994), Figure 1.

Although chemical oxidation of the 5′-hydroxyl of precursor **[1]**, (2-chloro-N-(2,2-diphenylethyl)-adenosine), using transition metal oxidants (such as $KMnO_4$), has been carried out in good chemical yields, this has posed considerable problems for scale up due to the heterogeneous nature of the reaction. The lack of regioselectivity of the oxidation reaction also necessitates the use of a protection/de-protection sequence to protect the 2′/3′-hydroxyl groups, thus introducing two extra steps into the route (Figure 1).

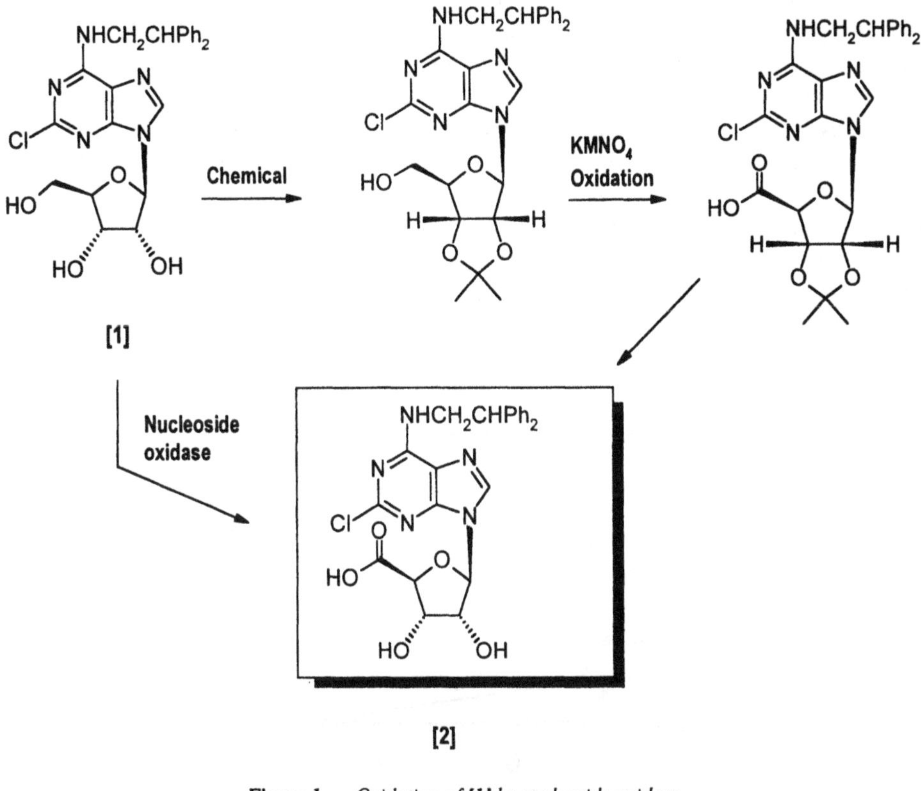

Figure 1 *Oxidation of [1] by nucleoside oxidase.*

There are also considerable environmental and handling implications with the use of this reagent. Enzymatic oxidation of a nucleoside precursor offers potential for an improved synthesis; it is environmentally clean and obviates the need for protection of other functional groups.

Nucleoside oxidase is produced by *Pseudomonas* species and related Gram-negative bacteria (Isono and Hoshino, 1992; Isono *et al.*, 1989a-b). Crude extracts of *Stenotrophomonas maltophilia* exhibiting nucleoside oxidase activity have been reported to convert the 5′-hydroxyl groups of natural purine and pyrimidine nucleosides to their corresponding carboxylic acids at analytical scale (Misaki *et al.*, 1983). The enzyme has been purified to homogeneity and catalyses a two step oxidation of a nucleoside *via* an aldehyde intermediate consuming one molecule of molecular oxygen (Isono *et al.*, 1989 a-b).

Small scale cultures of *Stenotrophomonas maltophilia* (FERM BP-2252), for production of nucleoside oxidase, were grown routinely at 25°C in 50 ml volumes of medium (yeast extract, 25 g/L; glucose, 30 g/L; K_2HPO_4, 1 g/L; KCl, 1 g/L; $MgSO_4$, 0.5 g/L) in shake flasks (250 rev/min, 5 cm throw). For larger scale cultivations, 450-500 L fermentations were inoculated from two 8-hr old 5-L seed fermenters grown as above (2% v/v).

Crude extracts of *S. maltophilia* were initially found to oxidise selectively the 5′-hydroxyl of [1] to generate [2] on a mg scale (Mahmoudian *et al.*, 1998). The crude extracts also cleaved the substrate slowly to release the purine base, presumably *via* a phosphorylase activity (Figure 2); this was, however, considered not to be significant when cells with a high nucleoside oxidase activity were used and crude extracts were found to be satisfactory for use in biotransformations without further purification. Crude extracts containing nucleoside oxidase were used in bioconversions with up to 20 g/L of substrate input. There was no evidence of substrate inhibition at high substrate concentrations; [1] was quickly oxidised to [2] and reactions had gone to completion within 24 hr. Oxidation of [1] was shown to go *via* a transient formation of the corresponding aldehyde intermediate to produce the carboxylic acid in high chemical yields (Figure 2).

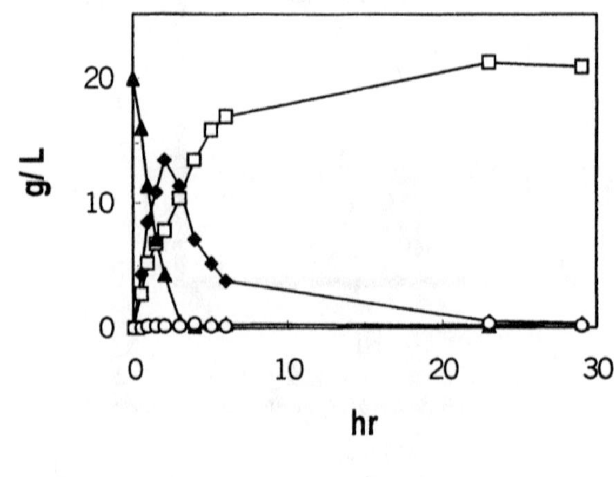

Figure 2 *Reaction profile using crude extracts of* Stenotrophomonas maltophilia. *1* (▲), *2* (□), *aldehyde* (◆), *base* (○). *The reaction was carried out at room temperature in a magnetically-stirred flask containing 3.5 ml of a clarified crude lysate (pH 6). [1] was added as a solid (70 mg) and stirred to obtain a homogeneous suspension. Periodically, samples were removed, diluted into the mobile phase and the clarified solution assayed by HPLC. [Reproduced from Mahmoudian et al. (1998) with the permission of Elsevier Science Ltd.].*

To simplify downstream processing and to allow the enzyme to be re-used nucleoside oxidase was immobilised directly from crude homogenates of *S. maltophilia* onto

Eupergit-C beads. Typically, up to 2 kg (wet weight) cell paste was suspended to give a 20% (w/v) suspension in lysis buffer (potassium phosphate, 50 mM; EDTA, 1 mM; DTT, 1 mM; PMSF, 1 mM; pH, 6-7) and disrupted by passing the suspension through a Manton-Gaulin homogeniser operating at 8,000 psi, or through a Dynomill (flow rate, 20 L/hr; speed, 3200 rev/min; bead size, 0.25-0.5 μm) to obtain the crude extract. Eupergit-C was added directly to the crude extract (10 g dry beads/ g protein) and the mixtures were left to stand at room temperature with occasional mixing to ensure adequate infiltration. After approximately 7 days, beads were washed with cold buffer (Tris-HCl, 100 mM; EDTA, 1 mM; DTT, 1 mM; NaCl, 0.5 M; ρ-hydroxybenzoic acid ethyl ester, 500 ppm; pH 7.5) then stored in the same buffer minus NaCl at 4°C.

In free enzyme reactions, when the enzyme to substrate ratio was low, oxidation was not only slow but of limited duration and stopped before all the substrate was used; this was also evident with the immobilised enzyme (data not shown). Nucleoside oxidase is known to be able to carry out a laccase reaction which is dependent on and stoichiometric with nucleoside oxidation (Isono and Hoshino, 1992). Working with crude extracts there was evidence that addition of a laccase substrate such as quinol increased the amount of nucleoside that could be oxidised before the reaction ceased. With the immobilised enzyme, although initial bioconversion rates were found to be significantly lower in the presence of 1 g/L of quinol, even at a substrate concentration of 20 g/L, reactions continued to completion in the presence of quinol, whereas the corresponding control reactions (without quinol) stopped at a product concentration around 15 g/L (Figure 3).

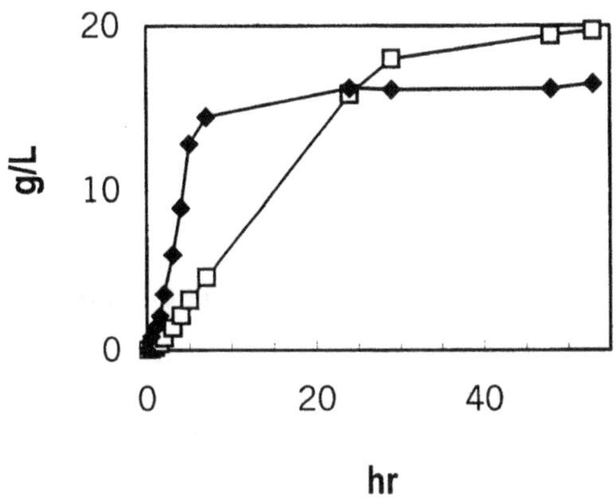

Figure 3 *Effect of quinol on bioconversions using the immobilised enzyme. The reaction (5 ml) was carried out in 50 mM potassium phosphate buffer (pH 6).* ***[1]****, (100 mg), was added as a solid and stirred at room temperature to obtain a homogeneous suspension. The reaction was started by the addition of washed immobilised beads at 40% (w/v). At intervals, samples were removed, cleared of enzyme beads and analysed by hplc. Reactions were carried out either in the presence (□) or absence (♦) of quinol at 1 g/L. [Reproduced from Mahmoudian* et al. *(1998) with the permission of Elsevier Science Ltd.].*

This was attributed to quinol having a protective role by stabilising the enzyme during bioconversions. It was found that the same batch of enzyme could now be re-used for 5 cycles and bioconversions were scaled up to produce larger quantities of [2] for further evaluation.

The substrate specificity of nucleoside oxidase towards natural nucleosides has been described previously (Isono and Hoshino, 1992; Isono *et al.*, 1989a-b). The enzyme accepts purine and pyrimidine nucleosides having ribose, deoxyribose or arabinose as a sugar moiety but does not oxidise the sugar in the absence of a base. We have extended these studies to investigate the effect of the enzyme on ribosides of unnatural purine bases and on carbocyclic nucleosides (Figure 4).

As the natural purine nucleosides (adenosine, guanosine, inosine, xanthosine) are all good substrates it appears that the enzyme is quite tolerant of changes in the purine base, accepting either amino or carbonyl functionality at both positions 2 and 6. It was, however, surprising to find that **1** (Figure 1), with the bulky diphenylethylamino group at the 6-position, was also a good substrate. We have shown further that the enzyme is tolerant of different functionality at the 2-position including chloro (**3**) or phenylethyl amino (**4**), Figure 4.

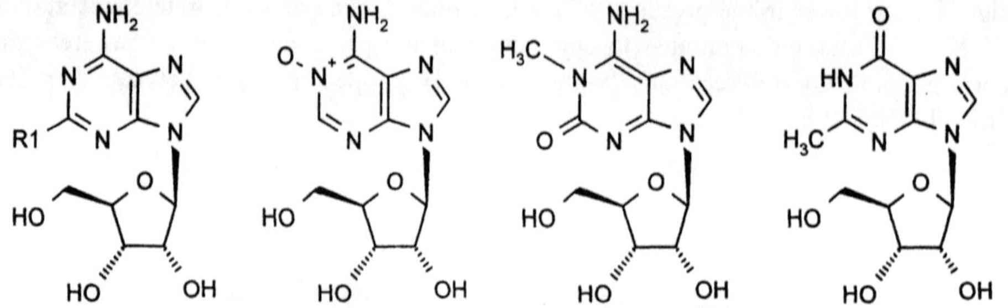

3 R1= Cl **5** **6** **7**

4 R1=NHCH$_2$CH$_2$(C$_6$H$_5$)

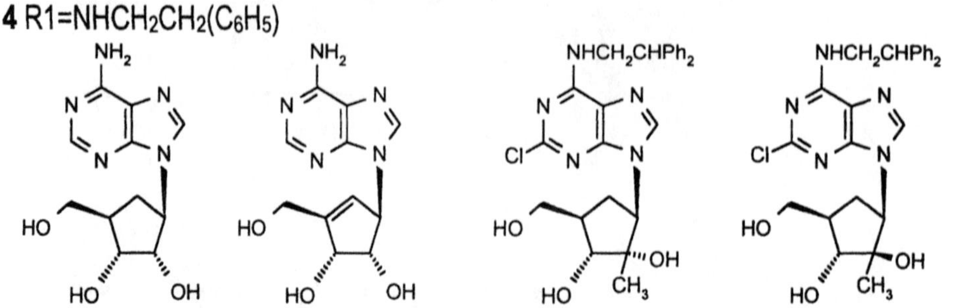

8 **9** **10** **11**

Figure 4 *Substrate specificity of nucleoside oxidase from* S. maltophilia. *[Reproduced from Mahmoudian* et al. *(1998) with the permission of Elsevier Science Ltd.].*

Furthermore, the nitrogen at the 1-position can be modified to the N-oxide (**5**) or methylated in the case of methyl isoguanosine (**6**). Interestingly, in the inosine series, with 2-methylinosine (**7**) as a substrate, the reaction tended to stop at the 5′-aldehyde intermediate and was not further oxidised.

Carbocyclic nucleosides such as aristeromycin (**8**) and neplanocin A (**9**) were also selectively oxidised to give the 5′-carboxylates (Figure 4). The enzyme did not, however, accept a methyl group at the 2′-position of the carbocyclic moiety (**10, 11**) or the 2′,3′ acetonides of any of the natural nucleosides.

The enzyme was found to have surprisingly wide substrate specificity toward unnatural nucleosides especially in the base moiety (Figure 4). This has formed the basis of a scaleable process for the generation of nucleoside 5′-carboxylic acid derivatives.

3. Abacavir (Ziagen™)

Abacavir [**12**] (Ziagen ™), a novel 2-aminopurine nucleoside analogue, is a selective and potent reverse transcriptase inhibitor which was recently approved by the FDA for the treatment of human immunodeficiency virus (HIV) and hepatitis B virus (HBV) infections in adults and children (Daluge, 1991), Figure 5.

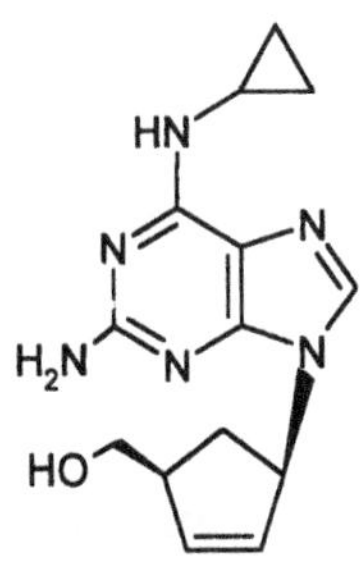

[12]

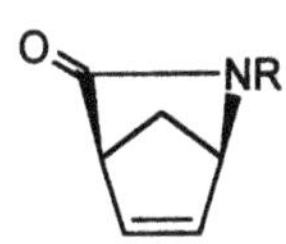

[15] R= $COOC(CH_3)_3$, [BOC]
[16] R= $COCH_3$, [acetyl]
[17] R= H

Figure 5 *Structures of abacavir and related* N-*substituted γ-lactams.*

Abacavir is lipophilic and water soluble, synergistic *in vivo* with protease and other reverse transcriptase inhibitors, as well as being well-tolerated and orally absorbed with significant CNS penetration (Daluge *et al.*, 1997). Abacavir is in the same class of agent as Epivir™ (3TC) and Retrovir™ (AZT). It provides a well-tolerated and compact dosing regimen (one tablet twice a day, with no dietary restrictions), which can be particularly important in improving patient adherence. Current triple combination regimens can be difficult to adhere to; patients may have to take between 10 to 20 pills a day, at different times of the day, and depending on the drugs involved, either with or without food and drink. The combination of abacavir and combivir™ (the fixed dose formulation of Epivir and Retrovir) constitutes two pills twice a day. This new simple dosing regimen therefore, addresses one of the most significant clinical challenges (patient adherence) faced in HIV treatment today (Glaxo Wellcome press release, 1998).

Prior to our involvement with abacavir at Glaxo Wellcome, we had spent a number of years investigating the potential of carbocyclic nucleosides, such as cabovir [(-)-**13**], as antiviral agents (Figure 6).

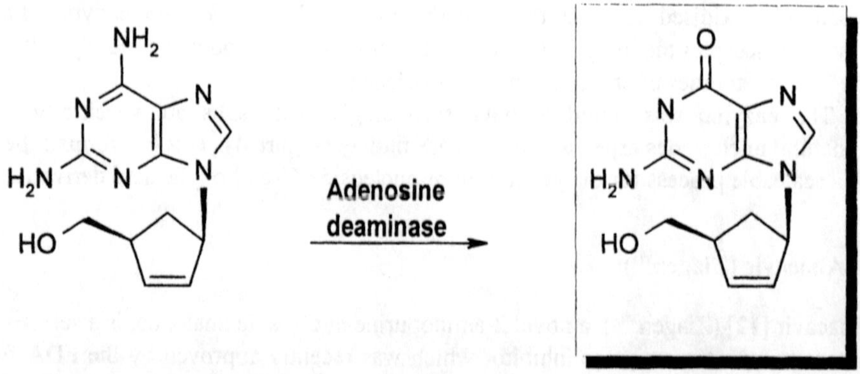

(-)-13 **(-)-14**

Figure 6 *Production of (-)-cabovir with adenosine deaminase.*

We had developed a chemoenzymatic route to produce (-)-**14** on a kilogram scale using adenosine deaminase (Mahmoudian and Dawson, 1997), Figure 6. In contrast to abacavir, however, carbovir (2',3'-didehydro-2',3'-dideoxycarbocyclic nucleoside, c-d4G), suffers form poor oral absorption, limited brain penetration, low aqueous solubility and potential for renal and cardiac toxicity (Mahmoudian and Dawson, 1997). Despite the structural similarity between abacavir and carbovir (Figures 5, 6), it is therefore, remarkable that an apparently small change in the molecule can have such a profound effect on its biological activity.

An important step in the manufacture of abacavir is the preparation of enantiomerically pure *N*-substituted γ-lactams [**15, 16**], Figure 5. The γ-lactam [**17**], (2-azabicyclo[2.2.1]hept-5-en-3-one), is a potentially useful intermediate, which may be used in the synthesis of abacavir. The Exeter and Chiroscience groups have developed a process for the resolution of racemic [17], using γ-lactamase containing micro-organisms such as *Pseudomonas solanacearum* NCIMB 40249 and *Rhodococcus sp.* NCIMB 40213 (Evans and Roberts, 1991; Evans *et al.*, 1992), Figure 7. These enzymes are, however, not commercially available for general use.

The *N*-substituted γ-lactam [**15**] is a key intermediate in the synthesis of abacavir [**12**]. We therefore, embarked on a programme to produce [**15**] in an optically pure form. It was argued that by activating the lactam ring with acyl protecting groups such as BOC or acetyl (Figure 5) we may be able to find a conventional hydrolytic enzyme, rather than needing a specialised γ-lactamase, that would hydrolyse the lactam bond of [**15**] and [**16**] enantioselectively.

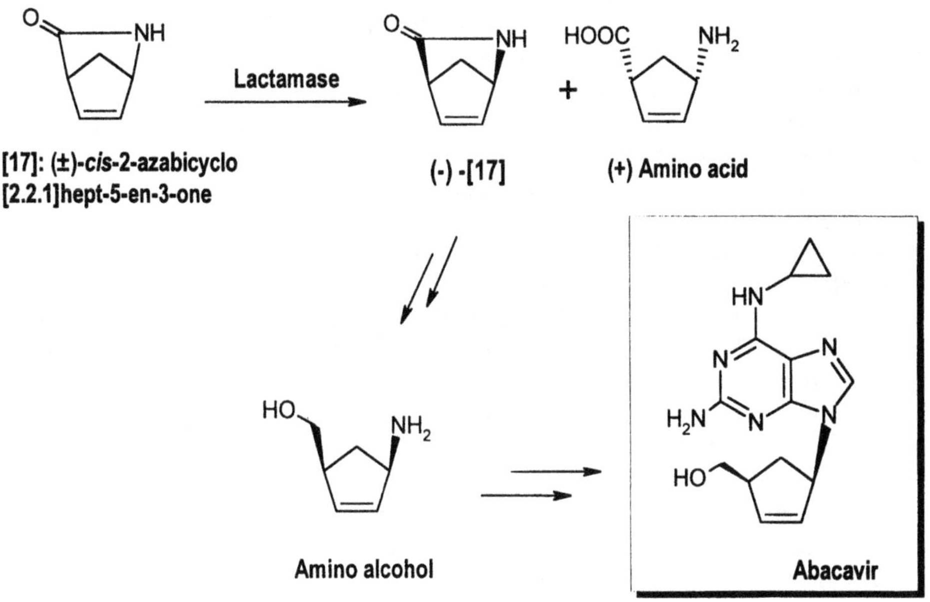

Figure 7 *Chemoenzymatic production of abacavir.*

A number of commercially available hydrolytic enzymes were screened for the ability to hydrolyse the lactam bond of racemic **[15]**, [(±)-*tert* butyl 3-oxo-2-azabicyclo(2.2.1)hept-5-ene-2-carboxylate], enantioselectively. There was substantial chemical hydrolysis of the *N*-BOC protecting group under aqueous conditions, but this could be minimised if reactions contained up to 50% (v/v) of organic solvents such as tetrahydrofuran.

Several enzymes were found to hydrolyse (+) 1*S*, 4*R* -**[15]** to the corresponding *N*-acyl amino acid leaving behind the residual (-) 1*R*, 4*S* -**[15]** of the correct absolute configuration for synthesis of abacavir (Figure 8).

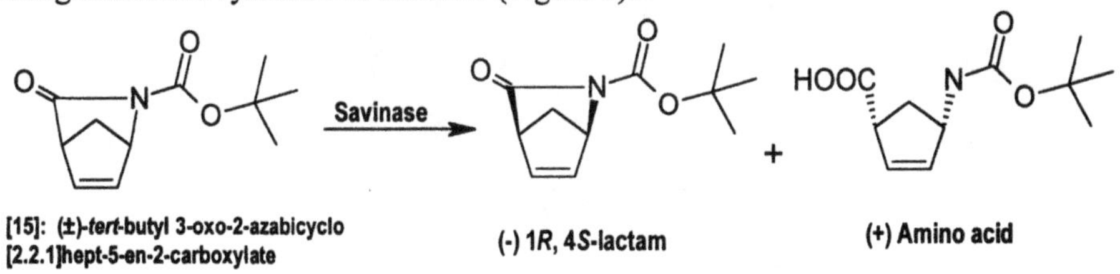

Figure 8 *Savinase-catalysed resolution of [15].*

These enzymes were pig liver esterase (ALTUS), *Bacillus* sp. protease (ALTUS), *Subtilisin carlsberg* (ALTUS), Neutrase (*Bacillus subtilis*, NOVO), Novozyme 243 (*Bacillus licheniformis*, NOVO), Alcalase (*Bacillus licheniformis*, NOVO), Savinase

(*Bacillus* sp. NOVO), porcine pancreas lipase (Biocatalysts), Flavorpro-192 (peptidase, Biocatalysts), Flavorpro-373 (glutaminase, Biocatalysts), Promod-TP (endopeptidase, Biocatalysts), lipase-CE (*Humicola lanuginosa*, Amano), protease-M (*Aspergillus* sp., Amano), prozyme-6 (*Aspergillus sp.*, Amano), lipase PGE (calf tongue root and salivary gland, Amano) and *Aspergillus sp.* acylase (Sigma). Savinase was found to be highly enantioselective (Figure 8). Typically, reactions were carried out at 30°C in phosphate buffer (pH 8.0) containing up to 50% (v/v) organic solvent such as tetrahydrofuran and up to 100 g/L of racemic [15]. Reaction mixtures were monitored by reverse-phase and chiral-hplc. Upon completion of reactions (50% conversion) the enantiomeric excess (ee) of (-)-[15] was better than 99%. The reaction mixture was filtered and (-)-[15] was isolated in good chemical yield (84% theory) by extraction into cyclohexane followed by evaporation of organic solvent.

Similarly, Savinase hydrolysed the lactam bond of racemic [**16**], (*cis*-2-acetyl-2-azabicyclo[2.2.1]hept-5-en-3-one), enantioselectively to afford (-) 1*R*, 4*S* -[**16**], ee> 99% (Figure 8). Interestingly, Savinase did not hydrolyse the unactivated racemic [**17**]. Savinase (Subtilisin, *EC 3.4.21.62*) is a serine-type protease, which is produced by submerged fermentation of a genetically modified alkalophilic *Bacillus* sp. This enzyme is inexpensive and available in bulk from Novo Nordisk for use in the detergent industry (Novo, 1996). This has formed the basis for development of a simple and scaleable process for preparation of optically pure *N*-substituted γ-lactams (Mahmoudian *et al.*, 1999a)

4. Production of the anti-leukaemic agent 506U78

506U78 (2-amino-9-β-*D*-arabinofuranoyl-6-methoxy-9*H*-purine) is being developed by Glaxo Wellcome for the treatment of leukaemia (Averett *et al.*, 1991; Lambe *et al.* 1995). 506U78 is a pro-drug of *ara*-G (9-β-*D*-arabinofuranosyl guanine), Figure 9.

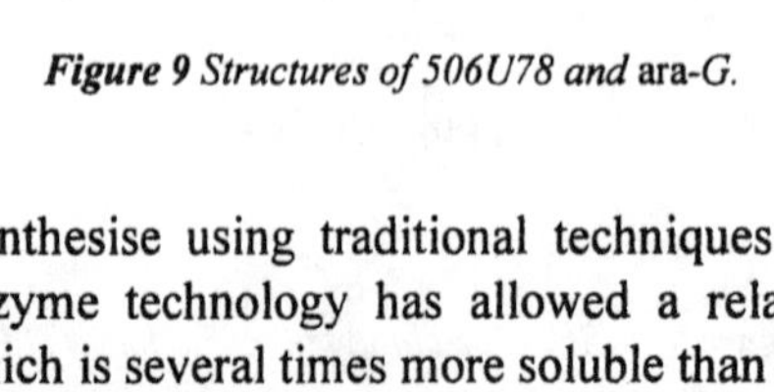

Figure 9 *Structures of 506U78 and* ara-*G.*

Ara-G is difficult to synthesise using traditional techniques and it is poorly water-soluble. The use of enzyme technology has allowed a relatively easy synthesis of 506U78, a compound which is several times more soluble than *ara*-G. 506U78 is rapidly demethoxylated *in vivo* by adenosine deaminase to *ara*-G (Krenisky *et al.*, 1996), and thus, many of the obstacles to the use of *ara*-G have been circumvented by the development of 506U78.

Uridine phosphorylase (Upase) and purine nucleoside phosphorylase (PNP) have been previously reported by Krenitsky *et al.* (1981) to catalyse the synthesis of purine arabinosides. Upase catalyses net transfer of arabinose from a pyrimidine to a purine base with retention of the β-*D* configuration. Our colleagues at the former Burroughs Wellcome extended these studies to the synthesis of 506U78 (Figure 10).

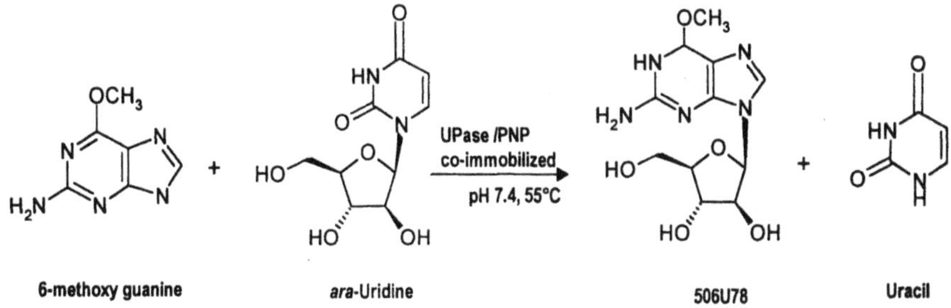

Figure 10 *Enzymatic production of 506U78.*

Each enzyme has been cloned and overexpressed in independent *E. coli* strains containing the corresponding genes on a multicopy plasmid. In the UK, to simplify regulatory approval, the preparation of organism banks and the fermentation process were carried out in media free of animal derived ingredients; for example tryptone could be replaced with soya peptone without compromising cell-yields and expression levels of each enzyme. Seed cultures (250 ml) of recombinant *E. coli* strains, for production Upase and PNP, (Glaxo Wellcome collection), were grown at 37°C in a medium containing soya peptone (10 g/L), yeast extract (5 g/L) and NaCl (5 g/L). All media were supplemented with tetracycline (20 mg/L final conc.) for Upase, or kanamycin (50 mg/L) for PNP strains. Cultures were transferred to a production medium (soya peptone, 10 g/L; yeast extract, 5 g/L; potassium phosphate, 13.1 g/L; sodium ammonium phosphate, 3.5 g/L; citric acid, 2 g/L; $MgSO_4.7H_2O$, 0.2 g/L; glycerol, 6.3 g/L post-autoclave) when growth was in early to mid log phase. Production stage flasks (2L) were used to inoculate fermentors. Fermentation conditions were optimised to attain best enzyme production.

Our colleagues at the former Burroughs Wellcome initially found that crude lysates of these enzymes not to be stable at the high reaction temperature (55°C). These could, however, be stabilised by direct co-immobilisation of the enzymes onto an ion-exchange support (DEAE-52); this also facilitated removal and recovery of the enzyme. Typically, bioconversions were carried out with up to 200 g/L substrate input and the co-immobilised enzymes could be re-used several times in bioreactors. Upon completion of each cycle, beads were filtered and washed with the reaction buffer (10mM potassium phosphate, pH 7.4). The crude product was crystallised from hot aqueous base (pH 11); this was followed by a final purification step that involved product precipitation from hot water, after a charcoal treatment, to remove any potential endotoxins. This has formed the basis of a scaleable process for production of 506U78. The process was transferred to factory to produce larger quantities for further evaluation.

Concurrent with this work, it was shown that esters of 506U78, in particular 5'-acetate, have a better water solubility and bioavailability (Moorman *et al.,* 1992). Chemical methods for its synthesis included selective acylation, or selective deacylation of the corresponding tri-acetate, all of which resulted in a poor selectivity (Eaddy, 1997). Furthermore such methods required chromatography, to remove other undesired acetates, which would not have been amenable to scale up. Preliminary work in the former Burroughs Wellcome indicated that an enzymatic approach would be feasible (Eaddy and Liu, 1994; Eaddy and Corona, 1998). In order to find a more selective enzyme, we embarked on a programme to screen hydrolytic enzymes that would regioselectively acylate the 5'-hydroxyl position, and to optimise conditions for its production.

Hydrolytic enzymes have been used extensively in industry for the production of chiral pharmaceutical intermediates. Lipases, in particular, are well known for their ability to catalyse acyl transfer reactions, in neat organic solvents, in the presence of suitable activated acyl donors such as vinylic esters (Stead *et al.*, 1996; Nishio *et al.*, 1989; Fukazawa & Hashimoto, 1993; Lauman *et al.*, 1989; Basavaiah & Krishna, 1994; Hoing & Seufer-Wasserthal, 1990).

Initially, bioconversions were carried out in *tert*-amyl alcohol, with up to 50% (v/v) vinyl acetate as the acyl donor and triethylamine (TEA) (10% v/v) as an acid scavenger. Reactions were monitored for up 4 days using an automated hplc system (Anachem SK233 Reaction Station) equipped with an online sample retrieval and analysis. Several enzymes were found to acylate 506U78 regioselectively at the 5'-position (Figure 11).

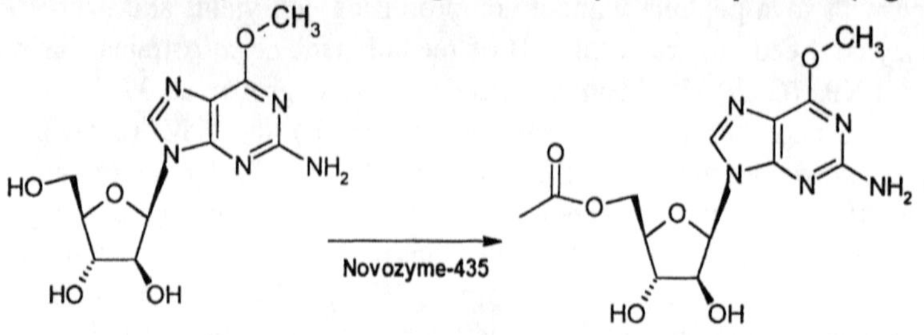

Figure 11 *Enzymatic acylation of 506U78.*

There were, however, other related impurities (up to 20% of the other acylated products) present. These enzymes were Novozyme-435 (*Candida antarctica* lipase, Novo), Alcalase (*Bacillus licheniformis* protease, Novo), Lipozyme IM (*Mucor miehei* lipase, Novo), CLEC-BL (*Bacillus licheniformis* protease, Altus), Savinase (*Bacillus sp.* protease, Novo), Novozyme-243 (*Bacillus licheniformis* protease, Novo), *Alcaligenes sp.* lipase (Altus) and Lipolase (Novo). Novozyme-435, an immobilised preparation of *Candida antarctica* lipase, was selected for further work.

Novozyme-435 is an immobilised preparation of a thermostable lipase. The enzyme is a triacylglycerol hydrolase (*EC 3.1.1.3*) and is manufactured by Novo Nordisk using recombinant DNA technology. The gene encoding for the lipase has been transferred from selected strains of *Candida antarctica* to the host organism, *Aspergillus oryzae.* The enzyme produced by this host organism is immobilised on a macroporous acrylic resin (Novo, 1997).

It was envisaged that the use of chromatography, to remove the undesired impurities, would not be a viable option for the scale up and we therefore, set out at this stage to improve the impurity (*ca.* 20%) profile of enzymatic reactions. Several parameters (solvents, acylating agents, temperature, and substrate concentration) were systematically investigated and reactions were optimised to minimise the impurity levels (Mahmoudian *et al.*, 1999b). For example, we found that that the choice of solvent had a profound effect on enzymatic acylations. Reaction rates were significantly lower in acetone, acetonitrile, cyclohexane, dimethyl sulphoxide, methanol, propan-2-ol, *tert*-methyl butyl ether, tetrahydrofuran, toluene or neat vinyl acetate compared with similar reactions in 1,4-dioxane, *N, N*-dimethylformamide (DMF) or *tert*-amyl alcohol (Table 1). Changing the acyl donor (vinyl acetate, isopropenyl acetate or acetic anhydride; up to 50% v/v), in combination with DMF or *tert*-amyl alcohol, also had a significant effect on reaction rates. For example vinyl acetate was the best acyl donor with DMF whereas, better conversions were obtained with isopropenyl acetate in *tert*-amyl alcohol (Table 1). Addition of TEA (10% v/v) to reaction mixtures also improved rates but this resulted in increased levels (up to 20%) of other related impurities. Lowering the concentration of TEA to 1% (v/v) did not, however, significantly affect either bioconversion rates or impurity profile (data not shown).

Since, bioconversions in 1,4-dioxane had generally gone to completion (Table 1) and had a better impurity profile than similar reactions in DMF or *tert*-amyl alcohol, we decided to re-examine two other bulk-available enzymes, Savinase and Lipolase (Novo), in 1,4-dioxane to compare reaction rates with Novozyme-435. Savinase is a proteolytic enzyme (Subtilisin) prepared by submerged fermentation of an alkalophilic *Bacillus sp.* whereas, Lipolase is a recombinant lipase of fungal origin (Novo 1996).

***Table 1** Effect of solvents on acylation of 506U78*

*✝ Moles of 5'-monoacylated product / moles of 506U78 (100). *Only 2-5% conversion was evident with isopropenyl acetate or acetic anhydride as the acyl donor. ✦Isopropenyl acetate was used as the acyl donor (43% conversion with vinyl acetate, whereas, 5% acetic anhydride). The reaction (5ml) was carried out in solvents containing vinyl acetate (2.5 ml) as the acyl donor. 506U78 (250 mg) was added as a solid and stirred at 30℃ to obtain a homogeneous suspension. The reaction was started by the addition of immobilised Novozyme-435 (500 mg). Reactions were analysed after 4d; samples were removed, cleared of enzyme beads and analysed by hplc. [Reproduced from Mahmoudian* et al. *(1999b) with the permission of Portland Press Ltd.].*

Solvents	% mole conversion ✝
Acetone	5
Acetonitrile	2
Cyclohexane	1

Table 1 Effect of solvents on acylation of 506U78

Solvents	% mole conversion ◆
Dimethyl sulphoxide	3
1,4-Dioxane	>99
Methanol	9
N, N-dimethylformamide †	54
Propan-2-ol	14
tert-Amyl alcohol ‡	74
tert-Methyl butyl ether	4
Tetrahydrofuran	40
Toluene	2
Vinyl acetate	3

◆ Moles of 5'-monoacylated product / moles of 506U78 (100). †Only 2-5% conversion was evident with isopropenyl acetate or acetic anhydride as the acyl donor. ‡ Isopropenyl acetate was used as the acyl donor (43% conversion with vinyl acetate, whereas, 5% acetic anhydride). The reaction (5ml) was carried out in solvents containing vinyl acetate (2.5 ml) as the acyl donor. 506U78 (250 mg) was added as a solid and stirred at 30°C to obtain a homogeneous suspension. The reaction was started by the addition of immobilised Novozyme-435 (500 mg). Reactions were analysed after 4d; samples were removed, cleared of enzyme beads and analysed by hplc. [Reproduced from Mahmoudian et al. *(1999b) with the permission of Portland Press Ltd.].*

Both enzymes are non-immobilised granulated preparations specially formulated for detergent industry. Bioconversion rates were much lower with either Savinase or Lipolase compared with similar reactions using Novozyme-435 (Table 2). This was in part attributed to the enhanced operational stability of Novozyme-435, in very polar organic solvents such as 1,4-dioxane (log P, -1.1), rendered by immobilisation onto a support. Interestingly as expected, the non-immobilised preparations of Savinase and Lipolase were less stable in 1,4-dioxane but worked better in a less polar solvent such as *tert*-amyl alcohol (log P, > +0.8), Table 2.

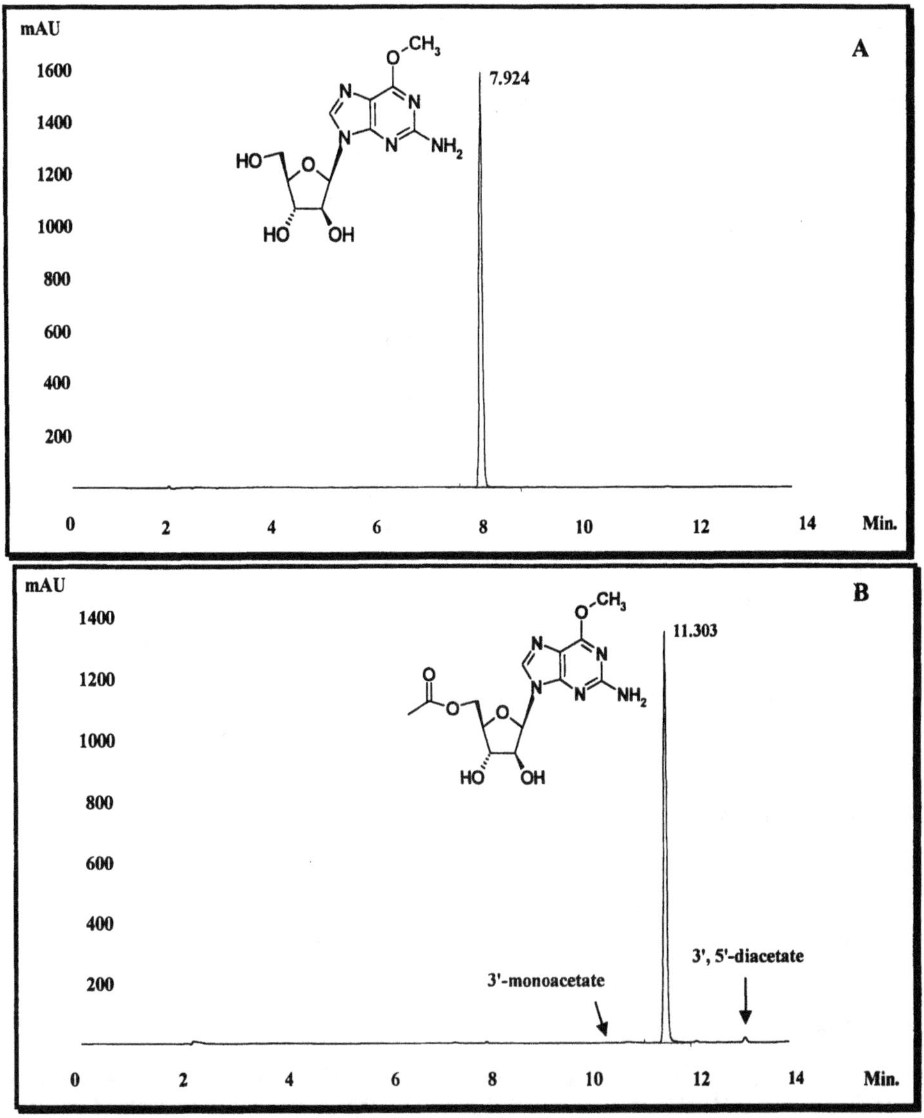

Figure 12A-B *Reaction profile for the regioselective acylation of 506U78 using Novozyme-435. The reaction (5ml) was carried out in anhydrous 1,4-dioxane containing vinyl acetate (1 ml) as the acyl donor. 506U78 (250 mg) was added as a solid and the reaction mixture was stirred at 30°C to obtain a homogeneous suspension. The reaction was started by the addition of immobilised enzyme (500mg). At intervals, samples were removed, cleared of enzyme beads and analysed by hplc.* ***A****: At the start of reaction.* ***B****: After 2d. [Reproduced from Mahmoudian* et al. *(1999b) with the permission of Portland Press Ltd.].*

Table 2 Comparison of bulk-available enzymes on acylation of 506U78

Reactions were carried out in anhydrous 1,4-dioxane with vinyl acetate as the acyl donor. Other conditions are as described in Table 1. ♦Reactions were carried out in tert-amyl alcohol and isopropenyl acetate. [Reproduced from Mahmoudian et al. (1999b) with the permission of Portland Press Ltd.].

Enzymes	% mole conversion
Savinase	5 (35♦)
Lipolase	10 (28♦)
Novozyme-435	>99 (74♦)

Preparative reactions were carried out under optimised conditions. Typically, bioconversions were carried out in anhydrous 1,4-dioxane, with vinyl acetate (20-50% v/v) as the acyl donor, and up to 100 g/L of 506U78. On completion of the reaction, the enzyme beads were removed by filtration and, after washing with neat methanol, were stored at 4°C prior to re-use. The immobilised enzyme was found to be stable when stored at 4°C and could be re-used for another reaction cycle. The level of related impurities was less than 0.5% and the reaction mixture could simply be evaporated to dryness to obtain the product in acceptable purity (Figure 12A-B). This formed the basis of a scaleable process for production of 5'-acetyl 506U78 (Mahmoudian *et al.*, 1999b).

Acknowledgements

This article represents contributions from a large team of scientists at Glaxo Wellcome Research and Development. The author gratefully acknowledges contributions of colleagues from Glaxo Wellcome Operations for the scale up of these processes.

References

Averett, D.R., Koszalka, G.W., Fyfe, J.A., Roberts G.B., Purifoy, D.J.M. and Krenitsky, T.A. (1991) 6-Methoxypurine arabinoside as a selective and potent inhibitor of varicella-zoster virus. *Antimicrob. Agents Chemother. 35*, 851-857.

Basavaiah, D. and Krishna, P.R. (1994) Pig liver powder as biocatalyst; enantioselective syntheis of *trans*-2-alkoxycyclohexan-1-ols. *Tetrahedron 50*, 10521-10530.

Daluge, S.M. Therapeutic nucleosides. (1991) *Eur. Pat. Appl.* 0 434 450.

Daluge, S.M., Good, S.S., Faletto, M.B., Miller, W.H., St. Clair, M.H., Boone, L.R., Tisdale, M., Parry, N.R., Peardon, J.E., Dornsife, R.E., Averett, D.R. and Krenitsky, T.A. (1997) 1592U89, a novel carbocyclic nucleoside analogue with potent, selective anti-HIV activity. *Antimicrob. Agents Chemother. 41*, 1082-1093.

Eaddy, J. (1997) Glaxo Wellcome R & D, USA. Unpublished work.

Eaddy, J. and Corona, J. (1998) Glaxo Wellcome R & D, USA. Unpublished work.

Eaddy, J. and Liu, W. (1994) Burroughs Wellcome Co. USA. Unpublished work.

Evans, C.T. and Roberts, S.M. (1991) Chiral azabicycloheptanone and a process for their preparation. *Eur. Pat. Appl.* 0424 064 B1.

Evans, C.T., Roberts, S.M., Shoberu, K.A. and Sutherland, A.G. (1992) Potential use of carbocyclic nucleosides for the treatment of AIDS: chemoenzymatic synthesis of the enantiomers of carbovir. *J Chem Soc Perkin Trans* (1), pp. 589-592.

Fukazawa, T. and Hashimoto, T. (1993) Synthesis of enantiomerically pure *R*-2-cycloalken-1-ols using highly enantioselective enzymatic transesterification. *Tetrahedron Asymm. 4*, 2323-2326.

Glaxo Wellcome press release. 1998 Abacavir (ziagen™).

Gregson, M., Ayres, B.E., Ewan, G.B., Ellis, F. and Knight, J. (1994) 2,6-Diaminopurine derivatives. *PCT Int. Appl., WO* 9417090.

Honig, H. and Seufer-Wasserthal, P. A general method for the separation of enantiomeric *trans*-2-substituted cyclohexanols. (1990) *Synthesis* pp. 1137-1140.

Isono, Y. and Hoshino, M. (1992) Nucleoside oxidase and assay method utilising same. *US Pat. Appl.* 5,156,955.

Isono, Y., Sudo, T. and Hoshino, M. (1989a) Purification and reaction of a new enzyme, nucleoside oxidase. *Agric. Biol. Chem., 53*, 1663-1669.

Isono, Y., Sudo, T. and Hoshino, M. (1989b) Properties of a new enzyme, nucleoside oxidase, frown *Pseudomonas maltophilia* LB-84. *Agric. Biol. Chem., 53*, 1671-1677.

Krenitsky, T.A., Averett, D.R., Moorman, A.R., Koszalka, G.W., Chamberlain S.D. and Wolberg, G. Method of treating T-Cell lymphoblastic leukaemia with *ara*-G nucleoside derivatives. (1996) *US Pat. Appl.* 5,492,897.

Krenitsky, T.A., Koszalka, G.W., Tuttle, J.V. and Elion, G.B. (1981) An enzymatic synthesis of purine *D*-arabinonucleosides. *Carbohydrate Res. 97*, 139-146.

Lambe, C.U., Averett, D.R., Paff, M.T., Reardon, J.E., Wilson, J.G. and Krenitsky, T.A. (1995) 2-Amino-6-methoxypurine arabinoside; an agent for T-cell malignancies. *Cancer Res. 55*, 3352-3356.

Lauman, K., Breitgoff, D. Seemayer, R. and Schneider, M.P. (1989) Enantiomerically pure cyclohexanols and cyclohexane-1,2-diol derivatives; chiral auxiliaries and substitutes for (-)-8-phenylmenthol. A facile enzymatic route. *J. Chem. Soc. Chem. Commun.* pp. 148-150.

Mahmoudian M, Dawson M.J. (1997) Chemoenzymic production of the anti-viral agent Epivir (3TC). In: *Biotechnology of Industrial Antibiotics*, 2nd edition. pp. 753-77. Edited by Strohl WR. Marcel Dekker, New York.

Mahmoudian, M., Lowdon, A., Jones, M.F., Dawson, M.J. and Wallis, C. (1999a) A practical enzymatic procedure for the resolution of *N*-substituted 2-azabicyclo[2.2.1]hept-en-3-one. *Tetrahedron Asymm. 10*, 1201-1206.

Mahmoudian, M., Eaddy, J. and Dawson M.J. (1999b) Enzymatic acylation of 506U78; a powerful new anti-leukaemic agent. *Biotechnol. Appl. Biochem. 29*, 229-233.

Mahmoudian, M., Rudd, B.A.M., Cox, B., Drake, C.S., Hall, R.M., Stead, P., Dawson, M.J., Chandler, M., Livermore, D.G., Turner, N.J. and Jenkins, G. (1998) A versatile procedure for the generation of nucleoside 5'-carboxylic acids using nucleoside oxidase. *Tetrahedron 54*, 8171-8182.

Misaki, H.; Ikuta, S.; Matsunia, K. (1983) Nucleoside oxidase and process for making same, and procrss and kit for using same. *US Pat. Appl.* 4,385,112.

Moorman, A.R., Chamberlain, S.D., Jones, L.A., Peoples, M.E., de Miranda, P., Reynolds, D.J. and Krenitsky, T.A. (1992) 5'-ester prodrugs of the vaicella-zoster antiviral agent, 6-methoxypurine arabinoside. *Antiviral Chem. Chemother 3*, 141-146.

Nishio, T., Kamimura, M., Murata, M., Terao, Y. and Achiwa, K. (1989) Production of optically active esters and alcohols from racemic alcohols by lipase-catalysed stereoselective transesterification in non-aqueous reaction system. *J. Biochem. 105*, 510-512.

Novo. (1996) Savinase 6T, Lipolase 100T. *Novo Nordisk enzyme data sheet.*

Novo. (1997) Novozyme-435, an immobilised preparation of *Candida antarctica* lipase. *Novo Nordisk enzyme data sheet.*

Stead, P., Marley, H., Mahmoudian, M., Webb, G., Noble, D., Ip, Y. T., Piga, E., Rossi, T., Roberts, S. and Dawson, M.J. (1996) Efficient procedures for the large-scale preparation of (1*S*, 2*S*)-*trans*-2-methoxycyclohexanol, a key chiral intermediate in the synthesis of tricyclic β-lactam antibiotics. *Tetrahedron Asymm.* 7, 2247-250

APOPTOSIS AND BIOPROCESS TECHNOLOGY

R.P. SINGH AND M. AL-RUBEAI*
School of Chemical Engineering, University Of Birmingham, Edgbaston, Birmingham, BJS 2TT UK. mailto:m.al-rubeai@bham.ac.uk

1. Introduction

The controlled elimination of cells by a genetically regulated suicide pathway is the primary mechanism of cell death in all multi-cellular organisms and some single-celled species. In higher organisms, this form of programmed cell death is termed apoptosis (Kerr et al., 1972), and is fully integrated with the mechanisms of cell division and differentiation. Together, these three fundamental processes combine to regulate the organisation and function of all tissues, and are essential for the survival and reproduction of the organism. Apoptosis research has had a tremendous impact on cell biology and biomedical science. These studies are now widely recognised as having profound implications for the treatment of many disease states, including cancer, neurological disorders and viral infections. Our growing understanding of the molecular pathways involved in the regulation and execution of the death pathway are yielding new and exciting strategies for therapeutic intervention (Evan and Littlewood, 1998; Hengartner, 1998; Green and Reed, 1998).

This newly acquired knowledge also has implications for process biotechnology. Whilst antibodies still remain the most important class of protein pharmaceuticals manufactured by the biotechnology industry, there are numerous other biopharmaceuticals currently on the market, with a total of around 1000 potential products undergoing clinical trials. This number is expected to increase substantially in the next few years as the human genome project and allied high-throughput and automation technologies revolutionise the drug development process. Consequently, optimisation of the technology for cost effective manufacture of these products has been a key area of research. Considerable progress has been made through the optimisation of culture medium formulations, process conditions and cell line development (Kioukia et al., 1992; MacQueen and Bailey, 1989; Bertheussen, 1993; Merten et al., 1994; Griffiths, et al. 1992). However, cellular growth kinetics during the commercial cultivation of mammalian cell lines in the bioreactor environment are a result of both controlled cell division and cell elimination by apoptosis (Singh, et al. 1994; Franek and Dolnikova, 1991; Mercille and Massie, 1994; Mitchell Logean and Murhammer, 1997;

A. Van Broekhoven et al. (eds.), Novel Frontiers in the Production of Compounds for Biomedical Use, 267–275.
© 2001 *Kluwer Academic Publishers. Printed in the Netherlands.*

Moore et al., 1995) rather than passive cell death by necrosis. As a result, cell culture optimisation strategies are no longer focused solely on the engineering of the culture environment to maximise cellular proliferation and maintain viability. Instead, apoptosis regulatory genes identified by basic biomedical research have been applied to the genetic engineering of cell lines in order to block death by apoptosis, and thereby increase cellular robustness and of course, ultimately, boost biopharmaceutical production. This is achieved by allowing the switching of culture resources from the generation of replacement cellular biomass for those cells that have "committed suicide", to the synthesis of additional product.

A detailed review of a field that has become as wide and varied as apoptosis would be impossible. Instead, in this article, we provide an over-view of the most active areas of research. We focus on the role of the caspase family of proteases that mediate the cell death signal and execute the death program and also describe the central role of mitochondria. Aspects of current practice within the biopharmaceutical industry that will benefit from the control of apoptotic cell death are described, focusing on the growing list of studies which clearly demonstrate the advantages of this approach to process optimisation.

2. Apoptosis: basic features of cell death

Cells in the early stages of apoptotic death are easily identifiable. Initially, the plasma membrane undergoes intensive blebbing that leads to the budding of cytoplasmic packages called apoptotic bodies. The loss of cytoplasmic material, together with osmotic changes, results in a reduction in cellular volume. The nucleus also undergoes highly characteristic changes. The chromatin condenses into the nuclear membrane and ultimately fragments into a number of highly spherical particles. In comparison, necrotic cell death is characterised by an increase in cellular volume followed by cellular lysis. Nuclear morphology remains relatively unchanged throughout the death process.

Our understanding of the molecular basis of the apoptotic process has developed extremely rapidly, being driven by the realisation that many disease states result from the disregulation of this phenomenon. For example, the failure of cells to undergo apoptosis at the end of their productive life-span is a key step in malignant transformation (for review see Evan and Littlewood, 1998). Increased cellular survival results in the expansion of cell numbers due to accumulation, although the rate of cell division is very similar to that of normal cells. Further mutations result in the de-regulation of the cell cycle, an increase in cellular proliferation and the development of the fully transformed phenotype. Many genes have now been identified which can enhance cellular life span in this way. By far the best studied of this ever growing list is bcl-2, a 24 kDa transmembrane protein associated primarily with the mitochondrial outer and nuclear membrane (Chen Levy et al., 1989). Over-expression of the bcl-2 gene has been demonstrated to result in a substantial improvement in cellular survival in a very wide range of cell types. Moreover, this protein has been highly conserved throughout evolution, with genes such as Ced 9 of the nematode *C. elegans,* showing substantial sequence homology as well as functional similarities with bcl-2 (Hengartner and Horovitz, 1994). Sequence homology has also been reported with pore-forming proteins found in *E. coli,* and seemed to indicate a mechanism of activity of the protein (Sattler et al., 1997). Other potential mechanisms of action are outlined below.

The Caspases

The Caspase enzymes act as both the initiators and effectors of apoptotic death. In viable cells, this group of about 13 proteases exist as non-active proenzymes which, following proteolysis, are activated and trigger the cell death pathway. These enzymes are highly specific, with cleavage after aspartic acid residues, and site recognition also involving four or more amino acids on the NH2-terminal side of the cleavage site. The composition of these four amino acids varies from one caspase family member to the next, and reflects the highly specific and regulated nature of the early events of apoptosis (Thornberry et al., 1997).

Activation of the caspases proceeds by two distinct mechanisms. In the first, known as the "caspase cascade", a previously active molecule will cleave an unprocessed pro-enzyme, thus leading to its activation, and thereby triggering a "chain reaction" which rapidly becomes amplified (Ashkenazi and Dixit, 1998). This in turn leads to the activation of downstream effector caspases directly involved in the execution of the cell. The second strategy is referred to as the "induced proximity mechanism". This is dependent upon the inherent low-level protease activity in caspase pro-enzymes (Muzio et al., 1998), and involves the elevation of local proenzyme concentrations through the activity of molecular chaperones.

The mechanism by which caspases carry out the execution of the cell is now beginning to emerge. Firstly, caspases are responsible for the cleavage and de-activation of enzymes which are responsible for suppressing the apoptotic pathway and thus keeping a cell viable. For example, DNA fragmentation occurs by the nuclease caspase-activated deoxyribosenuclease (CAD), which is inhibited by forming a complex with ICAD/DFF45. Deactivation of ICAD by caspases then leads to the release of active CAD and cleavage of DNA (Enari et al., 1998; Liu et al., 1997). Caspases also cleave bcl-2 family proteins that suppress the induction of apoptosis (Xue and Horvitz, 1997). Indeed, one of the products of this cleavage event acts as a positive regulator of apoptotic cell death. Cleavage of the lamins leads to the destruction of the nuclear lamina, a structure involved in chromatin organisation. The consequence of these changes is the condensation of chromatin which is a highly characteristic feature of apoptosis in many cell types (Orth et al., 1996). Caspases also act on targets involved in the organisation of the cytoskeleton. For example, cleavage of gelsolin, a protein that severs actin filaments, results in its constitutive activation and therefore, the collapse of the cytoskeleton (Kothakota et al., 1997). Caspases also target proteins involved in the cellular homeostasis, such as DNA repair, splicing and replication, thus ensuring that the process of death proceeds quickly and efficiently.

APOPTOSIS AND THE MITOCHONDRIA

The symbiotic relationship between the earliest ancestors of eukaryotic cells and the bacteria that become the protomitchondria was a perilous one in which the endosymbiont was able to regulate the death of the protoeukaryotic cell according to its lifecycle and the demands of the environment. Stabilisation of this relationship did not occur until the exchange of genetic material between the protomitochondria and the host cells. However, even after the relationship became an obligate symbiosis, the protoeukaryotic cells retained the cell death pathway as an important component of the life-cycle of the organism. Although initially the concept of an altruistic death pathway in single celled organisms was rejected on the basis that genes involved would clearly be lost from the gene pool, there are a growing number of examples of cellular suicide in modem single celled eukaryotic organisms and bacteria. For example, induction of cellular suicide in a bacterial colony following viral infection prevents the infection from spreading to other members of the colony, and thus provides a mechanism by which the cell death genes (in the surviving colony members) are selected for and thus increase in frequency within the gene pool. Induction of cell death also allows cell numbers to adjust and adapt to prevailing nutrient levels. Such mechanisms provided protoeukaryotic cells with a significant survival advantage, and is responsible for the ubiquity of cellular suicide in modem eukaryotes. Indeed, the importance of cellular suicide increased substantially with the evolution of multi-cellular organisms, were the controlled elimination of cells provided an important mechanism that has become central to embryogenesis and normal tissue function and turnover.

The role of mitochondria in the evolution of the cell death process is reflected in its central role in the apoptotic pathway of modem eukaryotes. In cell-free systems, apoptotic changes in nuclear morphology does not proceed in the absence of mitochondria (Newmeyer et al., 1994). Moreover, there is evidence to suggest that the role of the mitochondria is more critical than that of the caspases. Although caspase inhibitors can prevent the expression of some or all of the morphological and biochemical features of apoptosis, the cells do not maintain their clonogenic potential (as is the case when apoptosis is suppressed by, for example, anti-apoptosis genes such as bcl-2) and the cells die, although more slowly and, apparently, by a non-apoptotic mechanism. A number of pro-apoptotic signals, such as the over-expression of the bax gene, can result in cell death by mitochondrial damage in the absence of caspase activation (Green and Reed, 1998).

There are three possible mechanisms by which mitochondria may mediate cell death. The first and most obvious is through the disruption of the electron transport chain and energy metabolism. Several factors known to induce apoptosis appear to target electron transfer between cytochrome b-c l/cytochrome c. These include gamma irradiation and exposure to ceramide (Garcia-Ruiz et al., 1997). Clearly, inhibition of the electron transport chain would be expected to result in a cessation in ATP generation leading to cell death. However, a fall in ATP levels is only seen very late in the death process, and because of the active nature of the death process, it is unlikely that this in itself is an early event in cell death. The second mechanism involves the release of caspase inducing cytochrome c from the nucleus. The combination of cytosolic cytochrome c with the protein Apaf- 1 and procaspase-9 constitutes the formation of the vertebrate 'apoptosome", resulting in the activation of caspase 9 and the induction of other caspases, leading to the initiation of the cell death process (Yang et al., 1997; Li et al., 1997). The bcl-2 gene prevents the release of cytochrome c and thus results in the suppression of apoptosis. By contrast, caspase inhibitors do not inhibit the release of cytochrome c that results from a wide range of apoptosis inducers (Bossy-Wetzel et al., 1998; Jurgensmeier et at., 1998).

The consequences of cytochrome c release for the cell death mechanism is dependent upon the cell type. When cytochrome c is present in excess, disruption to the mitochondrion may be minimal, and collapse in ATP generation may not be seen in the early stages of cell death, thus providing the driving force for the induction of cell death. In the presence of caspase inhibitors and when the levels of cytochrome c are low, the suppression of the effectors of the death pathway results in a reduction in the rate at which cell death progresses, and the collapse of the electron transport chain ultimately leads to a necrotic death (Green and Reed, 1998).

The third and final route by which mitochondria might result in the induction of apoptosis is through the generation of free radicals. Even under ideal conditions, approximately 1-5% of electrons entering the transport chain will result in the generation of free radicals. Increased levels of super oxides have been reported following the induction of apoptosis in response to numerous inducers (Bredesen, 1995). However, the demonstration that apoptosis can proceed in the near complete absence of oxygen may suggest that free radical generation may not be an indispensable component of the apoptotic pathway (Jacobson and Raff, 1995). However, it should be noted that free radicals can be generated even under conditions that approach complete anaerobiosis (Degli Esposti and McLennan, 1998).

In many cases, the induction of apoptosis is associated with a loss of mitochondrial inner transmembrane potential (AΨm). This would indicate the opening of the mitochondrial PT pore which results in the equilibration of ions. One of the most important consequences of these changes is the expansion of the matrix leading to the rupture of the outer mitochondrial membrane and the release of caspase activating proteins into the cytosol. A number of factors that prevent opening of the PT pore also suppress the induction of apoptosis, thus providing evidence for the importance of this event in the apoptosis pathway. For example, cyclosporin and over-expression of the bcl-2 anti-apoptosis gene suppresses both apoptosis and PT opening. By contrast, over-expression of bax induces apoptosis and also results in pore opening, as do free radicals and increased levels of cytosolic calcium. However, as with many aspects of apoptosis research, there are examples which appear to undermine the suggestion that the PT opening is a key event in the induction of apoptosis. There are examples in which PT opening occurs downstream of caspase activation and cytochrome c release. However, it is possible that initial caspase activation leads to PT pore opening, which in-turn leads to caspase activation and amplification (Green and Reed, 1998)

3. Apoptosis and its control during industrial scale cell culture processes

Loss of culture viability during the production of biopharmaceuticals from mammalian cell lines has a negative impact on the culture process in a variety of ways. First and foremost, the loss of viable cells will reduce the number of "cellular factories" available for the production of the biopharmaceutical in question. Clearly, boosting culture viability will increase maximum product titres. High levels of cell death will also impact on culture productivity in more indirect ways. For example, the release of proteases from dead and dying cells will lead to product degradation and modification, the latter of which is clearly unacceptable from a regulatory standpoint. Moreover, the accumulation of cellular debris will complicate medium clarification and product purification and thereby add to the cost of the manufacturing process.

Electron microscopic studies led to the first suggestion that cell death in murine hybridoma cultures proceeded by apoptosis (Al-Rubeai and Emery, 1990). Further studies provided additional evidence for this and it is now clear that most if not all cell lines used in the biopharmaceutical industry will undergo apoptotic death under particular conditions, although the exact level of apoptosis can vary depending upon the cell line and the type of inducer (Singh, et al. 1994; Franek and Dolnikova, 1991; Mercille and Massie, 1994; Mitchell-Logean and Murhammer, 1997; Moore et al., 1995). Initial attention was focused on factors such as serum, glutamine and glucose deprivation. However, more recent studies indicate that deprivation of any essential amino acid can induce apoptosis in hybridoma cultures (Simspon et al., 1998). Oxygen limitation is also a critical factor in determining maximum cell numbers and culture viability, and also results in the induction of high levels of apoptosis (Simpson et al., 1997; Singh et al., 1997). There is also evidence to suggest that factors such ammonia toxicity and exposure to hydrodynamic stress can result in the induction of apoptosis (Singh et al., 1994; Westlund et al., 1998; Al-Rubeai et al., 1995).

These findings have important consequences for numerous aspects of the cell culture process. The effect of serum deprivation may explain the technical difficulties that have been associated with the development of effective protein/serum free media. The need to exclude components of animal origin from medium formulations used in the manufacture of human therapeutics has been the driving force behind protein free media development. However, these media tend to be highly expensive and cannot be applied generically. By developing strategies that suppress the induction of apoptosis, it should be possible to achieve better growth and product titres as well as more rapid adaptation to these media. A number of studies have clearly demonstrated that the suppression of apoptosis by bcl-2 enhances cellular robustness under conditions of serum deprivation. Moreover, this enhanced robustness leads to improved cell growth and viability on commercially available serum free media (Singh et al., 1996; Fassnacht et al., 1998).

Nutrient and oxygen limitation zones can arise in large scale bioreactors operated in batch and perfusion mode as well as in intensive culture systems. Thus, these systems are often characterised by high levels of cell death which are not completely alleviated by improved medium and bioreactor design. Suppression of apoptosis under these conditions has proved to be highly effective. For example, a substantial increase in batch culture duration has been reported following over-expression of the bcl-2 gene in a range of industrially important animal cell lines, including Burkitt's Lymphoma, NSO, CHO, Hybridoma and COS-l (Itoh et al., 1995; Singh et al., 1996; Simpson et al., 1997; Mitchell-Logean and Murhammer, 1997; Tey et al., 1999a). In addition, the Epstein Bar Virus 1 9kDa protein also protected NSO cells from starvation induced apoptosis (Mercille et al, 1999). However, in one particular study, it was reported that the over-expression of bcl-2 failed to protect NSO cells from apoptosis in batch cultures (Murray et al., 1996). The authors did, however, detect expression of endogenous bcl-xL, a related protein which also provides protection from the induction of apoptosis. Thus, it was suggested that the bcl-2 gene may be functionally redundant in this cell line. Clearly, this report indicates that the "genetic background" of a particular cell line will determine whether the genetic manipulation of the cell death pathway is successful or not.

The enhanced robustness of mammalian cells in batch cultures has not always translated into an improvement in maximum antibody titres, which obviously must be attributed to the exhaustion of key biosynthetic precursors in the latter stages of the culture process. However, feeding of fresh nutrients in perfusion and fed batch cultures has yielded more promising results. In fixed bed and hollow fibre reactors, over-expression of bcl-2 in hybridoma cultures resulted in a 100% improvement in maximum antibody titre (Fassnacht et al., 1999b). A similar improvement has been observed in spin-filter cultures of an NSO cell line expressing the anti-apoptosis Epstein Bar Virus 19kDa protein (Mercille and Massie, 1999). Additionally, studies in our laboratory demonstrated that fed batch cultures of bcl-2 transfected NSO cells resulted in a 100% increase in maximum cell numbers and a 250% increase in maximum antibody titre when compared to the control cell line (Tey et al., 1999b).

Induction of apoptosis appears to be an important cell defence mechanism following viral infection. The death of the infected cell prevents the replication of the virus and thereby stops the spread of the viral infection to the rest of the organism. However, many viral anti-apoptosis genes have now been identified which block the induction of the cell death pathway and thereby allow the progression of the infection. This finding has attracted the attention of a number of workers concerned with the development of virus-based expression systems for the production of recombinant proteins. These systems, such as those based on baculovirus, can achieve very high levels of protein expression. However, they are transient systems of short duration, and the viral infection eventually kills the cells. A number of studies have suggested that the use of cell lines over-expressing anti-apoptosis genes results in extended culture duration and therefore productivity of these systems (Mitchell-Logean and Murhammer, 1997 Mastrangelo and Betenbaugh, 1998).

4. Conclusion

The study of Apoptosis has had a far-reaching impact on many aspects of biological research. Given the fundamental importance of this phenomenon, it is perhaps not surprising that these studies also have important consequences for animal cell biotechnology. The induction and suppression of apoptosis in the bioreactor under laboratory conditions has now been well documented and can result in a substantial improvement in culture productivity. Studies are now underway to establish whether or not this strategy will lead to the same improvement in productivity under state of the art biopharmaceutical production conditions.

References

Al-Rubeai, M., Emery, A. N. (1990). Mechanisms and kinetics of monoclonal-antibody synthesis and secretion in synchronous and asynchronous hybridoma cell-cultures. Journal of Biotechnology 16:67-86.

Al-Rubeai, M., Singh, R. P., Goldman, M. H., Emery, A. N. (1995) Death mechanisms of animal cells under conditions of intensive agitation. Biotechnology and Bioengineering 45, 463-472.

Ashkenazi, A., Dixit, V. M. (1998) Death receptors: Signaling and modulation Science 281, 1305-1308.

Bertheussen, K. (1993) Growth of cells in a new defined protein-free medium. Cytotechnololgy 11:219-226.

Bossy-Wetzel, E., Newmeyer, D. D., Green, D. R. (1998) Mitochondrial cytochrome c release in apoptosis occurs upstream of DEVD-specific caspase activation and independently of mitochondrial transmembrane depolarization. EMBO Journal 17,37-49.

Bredesen, D. E. (1995) Neural apoptosis. Annals of Neurology 38, 839-851.

Chen-Levy, Z., Nourse, M., Cleary, M. L., (1989). The *bcl-2* candidate proto-oncogene product is a 24 kDa integral membrane protein highly expressed in lymphoid cell lines and lymphomas carrying the t(1 2:18) mutation. Molecular and Cellular Biology 9, 701-710.

Degli-Esposti, M., McLennan, H. (1998) Mitochondria and cells produce reactive oxygen species in virtual anaerobiosis: relevance to ceramide-induced apoptosis FEBS Letters 430, 338-342

Evan, G, Littlewood, T. (1998) A matter of life and death. Science 281, 13 17-1322.

Fassnacht, D., Rossing, S., Franek, F., Al-Rubeai, M., Portner, R. (1998) Effect of bcl-2 expression on hybridoma cell growth in serum-supplemented, protein-free and diluted media. Cytotechnololgy 26:219-225.

Franek, F. and Dolnikova., J. (1991). Nucleosomes occuring in protein-free hybridoma cell cultures. Evidence for programmed cell death. FEBS Letters 248:285-287.

Garcia-Ruiz, C., Colell, A., Man, M., Morales, A., Fernandez-Checa, J. C. (1997) Direct effect of ceramide on the mitochondrial electron transport chain leads to generation of reactive oxygen species . Role of mitochondrial glutathione. Journal of Biological Chemistry 272, 11369-11377.

Green, D. R., Reed, J. C. (1998) Mitochondria and Apoptosis. Science 281, 1309-1312.

Griffiths, J. B., et al. (1992) Maximization of perfusion systems and process comparison with batch-type cultures. Cytotechnology 9:3-12.

Hengartner, M. (1998) Death by Crowd Control. Science 281, 1298-1299.

Hengartner, M. 0., Horovitz, H. R. (1994) C. elegans cell survival gene ced-9 encodes a functional homolog of the mammalian proto-oncogene bcl-2. Cell 76, 665-676.

Jacobson, M. D., Raff, M. C. (1995) programmed cell-death and bcl-2 protection in very low oxygen Nature 374, 814-816.

Jurgensmeier, J. M., Xie, Z. H., Deveraux, Q., Ellerby, L., Bredesen, D., Reed, I. C. (1998) Bax directly induces release of cytochrome c from isolated mitochondria Proceedings Of The National Academy Of Sciences USA 95, 4997-5002.

Kerr, J. F. R., Wyllie, A. H., Currie, A. R. (1972) Apoptosis: A basic Biological phenomenon with wide rangeing implications in tissue kinetics. British Journal of Cancer 26, 3 9-257.

Kioukia, N., et al. (1992) The impact of fluid fluid dynamics on the biological performance of free suspension animal cell culture: further studies. Transactions of The Istitute of Chemical Engineers 70(C) :143-154.

Kothakota, S., Azuma, T., Reinhard, C., Klippel, A., Tang, J., Chu, K. T, McGarry, T. J., Kirschner, M. W., Koths, K., Kwiatkowski, D. J., Williams, L. T. (1997) Caspase-3-generated fragment of gelsolin: Effector of morphological change in apoptosis. Science 278, 294-298.

Li, P., Nijhawan, D., Budihardjo, I., Srinivasula, S. M., Ahmad, M., Alnemri, E. S., Wang, X. D. (1997) Cytochrome c and dATP-dependent formation of Apafl/caspase-9 complex initiates an apoptotic protease cascade. Cell 91, 479-489.

Liu, X. S., Zou, H., Slaughter, C., Wang, X. D. (1997) DFF, a heterodimeric protein that functions downstream of caspase-3 to trigger DNA fragmentation during apoptosis Cell 89, 175-184.

MacQueen, A. and Bailey, J. E. (1989) Influence of serum level, cell type and viscosity on flow-induced lysis of suspended mammalian cells. Biotechnology Letters 11:531-536.

Mastrangelo, A. J. and Betenbaugh, M. J. (1998) Overcoming apoptosis: new methods for improving protein-expression systems. Trends in Biotechnology 16, 8 8-95.

Mercille, S. and Massie, B. (1994). Induction of apoptosis in nutrient-deprived cultures of hybridoma and myeloma cells. Biotechnology and Bioengineering 44:1140-1154.

Mercille, S., Jolicoeur, P., Gervais, C., Paquette, D., Mosser, D. D., Massie, B. (1999) Dose-dependent reduction of apoptosis in nutrient-limited cultures of NS/0 mycloma cells transfected with the El B- 19K adenoviral gene. Biotechnology and Bioengineering 63, 5 16-528.

Mercille, S. and Massie, B. (1999) Apoptosis-resistant ElB-19K-expressing NS/0 myeloma cells exhibit increased viability and chimeric antibody productivity under perfusion culture conditions. Biotechnology and Bioengineering 63, 529-543.

Merten, O. W., Kierulef, J. V., Castignolles, N., Perrin, P. (1994) Evaluation of the new serum-free medium (MDSS2) for the production of different biologicals: use of various cell lines. Cytotechnology 14:47-59.

Mitchell-Logean, C. and Murhammer, D. W. (1997) Bcl-2 expression in Spodoptera frugiperda Sf9 and Trichoplusia ni BT-Tn-5B1-4 insect cell:effect on recombinant protein expression and cell viability. Biotechnology and Bioengeering 56:380-388.

Mitchell-Logean, C. and Murhammer, D. W. (1997) Bcl-2 expression in Spodoptera frugiperda Sf9 and Trichoplusia ni BT-Tn-5B 1-4 insect cell:effect on recombinant protein expression and cell viability. Biotechnology and Bioengeering 56:380-388.

Moore, A., Donahue, C. I., Hooley, J., Stocks, D. L., Bauer, K. D., Mather, J. P. (1995). Apoptosis in CHO cell batch cultures - examination by flow-cytometry. Cytotechnology 17:1-11.

Murray, K., Ang, C. E., Gull, K., Hickman, J. A., Dickson, A. J. (1996). NSO myeloma cell death: Influence of bcl-2 overexpression. Biotechnology and Bioengineering 51:298-304.

Muzio, M., Stockwell, B. R., Stennicke, H.R., Salvesen, G. S. Dixit, V. M. (1998) An induced proximity model for caspase-8 activation. Journal of Biological Chemistry 273, 2926- 2930.

Newmeyer, D. D., Farschon, D. M., Reed, J. C. (1994) cell-free apoptosis in xenopus egg extracts - inhibition by bcl- 2 and requirement for an organelle fraction enriched in mitochondria. Cell 79, 353-364.

Orth, K., Chinnaiyan, A. M., Garg, M., Froelich, C. J., Dixit, V. M. (1996) The CED3/ICE-like protease Mch2 is activated during apoptosis and cleaves the death substrate lamin A. Joumal of Biological Chemistry 271, 16443-16446.

Sakahira, H., Enari, M., Nagata, S. (1998) Cleavage of CAD inhibitor in CAD activation and DNA degradation during apoptosis. Nature 391, 96-99.

Sattler, M., Liang H, Nettesheim D, Meadows RP, Harlan JE, Eberstadt M, Yoon HS, Shuker SB, Chang BS, Minn AJ, Thompson CB, Fesik SWRabinder (1997) Structure of Bcl-x(L)-Bak peptide complex: recognition between regulators of apoptosis. Science 275, 983-986.

Simpson, N. H., Milner, A., Al-Rubeai, M. (1997) Prevention of hybridoma cell death by bcl-2 during sub-optimal culture conditions. Biotechnology and Bioengineering 54:1-16.

Simpson, N. H., Singh, R.P., Perani, A., Goldenzon, C., Al-Rubeai, M. (1998) In hybridoma cultures, deprivation of any single amino acid leads to the induction of apoptosis, which is suppressed by the *bcl-2* gene. Biotechnology and Bioengineering 59:90-98.

Singh, R. P., Al-Rubeai, M., Gregory, C.D., Emery, A. N. (1994) Cell death in Bioreactors: a role for apoptosis. Biotechnology and Bioengineering 44:720-726.

Tey, B. T., Singh, R.P., Piredda, L., Piacentini, M., Al-Rubeai, M. (1999a) Bcl-2 mediated suppression of apoptosis in an CHO cell line expressing a chimeric antibody. Biotechnology and Bioengineering. In press.

Tey, B. T., Singh, R.P., Piredda, L., Piacentini, M., Al-Rubeai, M. (1999b). Bcl-2 mediated suppression of apoptosis in an NSO cell line expressing a chimeric antibody. Journal of Biotechnology. In press.

Thornberry, N. A., Ranon, T. A., Pieterson, E.P., Rasper, D. M., Timkey, T., GarciaCalvo, M., Houtzager, V.M., Nordstrom, P.A., Roy, S., Vaillancourt, J. P., Chapman, K. T., Nicholson, D. W. (1997) A combinatorial approach defines specificities of members of the caspase family and granzyme B - Functional, relationships established for key mediators of apoptosis. Journal of Biological Chemistry 272, 17907-17911.

Westlund A., Haggstrom L. (1998). Ammonium ion transport by the Na(+)K(+)2C1(-)-cotransporter induces apoptosis in hybridoma cells. Biotechnology Letters 20:87-90.

Xue, D., Horvitz, H. R. (1997) Caenorhabditis elegans CED-9 protein is a bifunctional cell- death inhibitor. Nature 390, 305-308.

Yang, J., Liu, X S., Bhalla, K., Kim, C. M., Ibrado, A. M., Cai, J. Y., Peng, T. I., Jones, D. P., Wang, X. D. (1997). Prevention of apoptosis by Bcl-2: Release of cytochrome c from mitochondria blocked. Science 275, 1129-1132.

GRAM-POSITIVE BACTERIA AS HOST CELLS FOR HETEROLOGOUS PRODUCTION OF BIOPHARMACEUTICALS

Gram-positive bacteria as cell factories

LIEVE VAN MELLAERT AND JOZEF ANNÉ
Laboratory of Bacteriology, Rega Instituut, Katholieke Universiteit Leuven, Minderbroedersstraat 10, 3000 Leuven, Belgium

Abstract

The increasing demand of recombinant compounds in bioscience and bioindustries requires the further exploration and improvement of production systems including bacteria, fungi, insect and human cells. For compounds that do not require glycosylation for biological activity, microbial systems are most favourable hosts because of high level expression and relatively inexpensive culture systems. Traditionally, *Escherichia coli* was and still is most often the host of choice. However, the major drawbacks of this Gram-negative organism are the periplasmic location of the secreted proteins of interest due to the presence of an outer membrane and the frequent occurrence of cytoplasmically or periplasmically located inclusion bodies consisting of denaturated recombinant proteins. Gram-positive bacteria secrete the protein of interest directly into the culture media, thereby greatly facilitating downstream processing and protein recovery. As a consequence, they are being extensively explored for recombinant protein production.

This report reviews the possible applications of Gram-positive bacteria as host cells for the production of proteins of biopharmaceutical interest. It will also highlight eventual advantages of Gram-positive bacteria compared to Gram-negative organisms. Although successful in some cases, several bottlenecks in the secretion of heterologous proteins remain. Approaches undertaken to improve the yield of secreted recombinant proteins in these Gram-positive bacteria will be summarised. Finally, results obtained so far regarding the production of biopharmaceutical compounds in a soluble active form and with respect to the cell surface display of recombinant proteins by Gram-positive bacteria will be discussed.

A. Van Broekhoven et al. (eds.), Novel Frontiers in the Production of Compounds for Biomedical Use, 277–300.
© 2001 *Kluwer Academic Publishers. Printed in the Netherlands.*

1. Introduction

Much research has already been carried out to produce eukaryotic proteins of biopharmaceutical interest via heterologous expression systems at reasonable cost. In spite of its drawbacks, *Escherichia coli* is the most employed host, so far, for the production of recombinant proteins. This is largely because of its well-defined genetics and physiology and owing to the availability of a wide range of expression vectors. The membranaceous outer layer of Gram-negative bacteria, however, represents a barrier to the secretion of recombinant proteins into the culture medium. Recombinant polypeptides that are overproduced in *E. coli* often accumulate in the cytoplasm or periplasm as inclusion bodies, insoluble aggregates of denaturated proteins. Under these conditions protein recovery from *E. coli* requires breaking the cells and solubilising the aggregates with strong chaotropic agents followed by refolding of the polypeptide under appropriate conditions. An additional problem is the presence of endotoxin, a toxic lipopolysaccharide in the cell wall of Gram-negative bacteria, which is very difficult to remove during purification processes. Because of these problems encountered with *E. coli*, other systems are currently under investigation. Systems looked for are those that secrete the proteins in the culture medium at high yield in a soluble, biologically active form, thus avoiding inclusion bodies and accompanying problems of refolding. Gram-positive bacteria such as *Streptomyces* and *Bacillus* are promising hosts. They are widely exploited in industry for the production of homologous proteins extracellularly secreted in large amounts. However, the exploration of these bacteria for the production of heterologous (eukaryotic) proteins has shown a number of bottlenecks. These include impaired translation, inefficient translocation across the plasma membrane, retarded release of the secreted protein from the cell wall, incorrect folding and degradation of the recombinant protein by extracellular proteases. In recent years, much effort has been invested to characterise at the molecular level the secretion process in these organisms in order to be able to rationally modulate these bacterial cell factories for yield improvement of heterologous proteins. Other Gram-positive bacteria including lactic acid bacteria, *Staphylococcus* and others are also being tested as host for heterologous protein production, but with variable success.

In addition, bacteria are becoming attractive tools for the production of cell surface displayed heterologous proteins. Applications of this cell surface display technique are the use of bacteria as live vaccine delivery vectors, an alternative for the phage display technique to select peptides or recombinant antibody fragments from large libraries, and the use of enzyme-coated bacteria as novel biocatalysts. Gram-negative bacteria such as *E. coli* and *Salmonella* spp. were used first for this purpose as a consequence of the more extensive knowledge of their genetics and the wealth of available genetic tools for manipulating these organisms. However, both translocation through the cytoplasmic membrane and correct integration into the outer membrane are required for surface display in Gram-negative bacteria. In Gram-positive bacteria, the translocation through the cytoplasmic membrane is sufficient to achieve proper surface exposure of the heterologous protein. Therefore, *Mycobacterium*, *Listeria, Lactococcus*, *Lactobacillus*, *Staphylococcus* and *Streptococcus* are currently extensively investigated for their possible use as live antigen delivery system.

Obviously, the prerequisite for successful production and cell surface protein display of heterologous proteins is their efficient translocation. The increasing insight in the secretion processes of Gram-positive bacteria may help to overcome problems in secretion. In this overview, the state of the art of the secretion mechanisms in Gram-positive bacteria, and how this knowledge can help to overcome bottlenecks and thus improve secretion, will be described. We will discuss the results obtained with respect to the heterologous production of pharmaceutically important proteins in Gram-positive bacteria and focus on the increasing efforts to develop these bacteria as live vaccine delivery vectors. From these data, we may conclude that Gram-positive bacteria are most attractive hosts for heterologous protein production.

2. The general secretion pathway

At current, several secretion pathways, indicated type I, II or Sec-dependent secretion pathway, III, IV and the Tat-dependent secretion pathway are described for bacteria, each type specialised to translocate a specific set of proteins. Heterologous proteins are routed so far via the Sec-dependent secretion pathway. In this pathway proteins are synthesised as preproteins having a signal peptide as an N-terminal extension that is proteolytically removed by a signal peptidase upon translocation. Proteins to be secreted have to be kept in an unfolded or partially unfolded conformation in the cytoplasm prior to translocation. This requires the assistance of cytosolic chaperones or similar factors, which deliver the protein to the cytosolic membrane. Transport across this membrane occurs through the Sec translocation channel consisting of several membrane proteins. After translocation the proteins fold assisted by folding catalysts. The current knowledge of the general secretion pathway is essentially coming from studies on *E. coli*. Homologues of *E. coli* secretion pathway components have now been discovered in several bacteria. It is assumed therefore that secretion pathways are similar in all prokaryotes. In the next paragraphs, a short overview of the Sec-dependent secretion pathway and its elements with emphasis on Gram-positive bacteria will be given.

2.1. EARLY STAGE

Proteins to be secreted have a signal peptide that acts as address tag allowing secretory proteins to specifically recognise the translocase on the target membrane. This signal peptide consists of 3 domains: a positively charged N-terminal domain (n-region), a hydrophobic h-region and the C-terminal domain containing the signal peptidase cleavage site (c-region). The n-region of the signal peptide is believed to fulfil an important role in the secretion process by interacting electrostatically with the negatively charged heads of membrane phospholipids and by enhancing the interaction with SecA, a principal component of the translocase (see below). The h-region is the central hydrophobic core forming an α-helix and thus facilitates the insertion of the precursor into the hydrophobic interior of the membrane lipid bilayer (Lammertyn and Anné, 1998).

In the early stage of protein secretion, targeting of the preprotein to the membrane and its maintenance in a translocation-competent state, *i.e.* unfolded or partially folded, are crucial steps for an efficient secretion. In Gram-negative bacteria such as *E. coli*, the molecular chaperone SecB has an important role in targeting the preprotein to the membrane. This protein has an antifolding activity and shows an affinity for SecA, the translocation motor peripherally located in the plasma membrane. SecB is in *E. coli* required for efficient secretion of certain proteins. Very recently, CsaA exhibiting chaperone-like activities and suppressing the growth and secretion effects of *E. coli* SecA(ts) strains was described for *B. subtilis*. Since CsaA also suppresses the growth defects of *dnaK, dnaJ* and *grpE* mutants of *E. coli* and seems to interact specifically with SecA and preproteins, it might function as the *E. coli* SecB (Müller *et al.*, 2000). A counterpart for SecB has not yet been described in other Gram-positive bacteria.

Besides the Sec pathway, in several bacteria a Signal Recognition Particle (SRP) pathway has been identified. In this pathway a 4.5S RNA and Ffh having GTPase activity are involved. They are homologous to the eukaryotic SRP 7S RNA and 54kDa subunit, respectively. Unlike SecB, Ffh binds nascent polypeptides still associated with the ribosome. In *B. subtilis*, the for cell growth essential Ffh protein forms a complex with a small cytoplasmic RNA and Hbsu, a histone-like protein (Nakamura *et al.*, 1999). The ribosome-Ffh-nascent chain complex binds subsequently to an SRP-receptor, FtsY, which has been identified on basis of its homology with the α-subunit of the eukaryotic SRP-receptor. The preproteins transported in this way are released from the complex upon binding of Ffh to FtsY and eventually translocated by the Sec translocase (Valent *et al.*, 1998). Whereas in *E. coli* this targeting pathway seems mainly be used for targeting and insertion of inner membrane proteins, the SRP pathway in *Bacillus* is involved in the transport of most of the extracellular proteins (Hirose *et al.*, 2000). The requirement of SRP in localisation of membrane proteins in this bacterium has not yet been analysed.

For the maintenance of a translocation-competent folding state of precursors, cytosolic chaperones belonging to heat shock GroEL and DnaK families are mainly responsible. These molecular chaperones fulfil crucial roles under normal physiological conditions by assisting in the folding of newly synthesised proteins thus preventing protein denaturation, aggregation and misfolding. GroEL associated with GroES and DnaK co-operating with DnaJ and GrpE, interact with various structural elements in unfolded proteins. While in Gram-negative bacteria the *groE* and *dnaK* operons are under positive control of alternative sigma factors, in Gram-positive bacteria the expression depends on repressor systems. In some Gram-positive bacteria including *B. subtilis* both operons are repressed by binding of HrcA on CIRCE (controlling inverted repeats of chaperone expressing) motifs. Some other Gram-positive bacteria such as *Streptomyces* have a more complex regulatory system: the *groE* genes are regulated on the basis of CIRCE motifs while the *dnaK* operon is repressed by HspR through binding on HAIR (HspR associated inverted repeat) motifs. Homologues of this *hspR* gene were also found in *Mycobacterium* spp., *Helicobacter pylori* and *Aquifex aeolicus* (Bucca *et al.*, 1995; Grandvalet *et al.*, 1997). Other chaperones eventually involved in preprotein transport are Hsp18 and ClpB (Grandvalet *et al.*, 1999). In *B. subtilis*, TepA is another cytosolic factor required for efficient preprotein translocation. The exact function of this ClpP-like proteolytic protein has still to be resolved (Bolhuis *et al.*, 1999a).

2.2. MIDDLE STAGE

The translocation process through the membrane (middle stage of secretion) starts with the binding of the precursor eventually complexed with a targeting factor, to SecA, the peripheral subunit of the translocase. In Gram-negative bacteria, this translocase consists of a hexameric complex of SecYEGDF and YajC. In this complex SecYEG form together the translocation channel in the membrane and the integral membrane proteins SecDF are required for efficient secretion. The function in precursor translocation of YajC, co-immunoprecipitating with SecDF, is still unclear (Driessen *et al.*, 1998). Binding to SecA is facilitated by the signal peptide, but in addition, by the affinity for SecA of both the preprotein mature domain and of chaperones. On initiation of the translocation reaction the mature part of the preprotein is transferred onto SecA that is already bound to SecYEG at the membrane. Sources of energy for the translocase machinery are both ATP and the proton-motive force (PMF). SecA has ATPase activity that is activated by its interaction with translocation-competent precursor proteins, the SecYEG complex and acidic phospholipids. In the sequenced genome of *B. subtilis* (Kunst *et al.*, 1997), counterparts of these components are present. Exceptionally, mycobacteria contain two distinct SecA proteins which may have different substrate specificities (Economou, 1999). Genes of the Sec-dependent pathway, except a *yajC* homologue, were also described for *Streptomyces* (Lammertyn, 2000).

2.3. LATE STAGE

Upon translocation (late stage), membrane-bound type I signal peptidases (SPases) are responsible for the removal of the signal peptide from the preprotein resulting in the release of mature proteins from the membrane. In *E. coli*, only a single SPase, i.e. LepB, is present. Some Gram-positive bacteria, like *Mycobacterium*, also encode a sole SPase. On the other hand, *B. subtilis* and *Streptomyces lividans* contain multiple SPases (Bron *et al.*, 1998; Parro *et al.*, 1999). *Bacillus* contains 5 signal peptidase (*sip*) genes, which are scattered over the chromosome. In addition, two *sip* genes were found on *Bacillus* plasmids (Meijer *et al.*, 1995). In *Streptomyces* 4 *sip* genes are detected, which are clustered. *SipX, sipW, sipY* are located in an operon structure, while *sipZ* is the first gene of an adjacent operon with still three unrelated genes (Parro *et al.*, 1999). Although these SPases have partially overlapping substrate specificities, they clearly show substrate preferences (Bron *et al.*, 1998; N. Geukens, pers. comm.). A remarkable difference in membrane topology between type I SPases of Gram-positive versus Gram-negative bacteria is that SPases of the former contain 1 N-terminal transmembrane anchor, whereas in most instances SPases of Gram-negative bacteria have two membrane anchors. As a consequence, SPases of Gram-positive bacteria are often shorter than their homologues in Gram-negative bacteria (Dalbey *et al.*, 1997). In contrast to most other SPases, 3 out of the 4 *S. lividans* SPases have also a C-terminal anchor (Geukens *et al.*, 2000).

After removal by SPase processing the signal peptides are degraded by membrane-bound signal peptide peptidases (Spp). Such an Ssp of *B. subtilis*, SspA, was shown to be required for efficient processing of preproteins under conditions of hypersecretion (Bolhuis *et al.*, 1999a).

Folding of mature proteins to their native state is achieved by extracellular folding catalysts. These proteins accelerate rate-limiting steps along the folding pathway. These include thiol/disulphide oxidoreductases (Dsb proteins) and disulphide bond isomerases (PDI proteins) responsible for the formation and reshuffling of disulphide bonds, and by peptidyl prolyl *cis/trans* isomerases (PPIases) which catalyse the isomerisation of X-proline bonds. In *B. subtilis*, the lipoprotein PrsA, a putative PPIase, seems important for the high level secretion of certain secretory proteins. PrsA is believed to accelerate folding of the secreted protein rendering this protein resistant to proteases and, hence, increasing its stability in the secreted fraction (Kontinen and Sarvas, 1993). In *Lactococcus lactis*, the extracellular lipoprotein PrtM shows a high degree of similarity to *Bacillus* PrsA and it was described that PrtM is involved in maturation of lactococcal proteinase after translocation of its precursor across the membrane (Haandrikman *et al.*, 1989). Other foldases identified in *B. subtilis* are the thiol-disulphide oxidoreductases BdbA, B and C. Disruption of *bdbB* or *bdbC* significantly affects stability of secreted proteins with disulphide bonds (Bolhuis *et al.*, 1999c). In other Gram-positive bacteria these extracellular foldases remain to be examined.

In the late stage of secretion the cell wall can be a severe obstacle for secretion. The cell wall of Gram-positive bacteria is negatively charged through the presence of teichoic and teichuronic acids. Proteins with positive charges can be trapped in the cell wall by electrostatic interaction rendering the proteins more sensitive to membrane-bound proteases (Stephenson *et al.*, 1998). In the case of secretion of human pancreatic α-amylase (HPA) by *B. subtilis*, it was demonstrated that HPA fused to the signal peptide A2 was accumulated in the cell wall in an inactive form. The reason for the failure of A2-HPA to pass the cell wall was mainly related to its misfolding, rendering the protein inactive. This also suggests that the disulphide bond oxidoreductases of *B. subtilis* can not correctly form the disulphide bridge in HPA (Bolhuis *et al.*, 1999b). In addition, some secreted proteins may require cations for efficient folding and activity. As a consequence, their secretion can be dependent on the cation concentration at the membrane-cell wall interface as demonstrated for the secretion of α-amylase from *B. subtilis* (Stephenson *et al.*, 1998).

Finally, after translocation the protein can be degraded by proteases present as extracellular soluble, membrane- or cell wall-associated enzymes. This is a ubiquitous problem noticed in bacterial systems. For *Bacillus* several extracellular proteases were identified and *B. subtilis* strains in which up to seven proteases were knocked-out without loss of strain viability, were constructed (Wong, 1995). In this respect, *Bacillus brevis* is advantageous over *B. subtilis* for heterologous protein production due to lower extracellular protease activities (Udaka and Yamagata, 1993a). Lower protease activities were also noticed in *S. lividans,* the host of choice for heterologous protein production when using *Streptomyces.* Nevertheless, a number of extracellular proteases were identified (Taguchi *et al.*, 1995; Vitale *et al.*, 1999). These proteases may, of course, have an adverse effect on the quality and quantity of the heterologous protein produced in the bacterial host, when the protein is not properly folded and sensitive to proteases.

In *Staphylococcus carnosus*, a strain often used in the food industry for meat fermentation and almost completely devoid of soluble extracellular proteases, degradation of heterologous exported proteins could nevertheless be observed due to the presence of extracellular membrane- or cell wall-bound proteases (Meens *et al.*, 1997). Also *Bacillus* and *Streptomyces* spp. contain cell-associated proteases. In *Bacillus* spp., the cell wall-bound serine protease CWP52 is active at the site of preprotein translocation. It could even act on highly stable secreted proteins since depletion of this protease increased yield of extracellular α-amylase (Stephenson and Harwood, 1998). In *S. lividans,* Binnie *et al.* (1995) described two cell-associated proteinases, SlpD and SlpE. It is suggested that the function of these cell-associated proteases resembles that of *E. coli* DegP protease. This means that they might be involved in the selective degradation of unfolded or misfolded proteins trapped in the cell wall. An inverse correlation seems thus to exist between the efficiency of folding of newly translocated proteins into a conformation resistant to the hydrolytic action of cell-associated proteases, and their degradation (Meens *et al.*, 1997). Thus, proteases of the latter type are likely to be involved in the quality control of secreted proteins by removing incorrectly folded proteins in order to prevent jamming of the secretory pathway (Braun *et al.*, 1999).

2.4. INTRINSIC FEATURES OF SECRETORY PROTEINS

As indicated above, physico-chemical features of the heterologous protein to be translocated has a major influence on productive secretion. In the early stage of the translocation process, the interactions between the presecretory protein and the components of the secretion pathway are critical for efficient secretion. In first instance, the signal peptide influences the yield of the secreted protein. The positively charged n-region is of importance for interaction with the membrane and SecA, while the hydrophobicity of the h-region seems to influence the choice of targeting pathway. Not every signal peptide is as efficient for the translocation of a particular protein. Moreover, the mature part of the precursor can have characteristics that hamper efficient interaction with molecular chaperones, SecA or other translocase components. These inefficient interactions can cause aggregation of the newly synthesised precursor or lead to breakdown of unfolded or misfolded peptides by intracellular proteases.

Proteins containing transmembrane domains and sequences of hydrophobic residues are often inefficiently secreted from bacteria probably due to sticking in the cell membrane (Sjölander *et al.*, 1993; Nguyen *et al.*, 1995). It is noticed that secretory proteins with little or no net charge pass the cell wall unretarded, while those with an overall positive charge can be retained within the cell wall fraction as a consequence of electrostatic interactions between the protein and the negatively charged groups (Stephenson *et al.*, 1998).

Furthermore, for several Gram-positive bacteria, it has been demonstrated that the presence of a propeptide, a sequence located between the signal peptide and the mature protein which is removed after translocation, protects the secretory protein against extracellular breakdown by proteases in the late stage of secretion. It is believed that this propeptide acts as an intramolecular chaperone by guiding the post-translocational, correct folding of the protein in a protease-resistant form. Another explanation could be that the propeptide enhances the secretion efficiency by improving the release from the plasma membrane and/or the passage through the cell wall and, as a consequence, shortening the time span during which the secretory protein would be in the range of the cell-bound proteases (Meens *et al.*, 1997). Recently, Samuelson *et al.* (1999) investigated in more detail the relevance of the 207-aa lipase propeptide in efficient secretion by deleting this fragment and determining the effect on secretion. They concluded that the propeptide can be deleted in case the proteins are easily secretable but, in contrast, it seems to be advantageous for translocation of inefficiently secreted proteins like *e.g.* a 101-aa protein fragment derived from the G protein of human respiratory syncytial virus.

3. Improvement of secretion

Protein yield can be improved by affecting transcription and/or translation level of the heterologous gene (not discussed here), but also the secretion process itself offers ample possibilities to improve protein yield.

As outlined above, several factors have to be considered for achieving efficient secretion of heterologous proteins in Gram-positive bacteria. Each of these factors can act as a bottleneck in the secretion process. Modulation of these proteins can be envisaged in order to improve yield of secreted protein. In this chapter, we will review the bottlenecks identified and the approaches undertaken to enhance heterologous protein yield in Gram-positive bacteria. Data indicate that at each stage the process of secretion can be improved. Most efforts made to increase yield are protein-specific and can not be applied in general terms for all heterologous proteins. As a consequence, the efficient production of a protein of interest requires the construction of tailor-made host strains. It means that in first instance the specific bottleneck in the secretion of a particular protein needs to be identified, followed by a directed approach to solve the problem.

3.1. EARLY STAGE SECRETION IMPROVEMENT

A possible reason for the intracellular accumulation of a precursor protein during the early stage of secretion is the limited activity of chaperones. Hence, overproduction of molecular chaperones in order to keep preproteins in an optimal translocation-competent state and to minimise intracellular proteolysis can help to solve this problem. In *Bacillus* the secretion of antidigoxin single-chain antibodies (SCA) is hampered by their tendency to accumulate in inclusion bodies. Overproduction of the major intracellular molecular chaperones by inactivation of the repressor HrcA acting on the *groE* and *dnaK* operons, increased the secretion of an antidigoxin SCA fragment by about 60% (Wu *et al.*, 1998).

Another approach to influence secretion at the early stage is to manipulate the signal sequence of the precursor to be secreted. Lammertyn and Anné (1998) reviewed the research on signal peptide mutants and their effect on translocation in *Streptomyces*. Signal peptide mutants were tested in order to improve secretion of eukaryotic proteins. For example, decrease of the net positive charge in the n-region of the *Streptomyces venezuelae* subtilisin inhibitor (Vsi) signal peptide from wild type +3 to +2 resulted in a 3- to 10-fold enhanced secretion of mTNFα (Lammertyn *et al.*, 1997). Also in *Bacillus* several signal peptide mutants have been investigated. For example, the presence of a β-turn potential at the C-terminus of the *Bacillus amyloliquefaciens* neutral protease signal peptide provided a slight increase in secretion efficiency of a 34-aa fragment of human parathyroid hormone (Saunders *et al.*, 1991).

3.2. MIDDLE STAGE SECRETION IMPROVEMENT

SecA, the peripheral subunit of the translocase, directly recognises the signal sequence and unknown elements of the mature domain of the precursor. Leloup *et al.* (1999) showed that a difference exists in SecA dependency for native *B. subtilis* proteins expressed from their chromosomal genes. The secretion of α-amylase is affected only under conditions of strong SecA depletion, while that of levansucrase gradually varies with the SecA level in the cell. A 40% increase in levansucrase could be obtained in case the SecA was 7-fold overproduced. The researchers concluded that levansucrase has a lower affinity for the translocase compared to α-amylase. Overproduction of SecA or modification of the SecA component resulting in a higher affinity between this translocase subunit and the precursor of interest might thus be a possible way to improve protein yield. Also overproduction of other Sec components can be attempted to achieve higher secretion levels, but is not yet described for Gram-positive bacteria.

In *E. coli,* different *prl* (protein localisation) suppressor mutations in the *secA*, *secY*, *secE* and *secD* genes were identified. The *prl* mutations result, for instance, in enhanced affinity of SecA for the translocase due to loosened association of the SecYEG subunits in this complex. In this way, they disrupt the proof-reading activity, or broaden the specificity of the translocase allowing significant export of preproteins with a wide variety of defective (or even deleted) signal peptides or with a folded structure in the mature domain (Duong and Wickner, 1999). Taking advantage of the Prl phenotypes, the export efficiency of the pre-OmpA-human interleukin 6 fusion protein could be significantly raised in *E. coli* (Perez-Perez *et al.*, 1996). In Gram-positive bacteria, such suppressor mutations of *sec* genes have, so far, not been described.

In some cases the membrane itself forms a barrier for translocation. Nguyen *et al.* (1995) could improve cell surface display of the major glycoprotein (G protein) of human respiratory syncytial virus (RSV) on *Staphylococcus xylosus* by substitution or deletion of multiple hydrophobic phenylalanine residues. In contrast to the non-engineered RSV G protein, the engineered variants were efficiently targeted to the *S. xylosus* cell surface. If acceptable for the application, this hydrophobic engineering can be of general interest for secretion of proteins otherwise difficult to translocate through the bacterial cell membrane.

3.3. LATE STAGE SECRETION IMPROVEMENT

Most bottlenecks identified in heterologous protein secretion, so far, are related to the late stage of translocation. The problems encountered are associated to the SPase activity, proteolysis of the protein of interest by host-encoded proteases, protein folding or the cell wall passage.

3.3.1. SPase activity

Bacillus as well as *Streptomyces* contain several type I SPases. For *B. subtilis* it has been proven that these SPases have a different precursor preference. *B. amyloliquefaciens* pre-amylaseQ processing was significantly reduced in strains lacking SipT SPase, but was improved in the absence of SipS or SipU. This was explained by a competition between the different SPases with a variable processing efficiency for a particular substrate (Tjalsma *et al.*, 1997). In contrast, for other fusion proteins the processing rate was enhanced by overexpression of SipS. It was suggested that processing by SPase might be a secretion bottleneck for those proteins of which the mature form reaches the growth medium but on the other hand of which precursors are slowly processed (Bolhuis *et al.*, 1999b). Therefore, overexpression or depletion of certain SPases might be a way to enhance precursor processing.

Introduction of a nine-residue synthetic propeptide containing 2 negative charges at position +2 and +8 improved the secretion of *Staphylococcus aureus* nuclease (Nuc) 2- to 4-fold from *Lactococcus lactis*. Beside the positive effect on precursor conformation or interaction with SecA, this propeptide might optimise the charge balance around the signal peptidase cleavage site resulting in a better processing of the preprotein (Le Loir *et al.*, 1998). This 9-aa synthetic propeptide was also used for a successful translocation of a bovine coronavirus (BCV) epitope-Nuc fusion (Langella and Le Loir, 1999).

Also for *Streptomyces*, it was shown that the amino acids surrounding the signal peptidase cleavage site influence the final recombinant protein yield. Maintenance of two amino acids of the native Vsi sequence downstream the cleavage site allows a better processing of the fusion protein (Van Mellaert, unpublished data). A similar effect of importance of amino acids surrounding the signal peptidase cleavage site was noticed for the production of the D1D2 part of the soluble CD4 receptor. Keeping threonine, the first amino acid of the *S. longisporus* StiI mature protein, between the signal peptide and D1D2 already improved the secretion of D1D2 by *S. lividans* more than 85% compared to the direct fusion product (Fornwald *et al.*, 1993).

Under conditions of hypersecretion in *B. subtilis*, the signal peptide peptidase SspA is required (Bolhuis *et al.*, 1999a). Hence, overproduction of signal peptide peptidases might be beneficial for the abundant secretion of certain heterologous proteins.

3.3.2. Proteolytic breakdown

An important aspect in optimisation of heterologous protein yield in Gram-positive bacteria is the avoidance of proteolytic breakdown of the heterologous protein of interest, which could be attained by inactivation of the responsible proteases.

However, in *B. subtilis*, proteolytic breakdown of heterologous proteins, e.g. SCA, remained even when applying strains missing 6 or 7 extracellular proteases (Wu *et al.*, 1998; Bolhuis *et al.*, 1999b). Properly folded proteins are less susceptible to proteases. Hence, proteolytic activity could be circumvented by overproducing peptidyl-prolyl *cis-trans* isomerases that catalyses the isomerisation of prolyl residues, in case protein folding is a rate limiting step. For example, overproduction of PrsA increased the secretion yield of antidigoxin SCA fragment in *B. subtilis* by 60%. The proper folding of SCA owing to the presence of PrsA improved resistance to proteases and, hence, increased stability in the secreted fraction (Wu *et al.*, 1998).

The breakdown of heterologous proteins is less problematic in *S. lividans* than in *B. subtilis*. Nevertheless, limited proteolytic degradation of the N-terminus of mTNFα (Van Mellaert *et al.*, 1994) or hGM-CSF, hIL3 and hEPO (Krieger *et al.*, 1994) was observed. Deletion of the tripeptidyl aminopeptidase (*tap*) gene significantly reduced the removal of N-terminal tripeptides (Butler *et al.*, 1996). The tendamistat-proinsulin fusion protein was degraded to a product similar in size as tendamistat due to chymotrypsine-like protease activity. Addition of Ni^{2+} or Zn^{2+} significantly reduced this proteolytic activity (Aretz *et al.*, 1989). To optimise *S. lividans* as a production host, several protease-deficient strains have been produced. Construction of these and properties have been described in several patents (*e.g.* Patent US5856166). Furthermore, the growth medium used to cultivate the recombinant strains also influences the protease activity as demonstrated by Fornwald *et al.* (1993). By selecting the proper medium, also Pozidis *et al.* (2000) could bring the N-terminal degradation of mTNF produced by *S. lividans* to a minimum.

Gram-positive bacteria with low exoproteolytic activity such as *Staphylococcus carnosus* showed also degradation of heterologous proteins due to cell-associated proteases. This problem was circumvented in *S. carnosus* by fusion of *E. coli* OmpA to the pre-pro-part of the *Staphylococcus hyicus* lipase (Meens *et al.*, 1997). The same propeptide of lipase was used for efficient secretion of *E. coli* β-lactamase (Liebl and Götz, 1986), of REIv, the variable domain of an immunoglobulin κ light chain (Pschorr *et al.*, 1994), and of human antilysozyme fragments (Schnappinger *et al.*, 1995).

3.3.3. Protein folding

Different enzymatic activities might be required to achieve correct folding of heterologous proteins. However, the bacterial foldases are not always capable to correctly fold the foreign protein, especially when several disulphide bridges are involved (Bolhuis *et al.*, 1999b). When using a fungal protein disulphide isomerase (PDI) as fusion partner to secrete heterologous proteins in *B. brevis*, Kajino *et al.* (2000) observed an 8-fold and 12-fold increase of extracellular production of light chain of immunoglobulin G and geranylgeranyl pyrophosphate synthase, respectively. Linkage to PDI prevented aggregation of the proteins resulting in high-level accumulation of the fusion proteins in soluble and biologically active forms. The fused PDI did not only function as a folding catalyst, but also as a molecular chaperone.

4. Examples of Gram-positive bacteria as host cells for the production of heterologous proteins

Two major applications envisaged for the use of Gram-positive bacteria with respect to the production of heterologous proteins are: (1) Production of soluble bioactive pharmaceutical compounds in the culture medium in commercially attractive levels. If soluble active proteins could be isolated in sufficient quantities from the extracellular fermentation broth, this would greatly facilitate the protein purification carried out downstream from the fermentation process. *B. subtilis*, *B. brevis* and *S. lividans* are intensively investigated for this purpose. They have a proven excellent secretion capacity and organisms belonging to the *Bacillus* and *Streptomyces* genera are industrially well-known for the production of several types of hydrolytic enzymes secreted in large amounts. (2) Cell surface display of heterologous proteins, *e.g.* antigens or antibody fragments, in order to employ the recombinant organisms as live antigen systems or diagnostic tools. Gram-positive bacteria currently extensively investigated for their possible use as live antigen delivery systems are *Mycobacterium*, *Listeria, Lactococcus*, *Lactobacillus*, *Staphylococcus* and *Streptococcus*. The choice to use these bacteria arises from different observations. The attenuated strain of *Mycobacterium bovis*, named bacillus Calmette-Guérin (BCG), has already a long history as bacterial anti-tuberculosis vaccine. *Listeria monocytogenes* is a facultative intracellular bacterium that is extensively used to investigate cell-mediated immune response. The lactic acid bacteria *Lactococcus* and *Lactobacillus*, and certain *Staphylococcus* species are widely applied in the food industry for the production for fermented products and, therefore, are considered as GRAS (generally regarded as safe) organisms. Finally, among the *Streptococcus* and *Lactobacillus* species used as recombinant host cells there are human commensals giving the advantage that they can colonise human mucosal surfaces.

4.1. HIGH-LEVEL PRODUCTION OF BIOPHARMACEUTICAL COMPOUNDS

Both bacilli and streptomycetes are industrially well-known for the secretion of native enzymes in the growth medium in massive amounts exceeding 10 g/L under appropriate fermentation conditions. Notwithstanding this high secretion capacity for homologous proteins, attempts to produce heterologous proteins in these bacteria failed in many instances so far. On the other hand, positive results have also been obtained indicating the potentiality of these species for heterologous proteins when the correct conditions are met. Secretion of human epidermal growth factor (hEGF) in *B. brevis* yielded about 3 g/L (Udaka and Yamagata, 1993b). Remarkably, in *B. subtilis* hEGF is produced only at levels of 7 mg/L probably as a consequence of proteolytic breakdown (Lam *et al.*, 1998).

In *S. lividans*, production levels exceeding 300 mg/L could be obtained for some proteins including mTNFα (Lammertyn *et al.*, 1997) and sCD4 derivative D1D2 (Fornwald *et al.*, 1993). These proteins were produced using regulatory sequences of the *Streptomyces venezuelae* subtilisin inhibitor Vsi and the *Streptomyces longisporus* trypsin inhibitor StiI, respectively. Both inhibitors showing a high degree of similarity (Van Mellaert *et al.,* 1998) are abundantly secreted in their native hosts. Many other heterologous proteins, however, are secreted at much lower levels for reasons not yet understood. Extensive investigations are going on at present to identify the main bottlenecks that hamper high level secretion and how these can be overcome.

Other Gram-positive bacteria used for extracellular heterologous protein production are e.g. *Lactococcus lactis* and *Staphylococcus carnosus*, Bovine prochymosin and bovine plasmin were secreted in *L. lactis* using the signal sequence of the major *Lactococcus lactis* secreted proteinase (Prt) and of an abundant lactococcal extracellular protein (Usp45) of unknown function, respectively (De Vos *et al.*, 1989; Arnau *et al.*, 1997). In addition, *S. carnosus* lacking considerable exoproteolytic activity was tested for the production of antibody fragments, seemingly very sensitive for proteolytic breakdown in *Bacillus* (see above). A fusion protein, composed of an N-terminal portion of the prepropeptide of *Staphylococcus hyicus* lipase, a hexahistidine tag followed by the recognition sequence of IgA protease and REI_V (the variable domain of an immunoglobulin κ light chain) was secreted by *S. carnosus*. The fusion protein accumulated as a soluble protein in the culture medium yielding 5-10 mg/L. REI_V with 2 additional N-terminal amino acids was purified to homogeneity with a yield of more than 1 mg/L and was further used for physical characterisation (Pschorr *et al.*, 1994). In a similar way, human antilysozyme fragments were produced under induced conditions. When the light chain and the Fd fragment were co-secreted as a fusion protein separated by an IgA protease cleavage site, they accumulated in a structure capable of heterodimerisation after IgA protease cleavage.

Since heterodimerisation requires a specific tertiary structure, this indicated that both fragments were correctly folded (Schnappinger *et al.*, 1995). A selected overview of results obtained for the extracellular heterologous production of eukaryotic proteins in Gram-positive bacteria is given in Table 1.

Table 1: Examples of extracellular eukaryotic protein production in Gram-positive bacteria.

Host cell	Heterologous protein	Yield [a]	Reference
- *Bacillus*			
	hIL2	120	1
B. brevis			
	tune growth hormone	240	2
B. brevis			
	human growth hormone	200	3
B. brevis			
	human epidermal growth factor	3000	3
B. brevis			
	human α-amylase	60	3
B. brevis			
	anti-11-deoxycortisol light chain IgG	20	4
B. brevis			
B. subtilis	hIL1β	6.7	5
B. subtilis	human pancreatic secretory trypsin inhibitor.	~40	6
B. subtilis	human epidermal growth factor	7	7
B. subtilis	antidigoxin SCA	12	8
B. subtilis	antilysozyme SCA	0.25	9
- *Streptomyces*			
S. lividans	hIFNα1	1-2 x 10^5 [b]	10
S. lividans	hIL1β	3.8 x 10^6 [b]	11
S. lividans	proinsulin (from *Macaca fascicularis*)	20 -100	12

Table 1: Examples of extracellular eukaryotic protein production in Gram-positive bacteria.

Host cell	Heterologous protein	Yield [a]	Reference
S. lividans	hirudin (from *Hirudo medicinalis*)	0.2 -0.5	13
S. lividans	hIL2	0.03	14
S. lividans	hGM-CSF	10 -15	15
S. lividans	sCD4 derivative D1D2	>300	16
S. lividans	hTGFα	10	17
S. lividans	mTNFα	>300	18
S. lividans	hTNFα	120	Unpubl. data
S. lividans	hEPO-R	15	19
S. lividans	Fv of HyHEL10 (mAb)	1[c]	20
- *Lactococcus*			
L. lactis	bovine plasmin	<1	21
L. lactis	bovine prochymosin	NI	22
L. lactis	mIL2	0.9	23
L. lactis	mIL6	0.9	23
- *Staphylococcus*			
S. carnosus	REIv	5-10	24
S. carnosus	human lysozyme antibody D1.3	NI	25

a: Yield is in mg/L if not otherwise indicated; b: yield in U/ml; c: Yield after purification; NI: not indicated.

1) Takimura et al. *1997,* Biosci. Biotech. Biochem. 61, *1858-1861; 2) Sagiya* et al. *1994,* Appl. Microbiol. Biotechnol. 42, *358-363; 3) Udaka and Yamagata 1993,* Antonie van Leeuwenhoek 64, *137-143; 4) Kajino* et al. *2000,* Appl. Environ. Microbiol. 66, *638-642; 5) Bellini* et al. *1991,* J. Biotechnol. 18, *177-192; 6) Kim* et al. *1997,* Mol. Cells 7, *788-794; 7) Lam* et al. *1998,* J. Biotechnol. 63, *167-177; 8) Wu* et al. *1998,* J. Bacteriol.180, *2830-2835; 9) Bolhuis* et al. *1999,* Appl. Environ. Microbiol. 65, *2934-2941; 10) Noack* et al. *1988,* Gene 68, *53-62; 11) Lichenstein* et al. *1988,* J. Bacteriol. 170, *3924 –3929; 12) Koller* et al. *1989,* Bio/Technology 7, *1055-1059; 13) Bender* et al. *1990,* Appl. Microbiol. Biotechnol. 34, *203-207; 14) Bender* et al. *1990,* Gene 86, *227-232; 15) Malek* et al. *1990,* J. Cell. Biochem. Suppl. 14A, *127; 16) Fornwald* et al. *1993,* Bio/Technology 11, *1031-1036; 17) Taguchi* et al. *1995,* Gene 159, *239-243; 18) Lammertyn* et al. *1997,* Appl. Environ. Microbiol. 63, *1808-1813; 19) Binnie* et al. *1997,* Protein Expr. Purif. 11, *271-278; 20) Ueda* et al. *1993,* Gene 129, *129-134; 21) Arnau* et al. *1997,* Appl. Microbiol. Biotechnol. 48, *331-338; 22) de Vos* et al. *1989,* J. Dairy Sci. 72, *3398-3405; 23) Steidler* et al. *1998,* Infect. Immun. 66, *3183-3189; 24) Pschorr* et al. *1994,* Biol. Chem. Hoppe-Seyler 375, *271-280; 25) Schnappinger* et al. *1995,* FEMS Microbiol. Lett. 129, *121-128.*

4.2. GRAM-POSITIVE BACTERIA AS LIVE VACCINE DELIVERY SYSTEMS

Surface display of heterologous proteins in bacteria was first achieved in Gram-negative bacteria like *E. coli* and *Salmonella*. However, these engineered enteric bacteria can not be classified as safe and cell surface display in these bacteria requires translocation of the heterologous protein through the inner membrane and a correct integration in the outer membrane. These Gram-negative bacteria are moreover highly immunogenic, a disadvantage in use as live recombinant vaccine systems. Gram-positive bacteria are more favourable for cell surface display applications for following reasons. Translocation of the protein of interest across a single membrane is sufficient to achieve a proper surface exposure. Additionally, the surface receptors of Gram-positive bacteria seem to be more permissive for the insertion of extended sequence of foreign proteins, as compared to different Gram-negative receptors (Fischetti *et al.*, 1993). Furthermore, Gram-positive bacteria are more rigid due to the thicker cell wall making them less sensitive to cell lysis during certain laboratory procedures. Their use as vaccine delivery system has the additional advantage that many Gram-positive bacteria exhibit a natural immuno-adjuvanticity due to components in the peptidoglycan layer.

The use of Gram-positive bacteria as live vaccine vehicles is indeed gaining more and more interest. These bacteria are good candidates for this purpose since they respond to several of the prerequisites for a new vaccine. Ideally these vaccines should be given in a single dose early in life and as a cocktail for a variety of diseases. Furthermore, they should be safe, cheap, heat-stable, easy to administer and capable to induce broad immune responses with life-long memory both in infants and adults. Because of the increasing availability of genetic tools to manipulate a growing number of bacteria, in particular Gram-positive bacteria, these are being developed as antigen delivery system. In first instance, the antigens were expressed intracellularly and it was shown that the obtained recombinant bacteria could elicit an immune response. However, it was demonstrated that in general better immune responses could be obtained by secreting the antigen or by exporting the specific protein as a chimeric cell-associated protein. In respect to protection against pathogen challenge a protective immune response was mostly obtained only after secretion or cell surface display of the antigen.

4.2.1. Gram-positive bacteria used as antigen delivery systems

The best-known Gram-positive bacterium widely used for a long time in vaccination programmes is the attenuated *Mycobacterium bovis* strain, called Bacillus Calmette-Guérin (BCG). It is of particular interest since it has already been used during several decades as bacterial anti-tuberculosis vaccine as a result of several advantages.

Besides the low-cost production, the orally administration and the long-term immunity from a single dose given at, or any time after, birth, BCG expresses intrinsic adjuvant activities especially for the cell-mediated immunity and it has low incidence of severe side effects. Because of these features and resolved problems concerning the genetic manipulation of mycobacteria, BCG has become the last decade a very promising bacterium to express a range of bacterial or viral antigens.

In the prevention of veterinary diseases, the attenuated *Bacillus anthracis* Sterne strain has been used for 60 years as a vaccine against anthrax, a zoonosis caused by *B. anthracis*. However, due to toxin production this live vaccine Sterne strain can still elicit severe side effects such as fever, severe oedema or even death. Recently, the Sterne strain was manipulated in order to decrease the virulence of the vaccine and it was shown that attenuated *B. anthracis* can be used for the design of new live vaccines against other veterinary diseases by expressing a foreign antigen (Brossier *et al.*, 1999).

Very recently, *Listeria monocytogenes* has received increasing attention as a vaccine vector because of its ability to gain access to the host cell cytosol. It is able to enter phagocytic and nonphagocytic cells, to escape from endosomes by secreting virulence factor listeriolysin, to multiply within the host cell cytoplasm, and to spread directly to adjacent cells without exposure to extracellular environment (Tilney *et al.*, 1989). Proteins secreted by native or engineered *L. monocytogenes* are processed and presented by both major histocompatibility complex class I and class II pathways stimulating potent $CD8^+$ and $CD4^+$ T cell responses. At the same time, a cascade of cytokines is produced upon infection. These cell-mediated immune responses are considered to be very important for clearance of viruses, tumours, and intracellular infections (Weiskirch and Paterson, 1997). Because of pathogenicity, an attenuated *L. monocytogenes* strain impaired in intra- and intercellular movement and which is self-destructing intracellularly, was constructed (Dietrich *et al.*, 1998).

Although some attenuated strains have been successfully used as vaccines, the use of non-pathogenic food-grade or commensal bacteria have attracted attention as possible antigen vehicles. Since there is no risk of reversion to virulence, these GRAS bacteria are safe for immunisation purposes. In addition, it is unlikely that these food-grade and commensal bacteria should evoke a strong undesired immune response. On the contrary, the peptidoglycan of Gram-positive cell wall possesses intrinsic adjuvanticity. Especially, the use of commensals as vaccine delivery system is very advantageous since they colonise a specific mucosal niche (oral, intestinal, vaginal). As a consequence, recombinant commensals may be used to stimulate a mucosal immune response against a pathogen that enters the mammalian host at a mucosal site in a similar way. Upon colonisation of the recombinant commensals, an enhanced local IgA response to the foreign antigen as well as a systemic IgG response and T cell stimulation can be generated. Among the Gram-positive commensals exploited as vaccine vehicles, is *Streptococcus gordonii* isolated from the oral cavity. Also members of the *Lactobacillus* genus are found at mucosal sites including the gastrointestinal and female reproductive tracts and the oral cavity. They are maintained transiently or for a longer term via colonisation of the mucosa. Together with lactobacilli, lactococci belong to the lactic acid bacteria (LAB). These bacteria are extensively utilised in various food fermentations and may have a health promoting, "probiotic" effect. In contrast to some lactobacilli, *Lactococcus* species do not have the capacity to colonise the mucosa but it was shown that *Lactococcus lactis* (the model of LAB) retained its metabolic activity during transit in the digestive tract (Corthier *et al.*, 1998). Although LAB are known as low secretors, their potential as live bacterial vaccine delivery systems has already been proven.

Other candidate Gram-positive bacteria possibly to be used as vaccine delivery systems are the non-pathogenic *Staphylococcus xylosus* and *Staphylococcus carnosus.*

4.2.2.Approaches for antigen presentation

Antigens can be presented to the mammalian host in 3 different ways. (1) Intracellular expression of the heterologous protein under the influence of a prokaryotic promoter in the bacterial strain of choice. (2) Antigens can N-terminally be fused to a signal peptide resulting in the secretion of the antigen by the bacterium at a determined location in the mammalian host. (3) The antigen can simultaneously be fused N-terminally to a signal peptide and C-terminally to a cell surface-anchor in order to export the heterologous protein through the cytoplasmic membrane and bind it to the cell surface. Nearly all bacterial surface proteins contain indeed a highly conserved C-terminus responsible for cell attachment. This region consists of repetitive sequences of charged residues followed by a hexapeptide with the consensus sequence LPXTGX (with X any amino acid). Adjacent to this consensus is a hydrophobic segment of 15 to 20 amino acids sufficient to span the cytoplasmic membrane, finally followed by a positively charged C-terminus of up to 6 amino acids. The cell wall sorting of the exported protein is achieved by proteolytic cleavage between the threonine and glycine residues of the conserved hexapeptide, and subsequent covalent binding of the surface protein to the cell wall. The fusion of a heterologous protein to such a cell surface-anchor results in the foreign protein protruding the cell wall.

The cell anchors used in the surface display of heterologous proteins in Gram-positive bacteria are those coming from the *Staphylococcus aureus* Protein A (SPA), the cell-bound proteinase PrtP of *Lactobacillus casei*, and the fibrillar M6 protein of *Streptococcus pyogenes*. For the latter, it was demonstrated that it can be highly expressed and efficiently cell wall-anchored in various LAB unrelated to *S. pyogenes* (Piard *et al.*, 1997).

To display a foreign protein to the surface of *Mycobacterium* spp. a total distinct anchoring system was applied. Heterologous proteins were N-terminally fused to the 19-kDa mycobacterial membrane lipoprotein. As a consequence, the fusion protein is translocated through the membrane and, next, anchored into the outer leaflet of the cell membrane by the lipidated N-terminus of the 19-kDa protein.

The way of presenting the antigen to the mammalian hosts is important since obvious differences were observed in immune responses and protective effects depending on the manner of antigen presentation. A comparison was made between recombinant BCG strains expressing the antigen OspA from *Borrelia burgdorferi* or PspA from *Streptococcus pneumoniae*, as a cytoplasmic protein, a secreted (fusion) protein or a chimeric exported membrane-associated lipoprotein (Stover *et al.*, 1993; Langermann *et al.*, 1994). In the first study, it was shown that OspA as a membrane-bound lipoprotein elicited protective antibody responses which were 100- to 1000-fold higher than those resulting from immunisation with cytoplasmically expressed or secreted OspA. In the second case, the rBCG strains elicited comparable antibody titers against PspA. However, protective responses were only obtained with rBCG expressing PspA as a secreted or chimeric exported protein. Using *Lactococcus lactis* as antigen delivery system, it was also demonstrated that the cell wall-anchored tetanus toxin fragment C

(TTFC) was 10- to 20-fold more immunogenic than the alternatively presented TTFC (Norton *et al.*, 1996).

4.2.3. Heterologous antigens presented by Gram-positive bacteria

The last decade, a variety of antigenic determinants was expressed in different Gram-positive bacteria. Administration of these recombinant bacteria to experimental animal models could elicit immune responses against bacterial pathogens, viruses and tumour antigens. Although *in vivo* enhancement of immunoglobulins and/or proliferation of T cells were determined in several experiments, protection against pathogen challenge was only investigated in few cases. Examples of antigens produced by Gram-positive bacteria are given in table 2. From the results obtained so far it can be concluded that the mentioned Gram-positive bacteria are interesting candidates for live vaccine delivery systems and will undoubtedly become soon applicable vaccination tools in human health care.

Table 2: Selected examples of heterologous antigen expression in Gram-positive bacteria resulting in an immune response (IR) and eventual protection (P) against pathogen challenge in experimental animal models. The manner of antigen presentation to the mammalian host is indicated: intracellularly expressed (I), secreted (S) or cell surface displayed on the bacterium (CSD).

Host cell	Heterologous antigen	Source	I/S/CSD	IR/P	Ref.
- *Mycobacterium* spp.	TTFC	*Clostridium tetani*	I	IR	1
	outer surface protein	*Borrelia burgdorferi*	I/S/CDS	IR/P	2
	surface protein A	*Streptococcus pneumoniae*	I/S/CDS	IR/P	3
	surface proteinase p63	*Leishmania major*	S	P	4
	glutathione S-ransferase	*Schistosoma mansoni*	I	IR	5
	gp120	HIV1	I	IR	1
	Nef	HIV1	I	IR	6
-*Bacillus anthracis* Sterne	listeriolysin O	*Listeria monocytogenes*	S	IR/P	7
	Ib of iota toxin	*Clostridium perfringens*	S	IR/P	8
-*Listeria monocytogenes*	nucleoprotein	Influenza type A	S	IR/P	9
	idem (model tumour g)	*idem*	S	IR/P	10

Table 2: Selected examples of heterologous antigen expression in Gram-positive bacteria resulting in an immune response (IR) and eventual protection (P) against pathogen challenge in experimental animal models. The manner of antigen presentation to the mammalian host is indicated: intracellularly expressed (I), secreted (S) or cell surface displayed on the bacterium (CSD).

Host cell	Heterologous antigen	Source	I/S/CSD	IR/P	Ref.
	nucleoprotein	LCMV	S	IR/P	11
	E1 protein	CRPV	S	IR/P	12
- *Lactococcus lactis*	TTFC	Clostridium tetani	I	IR/P	13
- *Lactobacillus* spp.	TNP	(hapten)	CSD	IR	14
	TTFC	Clostridium tetani	CSD	IR	15
	myelin basic protein	human	S	IR	15
	haemagglutinin pitope	Influenza virus	I	IR	16
- *Streptococcus gordonii*	acid allergen Ag5.2	hornet	CSD	IR	17
	V3 of gp120	HIV1	CSD	IR	18
	E7 protein	HPV16	CSD	IR	18
- *Staphylococcus* spp.	epitope G lycoprotein	hRSV	CSD	IR	19

CRPV: cottontail rabbit papilloma virus; HIV1: human immunodeficiency virus type 1; HPV16: human papilloma virus type 16; hRSV: human respiratory syncytial virus; LCMV: lymphocytic choriomeningitis virus; TNP: trinitrophenyl; TTFC: tetanus toxin C fragment
1) Stover et al. *1991,* Nature 351, *456-460; 2) Stover* et al. *1993,* J. Exp. Med. 178, *197-209; 3) Langermann* et al. *1994,* J. Exp. Med. 180, *2277-2286; 4) Abdelhak* et al. *1995,* Microbiology 141, *1585-1592; 5) Kremer* et al. *1996,* J. Immunol. 156, *4309-4317; 6) Winter* et al. *1991,* Gene 109, *47-54; 7) Sirard* et al. *1997,* J. Immunol. 159, *4435-4443; 8) Sirard* et al. *1997,* Infect. Immun. 65, *2029-2033; 9) Ikonomidis* et al. *1997,* Vaccine 15, *433-440; 10) Pan* et al. *1995,* Nature Med. 1, *471-477; 11) Shen* et al. *1995,* Proc. Natl. Acad. Sci. USA 92, *3987-3991; 12) Jensen* et al. *1997,* J. Virol. 71, *8467-8474; 13) Norton* et al. *1997,* Vaccine 15, *616-619; 14) Claassen* et al. *1995,* Adv. Exp. Med. Biol. 371, *1553-1558; 15) Maassen* et al. *1999,* Vaccine 17, *2117-2128; 16) Pouwels* et al. *1996,* J. Biotechnol. 44, *183-192; 17) Medaglini* et al. *1995,* Proc. Natl. Acad. Sci. USA 92, *6868-6872; 18) Di Fabio* et al. *1998,* Vaccine 16, *485-492; 19) Nguyen* et al. *1995,* J. Biotechnol. 42, *207-219.*

4.3. OTHER APPLICATION AREAS

The surface display technique in Gram-positive bacteria has still other possible biopharmaceutical applications. The expression of functional single chain antibodies on the bacterial cell surface can create an inexpensive diagnostic tool. Gunneriusson *et al.* (1996) described the surface display of a functional anti-human IgE single chain Fv fragment in *S. xylosus* and *S. carnosus*. This means that also Gram-positive bacteria can be applied as "whole cell monoclonal antibodies" in different diagnostic tests.

Cell surface displayed libraries in Gram-positive organisms could also offer a powerful complement to phage display for screening of high-affinity ligands to target molecules. The possible major advantage of the bacterial display systems over the phage display technique lies in the ability to use fluorescence activated cell sorting (FACS) for selection of bacterial clones carrying the desired peptide. In contrast, FACS can not be used with phage displayed libraries because of the small size of bacteriophages. In the future, biopharmaceutical research laboratories will surely further explore cell surface display of foreign peptides on Gram-positive bacteria.

5. Conclusions and perspectives

The more extensive knowledge of *E. coli* genetics and subsequent availability of genetic tools made this Gram-negative bacterium the host of first choice for heterologous protein production. However, the presence of the outer membrane, which in addition contains toxic lipopolysaccharides, and the frequent occurrence of inclusion bodies are major drawbacks. Gram-positive bacteria are in this respect advantageous over Gram-negative organisms. The presence of a single membrane enables direct secretion to the extracellular medium and efficient cell surface display of heterologous proteins. They are more rigid compared to Gram-negative organisms due to the thicker cell wall. In addition, their peptidoglycan layer has a natural immuno-adjuvanticity, which is beneficial for vaccination applications. Moreover, the surface proteins of Gram-positive bacteria seem more permissive of the insertion of extended sequences from foreign proteins than the Gram-negative surface proteins. A potential drawback of the use of Gram-positive bacteria is their lower frequency of transformation compared to Gram-negative bacteria, a factor of importance in the creation of peptide libraries.

However, heterologous protein secretion in Gram-positive bacteria still encounters a number of problems, which in several instances limit high yield secretion. Therefore, several research groups are analysing the different components of the secretion pathway in order to modulate these compounds and to improve production yield. Since there is no general rule to optimise secretion of different heterologous proteins, tailor-made strains will be the answer to resolve bottlenecks in secretion of heterologous proteins. Nevertheless, from the recent results obtained with heterologous protein secretion in Gram-positive bacteria, we can conclude that these host cells have much to offer as cell factories for the production of biopharmaceutical compounds.

References

Aretz, W., Koller, K.P. and Riess, G. (1989) Proteolytic enzymes from recombinant *Streptomyces lividans* TK24, *FEMS Microbiol. Lett. 53*, 31-35.

Arnau, J., Hjerl-Hansen, E., and Israelsen, H. (1997) Heterologous gene expression of bovine plasmin in *Lactococcus lactis*, *Appl. Microbiol. Biotechnol. 48*, 331-338.

Binnie, C., Butler, M.J., Aphale, J.S., Bourgault, R., DiZonno, M.A., Krygsman, P., Liao, L., Walczyk, E. and Malek, T. (1995) Isolation and characterization of two genes encoding proteases associated with the mycelium of *Streptomyces lividans* 66, *J. Bacteriol. 177*, 6033-6040.

Bolhuis, A., Matzen, A., Hyyryläinen, H.-L., Kontinen, V.P., Meima, R., Chapuis, J., Venema, G., Bron, S., Freudl, R. and van Dijl, J.M. (1999a) Signal peptide peptidase- and ClpP-like proteins of *Bacillus subtilis* required for efficient translocation and processing of secretory proteins, *J. Biol. Chem. 274*, 24585-24592.

Bolhuis, A., Tjalsma, H., Smith, H.E., de Jong, A., Meime, R., Venema, G., Bron, S. and van Dijl, J.M. (1999b) Evaluation of bottlenecks in the late stages of protein secretion in *Bacillus subtilis*, *Appl. Environ. Microbiol. 65*, 2934-2941.

Bolhuis, A., Venema, G., Quax, W.J., Bron, S. and van Dijl, J.M. (1999c) Functional analysis of paralogous thiol-disulfide oxidoreductases in *Bacillus subtilis*, *J. Biol. Chem. 274*, 24531-24538.

Braun, P., Gerritse, G., van Dijl, J.-M. and Quax, W.J. (1999) Improving protein secretion by engineering components of the bacterial translocation machinery, *Curr. Opin. Biotechnol. 10*, 376-381.

Bron, S., Bolhuis, A., Tjalsma, H., Holsappel, S., Venema, G. and van Dijl, J.M. (1998) Protein secretion and possible roles for multiple signal peptidases for precursor processing in Bacilli, *J. Biotechnol. 64*, 3-13.

Brossier, F., Mock, M. and Sirard, J.-C. (1999) Antigen delivery by attenuated *Bacillus anthracis*: new prospects in veterinary vaccines, *J. Appl. Microbiol. 87*, 298-302.

Bucca, G., Ferina, G., Puglia, A.M. and Smith, C.P. (1995) The *dnaK* operon of *Streptomyces coelicolor* encodes a novel heat-shock protein which binds to the promoter region of the operon, *Mol. Microbiol. 17*, 663-674.

Corthier, G., Delorme, C., Ehrlich, S.D. and Renault, P. (1998) Use of luciferase genes as biosensors to study bacterial physiology in the digestive tract, *Appl. Environ. Microbiol. 64*, 2721-2722.

Dalbey, R.E., Lively, M.O., Bron, S. and van Dijl, J.M. (1997) The chemistry and enzymology of the type I signal peptidases, *Protein Sci. 6*, 1129-1138.

de Vos, W., Vos, P., Simons, G. and David, S. (1989) Gene organization and expression in mesophilic lactic acid bacteria, *J. Dairy Sci. 72*, 3398-3405.

Dietrich, G., Bubert, A., Gentschev, I., Sokolovic, Z., Simm, A., Catic, A., Kaufmann, S.H.E., Hess, J., Szalay, A.A. and Goebel, W. (1998) Delivery of antigen-encoding plasmid DNA into cytosol of macrophages by attenuated suicide *Listeria monocytogenes*, *Nature Biotechnol. 16*, 181-185.

Driessen, A.J.M., Fekkes, P. and van der Wolk, J.P.W. (1998) The Sec system, *Curr. Opin. Microbiol. 1*, 216-222.

Duong, F. and Wickner, W. (1999) The PrlA and PrlG phenotypes are caused by a loosened association among the translocase SecYEG subunits, *EMBO J. 18*, 3263-3270.

Economou, A. (1999) Following the leader: bacterial protein export through the Sec pathway, *Trends Microbiol. 7*: 315-320.

Fischetti, V.A., Medaglini, D., Oggioni, M and Pozzi, G. (1993) Expression of foreign proteins on Gram-positive commensal bacteria for mucosal vaccine delivery, *Curr. Opin. Biotechnol. 4*, 603-610.

Fornwald, J.A., Donovan, M.J., Gerber, R., Keller, J. Taylor, D.P., Arcuri, E.J. and Brawner, M.E. (1993) Soluble forms of the human T cell receptor CD4 are efficiently expressed by *Streptomyces lividans*, *Bio/Technology 11*, 1031-1036.

Geukens, N., Van Mellaert, L., Lammertyn, E., Schacht, S., Schaerlaekens, K., Parro, V., Mellado, R.P., Bron, S., Engelborghs, Y. and Anné, J. (2000) Membrane topology of the *Streptomyces lividans* type I signal peptidases (submitted)

Grandvalet, C., de Crécy-lagard, V. and Mazodier, P. (1999) The ClpB ATPase of *Streptomyces albus* G belongs to the HspR heat shock regulon, *Mol. Microbiol. 31*, 521-532.

Grandvalet, C., Rapoport, G. and Mazodier, P. (1997) Disruption of the *hspR*, the repressor gene of the *dnaK* operon in *Streptomyces albus* G, *Mol. Microbiol. 23*, 77-84.

Gunneriusson, E., Samuelson, P., Uhlén, M., Nygren, P.-A. and Stahl, S. (1996) Surface display of a functional single chain Fv antibody on staphylococci, *J. Bacteriol. 178*, 1341-1346.

Haandrikman, A.J., Kok, J., Laan, H., Soemitro, S., Ledeboer, A.M., Konings, W. and Venema, G. (1989) Identification of a gene required for maturation of an extracellular lactococcal serine protease, *J. Bacteriol. 171*, 2789-2794.

Hirose, I., Sano, K., Shioda, I., Kumano, M., Nakamura, K. and Yamane, K. (2000) Proteome analysis of *Bacillus subtilis* extracellular proteins: a two-dimensional protein electrophoretic study, *Microbiology 146*, 65-75.

Kajino, T., Ohto, C., Muramatsu, M., Obata, S., Udaka, S., Yamada, Y. and Takahashi, H. (2000) A protein disulfide isomerase gene fusion expression system that increases the extracellular productivity of *Bacillus brevis, Appl. Environ. Microbiol. 66*, 638-642.

Kontinen, V.P. and Sarvas, M (1993) The PrsA lipoprotein is essential for the secretion in *Bacillus subtilis* and sets a limit for high-level secretion, *Mol. Microbiol. 8*, 727-737.

Krieger, T.J., Bartfeld, D., Jenish, D.L. and Hadary, D. (1994) Purification and characterization of a novel tripeptidyl aminopeptidase from *Streptomyces lividans, FEBS Lett. 352*, 385-388.

Kunst, F. *et al.* (1997) The complete genome sequence of the Gram-positive bacterium *Bacillus subtilis, Nature 390*, 249-256.

Lam, K.H.E., Chow, K.C. and Wong, W K R (1998) Construction of an efficient *Bacillus subtilis* system for extracellular production of heterologous proteins, *J. Biotechnol. 63*, 167-177.

Lammertyn, E. (2000) Isolation and characterisation of a novel subtilisin inhibitor from *Streptomyces venezuelae* and evaluation of its regulatory sequences for heterologous protein secretion by *Streptomyces lividans*, Katholieke Universiteit Leuven, PhD thesis, pp. 147.

Lammertyn, E. and Anné, J. (1998) Modifications of *Streptomyces* signal peptides and their effects on protein production and secretion, *FEMS Microbiol. Lett. 160*, 1-101.

Lammertyn, E., Van Mellaert, L., Schacht, S., Dillen, C., Sablon, E., Van Broekhoven, A. and Anné, J (1997) Evaluation of a novel subtilisin inhibitor gene and mutant derivatives for the expression and secretion of mouse tumor necrosis factor alpha by *Streptomyces lividans, Appl. Environ. Microbiol. 63*, 1808-1813.

Langella, P. and Le Loir, Y. (1999) Heterologous protein secretion in *Lactococcus lactis*: a novel antigen delivery system, *Braz. J. Med. Biol. Res. 32*, 191-198.

Langermann, S., Palaszynski, S.R., Burlein, J.E., Koenig, S., Hanson, M.S., Briles, D.E. and Stover, C.K. (1994) Protective humoral response against pneumococcal infection in mice elicited by recombinant bacille Calmette-Guérin vaccines expressing pneumococcal surface protein A, *J. Exp. Med. 180*, 2277-2286.

Le Loir, Y., Gruss, A., Ehrlich, S.D. and Langella, P. (1998) A nine-residue synthetic propeptide enhances secretion efficiency of heterologous proteins in *Lactococcus lactis, J. Bacteriol. 180*, 1895-1903.

Leloup, L., Driessen, A.J.M., Freudl, R., Chambert, R. and Petit-Glatron, M.-F. (1999) Differential dependence of levansucrase and α-amylase secretion on SecA (Div) during the exponential phase of growth of *Bacillus subtilis, J. Bacteriol. 181*, 1820-1826.

Leloup, L., Haddaoui, E.A., Chambert, R. and Petit-Glatron, M.-F. (1997) Characterization of the rate-limiting step of the secretion of *Bacillus subtilis* α-amylase overproduced during the exponential phase of growth, *Microbiology 143*, 3295-3303.

Liebl, W. and Götz, F (1986) Studies on lipase directed export of *Escherichia coli* β-lactamase in *Staphylococcus carnosus, Mol. Gen. Genet. 204*, 166-173.

Meens, J., Herbort, M., Klein, M. and Freudl, R. (1997) Use of the pre-pro part of *Staphylococcus hyicus* lipase as a carrier for secretion of *Escherichia coli* outer membrane protein A (OmpA) prevents proteolytic degradation of OmpA by cell-associated protease(s) in two different gram-positive bacteria, *Appl. Environ. Microbiol. 63*, 2814-2820.

Meijer, W.J.J., de Jong, A., Wisman, G.B.A., Tjalsma, H., Venema, G., Bron, S. and van Dijl, J.M. (1995) The endogenous *Bacillus subtilis (natto)* plasmids pTA1015 and pTA1040 contain signal peptidase-encoding genes: identification of a new structural module on cryptic plasmids, *Mol. Microbiol. 17*, 621-631.

Müller, J.P., Bron, S., Venema, G. and van Dijl, J.M. (2000) Chaperone-like activities of the CsaA protein of *Bacillus subtilis, Microbiology 146*, 77-88.

Nakamura, K., Yahagi, S.-i., Yamazaki, T. and Yamane, K. (1999) *Bacillus subtilis* histone-like protein, HBsu, is an integral component of a SRP-like particle that can bind the *Alu* domain of small cytoplasmic RNA, *J. Biol. Chem. 274*, 13569-13576.

Nguyen, T.N., Gourdon, M.-H., Hansson, M., Robert, A., Samuelson, P., Libon, C., Adréoni, C., Nygren, P.-A. Binz, H., Uhlén, M. and Stahl, S. (1995) Hydrophobicity engineering to facilitate surface display of heterologous gene products on *Staphylococcus xylosus*, *J. Biotechnol. 42*, 207-219.

Norton, P.M., Brown, H.W.G., Wells, J.M., Macpherson, A.M., Wilson, P.W. and Le Page, R.W.F. (1996) Factors affecting the immunogenicity of tetanus toxin fragment C expressed in *Lactococcus lactis*, *FEMS Immunol. Med. Microbiol. 14*, 167-177.

Parro, V., Schacht, S., Anné, J. and Mellado, R.P. (1999) Four genes encoding different type I signal peptidases are organized in a cluster in *Streptomyces lividans* TK21, *Microbiology 145*, 2255-2263.

Perez-Perez, J., Barbero, J.L., Marquez, G. and Gutierrez, J. (1996) Different PrlA proteins increase the efficiency of periplasmic production of human interleukin-6 in *Escherichia coli*, *J. Biotechnol. 49*, 245-247.

Piard, J.-C., Hautefort, I., Fischetti, V.A., Ehrlich, S.D., Fons, M. and Gruss, A. (1997) Cell wall anchoring of the *Streptococcus pyogenes* M6 protein in various lactic acid bacteria, *J. Bacteriol. 179*, 3068-3072.

Pozidis, C., Lammertyn, E., Politou, A.S., Anné, J., Sianidis, G. and Economou, A. (2000) Protein secretion biotechnology: large-scale production of functional tumor necrosis factor α using *Streptomyces lividans* (submitted).

Pschorr, J., Bieseler, B. and Fritz, H.-J. (1994) Production of the immunoglobulin variable domain REI_v via fusion protein synthesized and secreted by *Staphylococcus carnosus*, *Biol. Chem. Hoppe-Seyler 375*, 271-280.

Samuelson, P., Cano, F., Robert, A. and Stahl, S. (1999) Engineering of a *Staphylococcus carnosus* surface display system by substitution or deletion of a *Staphylococcus hyicus* lipase propeptide, *FEMS Microbiol. Lett. 179*, 131-139.

Saunders, C.W., Pedroni, J.A. and Monahan, P.M. (1991) Optimization of the signal-sequence cleavage site for secretion from *Bacillus subtilis* of a 34-amino acid fragment of human parathyroid hormone, *Gene 102*, 277-282.

Schnappinger, D., Geissdörfer, W., Sizemore, C. and Hillen, W (1995) Extracellular expression of native human anti-lysozyme fragments in *Staphylococcus carnosus*, *FEMS Microbiol. Lett. 129*, 121-128.

Sjölander, A., Stahl, S., Lövgren, K., Hansson, M., Cavelier, L., Walles, A., Helmby, H., Wahlin, B., Uhlén, M., Berzins, K., Perlmann, P and Wahlgren, M. (1993) *Plasmodium falciparum*: The immune response in rabbits to the clustered-asparagine-rich-protein (CARP) after immunization in Freund's adjuvant or immunostimulating complexes (ISCOMs), *Exp. Parasitol. 76*, 134-145.

Steidler, L., Robinson, K., Chamberlain, L., Schofield, K.M., Remaut, E., Le Page, R.W.F. and Wells, J.M. (1998) Mucosal delivery of murine interleukin-2 (IL-2) and IL-6 by recombinant strains of *Lactococcus lactis* coexpressing antigen and cytokine, *Infect. Immun. 66*, 3183-3189.

Stephenson, K., Carter, N.M., Harwood, C.R., Petit-Glatron, M.-F. and Chambert, R. (1998) The influence of protein folding on late stages of the secretion of α-amylases from *Bacillus subtilis*, *FEBS Lett. 430*, 385-389.

Stephenson, K. and Harwood, C.R. (1998) Influence of cell-wall-associated protease on production of α-amylase by *Bacillus subtilis*, *Appl. Environ. Microbiol. 64*, 2875-2881.

Stover, C.K., Bansal, G.P., Hanson, M.S., Burlein, J.E., Palaszynski, S.R., Young, J.F., Koenig, S., Young, D.B., Sadziene, A. and Barbour, A.G. (1993) Protective immunity elicited by recombinant bacille Calmette-Guerin (BCG) expressing outer surface protein A (OspA) lipoprotein: a candidate lyme disease vaccine, *J. Exp. Med. 178*, 197-209.

Taguchi, S., Odaka, A., Watanabe, Y. and Momose, H. (1995) Molecular characterization of a gene encoding extracellular serine protease isolated from a subtilisin inhibitor-deficient mutant of *Streptomyces albogriseolus* S-3252, *Appl. Environ. Microbiol. 61*, 180-186.

Tilney, L.G. and Portnoy, D.A. (1989) Actin filaments and the growth, movement and spread of the intracellular bacterial parasite, *Listeria monocytogenes*, *J. Cell Biol. 109*, 1597-1608.

Tjalsma, H., Noback, M.A., Bron, S., Venema, G., Yamane, K. and van Dijl, J.M. (1997) *Bacillus subtilis* contains four closely related type I signal peptidases with overlapping substrate specificities, *J. Biol. Chem. 272*, 25983-25992.

Udaka, S. and Yamagata, H. (1993a) High-level secretion of heterologous proteins by *Bacillus brevis*, *Methods Enzymol. 217*, 23-33.

Udaka, S. and Yamagata, H. (1993b) Protein secretion in *Bacillus brevis*, *Antonie van Leeuwenhoek 64*, 137-143.

Valent, Q.A., Scotti, P.A., High, S., de Gier, J.W.L., von Heijne, G., Lentzen, G., Wintermeyer, W., Oudega, B. and Luirink (1998) The *Escherichia coli* SRP and SecB targeting pathways converge at the translocon. *EMBO J. 17*, 2504-2512.

Van Mellaert, L., Dillen, C., Proost, P., Sablon, E., Deleys, R., Van Broekhoven, A., Heremans, H., Van Damme, J., Eyssen, H. and Anné, J. (1994) Efficient secretion of biologically active mouse tumor necrosis factor α by *Streptomyces lividans*, *Gene 150*, 153-158.

Van Mellaert, L., Lammertyn, E, Schacht, S;, Proost, P., Van Damme, J., Wroblowski, B., Anné, J., Scarcez, T., Sablon, E., Raeymaeckers, J. and Van Broekhoven, A. (1998) Molecular characterization of a novel subtilisin inhibitor protein produced by *Streptomyces venezuelae* CBS762.70, *DNA Sequence 9*, 19-30.

Vitale, L. (1999) Aminopeptidases of the genus *Streptomyces*, *Food Technol. Biotechnol. 37*, 29-37.

Weiskirch, L.M. and Paterson, Y. (1997) *Listeria monocytogenes*: a potent vaccine vector for neoplastic and infectious desease, *Immunol. Rev. 158*, 159-169.

Wong, S.-L. (1995) Advances in the use of *Bacillus subtilis* for the expression and secretion of heterologous proteins, *Curr. Opin. Biotechnol. 6*, 517-522.

Wu, S.-C., Ye, R., Wu, X.-C., Ng, S.-C and Wong, S.-L. (1998) Enhanced secretory production of a single - chain antibody fragment from *Bacillus subtilis* by coproduction of molecular chaperones, *J. Bacteriol.180*, 2830-2835.

MULTIPLE PATHWAYS OF EXOPROTEIN SECRETION IN GRAM-NEGATIVE BACTERIA

ANTHONY P PUGSLEY
Unité de Génétique Moléculaire, Institut Pasteur, 25, rue du Dr Roux, 75724 Paris cedex, France

1. Introduction

Molecular studies on protein secretion in Gram-negative bacteria began in earnest a little over a decade ago when the first mutations preventing the appearance of extracellular proteins were constructed and analysed. Many of these studies were (and still are) fed by an interest in the function of the exoproteins in pathogenicity and were started at a time when the characterisation of protein export across the *Escherichia coli* inner membrane (the Sec system) by genetical and biochemical methods was already in full swing. What did we expect to find? Are we surprised that many different bacteria use basically the same pathway to secrete one or many different proteins or that so many totally different systems can function simultaneously in a single bacterial cell? What, exactly, are we all trying to figure out and how much have we achieved? How can we exploit the knowledge we have accrued to design more efficient systems for heterologous protein production and secretion in Gram-negative bacteria? Despite the very great differences between the secretion systems, the main questions being asked are very similar:

- How is an exoprotein recognised by its cognate machinery and where is/are the secretion signal(s)?
- Are exoproteins in folded or unfolded configurations before and during translocation through the envelope? Where does folding occur and what additional factors are required to assist the folding process? Are these processes specific to the exoprotein(s) under study or can they operate on heterologous proteins that are engineered to be secreted by the same pathway?
- Does secretion occur in one or two steps and is the Sec system directly involved?
- Is protein translocation energy-driven and if so, what is the source of energy and how is it coupled to translocation?
- How are components of the secretion machinery assembled? Is the assembled machinery stable or dynamic? Is energy involved in assembly?
- What prevents periplasmic proteins from leaking across the outer membrane?

A. Van Broekhoven et al. (eds.), Novel Frontiers in the Production of Compounds for Biomedical Use, 301–311.
© 2001 *Kluwer Academic Publishers. Printed in the Netherlands.*

- How is the peptidoglycan breached by the exoprotein or by the envelope-spanning-secretion apparatus?
- Can we exploit exoprotein secretion processes for biotechnological purposes?

Very few complete answers are available to any of these questions, and where they are, they probably only pertain to the particular system studied and might not be applicable to other systems. The basic features of the different secretion systems are illustrated in Fig. 1. Some of these systems are also found in Gram-positive bacteria, as illustrated, but this short overview will concentrate only on systems that are found in Gram-negative bacteria.

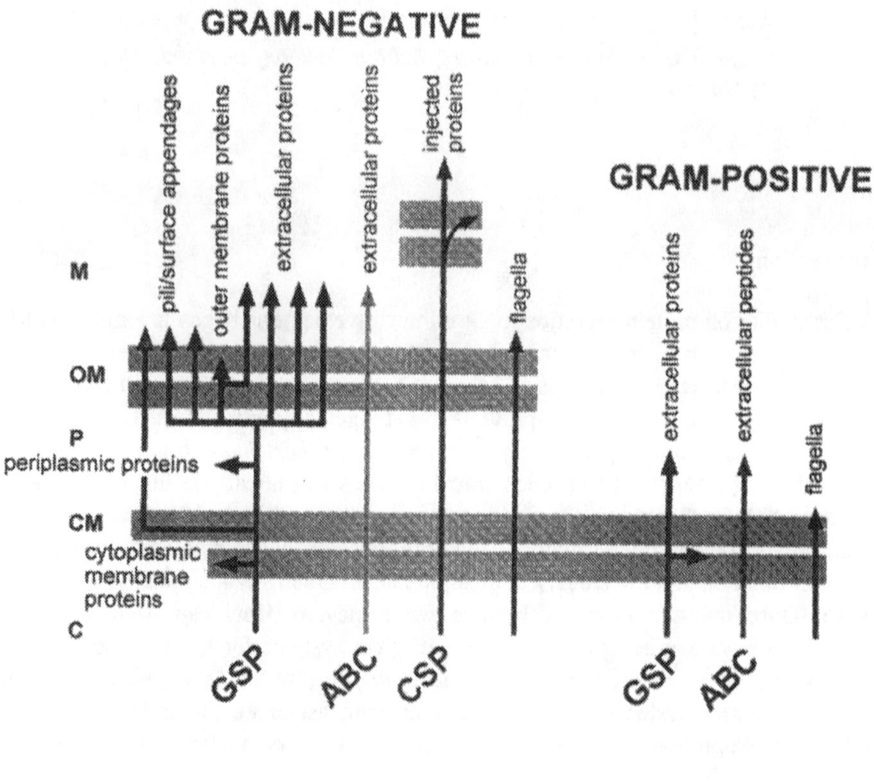

Fig. 1. Multiple protein secretion pathways in bacteria. GSP, general secretory pathway; ABC, ATP binding cassette pathway; CSP, contact secretion pathway (injection of proteins into and across the plasma membrane of target cells).

2. The General Secretory Pathway

The General Secretory Pathway (GSP) is the most complicated of the various pathways. It has numerous terminal branches involved in the insertion of proteins into the outer membrane or translocation of proteins through this membrane for assembly into surface organelles (pili) or for release as soluble proteins (Pugsley, 1993). All of these pathways are grouped together under the general designation of GSP because they all appear to have a common root, the Sec system, for transport across the inner membrane (Fig 2). All proteins secreted by the GSP transit via the periplasm and are therefore accessible to and probably subject to the action of periplasmic proteins that oxidise and reduce disulfide bonds, that catalyse the isomerisation of proline peptide bonds and that direct or assist the assembly and correct targeting of proteins and protein complexes (Missiakas & Raina, 1997).

The terminal branches of the GSP differ in their complexity. The simplest is the autosecretion pathway, of which the archetype is the *Neisseria gonorrhea* IgA protease secretion pathway (also sometimes called type IV). In this system, the protein that facilitates translocation across the outer membrane is the C-terminal segment of the pro-form of the exoprotein. Folding of the translocated N-terminal region of the pro- form apparently occurs after it has been translocated through the outer membrane, presumably through a barrel that is formed by the anti-parallel amphipathic b-strands of the C-terminal domain (Klauser *et al.*, 1992). This pathway can be used for the presentation of heterologous proteins on the bacterial cell surface.

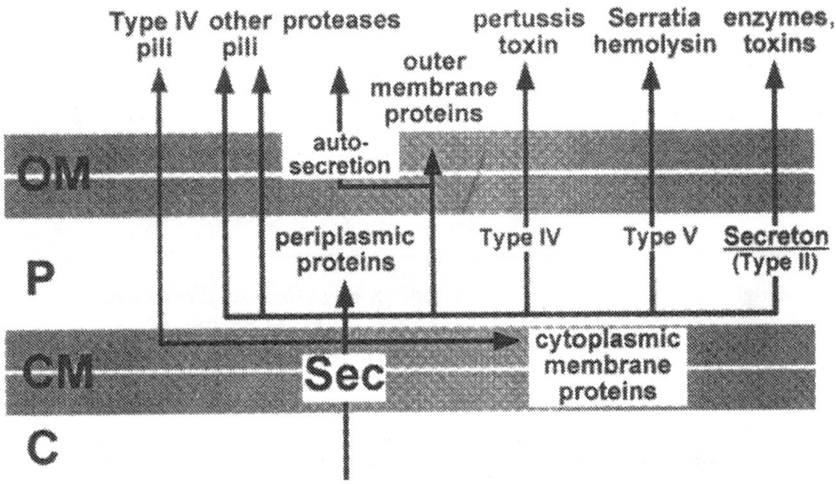

Fig. 2. Complete general secretory pathway in Gram-negative bacteria.

Table 1. *Examples of different exoprotein secretion pathways in Gram-negative bacteria.*

Pathway	Other name	Species (examples)	Exoproteins (examples)	Sec	Components	Energy requiremen t[3]
Secreton	II	*Klebsiella, Pseudomonas, Erwinia, Aeromonas*	Pullulanase, elastase, pectate lyase, aerolysin	+	>12	pmf and ATP
Type IV pilus		*Neisseria, Pseudomonas, Escherichia,*	Type IV pilins	?	>12	ATP
P pilus[1]		*Escherichia*	P pilins	+	2	
CS pilus		*Escherichia*	CS pilins	+	2	
Type IVa[2]		*Bordetella*	Pertussis toxin	+	>8	ATP
Type IVb[2]		*Legionella*	Unknown effectors	?	>8	ATP
Auto-secretion	IV	*Neisseria, Haemophilus, Serratia*	IgA protease, serine protease	+	0	
Type V		*Bordetella, Haemophilus, Serratia*	Filamentous haemagglutinin, HAP proteins, haemolysin	+	1	
ABC	I	*Escherichia, Erwinia, Pseudomonas*	alpha-haemolysin, metaloprotease	-	3	ATP, pmf
Contact	III	*Yersinia, Salmonella, Shigella, Xanthomonas, Pseudomonas, Escherichia*	Unknown ffector proteins	-	>10	ATP

1, together with related pili produced by a wide variety of Gram-negative bacteria by a similar mechanism.
*2, two categories distinguished by the extracellular release (pertussis toxin) or injection into target cells (*Legionella*).*
3, demonstrated requirement or requirement presumed on the basis of ATPase among components of secretion machinery.

The pathway typified by the *Serratia marcescens* haemolysin (Schiebel *et al.*, 1989) and the *Bordetella pertussis* filamentous haemagglutinin (FHA) (Guédin *et al.*, 1998) (type V) involves a single outer membrane component (in addition to the Sec proteins and a signal peptidase). This translocator protein might be equivalent to the C-terminal domain of pro-exoproteins secreted by the autosecretion system since there is evidence that it is only used in a single secretion cycle (Schiebel *et al.*, 1989).

Two components are needed for the assembly of most types of pili. In the archetypal Pap or P pilus system, these components are a periplasmic chaperone that protects and pilots pilin subunits to the second component of the system (Jacob-Dubuisson *et al.*, 1994), the outer membrane translocator called the usher that probably forms the exit conduit and assembly ring in the membrane (Thanassi *et al.*, 1998). Type IV pili are completely different, however. Indeed, it has not been conclusively established whether pilin subunits are actually exported across the inner membrane by the Sec system or by components of the type IV pilus secretion system. Nevertheless, the precursors of type IV pilins have N-terminal hydrophobic segments that resemble signal sequences recognised by the Sec pathway, and can be exported, presumably via this pathway, in cells devoid of the pilus secretion machinery (Strom & Lory, 1993).

The type IV pilus secretion system is particularly interesting because of its extensive similarity to the secreton (main terminal branch of the GSP or type II secretion system). Indeed, most of the components of the secreton, of which there are usually between 12 and 14 depending on the bacterium, have homologues in the type IV pilus system, and the former is sometimes regarded as a variant of the latter in which a putative rudimentary pilus (called the pseudopilus) is proposed to be specifically adapted to protein translocation (Pugsley, 1993). Both systems include a prepilin peptidase enzyme that cleaves the short *N*-terminal extension on the pilin/pseudopilin subunits, a membrane-anchored protein with predicted ATPase activity, a multimeric outer membrane protein called secretin, and a co-factor that protects secretin from proteolysis, that pilots it to the outer membrane and, in one case at least, remains associated with the secretin once it is embedded in the outer membrane (see below). The secreton is probably the most widely-spread exoprotein secretion pathway in Gram-negative bacteria; it is the only exoprotein secretion pathway naturally present in *Escherichia coli* K-12 (where its function is presently unknown) and it also exists in many other bacteria where it is used for enzyme and toxin secretion. The secreton is also probably one of the most interesting pathways from the biotechnology standpoint because it has been established that proteins fold in the periplasm before they are translocated across the outer membrane (Hardie *et al.*, 1995; Hirst & Holmgren, 1987; Pugsley, 1992) and that secretion signals from exoproteins can be grafted onto normally periplasmic β-lactamase to promote its extracellular release (Lu & Lory, 1996; Sauvonnet & Pugsley, 1996). Whether or not this secretion system can be used to secrete a wide range of heterologous proteins is currently under investigation (see section 6).

Finally, the protein secretion systems involved in *Bordetella pertussis* toxin secretion (Weiss *et al.*, 1993) and probably also in translocation of the *Legionella pneumophila* effector proteins into host cells (Segal *et al.*, 1998; Vogel *et al.*, 1998) (types IVa and IVb) are superficially similar to the secreton system, since exoproteins appear to be exported by the Sec pathway, assembled in the periplasm and then translocated across the outer membrane by a dedicated terminal branch of the GSP. However, the components of this pathway and the secreton are completely different and the former share intriguing similarities and even functional overlap with components of the plasmid conjugation (Segal & Shuman, 1998).

It is worth noting that there are other terminal branches to the GSP that are still poorly characterised. One of these is involved in the assembly of CS I and related pili in *E. coli* (Sakellaris *et al.*, 1996). Furthermore, homologues of genes known to be involved in different GSP terminal branches have been identified in numerous other bacteria that are not known to secrete proteins or in which the gene products have another function. For example the type IV pilus secretion genes and type IV pilin-like proteins are required for DNA uptake in the Gram-positive bacteria *Bacillus subtilis* and *Streptococcus pneumoniae* (Chung *et al.*, 1998; Lunsford & Roble, 1997).

3. The Type I or ABC secretion pathway

The ABC secretion pathway (type 1) is one of the least complicated and the best characterised secretion systems in Gram-negative bacteria. Only 3 components are needed, an integral inner membrane protein that carries an ATP binding cassette motif similar to that of human MDR protein and other bacterial ABC proteins involved in solute transport, a "membrane fusion protein", and an outer membrane. In the α-haemolysin secretion system of *E. coli,* these proteins are called HlyB, HlyD and TolC, respectively. In this and related systems, substrate induces assembly of the translocator into a complex that spans the entire cell envelope, ensuring that the exoprotein is translocated directly from the cytoplasm to the external milieu, without passing through the periplasm and, presumably without adopting an appreciable amount of tertiary structure (Létoffé *et al.*, 1996; Thanabalu *et al.*, 1998). Secretion is driven both by ATP hydrolysis (by the ABC protein) and by the proton motive force (pmf) (Delepelaire, 1994; Koronakis *et al.*, 1991). Specific recognition of α-haemolysin and of many other exoproteins secreted by the ABC pathway depends on a secretion signal located within the last 50 amino acids that is probably recognised by the ABC protein. This secretion signal can be grafted onto other proteins to promote their extracellular secretion (Létoffé & Wandersman, 1992; Spreng *et al.*, 1999), but the utility of the system remains to be tested because it is by no means clear that heterologous proteins secreted by this system can fold correctly in the external medium.

4. The type III or Contact Secretion Pathway

The archetypal example of the contact or type III secretion system is the YOP secretion pathway of *Yersinia.* The name "contact" was adopted for this system to indicate that the exoproteins are normally secreted when the bacteria are in close proximity to the cells they infect (usually epithelial cells). Indeed, most YOPS and other exoproteins secreted by this pathway by other bacteria are injected into the host cell; in other words, the exoproteins are translocated across three membranes, two in the producing cell and one in the target cell (Cornelis & Wolf-Watz, 1997). In some cases, the translocated protein also carries a nuclear localisation signal that directs it into the nucleus of the target cell (Skrzypek *et al.*, 1998; Van den Acherveken *et al.*, 1997). In many cases, secretion and exoprotein production are stimulated by contact with the host cell (or by conditions that, fortuitously, mimic such contact), which avoids the wasteful production of exoproteins under conditions where they are not required.

The contact secretion pathway is complex and there are both marked and subtle differences between pathways in different bacteria. For example, the YOP system relies in part on a secretion signal corresponding to the first 15 codons of the exoprotein mRNA that presumably targets nascent chains to the translocator (Anderson & Schneewind, 1997). A second signal, located within the exoprotein sequence just downstream from the first 15 amino acids, is required for secretion and injection into the host cell. This signal is recognised by chaperones that, presumably, target the exoproteins to the translocator as well as protecting them from proteolysis or premature association (Cornelis & Wolf-Watz, 1997). The fusion of secretion signals to reporter proteins such as calmodulin-dependent adenylate cyclase (Cornelis & Wolf-Watz, 1997) or protein A (Galyov *et al.*, 1997) leads to their injection into target cells. Thus, this delivery system might be exploited to target specific proteins into cells to which bacteria can adhere.

Studies on the integral membrane components of the contact translocator are still in their infancy. One remarkable feature that is common to this pathway and to the secreton and type IV pilus secretion pathways is the outer membrane, multimeric protein secretin and its cofactor. Secretins presumably form the conduit by which exoproteins cross the outer membrane. Secretins from different secretion pathways share many features that distinguish them from all other outer membrane proteins, and have been visualised as multimeric complexes that resemble cylinders with a large central channel (Bitter *et al.*, 1998; Crago & Koronakis, 1998; Koster *et al.*, 1997; Linderoth *et al.*, 1997). In the case of the secreton, these channels are large enough to accommodate exoproteins that adopt their final quaternary structure in the periplasm. In the contact system, however, there is no evidence that exoproteins transit the periplasm. Hence, if they do fold, they must do so in the cytoplasm and must cross all three membranes in a folded state.

At least some of the components of the contact translocator used for transport of exoproteins across the outer membrane of the target cell must be made by the bacterium itself and are secreted by the contact secretion pathway. There are striking superficial similarities between this phenomenon and conjugal DNA transfer and the related effector protein secretion system of *Legionella.* These similarities are strengthened by the observation of pilus- and/or rudimentary flagellum-like organelles on the surfaces of bacteria in the process of contact secretion and injection (Ebel *et al.*, 1998; Roine *et al.*, 1997). For example, many of the components of the contact secretion system of *Salmonella typhimurium* pathogenicity island I are strikingly similar, both in sequence and structure, to the membrane-proximal parts of the bacterial flagellum (Kubori *et al.*, 1998). Do this flagellum-like structure and the pilus-like structures observed in other bacteria mediate adherence of the secreting bacteria to their target cell or do they form part of a continual conduit that leads from the secretin in the bacterial outer membrane to the surface of the target cell? Interestingly, plant pathogens must breach the thick plant cell wall in order to inject their effector proteins into the target cell. The long pilus-like organelle observed in these bacteria (Roine *et al.*, 1997) might be the conduit that allows the exoproteins to be properly targeted in this and other circumstances. Is it conceivable that exoproteins are translocated through these tubes in a folded configuration?

5. Progress and technical problems

Considerable progress has been made in the last decade or so on identifying components of the various extracellular protein secretion machineries of Gram-negative bacteria. We are now addressing the more challenging question of their individual roles in secretion. In time, we should be able to produce a three-dimensional image of the assembled secretion machinery, to localise protein-protein interfaces and to map sites of exoprotein recognition. The major stumbling blocks to advances in these areas of research are the extremely small quantities of the secretion factors present in the cell envelope and consequent inability to reconstitute the process *in vitro*. Furthermore, disturbance of the normal stoichiometry of secretion machinery components often prevents the machine from functioning, presumably because other components become trapped in non-functional complexes. Nevertheless, some important progress has recently been made, notably in the visualisation of assembled contact secretion machineries and of secretin-cosecretin complexes by electron microscopy (Bitter *et al.*, 1998; Crago & Koronakis, 1998; Koster *et al.*, 1997; Kubori *et al.*, 1998; Linderoth *et al.*, 1997).

6. A specific example: the *Klebsiella oxytoca* pullulanase secreton

The most widespread system for the secretion of extracellular proteins in Gram-negative bacteria is the secreton, a complex of up to 14 proteins that permits the translocation of selected proteins from the periplasm through the outer membrane and into the surrounding medium. As already mentioned, the secreton, also called the type II secretion system, is the main terminal branch of the general secretory pathway (GSP) The first part of the GSP is the Sec system, which recognises and translocates signal-peptide bearing precursor proteins across the cytoplasmic membrane into the periplasm. Signal peptide processing releases the free polypeptide into the periplasm, where its folding and multimerisation is catalysed by a variety of ATP-independent enzymes and other proteins with disulphide oxydo-reductase, prolyl-peptidyl isomerase, general or specific chaperone and piloting activities.

The secreton system selects specific proteins from among this periplasmic pool and translocates them across the outer membrane. The biotechnological potential of such a system is obvious because the periplasm provides an ideal environment in which the correct folding of the recombinant proteins can be achieved before they are secreted into the growth medium. From a more fundamental viewpoint, the secreton is of particular interest because it is one of the very few systems that can translocate folded proteins through a lipid bilayer membrane. From the fundamental standpoint, it is essential to understand how the system function. From the applied standpoint, it should be possible to exploit better basic knowledge of the secreton to see how it can be adapted for the secretion of recombinant proteins.

There are two main elements to the secreton that are worthy of particular mention. The first is how exoproteins are specifically recognised by the secreton. If the exoproteins carry specific recognition (secretion signals), then it might be possible to incorporate them into recombinant proteins together with a signal peptide and thereby "trick" the secreton into secreting them. Using the secreton of *Klebsiella oxytoca* as a model system, we have identified two regions of the exoprotein pullulanase that, when fused to certain other proteins, can promote their secretion (Sauvonnet & Pugsley, 1996). The efficiency of recombinant protein secretion induced by these "secretion systems" varies depending to the recombinant protein and the signal peptide used (O. Francetic, unpublished data). Attempts are now being made to define the signals more precisely, to identify key residues within them and, thereby, to increase their efficiency.

The existence of secretion signals in the exoproteins implies that there are "receptors" to which they bind. Bacteria with different secretons secrete different proteins and are unable to secrete proteins that are normally secreted by another secreton. This implies that at least some secreton components are specific to a particular system, and therefore to one or a limited number of substrates. In an attempt to identify these receptors, we have studied the ability of secreton components from one bacterium to replace the corresponding component in another bacterium. These studies show that all but two of the secreton components are at least sometimes interchangeable between different systems (I. Guilvout, O. Possot and G. Vignon, unpublished data). One of these components is an integral outer membrane protein that forms large (Hardie *et al.*, 1996), barrel-like complexes with ion-conducting activity in artificial lipid bilayers (N. Nouwen, A. Engel, H. Saibil and A. Chazi, unpublished data). This protein, called secretin, is likely to be the channel via which exoproteins cross the outer membrane. However, genetic and biochemical studies with the secretin from the *K. oxytoca* secreton seem to rule out the possibility that pullulanase binds to it (I. Guilvout, unpublished data). We are now beginning studies on the second non-interchangeable secreton component to determine whether it has pullulanase-binding activity (O. Possot, M. Gérard-Vincent, unpublished data).

Another problem of fundamental interest is how the channel in the outer membrane is gated. The secretin channel has an internal diameter of 7-8 nm, sufficiently wide to permit the diffusion of a globular 100 kDa polypeptide (N. Nouwen, A. Engel and H. Saibil, unpublished data). However, the outer membrane of secreton-bearing bacteria is not permeable to other molecules of anything like this size, implying that the channel is normally in a closed configuration. Studies underway should determine whether the substrate, other secreton components and energy are required to signal channel opening.

Studies on the secreton in our laboratory and in others is now aimed at determining the role of its individual components, at studying their interactions, at isolating the entire secreton or sub-complexes thereof for molecular analysis, and at reconstituting different steps in the secretion process *in vitro*. These studies should provide a greater understanding of the molecular details of this fascinating secretion system.

References

Anderson, D.M. & Schneewind, O. (1997) A mRNA signal for the type III secretion of Yop proteins by *Yersinia enterocolitica*. *Science* **278:** 1140-1143.

Bitter, W., Koster, M., Latijnhouwers, M., de Cock, H. & Tommassen, J. (1998) Formation of oligomeric rings by XcpQ and PilQ, which are involved in protein transport across the outer membrane of *Pseudomonas aeruginosa*. *Mol. Microbiol.* **27:** 209-219.

Chung, Y.S., Breidt, F. & Dubnau, D. (1998) Cell surface localisation and processing of the ComG proteins, required for DNA binding during transformation of *Bacillus subtilis*. *Mol Microbiol* **29:** 905-913.

Cornelis, G.R. & Wolf-Watz, H. (1997) The *Yersinia* Yop virulon: a bacterial system for subverting eukaryotic cells. *Mol Microbiol* **23:** 861-867.

Crago, A.M. & Koronakis, V. (1998) *Salmonella* InvG forms a ring-like multimer that requires the InvH lipoprotein for outer membrane localization. *Mol Microbiol* **30:** 47-56.

Delepelaire, P. (1994) PrtD, the integral membrane ATP-binding cassette component of the *Erwinia chrysanthemi* metalloprotease secretion system, exhibits a secretion signal-dependent ATPase activity. *J Biol Chem* **269:** 27952-27957.

Ebel, F., Podzadel, T., Rohde, M., Kresse, A.U., Krämer, S., Deibel, C., Guzman, C.A. & Chakraborty, T. (1998) Initial binding of Shiga toxin-producing *Escherichia coli* to host cells and subsequent induction of actin rearrangements depend on filamentous EspA-containing surface appendages. *Mol Microbiol* **30:** 147-161.

Galyov, E.E., Wood, M.W., Rosqvist, R., Mullan, P.B., Watson, P.R., Hedges, S. & Wallis, T.S. (1997) A secreted effector protein of Salmonella dublin is translocated into eukaryotic cells and mediates inflammation and fluid secretion in infected ileal mucosa. *Mol Microbiol* **25:** 903-912.

Guédin, S., Willery, E., Locht, C. & Jacob-Dubuisson, F. (1998) Evidence that a globular conformation is not compatible with FhaC-mediated secretion of the *Bordetella pertussis* filamentous haemagglutinin. *Mol Microbiol* **29:** 763-774.

Hardie, K.R., Lory, S. & Pugsley, A.P. (1996) Insertion of an outer membrane protein in *Escherichia coli* requires a chaperone-like protein. *EMBO J* **15:** 978-988.

Hardie, K.R., Schulze, A., Parker, M.W. & Buckley, J.T. (1995) *Vibrio spp.* secrete proaerolysin as a folded dimer without the need for disulphide bond formation. *Mol Microbiol* **17:** 1035-1044.

Hirst, T.R. & Holmgren, J. (1987) Conformation of protein secreted across bacterial outer membanes: a study of enterotoxin translocation from *Vibrio cholerae*. *Proc Natl Acad Sci USA* **84:** 7418-7422.

Jacob-Dubuisson, F., Pinkner, J., Xu, Z., Striker, R., Padmanhaban, A. & Hultgren, S. (1994) PapD chaperone function in pilus biogenesis depends on oxidant and chaperone-like activities of DsbA. *Proc Natl Acad Sci USA* **91:** 11552-11556.

Klauser, T., Pohler, J. & Meyer, T.F. (1992) Selective extracellular release of cholera toxin B subunit by *Escherichia coli*: dissection of *Neisseria* Iga$_{\beta}$-mediated outer membrane transport. *EMBO J* **11:** 2327-2335.

Koronakis, V., Hughes, C. & Koronakis, E. (1991) Energetically distinct early and late stages of HlyB/HlyD-dependent secretion across both *Escherichia coli* membranes. *EMBO J* **10:** 3263-3272.

Koster, M., Bitter, W., de Cock, H., Allaoui, A., Cornelis, G. & Tommassen, J. (1997) The outer membrane component, YscC, of the Yop secretion machinery of *yersinia enterocolitica* forms a ring-shaped multimeric complex. *Mol Microbiol* **26:** 789-797.

Kubori, T., Matsushima, Y., Nakamura, D., Uralil, J., Lara-Tejero, M., Sukhan, A., Galan, J.E. & Aizawa, S.-I. (1998) Supramolecular structure of the *Salmonella typhimurium* type III protein secretion system. *Science* **280:** 602-605.

Létoffé, S., Delepelaire, P. & Wandersman, C. (1996) Protein secretion in Gram-negative bacteria: assembly of the three components of ABC protein-mediated exporters is ordered and promoted by substrate binding. *EMBO J* **15:** 5804-5811.

Létoffé, S. & Wandersman, C. (1992) Secretion of CyaA-PrtB and HlyA-PrtB fusion proteins in *Escherichia coli*: involvement of the glycine-rich repeat domain of *Erwinia chrysanthemi* protease B. *J Bacteriol* **174:** 4920-4927.

Linderoth, N.A., Simon, M.N. & Russel, M. (1997) The filamentous phage pIV multimer visualized by scanning transmission electron microscopy. *Science* **278:** 1635-1638.

Lu, H.-M. & Lory, S. (1996) A specific targeting domain in mature exotoxin A is required for its extracellular secretion in *Pseudomonas aeruginosa*. *EMBO. J.* **15:** 429-436.

Lunsford, R.R. & Roble, A.G. (1997) *comYA*, a gene similar to *comGA* of *Bacillus subtilis*, is essential for competence-factor-dependent DNA transformation in *Streptococcus gordonii*. *J Bacteriol* **179:** 3122-3126.

Missiakas, D. & Raina, S. (1997) Protein folding in the bacterial periplasm. *J Bacteriol* **179:** 2465-2471.

Pugsley, A.P. (1992) Translocation of a folded protein across the outer membrane via the general secretory pathway in *Escherichia coli*. *Proc. Natl. Acad. Sci. USA* **89:** 12058-12062.

Pugsley, A.P. (1993) The complete general secretory pathway in gram-negative bacteria. *Microbiol Rev* **57:** 50-108.

Roine, E., Wei, W., Yuan, J., Nurmiaho-Lassila, E.L., Kalkkinen, N., Romantschuk, M. & He, S.Y. (1997) Hrp pilus: an hrp-dependent bacterial surface appendage produced by *Pseudomonas syringae* pv. tomato DC3000. *Proc Natl Acad Sci USA* **94:** 3459-3464.

Sakellaris, H., Balding, D.P. & Scott, J.R. (1996) Assembly proteins of CS1 pili of enterotoxigenic *Escherichia coli*. *Mol Microbiol* **21:** 529-541.

Sauvonnet, N. & Pugsley, A.P. (1996) Identification of two regions of *Klebsiella oxytoca* pullulanase that together are capable of promoting ß-lactamase secretion by the general secretory pathway. *Mol Microbiol* **22:** 1-7.

Schiebel, E., Schwartz, H. & Braun, V. (1989) Subcellular location and unique secretion of the haemolysin of *Serratia marcescens*. *J Biol Chem* **264:** 16311-16320.

Segal, G., Purcell, M. & Shuman, H.A. (1998) Host cell killing and bacterial conjugation require overlapping sets of genes within a 22-kb region of the *Legionella pneumophila* genome. *Proc Natl Acad Sci USA* **95:** 1669-1674.

Segal, G. & Shuman, H.A. (1998) Intracellular multiplication and human macrophage killing by *Legionella pneumophila* are inhibited by conjugal components of IncQ plasmid RSF1010. *Mol Microbiol* **30:** 197-208.

Skrzypek, E., Cowan, C. & Straley, S.C. (1998) Targeting of the *Yersinia pestis* YopM protein into HeLa cells and intracellular trafficking to the nucleus. *Mol Microbiol* **30:** 1051-1065.

Spreng, S., Dietrich, G., Goebel, W. & Gentschev, I. (1999) The *Escherichia coli* haemolysin secretion apparatus - a potential universal antigen delivery system in Gram-negative bacterial vaccine carriers. *Mol Microbiol. in press.*

Strom, M.S. & Lory, S. (1993) Structure-function and biogenesis of the type IV pili. *Annu Rev Microbiol* **47:** 565-596.

Thanabalu, T., Koronakis, E., Hughes, C. & Koronakis, V. (1998) Substrate-induced assembly of a contiguous channel for protein export from *E. coli*: reversible bridging of an inner-membrane translocase to an outer membrane exit pore. *EMBO J* **17:** 6487-6496.

Thanassi, D.G., Saulino, E.T., Lombardo, M.-J., Roth, R., Heuser, J. & Hultgren, S.J. (1998) The PapC usher forms an oligomeric channel: Implications for pilus biogenesis across the outer membrane. *Proc Natl Acad Sci USA* **95:** 3146-3151.

Van den Acherveken, G., Marois, E. & Bonas, U. (1997) Recognition of the bacterial avirulence protein AvrBs3 occurs inside the host plant cell. *Cell* **87:** 1307-1313.

Vogel, J.P., Andrews, H.L., Wong, S.K. & Isberg, R.R. (1998) Conjugative transfer by the virulence system of *Legionella pneumophila*. *Science*: 873-876.

Weiss, A., Johnson, F.D. & Burns, D.L. (1993) Molecular characterization of an operon required for pertussis toxin secretion. *Proc Natl Acad Sci USA* **90:** 2970-2974.

ALTERATIONS OF METABOLIC FLUX DISTRIBUTIONS IN RECOMBINANT *ESCHERICHIA COLI* IN RESPONSE TO HETEROLOGOUS PROTEIN PRODUCTION

JAN WEBER AND URSULA RINAS
GBF, National Research Centre for Biotechnology, Biochemical Engineering Division, Mascheroder Weg 1, 38124 Braunschweig, Germany

Summary

Heterologous protein production in *Escherichia coli* can perturb cellular functions at many levels. The impact on the host carbon and energy metabolism may be severe and may implicate lowered yields of the recombinant product. This article is focused on the alterations of metabolic pathway utilisation during heterologous protein production in high-cell density fed-batch cultures of *E. coli*. The effect of the production of different recombinant proteins from different expression systems on the host metabolism is investigated by a stoichiometric metabolic model; general tendencies are demonstrated.

1. Introduction

By the use of recombinant microorganisms heterologous proteins can be produced in large quantities in bioreactors. Successful commercialisation of a heterologous protein requires a genetically stable, high-yielding recombinant culture and a productive cultivation process. *Escherichia coli* is the most important prokaryotic host organism for heterologous protein production. This bacterium is well characterised at the molecular level and foreign genes can be easily introduced. In addition, *E. coli* offers the advantages of fast growth, cultivation to high cell densities and the ability to utilise inexpensive carbon sources such as glucose. Although *E. coli* can not be used to produce glycosylated proteins, quite a number of other proteins of human interest, *e.g.* interferons, interleukins, human-insulin and growth hormones have been successfully produced by *E. coli* and the market volume of such proteins is expected to increase.

The development of a cost-effective cultivation process aims at a maximum of volumetric productivity. This can be achieved by increasing the biomass concentration in the bioreactor. For *E. coli*, fed-batch processes are usually applied to obtain cell

A. Van Broekhoven et al. (eds.), Novel Frontiers in the Production of Compounds for Biomedical Use, 313–337.
© 2001 *Kluwer Academic Publishers. Printed in the Netherlands.*

densities of more than 100 g l^{-1} (Knorre et al., 1991; Korz et al., 1995). Different feeding strategies have been developed for this purpose (see review (Lee, 1996). The main task of all feeding strategies is to supply sufficient oxygen and to prevent the accumulation of toxic levels of acetic acid which is formed in response to incomplete substrate oxidation due to oxygen limitation or excess carbon.

The volumetric productivity can further be increased by increasing the specific productivity (*i.e.* volumetric productivity per cell concentration) which is dependent on the gene construct and on the metabolic properties of the host organism. In the past, research focused mainly on the development of efficient expression vectors (see review (Hannig and Makrides, 1998) while minor attention has been spent on the host cell metabolism.

The presence of a plasmid and the overproduction of a particular protein often disturbs the balance of metabolic processes in the host organism (Birnbaum and Bailey, 1991; Snow and Hipkiss, 1987; Ramirez and Bentley, 1993). The host resources are used by means of energy and raw material often described as "metabolic burden" (Glick, 1995). Moreover, the appearance of a recombinant protein causes several side-effects like the expression of stress proteins as found in several different recombinant strains of *E. coli* (Parsell and Sauer, 1989; Rinas, 1996). This can lower the growth rate and even lead to cell death (Dong et al., 1995; Kurland and Dong, 1996). To optimise protein production it would be desirable to impair the metabolic homeostasis to a lesser extent or, at least, to maintain the cells in the productive stage for a longer time. In this respect, information about the metabolic properties of the recombinant organism is essential, first, for the design of an optimal cultivation process and second, for the construction of a superior host organism in terms of metabolic engineering.

For quantitative evaluation of the metabolic characteristics, the analysis of metabolic fluxes has emerged as an important tool in the field of metabolic engineering (Vallino and Stephanopoulos, 1990). Quantification of metabolic fluxes provides access to relevant information on metabolic pathway utilisation and potential limitations (Holms, 1986; Stephanopoulos and Vallino, 1991). Information on metabolic fluxes can be obtained from a metabolic model and a set of measured fluxes, typically the uptake rates of substrates and secretion rates of metabolites (Vallino and Stephanopoulos, 1990); (Varma and Palsson, 1994). The analysis relies on the stoichiometry of cellular pathways, the metabolic demands for growth and optimisation principles to estimate the intracellular carbon flux within a defined network. The flux distributions obtained are based on a pseudo-steady state assumption for intracellular metabolites. Stoichiometry-based metabolic models have been applied to a variety of organisms, *e.g.* hybridoma (Bonarius et al., 1996), *Saccharomyces cerevisiae* (Nissen et al., 1997); (van Gulik and Heijnen, 1995), *Penicillium chrysogenum*, (Jorgensen et al., 1995), *Corynebacterium glutamicum* (Vallino and Stephanopoulos, 1994a; Vallino and Stephanopoulos, 1994b; Vallino and Stephanopoulos, 1993), *Bacillus subtilis* (Sauer et al., 1996) and *E. coli* (Varma and Palsson, 1993; Pramanik and Keasling, 1996).

The actual metabolic flux distributions being examined in this study concern industrial-like production processes for the production of heterologous proteins by recombinant strains of *E. coli.* These proteins comprise the human fibroblast growth factor (hFGF-2), human insulin fusion protein and a stable and unstable version of the antigenic C-terminus of the VP60 capsid protein from the rabbit haemorrhagic disease virus. In these strains, the expression of hFGF-2 and insulin is controlled by the strong and highly repressible, temperature-sensitive λ promoter/repressor system. This promoter, widely used in large-scale protein production, is induced by temperature shift from 30°C to 42°C. Since the elevated temperature causes a heat-shock response in the host organism, a chemically inducible system was additionally investigated. The VP60 capsid protein is synthesised by inducing the T7 promoter system with isopropyl-β-D-thiogalactopyranoside (IPTG). This protein is also a good example of how the stability of a protein can affect the host' metabolism. The data for flux calculations were available from carbon-limited high-cell density cultivations carried out in a pilot-scale bioreactor.

2.Methodology

2.1 METABOLIC FLUX ANALYSIS

In a protein production process, the estimation of the carbon fluxes through the central metabolic pathways and the amino acid synthesising pathways is the desired goal. The latter is particularly of interest in the case when the amino acid composition of the heterologous protein differs significantly from an average bacterial protein. The basis for the calculation of metabolic fluxes is a stoichiometric metabolic network as illustrated in Figure 1. Stoichiometrically derived mass balances are used to estimate the carbon flux through the central metabolic pathways (Vallino and Stephanopoulos, 1990; van Gulik and Heijnen, 1995). The biochemical reactions and considered metabolites form a set of linear equations which can be expressed in matrix notation as:

$$A \cdot x = r \tag{1}$$

by introducing the stoichiometric ($\mathbf{m}\times\mathbf{n}$) matrix $\mathbf{A}$, the ($\mathbf{n}\times 1$) vector of unknown fluxes $\mathbf{x}$, and the ($\mathbf{m}\times 1$) vector $\mathbf{r}$ of the accumulation rates. The vector $\mathbf{r}$ can be split into $\mathbf{r}_i$ and $\mathbf{r}_e$ corresponding to intra- and extracellular metabolites, respectively. Assuming pseudo-steady state for intracellular metabolites, $\mathbf{r}_i$ can be set to zero. For the metabolites j that are exchanged with the environment, *e.g.* substrates, biomass and by-products, the (volumetric) accumulation rates $\mathbf{r}_e$ are calculated in fed-batch processes as follows:

$$r_e = \frac{dc_j}{dt} + (c_j - c_{j,f}) \cdot \frac{F(t)}{V(t)} \tag{2}$$

F is the time-varying feed F (= dV/dt) containing substrates at a constant concentration c_f, and F(t)/V(t) denotes the dilution rate with the reactor volume V.

2.2 UNDERDETERMINED NETWORKS

The generated bioreaction network for *E. coli* is underdetermined (see below). Underdetermined networks are rank-deficient, *i.e.* the number of linear independent metabolite balances is not sufficient to determine the unknown intracellular fluxes uniquely. Underdetermined networks occur for the following two reasons:

- The number of reactions is larger than the number of metabolites. With regard to the stoichiometric matrix $\mathbf{A}(\mathbf{m} \times \mathbf{n})$, $\mathbf{m}$ is smaller than $\mathbf{n}$.
- Reactions are linear dependent. These singularities occur often in complex networks *e.g.* when cyclic pathways are included. Then, the rank of the stoichiometric matrix $\mathbf{A}$, $R(\mathbf{A})$, is smaller than the number of unknown fluxes $\mathbf{n}$.

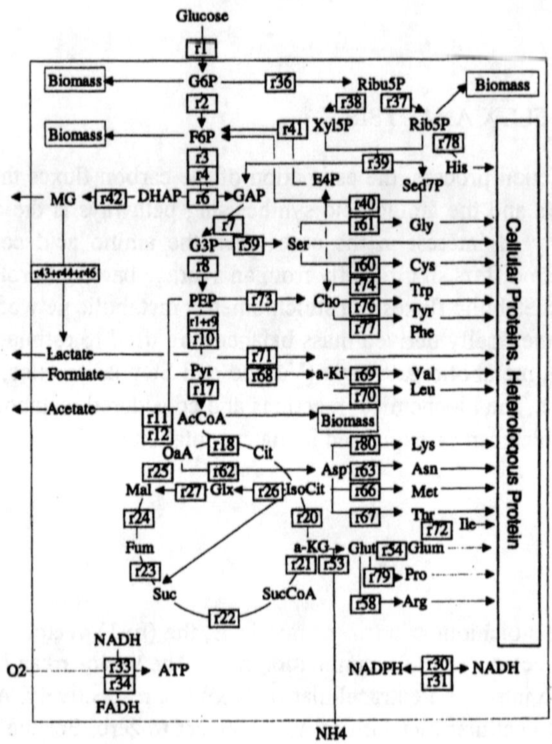

Figure 1: *Metabolic network of* E. coli *grown on glucose. Reactions are represented by numbers as given in the appendix. Not all reactions are displayed. Biomass comprises all components (RNA, DNA, lipids, lipopolysaccharide, peptidoglucan and glycogen) except protein.*

In both cases, an infinite number of solutions exists. To obtain a unique solution for all unknown fluxes additional constraints can be added by applying linear programming. Constraints are specified by the thermodynamics determining the direction of a reaction, and by formulating an objective function. A linear programming problem can be expressed as follows:

$$\text{minimise } Z = c \cdot x \tag{3}$$

which is subject to the stoichiometry (Equation 1). In this study, the objective function maximise biomass yield was used. In this case, the coefficient c for biomass production is set to -1, whereas the other factors are set to zero. Linear programming was carried out by using a two-phase simplex algorithm (MATLAB, Version 4.2c1, The MathWorks, Inc., Modelling and Simulation Toolbox, Version 1.5b GBF, kindly provided by Klaus Gollmer, FH Trier).

2.3 WHY LINEAR PROGRAMMING?

Although thousands of reactions take place within a cell, many of them can be lumped together. It is dependent on the application, which reactions are considered in more detail and which are lumped together. However, one must be aware that increasing the number of reactions does not simultaneously increase the observability of the overall network, *e.g.* linear reaction sequences do not provide any additional information. In the literature, networks from 6 (*e.g.* in (Stephanopoulos,) to more than 700 reactions (Schilling et al., 1999) can be found.

In small networks the biosynthetic routes are usually collected in one single reaction based on the macromolecular composition of the biomass. The dimension of the network can further be reduced by considering only those metabolites that comprise branch points in the network. Those networks are often constructed in a way to obtain a determined or overdetermined stoichiometric matrix; the solution can be found by computing the inverse (by Gaussian elimination) or the pseudoinverse of **A**, respectively. These networks are suitable when the biochemical routes are known under the given conditions. However, for a new physiological situation the network must be reconstructed. It may also happen that no solution can be found, *e.g.* if we consider growth of *E. coli* on acetate. Acetate induces enzymes of the glyoxalate shunt (Holms, 1986). Since inclusion of the corresponding reactions causes a singularity (*i.e.* linear dependent reactions), the system is not solvable without additional information (*e.g.* measured split ratios).

Metabolic networks are by nature underdetermined. The genetic basic set of a cell allows the synthesis of enzymes for alternative reactions, alternative pathways and cycles. Thus, it may be desired to construct more universal networks which represent the flexible nature of the bacterial metabolism better. The resulting networks are usually underdetermined requiring additional intracellular measurements or linear programming techniques. These stoichiometric models can be constructed on basis of metabolic databases such as EcoCyc[1], WIT[2] or KEGG[3]. The most universal approach would be the

[1] http://ecocyc.PangeaSystems.com/ecocyc/ecocyc.html
[2] http://wit.mcs.anl.gov
[3] http://kegg.genome.ad.jp/kegg

construction of the stoichiometric model from genome sequence databases taking benefit of the rapidly growing information on annotated genomes (Schilling et al., 1999). This approach is, in contrary to determined/overdetermined systems, deductive; from the universal a particular case is deduced, namely the virtually utilised pathways.

When applying linear programming, it may be difficult to find an appropriate objective function. Often it is attempted to use biologically meaningful objective functions. For *E. coli*, the objective "maximise growth" revealed good agreement with experimental data (Varma and Palsson, 1993), while this was the case for hybridoma cells using "maximise NADPH" or "maximise ATP" (Bonarius et al., 1998) (see also review (Bonarius et al., 1997). Another uncertainty in linear programming concerns the energetic parameters, P/O ratio (ATP gain per atom oxygen reduced in the respiration chain), ATP requirements for biomass formation ($Y_{ATP/X}$), and maintenance requirements (m_{ATP}). In aerobic cultures, these parameters can neither be measured directly nor evaluated independently. However, from the observed biomass yield, $Y_{X/S}$, a combination of these parameters can be determined and the resulting flux distributions are independent on the parameter combination as long as $Y_{X/S}$ is matched (Vanrolleghem et al., 1996).

In summary, underdetermined networks require constraints and an objective function to be defined. The expenditure of computer power and computing time is bigger, but this approach is more universal and flexible and represents the nature of the metabolism better. The use of determined/overdetermined networks, on the other hand, is appropriate if the biochemical routes are known under the given conditions. The computational requirements are relatively low allowing also the use for online-monitoring of metabolic fluxes during a cultivation process.

2.4 PROPERTIES OF THE METABOLIC NETWORK

The metabolic network constructed for the protein production process is based on the *E. coli* database EcoCyc (Karp *et al.*,1999). It includes the reactions of the central metabolic pathways, *i.e.* the Embden-Meyerhof-Parnas (EMP) pathway, the Krebs or tricarboxylic acid (TCA) cycle, the pentose phosphate (PP) pathway, the methylglyoxal (MG) pathway and the glyoxylate shunt. Also, a pyridine dinucleotide transhydrogenase for the interconversion of NADPH and NADH is included. The formation of NADPH by transhydrogenation was assumed to be energy dependent, whereas the backwards reaction was considered to be energy independent (Zahl et al., 1978; Voordouw et al., 1983). Some reactions in linear metabolic routes were lumped together, *i.e.* anabolic reactions concerning the nucleotide and lipid metabolism and the synthesis of several amino acids. The biomass equation was adapted to the macromolecular composition of *E. coli* consisting of 62 %(w/w) protein, 11.5 %(w/w) RNA (values obtained from own measurements, respectively) and 3.1 %(w/w) DNA, 9.1 %(w/w) lipids, 3.4 %(w/w) lipopolysaccharide, 2.5 %(w/w) peptidoglucan and 1.3 %(w/w) glycogen following the analysis of (Ingraham et al., 1983). The biomass composition is assumed to be constant in the course of a cultivation since the impact of changes in the biomass composition on the intracellular flux distributions is small as shown by a sensitivity analysis (Pramanik and Keasling, 1997); (Daae and Ison , 1999). The reaction for the synthesis of the recombinant protein is formulated on basis of its amino acid composition. The amino acid composition can deviate significantly from an average cellular protein of *E. coli* as shown for hFGF-2 in Figure 2, and therefore is considered in the model. The final metabolic network for the protein production process contains 103 reactions and 91 metabolites and is listed in the Appendix. A P/O ratio of 1.75 was assumed as determined for *Enterobacteria* (Zeng et al., 1991). For growth associated ATP consumption, $Y_{ATP/X}$, a value of 97 mmol g^{-1} h^{-1} was used as determined from chemostat cultures of *E. coli* TG1 (Kayser, 1998). Non-growth associated ATP consumption, m_{ATP}, was used to represent the overall additional energy requirements, used for *e.g.* maintenance and turnover of macromolecules, and was adjusted to the measured biomass yield.

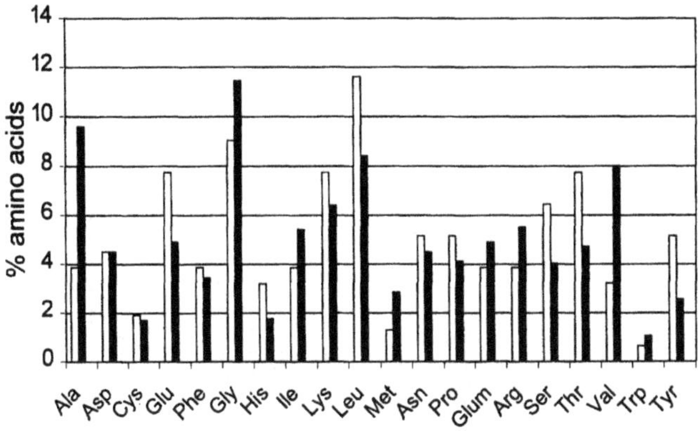

Figure 2: Comparison of the amino acid composition of hFGF-2 and an average E. coli *protein. hFGF-2: white columns, biomass protein: black columns.*

2.5 OPTIMAL AMINO ACID DRAIN FOR PROTEIN PRODUCTION

The first application of the stoichiometric network is the calculation of the optimal theoretical supply of amino acids for the synthesis of a heterologous protein. It represents the desired "best" flux distribution assuming that all of the carbon source is channelled to the protein. This flux distribution can then be compared with the flux estimations obtained from cultivation data. For this purpose, we set the substrate uptake rate to 100 and maximise the production of in this case hFGF-2. From this calculation, the maximum yield is 89 % mol mol^{-1} (= 54 % g g^{-1}) hFGF-2 per glucose. It is seen that the source of NADPH required for hFGF-2 synthesis is mainly the PP shunt (Figure 3). The TCA cycle as an alternative source (*i.e.* isocitrate dehydrogenase, r20) shows relatively low contribution (see Figure 6 for comparison with a non-producing cell). Remarkable is the pyruvate branch point. An excess of pyruvate is produced by the phosphotransferase (PTS) system and can not be consumed by reactions further downstream. It is therefore degraded by phosphoenolpyruvate (PEP) synthetase.

3.Applications

3.1 PRODUCTION OF HFGF-2 BY TEMPERATURE SHIFT

As a first example the production of hFGF-2 in a high-cell density cultivation process is described. The strain *E. coli* TG1 harbours the plasmid pλFGFB that carries the hFGF-2 structural gene behind the $\lambda P_R P_L$ promoter. The repressor cI_{857} of the λ promoter is thermolabile and is destroyed at high temperatures, so induction is achieved by temperature shift from 30°C to 42°C. Carbon-limited high-cell density cultivations were carried out in minimal medium using an exponential feeding strategy (Seeger *et al.*, 1995).

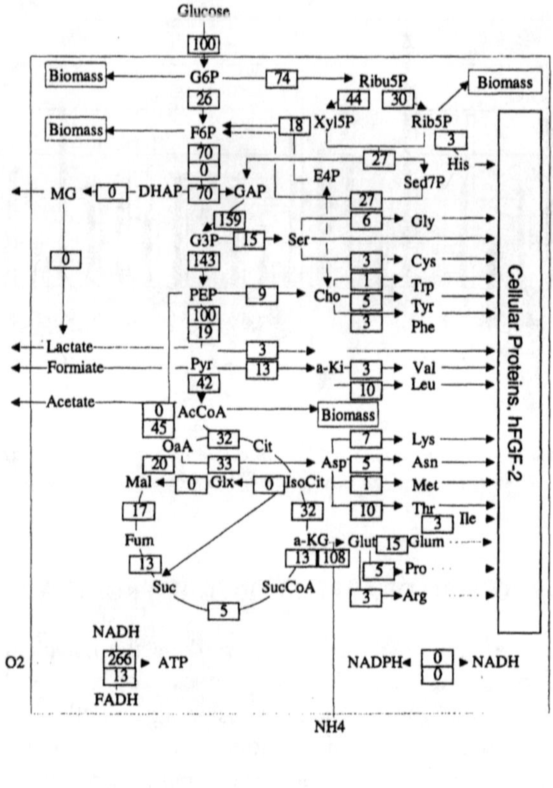

***Figure 3:** Optimal flux distribution for the production of hFGF-2 from glucose estimated from the stoichiometric model using the objective function "maximise hFGF-2".*

The process can be divided into three separate phases. The first one is a batch phase with cells growing at maximum growth rate (μ_{max}=0.46 h^{-1}) followed by a feeding phase with reduced growth rate at μ_{set} = 0.12 h^{-1} (second phase). In the third phase, hFGF-2 synthesis is induced by temperature shift from 30°C to 42°C while the feeding rate is further reduced ($\mu_{set} = 0.08h^{-1}$).

In Figure 4, profiles of the biomass, glucose and hFGF-2 concentration during a typical high-cell density cultivation of *E. coli* TG1:pλFGFB are shown. hFGF-2 accumulated to concentrations of app. 2.6 g l^{-1}. This corresponds to a specific concentration of 45mg g^{-1} and a yield on glucose of 11 % mol mol^{-1}. Nearly half of the recombinant protein is formed as inclusion bodies. Upon temperature shift, growth of TG1:pλFGFB nearly ceased. The specific growth rate was with in average $\mu = 0.03\ h^{-1}$ lower than the set value of $\mu = 0.08\ h^{-1}$. The biomass yield on glucose, $Y_{X/S}$, dropped from 0.50 g g^{-1} in the first fed-batch phase at 30°C to 0.18 g g^{-1} at 42°C. The metabolic perturbation becomes also evident by means of the specific oxygen uptake rate (OUR) and the carbon dioxide evolution rate (CER) which increased immediately upon temperature shift (Figure 5).

In Figure 5, also the online glucose uptake rate, r_s, and the online biomass formation rate, r_x, are shown. The glucose uptake rate was calculated from the actual mass flow rate of the feeding solution; the online biomass formation rate was calculated from a linear correlation of the ammonia consumption and biomass concentration (for details on the real-time calculations see Schmidt *et al.*, 1999b).

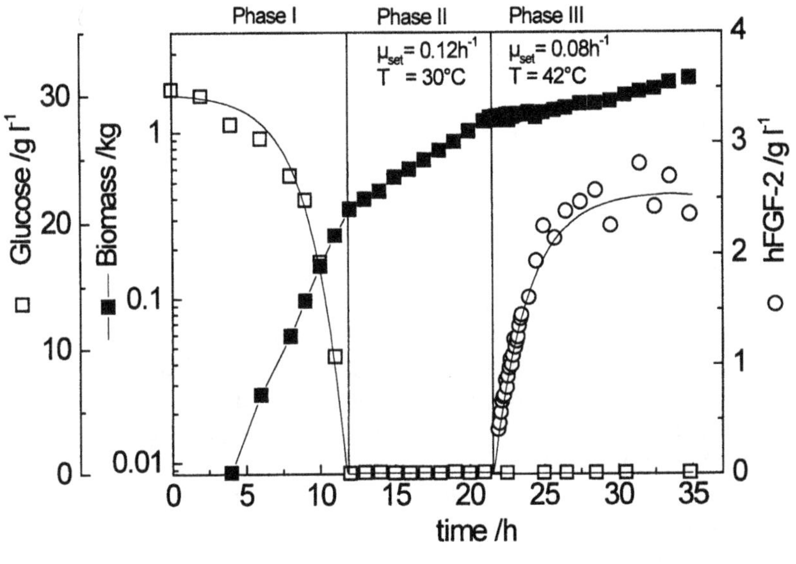

Figure 4: *Temperature-induced production of hFGF-2 in high-cell density culture of* E. coli *TG1:pλFGFB in a 50L-bioreactor. Product formation was induced by temperature shift from 30°C to 42°C. The time course of biomass, glucose concentration and hFGF-2 concentration is shown.*

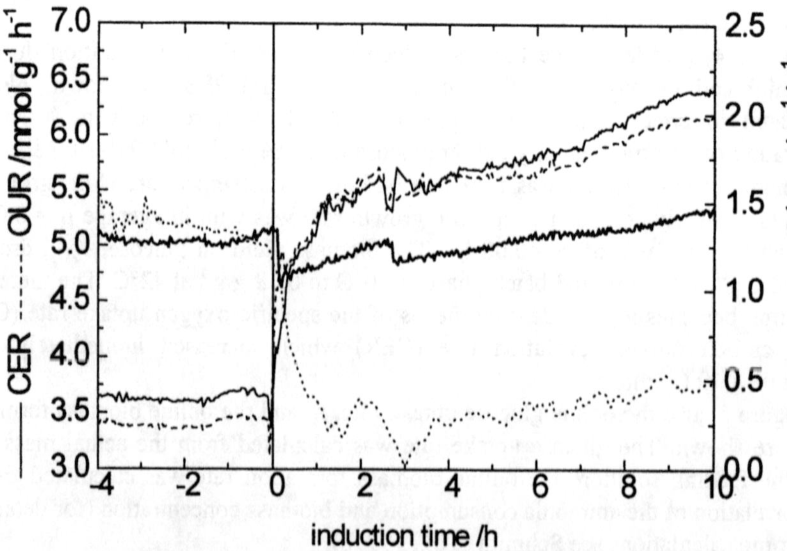

***Figure 5:** Time course of the specific oxygen uptake rate (OUR), carbon dioxide evolution rate (CER), glucose uptake rate (r_S) and biomass formation rate (r_X) during the induction period of a temperature-induced high-cell density culture of* E. coli *TG1:pλFGFB. The online glucose uptake rate is calculated from the feed rate; the online biomass formation rate is calculated from the ammonia consumption used for pH control (for details on the correlation see Schmidt* et al.*, 1999b); the molecular weight taken for biomass was 100 g mol^{-1} (Kayser et al., 2000).*

Fluxes were calculated from the glucose uptake rate and the hFGF-2 production rate. The biomass formation rate was maximised as objective function. It was adjusted to the measured values by adjusting m_{ATP}. For two specified time points, 2 h before and 0.6 h after the temperature shift, flux distribution maps are shown in Figure 6 and 7. In Figure 8, the major pathway fluxes are shown as a function of time to facilitate comparison of the alterations of metabolic flux distributions using other expression systems in high-cell density cultivation (see below).

The calculated fluxes are normalised with respect to the glucose uptake rate to facilitate comparison. Major changes happened at the branch point between PP pathway and EMP pathway. Under unperturbed growth conditions at 30°C in the fed-batch phase I, one third of the glucose was processed *via* the PP pathway and the remaining entered the EMP pathway (Figure 6). After the temperature shift the EMP pathway was fully active (Figure 7). With decreasing biomass yield, enhanced TCA cycle activity was enabled by reduced precursor drain for biosynthesis from the EMP pathway. The flux into the TCA cycle increased by 100%. Therefore, the increasing respiratory activity can be assigned to the increased TCA cycle flux, which is a major source of carbon dioxide. At high TCA cycle fluxes, an excess of NADPH was produced by isocitrate dehydrogenase of the TCA cycle. The electrons were transferred to NADH by transhydrogenase, which was not active in the fed-batch phase I, and later transferred into the respiratory chain to yield energy in the form of ATP.

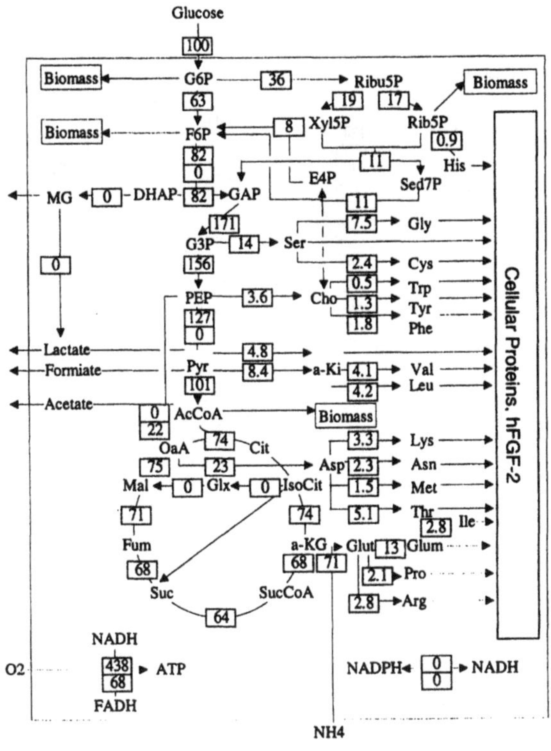

Figure 6: *Intracellular flux distribution of* E. coli *TG1:pλFGFB in high-cell density cultivation for the fed-batch phase I at 30°C (at -2 h of induction time). In this phase, no hFGF-2 is produced. All fluxes are given as molar metabolite fluxes normalised with respect to the glucose uptake rate (r_S = 1.29 mmol g^{-1} h^{-1}).*

The reduced biomass yield is caused by an increasing energy demand. The parameter m_{ATP} estimated from flux analysis can be used to assess the multiple metabolic loads and thus to capture the energetic characteristics of the cells quantitatively. Before induction, m_{ATP} was 2.93 mmol g^{-1} h^{-1}, 0.6 h after the temperature shift, m_{ATP} increased to 15.8 mmol g^{-1} h^{-1}. m_{ATP} represents maintenance requirements, *e.g.* for the turnover of macromolecules. Maintenance requirements are known to be very temperature sensitive (Wallace and Holms, 1986; Nielsen, 1994). Furthermore, it is known that the protein synthesis rate increases with temperature (Farewell and Neidhardt, 1998). Protein synthesis is the major energy sink of the cell, as shown by stoichiometric analysis (Stouthammer, 1973). For the system at hand, it has been shown in a previous work that the additional protein synthesis after the temperature shift can be attributed mainly to heat shock proteins (Rinas, 1996). The increased energy demand during temperature-induced recombinant protein production is manifested by the increased TCA cycle flux since in the TCA cycle the NADH produced is finally converted to ATP in the respiratory chain.

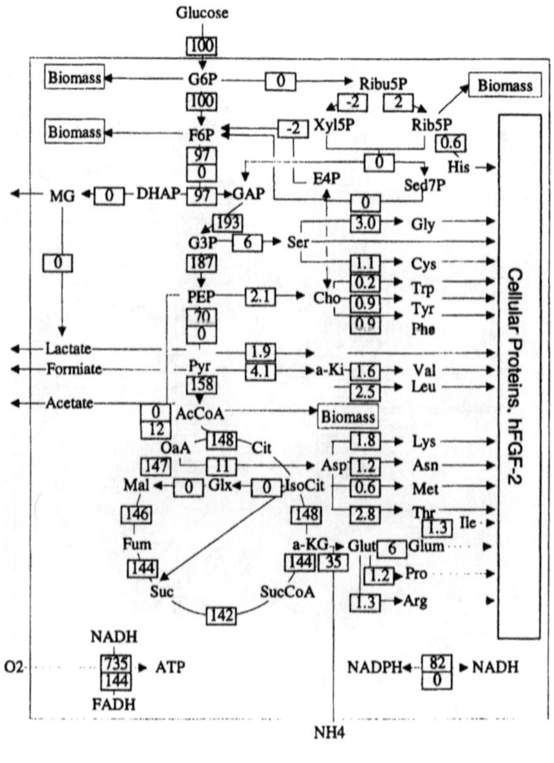

Figure 7: Intracellular flux distribution of E. coli *TG1:pλFGFB in high-cell density cultivation for the fed-batch phase II at 42°C (0.6 h after induction). All fluxes are given as molar metabolite fluxes normalised with respect to the glucose uptake rate (r_S = 1.12 mmol g^{-1} h^{-1}).*

A comparison of the flux distributions during growth and hFGF-2 production with the theoretical optimum flux distribution for maximum hFGF-2 synthesis shows entirely different distributions. In Figure 9, selected fluxes from the flux distribution maps (Figures 3, 6 and 7) are compared. In contrast to the cultivation process, the EMP pathway shows rather low activity under optimal conditions since a substantial flux through the PP pathway for NADPH formation is required. Correspondingly, the TCA cycle as an alternative source reveals low contribution. The flux distribution during hFGF-2 production shows even more deviations from the optimum flux distribution than the one estimated for an at 30°C growing cell during the feeding phase. The main reason being for these differences that during temperature induced hFGF-2 synthesis the substrate is used very inefficiently. Due to the additional energy requirements during hFGF-2 production most of the carbon is lost in form of carbon dioxide (> 75 % c-mol c-mol^{-1}).

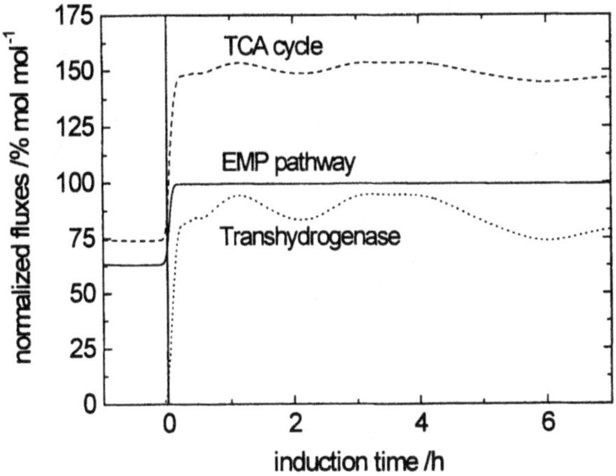

Figure 8: *Fluxes through the TCA cycle, the EMP pathway and the transhydrogenase reaction upon temperature induction of a* E. coli *TG1:pλFGFB culture producing the growth factor hFGF-2. Fluxes were calculated at many time points and interpolated by a cubic spline function. The fluxes are normalised with respect to the glucose uptake rate.*

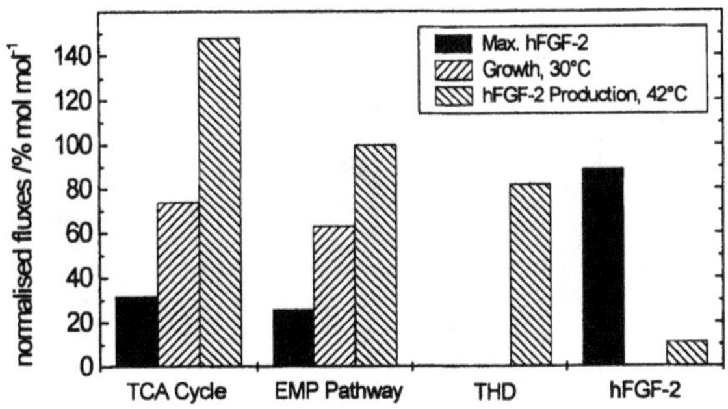

Figure 9: *Selected fluxes from a high-cell density culture of* E. coli *TG1:pλFGFB during growth at 30°C (refer to Figure 6) and hFGF-2 production at 42°C (refer to Figure 7), and from the theoretical optimal hFGF-2 production (Figure 3). The TCA cycle, EMP pathway, transhydrogenase reaction (THD) and hFGF-2 production are compared.*

3.2 TEMPERATURE-INDUCED PRODUCTION OF HUMAN INSULIN

Another example for the production of a recombinant protein using a temperature inducible expression system is human insulin by *E. coli* TG1. Herein, the chain combination approach is applied to produce native insulin. For more details on the process see (Schmidt et al., 1999a). In this approach, the A- and B-chain are produced separately and fused together after purification. Both proteins are attached to fusion protein partners to prevent intracellular degradation. The expression vectors encoding the genes for the insulin A- and B-chain fusion proteins (PMYW-A and PMYW-B) are under control of the λ-P_L promoter. In a typical high-cell density cultivation the insulin B-chain fusion protein accumulated to about 4.6 g l^{-1} (specific concentration 65 mg g^{-1}) corresponding to a concentration of about 800 mg l^{-1} of the insulin B-chain (18 % of the fusion protein represents the B-chain) (Figure 10). Similar results were found for the A-chain. The fusion proteins in both cases were found exclusively in form of inclusion bodies.

Similar to the hFGF-2 production process, the biomass yield $Y_{X/S}$ dropped from 0.48 g g^{-1} in the growth phase to an average of 0.15 g g^{-1} in the production phase. The specific growth rate was with 0.03 h^{-1} below the set value of 0.08 h^{-1}. The other specific rates, *i.e.* OUR, CER, glucose uptake and biomass formation rate are displayed on a molar basis in Figure 11. They resemble the rates of the hFGF-2 cultivation considerably. Thus, with respect to the metabolic flux distributions virtually the same results are obtained compared to the hFGF-2 process. During the temperature induced synthesis of the insulin B-chain fusion protein the glucose is mainly utilised in the fuelling reactions, *i.e.* the EMP pathway and the TCA cycle, for energy generating purposes. In Figure 12, the major pathway fluxes in the course of the cultivation are shown.

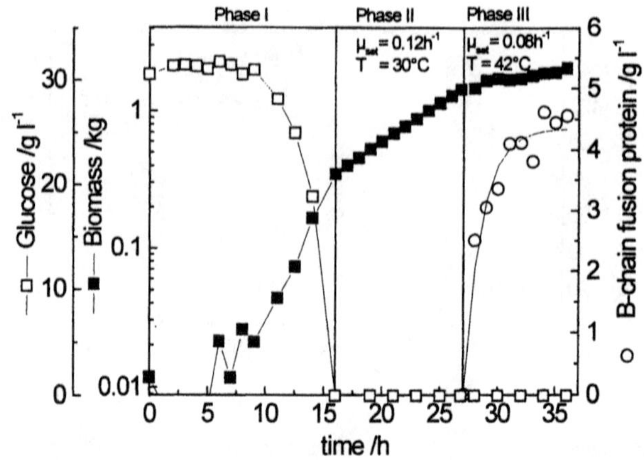

Figure 10: *Temperature-induced (30°C to 42°C) production of the recombinant B-chain insulin fusion protein in high-cell density culture of* E. coli *TG1:pMYW-B. The time course of biomass, glucose concentration and B-chain fusion protein concentration is shown (permission for reprint from (Schmidt et al., 1999a), on the way).*

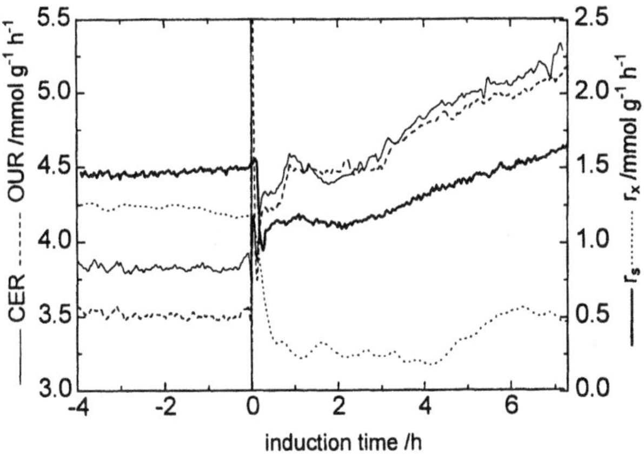

Figure 11: *Time course of the specific oxygen uptake rate (OUR), carbon dioxide evolution rate (CER), glucose uptake rate (r_S) and biomass formation rate (r_X) during the induction period of a temperature-induced high-cell density culture of* E. coli *TG1:pMYW-B producing the B-chain insulin fusion protein.*

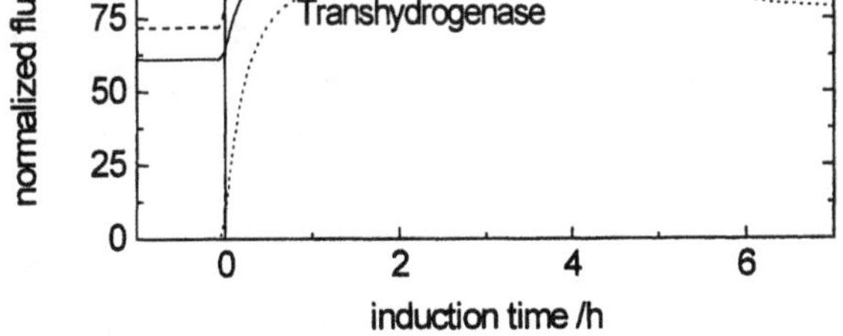

Figure 12: *Fluxes through the TCA cycle, the EMP pathway and the transhydrogenase reaction upon temperature induction of the* E. coli *TG1:pMYW-B culture producing the recombinant B-chain insulin fusion protein. The fluxes are normalised with respect to the glucose uptake rate.*

3.3 EFFECT OF A STABLE AND UNSTABLE RECOMBINANT PROTEIN ON THE HOST METABOLISM

The previous two examples showed the effect of both the recombinant protein production and the elevated temperature on the host metabolism. It is therefore of interest to see how an temperature independent expression system affects the host metabolism. As an example, the *E. coli* strain BL21(DE3) containing a chromosomal copy of the T7 RNA polymerase gene under control of the IPTG inducible *lac*UV5 promoter was used as host strain. Two closely related plasmids were constructed, pETRE9H and pETRE3H (Viaplana et al., 1997), encoding the antigenic C-terminus of the VP60 capsid protein from the rabbit haemorrhagic disease virus fused to a C-terminal histidine tag. pETRE3H contains additionally a signal sequence for the export into the periplasm. This protein is sensitive for proteolytic degradation. pETRE9H does not carry a signal sequence for protein export, but a control peptide tag of 13 amino acids (see (Viaplana et al., 1997), for further details on the constructs). pETRE9H forms a proteolytically stable recombinant protein in the cytoplasm in form of inclusion bodies.

In Figure 13, the estimations of OUR, CER, glucose uptake rate and biomass formation rate are shown in the induction period of a high-cell density cultivation (for details on the process see Schmidt *et al.*, 1999b). Both strains exhibited nearly the same performance in the growth phase prior to IPTG induction of recombinant protein synthesis. After induction, however, significant differences between the two strains became obvious. In the culture of BL21(DE3):pETRE3H producing the unstable protein the biomass formation rate as well as the biomass yield dropped markedly while the respiratory activity increased. In BL21(DE3):pETRE9H producing the stable protein these effects were observed merely slightly in response to IPTG addition.

The energy and material required for the formation of biomass and for the stable and proteolytically degraded recombinant protein is expected to be equal in both processes. Both proteins are identical except for minor variations in the N-terminal sequence. BL21(DE3):pETRE3H, however, seem to utilise the carbon substrate more for energy generation as seen by the increased respiration. Accordingly, flux analysis shows an increased activity of the fuelling pathways, *i.e.* the EMP pathway and the TCA cycle, compared to BL21(DE3):pETRE9H producing the stable protein (Figure 14). After 1 h, the energy surmounts the need of NADPH for biomass and recombinant protein synthesis and the stoichiometric model predicts activity of a transhydrogenase to convert NADPH to NADH which finally yields energy in the respiratory chain. The parameter m_{ATP} estimated from flux analysis allows comparison of the changes of the energetic requirements during the synthesis of the stable and unstable heterologous protein in the course of the cultivation (Figure 15). It can be seen that after 2 h of induction, more than 3 fold of energy is used during the production of the unstable protein. These differences may be a result of the higher energy requirements for the proteolytic degradation of the recombinant protein. Intracellular proteases, *e.g.* ClpXP, HflB, Lon (which is not active in BL21(DE3), are known to be energy-dependent (Gottesmann, 1996). It has also been shown that the ATP-dependent proteolytic activity and the synthesis rates of heat-shock proteins are elevated when a proteolytically unstable protein is synthesised (Kosinski et al., 1992a). In addition to these effects it may also be possible that additional energy is required for maintaining the general cell viability and for other ATP-consuming processes such as futile cycling.

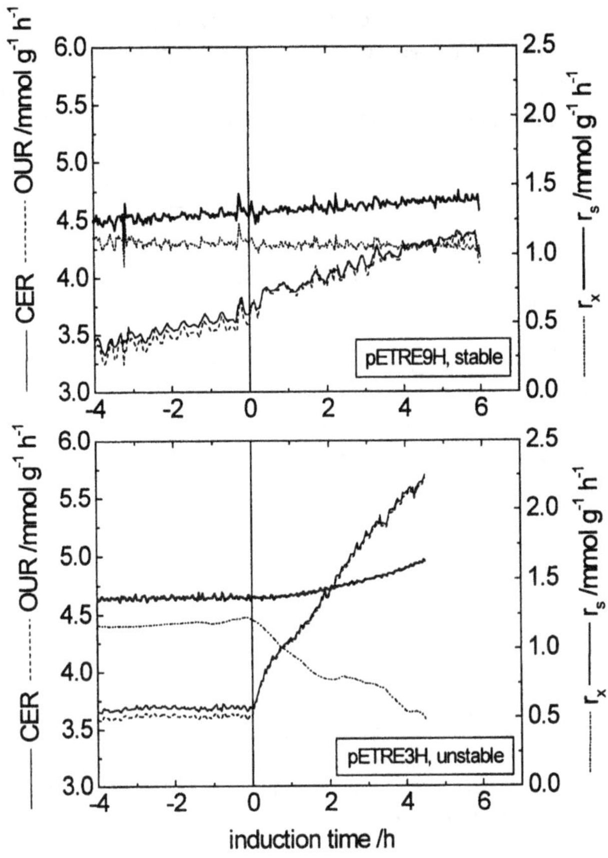

Figure 13: *High-cell density cultures of* E. coli *BL21(DE3) carrying either the plasmid pETRE9H encoding the stable recombinant protein (top) or pETRE3H encoding the proteolytic-sensitive recombinant protein (bottom). Induction was performed by addition of IPTG. The online estimations of the specific oxygen uptake rate (OUR), carbon dioxide evolution rate (CER), glucose uptake rate (r_S) and biomass formation rate (r_X) are shown.*

4. Summary and Concluding Remarks

The impact of the production of a heterologous protein on the *E. coli* metabolism has been examined using a stoichiometric modelling approach combined with linear programming techniques. The experimental data were obtained from carbon-limited high- cell density fed-batch cultivations. Feeding was carried out to increase the cell mass concentration exponentially in the bioreactor controlling biomass accumulation at growth rates which do not cause the formation of acetic acid. By comparing different expression systems and different heterologous proteins it could be shown that the metabolisms responds in a similar matter when the cell is faced to stress situations. Stress appeared when the recombinant protein was produced under heat shock conditions using the temperature-sensitive λ promoter/repressor system. Stress was also caused by a heterologous protein sensitive to proteolytic degradation. The same protein produced in a stable form in the cytoplasm did not reveal a significant stress response.

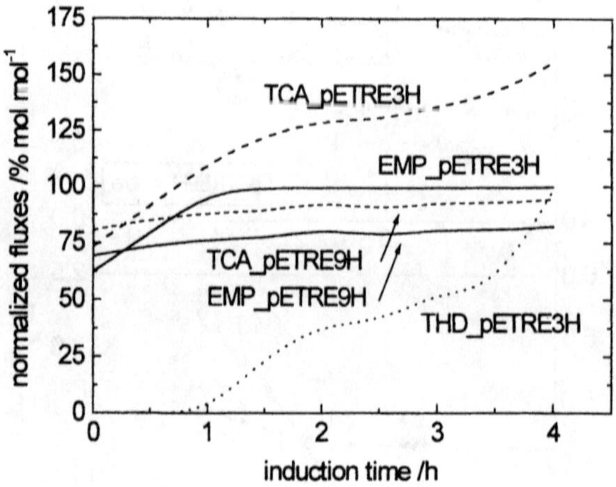

Figure 14: Fluxes of the TCA cycle, the EMP pathway and the transhydrogenase (THD) in BL21(DE3) pETRE3H encoding the proteolytic-sensitive recombinant protein and BL21(DE3) pETRE9H encoding the stable recombinant protein. The fluxes are normalised with respect to the glucose uptake rate. In BL21(DE3) pETRE9H the flux through the transhydrogenase reaction was zero.

The metabolic perturbation of the host organism was manifested by increasing fluxes through the energy generating pathways: The EMP pathway flux increased at the cost of the PP pathway. Due to the reduced biomass yield upon recombinant protein synthesis the precursor drain from the EMP pathway was reduced as well, thus increasing the flux through the TCA cycle (see Figure 15 for comparison of all strains investigated).
The elevated TCA cycle flux produced an excess of NADPH by isocitrate dehydrogenase. NADPH was converted to NADH by a transhydrogenase for additional energy generation as predicted by the stoichiometric model.

The overall additional energy demand was reflected by the non-growth associated ATP consumption parameter m_{ATP} which is summarised in Figure 15 (bottom) for all strains investigated. The regime of the curves is nearly identical to the corresponding TCA cycle fluxes (Figure 15, top) revealing the close correlation between the energy burden and TCA cycle activity. As a consequence, the increasing energy demand causes the cell to completely redirect the fluxes through the central pathways resulting in a very inefficient utilisation of the carbon substrate; the major portion is converted to carbon dioxide, not to biomass or protein. To improve recombinant protein synthesis the strategy might be to alter the cellular properties in a way to maintain the cellular homeostasis as best as possible. For instance, in terms of metabolic engineering one could think of the construction of an *E. coli* strain carrying instead of the NADP-specific isocitrate dehydrogenase a NAD-specific one which is known to exist in *Saccharomyces cerevisiae.* Simulations with the metabolic model show a significant less disturbed cell since NADPH generation requires an active PP pathway.

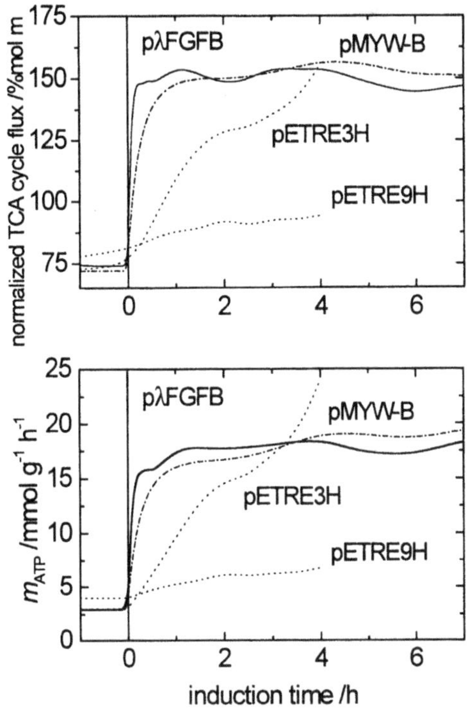

Figure 15: *Top: Comparison of the TCA cycle fluxes of the recombinant* E. coli *strains investigated upon induction of heterologous protein production in high-cell density cultivations. Bottom: Non-growth associated energy demand* m_{ATP} *of the recombinant* E. coli *strains investigated.* m_{ATP} *was adjusted to the measured biomass yields with the assumptions* $P/O = 1.75$ *and* $Y_{ATP/x} = 97$ *mmol* g^{-1}.

Appendix

Note: "=" points to a reversible, "= >" to an irreversible reaction.

BIOREACTION NETWORK OF *E. COLI*

Phosphotransferase System

r1) GLC + PEP = > GLC6P + PYR

Embden-Meyerhof-Parnas Pathway

r2) GLC6P = FRU6P
r3) FRU6P + ATP = > FRU16P
r4) FRU16P = > FRU6P
r5) FRU16P = GAP + DHAP
r6) DHAP = GAP
r7) GAP = NADH + G3P + ATP
r8) G3P = PEP
r9) PEP = > ATP + PYR
r10) PYR + ATP = > PEP

PEP Carboxykinase and PEP Carboxlyase

r11) OAA + ATP = > PEP + CO2
r12) PEP + CO2 = > OAA

By-products

r13) PYR = ACCOA + FORM
r14) ACCOA = > AC + ATP
r15) 2 ATP + AC = > ACCOA
r16) ACCOA + 2 NADH = > ETOH
r17) PYR = > ACCOA + CO2 + NADH

TCA Cycle

r18) ACCOA + OAA = > CIT
r19) CIT = ISOCIT
r20) ISOCIT = AKG + NADPH + CO2
r21) AKG = > SUCCOA + CO2 + NADH
r22) SUCCOA = ATP + SUC
r23) SUC = > FADH + FUM
r24) FUM = MAL
r25) MAL = > OAA + NADH

Glyoxylate shunt

r26) ISOCIT = > SUC + GLX
r27) ACCOA + GLX = > MAL
r28) MAL = > PYR + CO2 + NADPH
r29) MAL = > PYR + CO2 + NADH

Transhydrogenase

r30) NADPH = > NADH

r31) NADH + PO ATP = > NADPH

Oxidative Phosphorylation

r32) NADH + 0.5 O2 = > PO ATP
r33) NADH + 0.5 O2 = > 2 PO ATP
r34) FADH + 0.5 O2 = > PO ATP
r35) FORM + 0.5 O2 = > PO ATP + CO2

Pentose Phosphate Pathway

r36) GLC6P = > RIBU5P + CO2 + 2 NADPH
r37) RIBU5P = RIB5P
r38) RIBU5P = XYL5P
r39) XYL5P + RIB5P = SED7P + GAP
r40) SED7P + GAP = FRU6P + E4P
r41) XYL5P + E4P = FRU6P + GAP

Methylglyoxal Pathway

r42) DHAP = > MG
r43) MG + NADPH = LACALD + NADP
r44) MG + GLUTH = LACTGL
r45) LACTGL = DLAC + GLUTH
r46) MG = > DLAC
r47) PYR + NADH = > DLAC
r48) DLAC = > PYR + FADH
r49) LACALD = > LLAC + NADH
r50) PYR + NADH = > LLAC
r51) LLAC = > PYR + FADH

Ammonium, Glutamate and Glutamine

r52) NH3EX + 0.5 ATP = NH3 + 0.5 ADP
r53) NH3 + AKG + NADPH = GLUT + NADP
r54) GLUT + NH3 + ATP = GLUM + ADP

Amino Acids

r55) NADPH + ATP + 2 GLUT + ACCOA = > AKG + AC + ORN
r56) CO2 + GLUM + 2 ATP = GLUT + CAP + 2 ADP
r57) CAP + ORN = CR
r58) ATP + ASP + CR = FUM + ARG
r59) GLUT + G3P = NADH + AKG + SER
r60) SER + ACCOA + H2S = CYS + AC
r61) SER = NNMTHF + GLY
r62) OAA + GLUT = ASP + AKG
r63) 2 ATP + GLUM + ASP = ASN + GLUT
r64) 2 NADPH + ATP + ASP = HSER
r65) SUCCOA + CYS + HSER = SUC + PYR+ NH3 + HCYS
r66) NMTHF + HCYS = MET
r67) ATP + HSER = THR
r68) NADPH + 2 PYR = CO2 + AKI
r69) AKI + GLUT = AKG + VAL
r70) AKI + ACCOA + GLUT = LEU + AKG + NADH + CO2
r71) GLUT + PYR = AKG + ALA
r72) NADPH + PYR + GLUT + THR = CO2 + NH3 + AKG + ILE

r73) ATP + NADPH + E4P + 2 PEP = > CHOR
r74) CHOR + GLUM + SER + PRPP = > TRP + GLUT + PYR + GAP + CO_2
r75) CHOR = PREPH
r76) PREPH + GLUT = > CO_2 + AKG + TYR + NADH
r77) GLUT + PREPH = > CO_2 + AKG + PHA
r78) GLUM + ATP' + PRPP = > 2 NADH + AICAR + AKG + HIS
r79) GLUT + ATP + 2 NADPH = > PRO
r80) ASP + PYR + 2 NADPH + SUCCOA + GLUT + ATP = > SUC + AKG + CO_2 + LYS

Protein

r81) 1.18 ATP + .042 VAL + .005 TRP + .013 TYR + .024 THR + .021 SER + .021 PRO + .018 PHA + .015 MET + .033 LYS + .043 LEU + .028 ILE + .009 HIS + .059 GLY + .025 GLUM + .025 GLUT + .009 CYS + .023 ASP + 0.023 ASN + .028 ARG + .049 ALA = PROTEIN

Nucleotides

r82) NFTHF + AICAR = IMP
r83) IMP + ASP + 3 ATP = ATP' + MAL
r84) IMP + GLUM + 4 ATP = GTP + GLUT + NADH
r85) 4 ATP + GLUM + ASP + PRPP = UTP + GLUT + NADH
r86) ATP + GLUM + UTP = GLUT + CTP

RNA

r87) 0.0165 ATP' + 0.0203 GTP + 0.0136 UTP + 0.0126 CTP + 0.0256 ATP = RNA

DNA

r88) 0.00247 ATP' + 0.00247 UTP + 0.00254 GTP + 0.00254 CTP + 0.015 ATP = DNA

Lipids

r89) 0.0129 PAL + 0.0129 OL + 0.0129 GAP + 0.0129 SER + 0.0258 ATP = LIPID
r90) 8 ACCOA + 7 ATP + 13 NADPH = PAL
r91) 9 ACCOA + 9 ATP + 15 NADPH = OL

Lipopolysaccharide

r92) 0.00509 GLC6P + 0.0481 ATP + 0.0376 NADPH + 0.033 ACCOA + 0.0023 RIB5P + 0.0023 PEP + 0.00392 GLUT + 0.00235 G3P = LPS + 0.0023 NADH + 0.00392 AKG

Peptidoglucane

r93) 0.00276 FRU6P + 0.0055 ACCOA + 0.00276 PEP + 0.00276 PYR + 0.00276 OAA + 0.02484 ATP + 0.0193 GLUT + 0.0193 NADPH = PG + 0.0138 AKG

Glycogen

r94) 0.0154 GLC6P + 0.0154 ATP = GLYC

One-carbon unit and Polyamine

r95) 0.00485 SER = C1
r96) 0.0119 ATP + 0.01779 NADPH + 0.0119 GLUT = PA + 0.01397 AKG

Biomass

r97) 1.12 PROTEIN + 0.56 RNA + LIPID + LPS + 0.52 GLYC + PG + DNA + C1 + PA + YATP ATP = BIOMASS

Miscellaneous

r98) NNMTHF + NADH = NMTHF
r99) GLY = NADH + CO2 + NH3 + NNMTHF
r100) 2 ATP + RIB5P = PRPP
r101) PRPP + GLY + ASP + NFTHF + CO2 + 4 ATP = AICAR + MAL + 2 GLUT
r102) NNMTHF = NADPH + NFTHF

Fibroblast Growth Factor hFGF-2

r103) 0.0387 ALA + .04516 ASP + .01935 CYS + .07742 GLUT + 0.0387 PHA + 0.0903 GLY + .03226 HIS + .03871 ILE + .07742 LYS + .11613 LEU + .012903 MET + 0.05161 ASN + 0.05161 PRO + 0.0387 ARG + 0.06452 SER + 0.07742 THR + 0.03226 VAL + 0.00645 TRP + 0.05161 TYR + 0.0387 GLUM + 4.3 ATP = BFGF

Objective

r104) BIOMASS

References

Birnbaum, S., Bailey, J.E. 1991. Plasmid presence changes the relative levels of many host cell proteins and ribosome components in recombinant *Escherichia coli.* Biotechnol. Bioeng. 37, 736-745.

Bonarius, H.-P.J., Hatzimanikatis, V., Meesters, K.-P. H., de-Gooijer, C.-D., Schmid, G., Tramper, J. 1996. Metabolic flux analysis of hybridoma cells in different culture media using mass balances. Biotechnol. Bioeng. 50, 299-318.

Bonarius, H.-P.J., Schmid, G., Tramper, J. 1997. Flux analysis of underdetermined metabolic networks: the quest for the missing constraints. Trends Biotechnol. 15, 308-314.

Bonarius, H.-P.J., Timmerarends, B., de-Gooijer, C.-D., Tramper, J. 1998. Metabolite-balancing techniques vs. 13C tracer experiments to determine metabolic fluxes in hybridoma cells. Biotechnol. Bioeng. 58, 258-262.

Daae, E.B., Ison , A.P. 1999. Classification and sensitivity analysis of a proposed primary metabolic reaction network for *Streptomyces lividans.* Metabol. Eng. 1, 153-165.

Dong, H., Nilsson, L., Kurland, C.G. 1995. Gratuitous overexpression of genes in *Escherichia coli* leads to growth inhibition and ribosome destruction. J. Bacteriol. 177, 1497-1504.

Farewell, A., Neidhardt, F.C. 1998. Effect of temperature on *in vivo* protein synthetic capacity in *Escherichia coli.* J. Bacteriol. 180, 4704-4710.

Glick, B.R. 1995. Metabolic load and heterologous gene expression. Biotechnol. Adv. 13, 247-261.

Gottesman, S. 1996. Proteases and their targets in *Escherichia coli.* Annu. Rev. Genet. 30, 465-506.

Hannig, G., Makrides, S.C. 1998. Strategies for optimizing heterologous protein expression in *Escherichia coli.* Trends. Biotechnol. 16, 54-60.

Holms, W.H. 1986. The central metabolic pathways of *Escherichia coli*: Relationship between flux and control at a branch point, efficiency of conversion to biomass, and excretion of acetate. Curr. Top. Cell. Reg. 28, 69-105.

Ingraham, J.L., Maaloe, O., Neidhardt , F.C. 1983. Growth of the bacterial cell. Sinauer Associates, Sunderland/MA, pp. 3-173.

Jorgensen, H., Nielsen, J., Villadsen, J., Mollgard, H. 1995. Metabolic flux distributions in *Penicillium chrysogenum* during fed-batch cultivations. Biotechnol. Bioeng. 46, 117-131.

Karp, P.D., Riley, M., Paley, S.M., Pellegrini-Toole, A., Krummenacker, M. 1999. EcoCyc: Encyclopedia of *Escherichia coli* genes and metabolism. Nucleic Acids Res. 27, 55-58.

Kayser, A., Weber, J., Rinas, U. 2000. Steady state in continuous culture of *Escherichia coli*: energetics and metabolic fluxes. submitted.

Korz, D.J., Rinas, U., Hellmuth, K., Sanders, E.A., Deckwer, W.-D. 1995. Simple fed-batch technique for high cell density cultivation of *Escherichia coli*. J. Biotechnol. 39, 59-65.

Kosinski, M.J., Rinas, U., Bailey, J. E. 1992. Proteolytic response to the expression of an abnormal beta-galactosidase in *Escherichia coli*. Appl. Microbiol. Biotechnol. 37, 335-341.

Kurland, C.G., Dong, H. 1996. Bacterial growth inhibition by overproduction of protein. Mol. Microbiol. 21, 1-4.

Lee, S.Y. 1996. High cell-density culture of *Escherichia coli*. Trends Biotechnol. 14, 98-105.

Nielsen, J., Villadsen J. 1994. Bioreaction Engineering Principles. New York, Plenum Press, pp. 61.

Nissen, T.L., Schulze, U., Nielsen, J., Villadsen, J. 1997. Flux distributions in anaerobic, glucose-limited continuous cultures of *Saccharomyces cerevisiae*. Microbiology 143, 203-218.

Parsell, D.A., Sauer, R.T. 1989. Induction of a heat shock-like response by unfolded protein in *Escherichia coli*: dependence on protein level not protein degradation. Genes Dev. 3, 1226-1232.

Pramanik, J., Keasling, J.D. 1997. Stoichiometric model of *Escherichia coli* metabolism: Incorporation of growth-rate dependent biomass composition and mechanistic energy requirements. Biotechnol. Bioeng. 56, 398-421.

Ramirez, D.M., Bentley, W.E. 1993. Enhancement of recombinant protein synthesis and stability via coordinated amino acid addition. Biotechnol. Bioeng. 41, 557-565.

Riesenberg, D., Schulz, V., Knorre, W.A., Pohl, H.D., Korz D., Sanders, E.A., Ross, A., Deckwer, W.-D. 1991. High cell density cultivation of *Escherichia coli* at controlled specific growth rate. J. Biotechnol. 20, 17-27.

Rinas, U. 1996. Synthesis rates of cellular proteins involved in translation and protein folding are strongly altered in response to overproduction of basic fibroblast growth factor by recombinant *Escherichia coli*. Biotechnol. Prog. 12, 196-200.

Sauer, U., Hatzimanikatis, V., Hohmann, H.P., Manneberg, M., van-Loon, A.-P., Bailey, J.E. 1996. Physiology and metabolic fluxes of wild-type and riboflavin-producing *Bacillus subtilis*. Appl. Environ. Microbiol. 62, 3687-3696.

Schilling, C.H., Edwards, J.S., Palsson, B.O. 1999. Toward metabolic phenomics: Analysis of genomic data using flux balances. Biotechnol. Prog. 15, 288-295.

Schmidt, M., Babu, K.R., Khanna, N., Marten, S., Rinas, U. 1999a. Temperature-induced production of recombinant human insulin in high-cell density cultures of recombinant *Escherichia coli*. J. Biotechnol. 68, 71-83.

Schmidt, M., Viaplana, E., Hoffmann, F., Marten, S., Villaverde, A., Rinas, U. 1999b. Secretion-dependent proteolysis of heterologous protein by recombinant *Escherichia coli* is connected to an increased activity of the energy-generating dissimilatory pathway. Biotechnol. Bioeng. 66, 61-67.

Snow, A., Hipkiss, A.R. 1987. Stability of uragastrone and some fusion derivatives and the induction of stress proteins in *Escherichia coli*. Biochem. Soc. Trans. 15, 965-966.

Stephanopoulos, G., Aristidou, A.A., Nielsen, A. 1998. Metabolic engineering. Principles and methodologies. Academic Press, San Diego, pp. 261.

Stephanopoulos, G., Vallino, J.J. 1991. Network rigidity and metabolic engineering in metabolite overproduction. Science 252, 1675-1681.

Stouthammer, A.H. 1973. A theoretical study on the amount of ATP required for synthesis of microbial cell material. Antonie van Leeuwenhoek 39, 545-565.

Vallino, J.J., Stephanopoulos, G. 1990. Flux determination in cellular bioreaction networks: applications to lysine fermentations, in Sikdar, S.K., Bier, M., Todd, P. (eds), Frontiers in bioprocessing, CRC Press, Boca Raton/Fl., pp.205-219.

Vallino, J.J., Stephanopoulos, G. 1993. Metabolic flux distributions in *Corynebacterium glutamicum* during growth and lysine overproduction. Biotechnol. Bioeng. 41, 633-646.

van Gulik, W.M., Heijnen, J.J. 1995. A metabolic network stoichiometry analysis of microbial growth and product formation. Biotechnol. Bioeng. 48, 681-698.

Vanrolleghem, P.A., de Jong-Gubbels, P., van Gulik, W.M., Pronk, J.T., van Dijken, J.P., Heijnen, S. 1996. Validation of a metabolic network for *Saccharomyces cerevisiae* using mixed substrate studies. Biotechnol. Prog. 12, 434-448.

Varma, A, Palsson, B.O. 1993. Metabolic capabilities of *Escherichia coli* II. Optimal growth patterns. J. Theor. Biol. 165, 503-522.

Varma, A., Palsson, B.O. 1994. Metabolic flux balancing: basic concepts, scientific and practical use. Bio/Technology 12, 994-998.

Viaplana, E., Rebordosa, X., Pinol, J., Villaverde, A. 1997. Secretion-dependent proteolysis of recombinant proteins is associated with inhibition of cell growth in *Escherichia coli.* Biotechnol. Lett. 19, 373-377.

Voordouw, G., van der Vies, S. M., Themmen, A.P.N. 1983. Why are there two different types of pyridine nucleotide transhydrogenase found in living organsims? Eur. J. Biochem. 131, 527-533.

Wallace, R.J., Holms, W.H. 1986. Maintenance coefficients and rates of turnover of cell material in *Escherichia coli* ML 308 at different growth temperatures. FEMS Microbiol. Lett. 37, 317-320.

Zahl, K.J., Rose, C., Hanson, R.L. 1978. Isolation and partial charterization of a mutant of *Escherichia coli* lacking pyridine nucleotide transhydrogenase. Arch. Biochem. Biophys. 190, 598-602.

Zeng, A.P., Ross, A., Deckwer, W.-D. 1990. A method to estimate the efficiency of oxidative phosphorylation and biomass yield from ATP of a facultative anaerobe in continuous culture. Biotechnol. Bioeng. 36, 965-969.

DYNAMICS OF PROTEOLYSIS AND ITS INFLUENCE ON THE ACCUMULATION OF INTRACELLULAR RECOMBINANT PROTEIN

ROZKOV A., YANG S. AND ENFORS S.-O.
Department of Biotechnology, Royal Institute of Technology, Stockholm, Sweden.
E-mail: enfors@biotech.kth.se .Fax: +46 8 723 1890

Summary

The degradation rates of several recombinant proteins were determined throughout fed-batch cultivation, enabling to calculate synthesis rates of the proteins from their accumulation data. The simulations demonstrated the impact of proteolysis on accumulation rate and the final yield of the products. The method to determine proteolysis was also used to analyse the reason for decreased productivity of recombinant proteins towards the end of experiment.

1.Introduction

Higher yield of recombinant proteins is important in production optimisation. Although this could be done in some cases by increasing the expression by genetic methods, such as maintenance of high copy number (Nilsson and Skogman, 1986), using a stronger promoter (Brosius and Lupski, 1988), increasing an efficiency of translation (Lee et al., 1987) and the stabilisation of mRNA (Chan et al., 1991), sometimes the expected increase of productivity is not achieved.

Stability of recombinant proteins can be a significant factor influencing productivity. Proteolysis is one of the mechanisms used by the cell to dispose off proteins which are unwanted, misfolded or incorrectly synthesised. Recombinant proteins are often regarded by a cell as foreign and, therefore, often degraded much faster than most bulk proteins. Poor product accumulation is often attributed to the lack of expression, when in reality high expression rate is counteracted by high proteolysis rate.

The commonly used technique for production of recombinant proteins is a fed-batch cultivation with exponentially increasing feeding rate to control the specific growth rate. Since the maximum feeding rate is limited by the oxygen transfer capacity of bioreactor the exponential feeding is eventually turned into a constant feeding to control the oxygen consumption rate, leading to a decrease in a specific growth rate. Decrease of

A. Van Broekhoven et al. (eds.), Novel Frontiers in the Production of Compounds for Biomedical Use, 339–347.
© 2001 *Kluwer Academic Publishers. Printed in the Netherlands.*

specific growth rate below 0.05 h^{-1} may lead to dramatic changes in the *E. coli* cell physiology, such as a considerable loss of ability of cells to divide and induction of stringent response (Andersson et al., 1996). Stringent response is known to increase proteolysis, either directly (Voellmy and Goldberg, 1980) or by positive regulation of the Lon protease (Chung and Goldberg, 1981). Apparently, the increase of proteolysis when nutrients are exhausted (Mandelstam, 1960) is one of the defence responses triggered in order to provide amino acids for synthesis of "starvation proteins" essential for survival of a cell (Groat et al., 1986). Although the influence of cultivation conditions, in particular, nutrient availability, on production of recombinant proteins has been studied quite intensively, these studies seldom involved quantification of proteolysis.

Yang and co-workers (Yang and Enfors, 1995) showed that the intracellular degradation of several recombinant proteins can be described by first order kinetics. Thus, the accumulation of intracellular proteins with concentration P ($g \cdot g^{-1}$) is a balance between its specific rate of synthesis (q_p $g \cdot g^{-1} \cdot h^{-1}$), degradation with a rate constant K_{deg} (h^{-1}) and dilution by cell growth μ (h^{-1}) according to:

$$dP \cdot dt^{-1} = q_P - k_{deg} \cdot P - \mu \cdot P \quad (1)$$

Wong and co-workers (Wong et al., 1998) showed that the product accumulation is sensitive to feeding method after induction. Too low or too high feeding rate resulted in a lower product concentration than in the case of optimal feeding. Cultivation with complex medium usually gives higher product yields (Tsai et al., 1987; Zabriskie et al., 1987), but it is not clear whether this was achieved by increase of synthesis or reduced proteolysis.

Not only the nutrient limitation or starvation increases proteolysis, but also induction of the protein itself or the induction method may influence proteolysis (Kosinski et al., 1992). Since recombinant proteins often are regarded by cell as abnormal their presence in a cell may provoke stress responses (Aris et al., 1998; Goff and Goldberg, 1985; Rinas, 1996). No connection between the SOS response and proteolysis of recombinant proteins has been found yet. Protein degradation can be increased by heat shock because Lon and Clp proteases are induced as a part of it (Goff and Goldberg, 1985; Kroh and Simon, 1990).

Most degradation of the recombinant proteins is done by ATP-dependent proteases with initial steps of cleavage limiting the overall rate (Goldberg, 1992; Maurizi, 1992). Therefore, relative amount of degradation bands of the product on a polyacrylamide gel gives no information on the rate of proteolysis. Inhibition of protein synthesis by chloramphenicol or other suitable antibiotic enables to quantify rates of protein degradation (Yang and Enfors, 1995). We used this method to measure proteolysis rate throughout the production of recombinant proteins and then used obtained rate constants in simulations, quantifying the influence of proteolysis on the product accumulation rate. Since accumulation of proteolytically susceptible protein is a function of both synthesis and degradation rates, a minimisation of proteolysis is an obvious strategy to achieve a higher productivity, but quantification of the proteolysis is needed to predict the outcome of this strategy.

2.Materials and Methods

2.1 MICROORGANISM

E. coli RR1ΔM15 containing the plasmid pRITcI857, which encodes for the temperature-sensitive λ repressor c1857 and contains a kanamycin-resistance gene. SpA and the fusion protein SpA-β-gal are encoded by plasmids pRIT2 and pRIT1, respectively. Both plasmids contain an ampicillin resistance gene.

E. coli strain W3110 containing the plasmid pRIT44T2 was used in the studies involving protein ZZT2. Plasmid pRIT44T2 encodes protein ZZT2 and contains ampicillin and tetracycline resistance genes. The expression of protein ZZT2 is controlled by the *trp* promoter, which is induced by indole-3-acrylic acid (IAA).

2 .2 MEDIA AND CULTIVATION:

E. coli RR1ΔM15 was grown on the following medium (per litre); in shake flasks: 2 g $(NH_4)_2SO_4$, 1.6 g KH_2PO_4, 6.6 g $Na_2HPO_4{\cdot}2H_2O$, 0.5 g $(NH_4)_2$-H-citrate, 20 g casamino acids, 20 g glucose, 70 mg ampicillin, 20 mg kanamycin, 2 ml of a 1 M $MgSO_4$ and 2 ml of a trace element solution (Holme et al., 1970); bioreactor medium the same as above except for 10 g casamino acids and 5 g glucose per litre; feeding solution the same as above except for 100 g casamino acids and 200 g glucose. Production of SpA and SpA-β-gal in *E. coli* RR1ΔM15 was performed in bioreactor initially contained 5 litres of medium. Feeding with a flow rate of 74 $mL{\cdot}h^{-1}$ was started when the glucose was consumed at a cell density of about 3.3. Cells were grown at 30°C until cell density 3.5 $g{\cdot}L^{-1}$ when SpA or SpA-β-gal is induced by temperature shift to 42°C.

E. coli W3110 was grown at 35°C on the following medium (per litre): 2 g Na_2SO_4, 2.68 g $(NH_4)_2SO_4$, 0.5 g NH_4Cl, 14.6 g K_2HPO_4, 3.6 g $NaH_2PO_4{\cdot}H_2O$, 1 g $(NH_4)_2$-H-citrate. 2 ml of a 1M solution of $MgSO_4$, 3 ml of a trace element solution, 100 mg ampicillin, 100 mg tryptophane, 100 mg thiamine and 10 g of glucose. Feeding solution contained 4 g Na_2SO_4, 5.36 g $(NH_4)_2SO_4$, 1 g NH_4Cl, 29.2 g K_2HPO_4, 7.2 g $NaH_2PO_4{\cdot}H_2O$, 2 g $(NH_4)_2$-H-citrate, 2 ml of a 1M solution of $MgSO_4$, 3 ml of a trace element solution, 100 mg ampicillin, 100 mg tryptophane, 100 mg thiamine and 500 g glucose. Production of ZZT2 protein was performed in bioreactor containing initially 6.6 litres of medium. Exponential feeding started after 1.6 h according to the profile described by equation (2):

$$F(t)=F_0{\cdot}e^{k{\cdot}t}, \qquad (2)$$

where F - current feeding rate, F_0 - initial feeding rate (2.96 $mL{\cdot}h^{-1}$), k - exponential rate constant (0.31 h^{-1}), t - time

2.3 PRODUCT CONCENTRATIONS

The concentration of SpA-β-gal was estimated by assay of the β-galactosidase activity in crude cell extracts (Veide et al., 1983). The concentration was calculated with a specific activity of 390 U/mg (Strandberg and Enfors, 1991). SpA and the ZZT2 were purified by IgG-affinity chromatography as described elsewhere (Rozkov and Enfors, 1999). The concentrations of protein A or ZZT2 were determined from the absorbance data at 280 nm using an extinction coefficients 0.33 or 1.389 $mL \cdot mg^{-1} \cdot cm^{-1}$, respectively. The eluted fraction was collected and freeze-dried. SDS-PAGE was performed with 3.5% stacking gel and separation gel with 16% of acrylamide (for ZZT2) or 12.5% (for SpA) according to Laemmli (1970). Proteins were stained by Coomassie Brilliant Blue R-250. Analysis of proteins A or ZZT2 by Western Blotting was performed as described by Yang and Enfors (1995). Membranes were scanned and evaluated by computer software (ImageMaster, Pharmacia Biotech). Proteins A or ZZT2 in this case was quantified by integration of its density peak and expressed as percentage of the initial concentration.

2.4 DETERMINATION OF THE PROTEOLYSIS RATE CONSTANT

Culture medium was taken from bioreactor and was diluted by fresh medium to optical density 3-4 Units in shake flasks. 100 mg/L chloramphenicol was added to stop protein synthesis. Flasks were incubated on a orbital shaker (200 RPM) at the temperature of induction for 60 minutes. 0.1 ml aliquots of cell suspension were withdrawn at regular intervals and resuspended in SDS-buffer. Samples were stored at -20°C prior to analysis by Western Blotting.

3. Results and Discussion

Staphylococcal protein A expressed in *E. coli* is degraded with a half-life of 15-30 minutes (Kandror et al., 1994; Yang and Enfors, 1995; Yang and Enfors, 1996), but is stabilised when it is fused with β-galactosidase (Yang et al., 1995). Since these two proteins share the same promoter and are expressed in the same host strain and at the same conditions we assume that rate of translation (on the molar basis) of SpA and SpA-β-gal is the same. Therefore, any difference in accumulation of these proteins in cells should be due to the proteolysis of SpA. We used SpA and SpA-β-gal to quantify the effect of proteolysis on accumulation of the product.

The comparison of accumulated levels of SpA and SpA-β-gal is shown in Figure 1. The accumulation rate of SpA-β-gal was much higher compared to SpA and the final SpA-β-galactosidase yield reached 13.8% in the end of cultivation, compared with only 0.89% of SpA yield, which even declined to about half that value during the last hours of cultivation. The bacterial growth in both cultures was identical during batch phase and fed-batch phase before induction (Fig. 1). After induction the growth rate declined, with SpA-β-gal-producing culture being more inhibited than SpA probably because higher concentration of SpA-β-galactosidase, which is toxic to a cell.

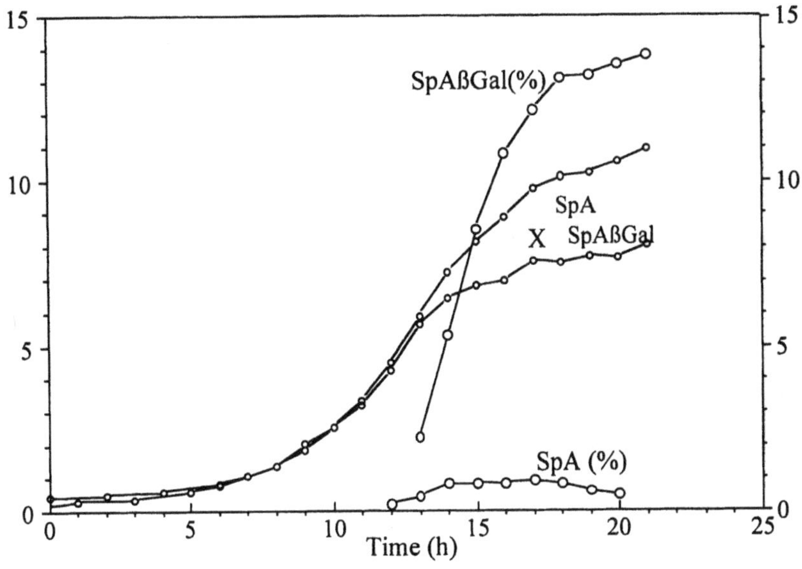

Figure 1. Production of SpA and SpA-β-gal proteins in fed-batch culture.

As described in equation 1 the accumulation of product in cell is a combined effect of synthesis and proteolysis. Since synthesis rates of these proteins are the same, the significant difference between obtained yields of SpA and SpA-□-gal can be only explained by continuous degradation of SpA. The proteolysis rate constant for SpA was determined at regular intervals throughout cultivation. The results show that the rate of proteolysis, expressed as first order kinetics rate constant, is highest immediately after induction - 5 h^{-1} and decreases to 2 h^{-1} in the end of cultivation (Figure 2).

To demonstrate the impact of proteolysis on productivity we calculated the theoretical accumulation rate of protein A using data on the rates of synthesis q_p of SpA-β-gal and degradation of SpA in Figure 2. A simulation of SpA accumulation is shown in Figure 3. Since proteolysis rate does not increase, the decrease of SpA yield after 18 hours should be attributed to decrease of synthesis.

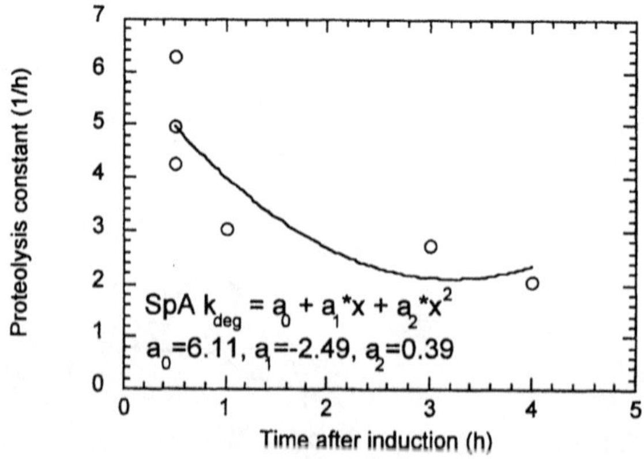

Figure 2. Dynamics of proteolysis of SpA after induction by temperature

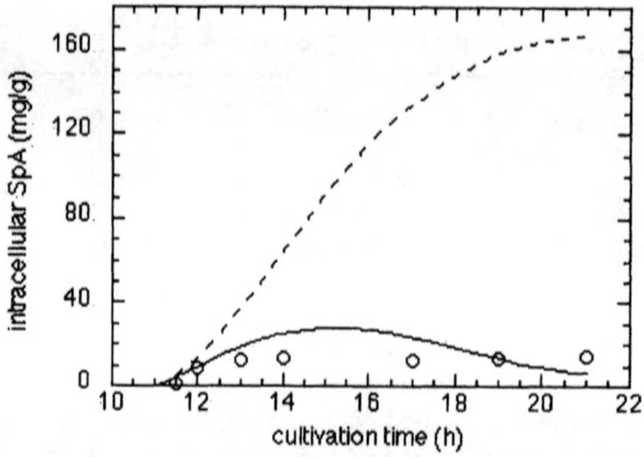

Figure 3. The comparison between SpA accumulation results obtained in experiment (points) and simulated SpA accumulation rate assuming proteolysis rate a) measured experimentally (solid line); b) equals zero (dashed line)

The thermal induction used in the experiments described above induces SOS (Aris et al., 1998) and heat shock responses, that could possibly affect dynamics of proteolysis. We, therefore, ran similar experiments with an other expression system, which does not use thermal induction. The proteolysis data of protein ZZT2 are shown in Figures 4 and 5. We present the obtained proteolysis data as both zero-order and first order kinetics rate constants since data fit to both models equally well. Proteolysis rate expressed as amount of ZZT2 protein degraded per biomass unit per hour (zero order kinetics) shows that degradation of ZZT2 increased after induction from 5-7 to 22 $mg \cdot g^{-1} \cdot h^{-1}$ and fluctuated between 14 and 23 $mg \cdot g^{-1} \cdot h^{-1}$ until the end of cultivation (Fig. 5). The same results expressed as the rate constants of first order kinetics shows that degradation rate constant decreases from the highest value 2.8 h^{-1} 0.5 hour after induction to 0.4 h^{-1} 10.5 hours after induction (Fig. 4). The low rate of proteolysis before induction can be explained by low amount of ZZT2 (2-5 $mg \cdot g^{-1}$), which makes its degradation substrate-limited. The accumulation rate of ZZT2 protein was close to linear after induction and specific proteolysis rate also increased after induction and quickly established within 3 hours at the higher value 22 $mg \cdot g^{-1} \cdot h^{-1}$ when protein yield was 15 $mg \cdot g^{-1}$ (Fig. 5). At the end of the experiment ZZT2 yield reached 39 $mg \cdot g^{-1}$, but the specific proteolysis rate (amount of protein degraded per unit of biomass per hour) remained on approximately the same level as after 3 hours induction. This indicates that proteolysis after 3 hours did not depend on the ZZT2 concentration and, therefore, was probably limited by protease activity. The specific synthesis rate increases after induction, reaches maximum 2.5 hour after induction (28 $mg \cdot g^{-1} \cdot h^{-1}$) and slowly decreases to the end of cultivation (Fig. 6). Therefore, decrease of accumulation rate of recombinant protein in the end of cultivation can be explained by decrease of synthesis rate rather than increase of proteolysis.

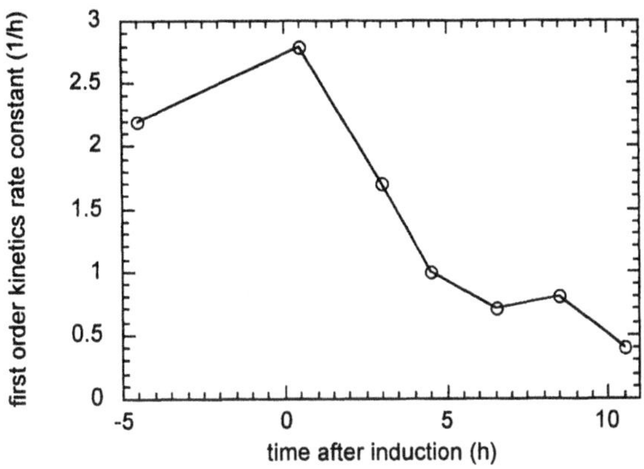

Figure 4. First order kinetics rate constant of proteolysis of ZZT2 protein

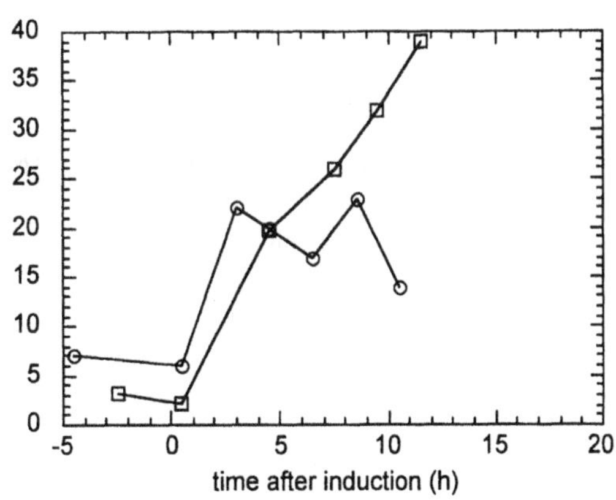

Figure 5. Specific rate of proteolysis (zero order kinetics) of ZZT2 protein (mg/g.h) (o) and yield (□) of protein ZZT2 (mg/g of dw) during cultivation.

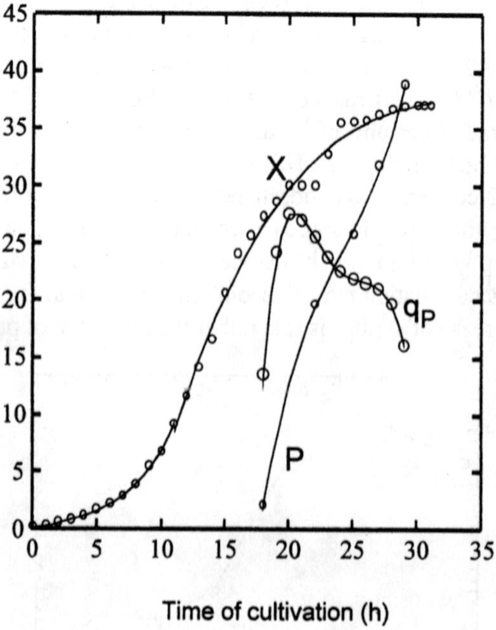

Figure 6. Calculation of specific synthesis rate of ZZT2 protein (q_P) ($mg \cdot g^{-1} \cdot h^{-1}$). Biomass curve (X) ($g\ L^{-1}$) and ZZT2 yield (P) ($mg\ g^{-1}$) are also shown.

4. Conclusions

Proteolysis has a significant impact on production of proteolytically sensitive recombinant proteins. Because of proteolysis the accumulation rate and the final yield of SpA protein decreased several fold compared to the stable fusion protein SpA-□-galactosidase.

The results from cultivations with three recombinant proteins showed that proteolysis reaches its maximum value during first hours of induction and does not increase to the end of the process. The decreasing accumulation rate of recombinant protein is explained by decreasing its synthesis rate.

References

Andersson, L., Yang, S., Neubauer, P. , Enfors, S.-O. 1996. Impact of plasmid presence and induction on cellular responses in fed-batch cultures of *Escherichia coli*. J. Biotechnol. **46:** 255-263.

Aris, A., Corchero, J.L., Benito, A., Carbonell, X., Viaplana, E. , Villaverde, A. 1998. The expression of recombinant genes from bacteriophage lambda strong promoters triggers the SOS response in *Escherichia coli*. Biotechnol. Bioeng. **60:** 551-559.

Brosius, J. , Lupski, J.R. 1988. Plasmids for the selection and analysis of prokaryotic promoters. Methods Enzymol. **153:** 54-68.

Chan, W.K.Y., Belfort, G. , Belfort, M. 1988. Stability of group I intron DNA in *Escherichia coli* and its potential application in a novel expression vector. Gene **73:** 295-304.

Chung, C.H. , Goldberg, A.L. 1981. The product of the lon (capR) gene in *Escherichia coli* is the ATP-dependent protease, protease La. Proc. Natl. Acad. Sci. **78:** 4931-4935.

Goff, S.A. , Goldberg, A.L. 1985. Production of abnormal proteins in E. coli stimulates transcription of *lon* and other heat shock genes. Cell **41:** 587-595.

Goldberg, A.L. 1992. The mechanism and functions of ATP-dependent proteases in bacterial and animal cells. Eur. J. Biochem. **203:** 9-23.

Groat, R.G., Schultz, J.E., Zychlinsky, E., Bockman, A. , A., M. 1986. Starvation proteins in *Escherichia coli*: Kinetics of synthesis and role in starvation survival. J. Bacteriol. **168:** 486-493.

Holme, T., Arvidson, S., Lindholm, B. , Pavlu, B. 1970. Enzymes - laboratory-scale production. Process Biochem. **5:** 62-66.

Kandror, O., Busconi, L., Sherman, M. , Goldberg, A.L. 1994. Rapid degradation of an abnormal protein in E.coli involves the chaperones GroEL and GroES. J. Biol. Chem. **269:** 23575-23582.

Kosinski, M.J., Rinas, U. , Bailey, J.E. 1992. Proteolytic response to the expression of an abnormal ß-galactosidase in *Escherichia coli.* Appl. Microbiol. Biotechnol. **37:** 335-341.

Kroh, H.E. , Simon, L.D. 1990. The ClpP component of Clp protease is the s32-dependent heat shock protein F21.5. J. Bacteriol. **172:** 6026-6034.

Laemmli, U.K. 1970. Cleavage of structural proteins during the assembly of the head of bacteriophage T4. Nature **227:** 680-685.

Lee, N., Zhang, S.-Q., Cozzitorto, J., Yang, J.-S. , Testa, D. 1987. Modification of mRNA secondary structure and alteration of expression of human interferon α1 in *Escherichia coli.* Gene **58:** 77-86.

Mandelstam, J. 1960. The intracellular turnover of protein and nucleic acids and its role in biochemical differentiation. Bacteriol. Rev. **24:** 289-308.

Maurizi, M.R. 1992. Proteases and protein degradation in *Escherichia coli.* Experientia **48:** 178-201.

Nilsson, J. , Skogman, S.G. 1986. Stabilization of *Escherichia coli* tryptophan-production vectors in continuous cultures: a comparison of three different systems. Bio/Technology **4:** 901-903.

Rinas, U. 1996. Synthesis rates of cellular proteins involved in translation and protein folding are strongly altered in response to overproduction of basic fibroplast growth factor by recombinant *Escherichia coli.* Biotechnol. Prog. **12:** 196-200.

Rozkov, A. , Enfors, S.-O. 1999. Stabilization of a Proteolytically Sensitive Cytoplasmic Recombinant Protein During Transition to Downstream Processing. Biotechnol. Bioeng. **62:** 730-738.

Strandberg, L. , Enfors, S.-O. 1991. Factors influencing inclusion body formation in the production of a fused protein in *Escherichia coli.* Appl. Environ. Microbiol. **57:** 1669-1674.

Tsai, L.B., Mann, M., Morris, F., Rotgers, C. , Fenton, D. 1987. The effect of organic nitrogen and glucose on the production of recombinant human insulin-like growth factor in high cell density *Escherichia coli* fermentations. J. Ind. Microbiol. **2:** 181-187.

Veide, A., Smeds, A.-L. , Enfors, S.-O. 1983. A process for large-scale isolation of β-galactosidase in an aqueous two-phase system. Biotechnol. Bioeng. **25:** 1789-1800.

Voellmy, R. , Goldberg, A.L. 1980. Guanosine-5′-diphosphate-3′-diphosphate (ppGpp) and the regulation of protein breakdown in *Escherichia coli.* J. Biol. Chem. **255:** 1008-1014.

Wong, H.H., Kim, Y.C., Lee, S.Y. , Chang, H.N. 1998. Effect of post-induction nutrient feeding strategies on the production of bioadhesive protein in *Escherichia coli.* Biotechnol. Bioeng. **60:** 271-276.

Yang, S. , Enfors, S.-O. 1995. The influence of energy sources on proteolysis of a recombinant protein A in *Escherichia coli.* Eur. J. Biochem. **233**: 969-975.

Yang, S. , Enfors, S.-O. 1996. Process conditions affecting proteolysis of recombinant proteins in Escherichia coli. Biotechn. Lett. **18:** 7-12.

Yang, S., Veide, A. , Enfors, S.-O. 1995. Proteolysis of fusion proteins: stabilisation and destabilisation of staphylococcal protein A and E.coli ß-galactosidase. Biotechnol. Appl. Biochem. **22:** 145-159.

Zabriskie, D.W., Wareheim, D.A. , Polansky, M.J. 1987. Effects of fermentation feeding strategies prior to induction of expression of a recombinant malaria antigen in *Escherichia coli.* J. Ind. Microbiol. **2:** 87-95.

Part 5

ARTIFICIAL ORGANS AND XENOGRAFTING

THE IMPACT OF TRANSGENESIS AND CLONING ON CELL AND ORGAN XENOTRANSPLANTATION TO HUMANS

LOUIS-MARIE HOUDEBINE [1] , BERNARD WEILL [2]

[1] *Unité de Biologie du Développement et de Biotechnologie, Institut National de la Recherche Agrononomique, 78352 Jouy-en-Josas Cedex France*

[2] *Laboratoire d'Immunologie, AP-HP, CHU Cochin, Université Paris 5, 75679 Paris Cedex 14 France*

Summary

Grafting cells or organs of animal origin into humans has been envisioned for a long time. Yet, no success has ever been met so far. The major hurdle, which still remains, is the strong immunological rejection of animal cells and organs by human hosts. The metabolic and functional compatibilities between the animal cells and the host will pose other problems varying according to the cells or organs involved. In principle, an animal close to humans should be the best potential donor. Although primates are the best candidates, they have been disqualified for ethical, biosafety and cost reasons. The pig is now considered as the best compromise. Pig organs are successively subjected to hyperacute rejection, delayed vascular rejection and finally to acute rejection. The latter seems essentially similar to the rejection occurring after allotransplantation. The mechanisms of hyperacute rejection have been deciphered. They imply the activation of the recipient's complement by preformed antibodies bound to endothelial cells. Those cells are the major targets of the rejection process. Their activation is followed by thrombosis and local inflammation. This activation in primate hosts can be prevented to some extent by the presence of human DAF, CD59 and MPC molecules at the surface of the graft and chiefly of enthelial cells. Pig hearts and kidneys expressing those molecules and grafted to primates can survive for several weeks without being damaged by complement attack. Studies using transgenic animal models (mouse, rat, rabbit and pig) are being carried in order to better clarify the rejection mechanisms. Recent progresses in gene transfer and animal cloning have opened new avenues for xenografting. Gene addition and replacement have become easier in several species. Further progress is needed before these techniques can be easily extended to the pig. The cloning technique virtually offers the possibility to generate animal totipotent embryonic

A. Van Broekhoven et al. (eds.), Novel Frontiers in the Production of Compounds for Biomedical Use, 351–363.
© 2001 *Kluwer Academic Publishers. Printed in the Netherlands.*

cells in which transgenes have been added or in which endogenous genes have been specifically inactivated. In theory, those cells can be differentiated in vitro and transplanted as stem cells to humans for treatment of Parkinson, Huntington or Alzheimer diseases. Biosafety problems, namely the transfer of viruses (essentially retroviruses) from the grafted organs to the hosts, represent an additional hurdle before xenografting can become acceptable for humans.

1. Why xenografting ?

The project of using organs from animals to humans is not new. The first attempts were performed at the beginning of the 20 th century. In 1905, Jaboulay grafted for the first time a goat kidney to a human being. The surgery was successful but the organ was rapidly destroyed. Various other attempts throughout the world were followed by rapid failures. The discovery of more and more effective immunosuppressive drugs led to an increasing success in allo transplantation. It is only at the beginning of the eighties that the growing shortage of human donors renewed the interest of scientists in xenotransplantation. Indeed, organs of human origin are chronically lacking and, everyday, a number of people die each year while being on waiting lists. Even an optimised collect of human organs would not significantly improve the situation. The number of patients waiting for an organ transplantation is considered as five times higher than that of available human organs (Persidis 1999; Gibbons et al 2000). Thus, the replacement of human organs by animal counterparts appears to be an inescapable solution to that problem. It will offer the alternative between bioartificial organs and xenotransplants.

Actually, cells or organoids of animal origin are needed for xenografting as much as functional organs. It is generally admitted that only a few organs from animals might be used. This is expected to be the case for heart and kidney. Transplanting animal livers does not seem realistic to most surgeons. Indeed, this organ is too complex and its compatibility with human metabolism has little chance to occur. Among the animal cells that might be grafted into humans, are Langerhans islets for diabetic patients, neurone cells for people suffering from Parkinson, Huntington or Alzheimer diseases, and liver cells for people with fulminant hepatitis.

Certain diseases require that the grafted cells are stem cells able to survive for long periods of time, to divide and to differentiate in order to functionally replace the defective organ. This is the case for Parkinson disease. In number of situations, grafted cells may potentially exert significant therapeutic effects after being maintained for various period of times in patients (Gurdon and Colman 1999; Lauza et al 1999b; Heath 2000). If stem cells cannot be easily obtained from adult humans, they can be obtained from animal foetuses. In addition, a few experiments have demonstrated that foetal cells from human brain can improve Parkinson and Huntington diseases. Stem cells can be obtained easily from animal foetuses. The recent progress in the cloning of embryos by nuclear transfer suggests that totipotent embryonic cells can be obtained from human somatic cells used as sources of nuclei (Wilmut et al 1998). Interestingly, mouse cloning has been obtained from established lines of embryonic stem cells (ES) (Wakayama et al

1999). These cells might allow repeated cloning generating genetically identical organ stem cells. The differentiation of human totipotent cells into fully functional cells has been obtained in vitro (Thomson et al 1998; Shamblott et al 1998; Zigowa and Sauberg 1998; Gearhart 1998; Lauza et al 1999a; Gurdon and Colman 1999; Lauza et al 1999b; Salter and Gearhart 1999; Bloom 1999). However, this protocol raises serious technical and ethical problems and cannot be considered as a realistic approach for human therapy in a near future (Annas et al 1999; Fox 1999; Vogel 1999). Indeed, the therapeutic cloning imagined so far implies that potentially healthy human embryos are generated only to be the source totipotent cells. A possible alternative might be to generate chimeric embryos by introducing human nuclei into bovine enucleated oocytes (Dominko et al 1999). Recent and preliminary experiments have shown that totipotent cell lines can be derived from the blastocysts obtained in this way. It is not presently known if these cells may be used for cell therapy in human. Most likely, the chimeric human-cow embryos are not viable and they stop their development at early stages. If so, no living organisms can be generated by this procedure and the ethical problems raised by the use of these chimeric embryos are greatly attenuated.

A simple approach not involving cloning to obtain organ stem cells might emerge in the coming years. Several works have shown that nervous stem cells extracted from brain can spontaneously be transformed into functional hematopoietic stem cells when introduced in the bone marrow of irradiate mice. These newly generated hematopoietic cells were able to generate red and blood cells (Bjornson et al 1999). Similarly muscular stem cells can be converted to some extent into hematopoietic cells (Gussoni et al 1999; Jackson et al 1999). Human bone marrow stromal cells can be stably implanted into brain and be active (Azizi et al 1998). Clearly, stem cells functions are more flexible than imagined and their potential use must be reconsidered (Lemischka 1999). If successful, the implantation of specific stem cells from an organ into a quite different organs would raise not particulary ethical problems. It remains difficult to evaluate the impact of organ stem cells for grafting. Indeed, only a limited number of stem cell types can be obtained in vitro from totipotent cells so far. On the other hand, it is not known how stem cells would be functional when introduced into an organ. It is difficult to predict how transplanted muscle stem cells will be capable of colonizing the organ and participate in an appropriate manner to muscle contraction. A success seems more accessible for diseases, such as Parkinson disease or diabetes mellitus which require a limited number of transplanted cells.

In most cases, the cells collected in humans or animals will require to be cultured before being transplanted into a patient. During the culture, foreign genes can be theoretically transferred into the cells and stably integrated into their genome. These genes can code for therapeutic proteins. The cell graft is then a cell therapy but also a gene therapy.

In the first xenotransplantations, organs from non human primates were frequently used. Indeed, it seemed consistent to use material from closely related species. However, primates are presently no longer considered as likely sources of organs and cells for humans for three main reasons. The first in infection. Indeed, there is a potential risk of contamination of human recipients by viruses derived from monkeys. The second reason is of ethical order. Several species of primates are protected. The third reason is just economical. The high cost of primates would certainly restrict xenotransplantation to a limited number of patients.

For all those reasons, another donor, the pig, has been considered as the best potential donor. The pig is, as the human, omnivorous. Its physiology is not very different from human physiology in general. As far as we presently know, bacterial and parasitical diseases that can be transmitted from pig to human are few and curable. Moreover, breeding conditions allowing the preparation of organs and cells free of bacterial and parasitic pathogens are well known. However, uncertainties remain concerning possible viral contaminations that could occur under clinical circumstances such as xenotransplantation.

3. The mechanisms of xenograft rejection

The mechanisms responsible for the rejection of xenografts are obviously more complex and more potent than those involved in allografting. The rejection of xenografts from phylogenetically distant species develops in three steps and can be described as follows : the first step is a rapid and catastrophic event named hyperacute rejection; a delayed vascular rejection follows after 36-48h; an acute rejection similar to that observed after allografting appears after 7-10 days (Bach et al 1977; Platt 1998).

During the last decade, many studies led to a better understanding of the hyperacute rejection. This phenomenon occurs in vascularized organs, and endothelial cells are its major targets. Hyperacute rejection starts as soon as the graft is revascularized. The graft vessels are then subjected to an intense vasoconstriction followed by platelet aggregation and a massive thrombosis which interrupts blood circulation and leads to a rapid necrosis of the graft.

These phonomena are induced by the activation of the recipient's complement. Complement is a defence system which is activated by foreign molecules or microorganisms. It generally implies the initial recognition of the foreign material by circulating preformed antibodies. The complement is thus activated very rapidly, without any other induction of specific immunological agents. Complement activation is a complex process involving a cascade of events leading to the formation of a membrane attack complex which degrades the membrane of foreign cells. The endothelial cells of the xenograft are thus the first targets of activated complement.

Several components of the complement cascade bind to endothelial cells which become themselves activated. Under the action of complement components, endothelial cells retract and the endothelium becomes leaky. The endothelial cells loose most of their anticoagulant factors. They synthesize receptors to which lymphoid cells bind. This binding induces a local and intense inflammation which contributes to local thrombosis and endothelium disruption.

The delayed vascular rejection occurs during the days following xenotransplantation. It has not been described as precisely as the hyperacute rejection. It cannot be observed so easily because some of its manifestations are interfering with those of the hyperacute rejection. The delayed vascular rejection involves NK cells and the phenomen on of antibody - dependent cellular cytotoxity (ADCC) that leads to the apoptotic death of endothelial cells. In addition, the production of tissue factor by the activated endothelium enhances thrombogenesis.

The acute rejection is clinically similar to the classical allograft rejection since it occurs 7-10 days postransplantation. However it differs by two main features. First, xenoantigens are essentially presented to the recipient's T cells by recipient's antigen-presenting cells and not by the graft itself. Second, the intragraft infiltrate associated with the rejection is comprised of macrophages in higher proportion than in classical allografts.

4. The preventive treatments of xenograft rejection

The mechanisms of xenograft rejection are obviously multiple and complex. The control of these mechanisms cannot be established without a deep understanding of their essential steps. The study of those mechanisms implies the use of cultured cells but also of transgenic animals. Indeed, the cell systems cannot mimick all the events which occur during xenograft rejection. Immortalized pig endothelial cells lines have been recently obtained. Interestingly, these cells have retained most of the characteristics of the primary cells (Malassagne et al 1996). These studies must be carried out by immunologists, biochemists and surgeons working in close collaboration.

A temporary inhibition of hyperacute rejection can be obtained by different ways. To prevent the rapid action of the natural antibodies, those molecules can be removed from the blood circulation of the recipient. This can be achieved by specifically absorbing the antibodies during extracorporeal blood circulation. It is also possible to destroy several constitutive elements of the complement with substances such as cobra venom factor. Under these conditions, the hyperacute rejection effectively does not occur anymore. If this treatment is of short duration, the natural antibodies and the complement return to their initial level and the xenografts are again subjected to rejection. If the treatment is prolonged, the graft is not rejected despite the normalization of antibodies and complement. This phenomenon has been named accommodation. Its mechanism is poorly understood. Yet, a good correlation between the capacity of an accommodated xenograft to resist hyperacute rejection and the expression of anti-apoptotic genes has been observed.

Another possible way to inhibit or a least to abrogate the hyperacute rejection consists in withdrawing the antigens present at the surface of the xenograft endothelial cells. The dominant antigen on pig cells, recognized by primate preformed antibodies is comprised of a gal α1,3 gal motifs linked to membrane proteins. It is possible to mask this epitope by adding a fucose using a fucosyl transferase by transgenesis. Indeed, the newly created motif is commonly found at the surface of most human epithelial cells. The pig cells having a fucose added to their gal α1,3 gal structure are more resistant to human complement attack (Fodor 1999). The genes coding for α-galactosidase (Osman et al 1997) and glycosyltransferase (Miyagawa et al 1999) may also contribute to mask the α 1,3-gal epitope in pig cells. It is also theorically possible to prevent the synthesis of the gal α1,3 Gal epitope by inhibiting the cellular galactosyltransferase responsible for protein galactosylation. This can be theoretically achieved by different ways. Antisense RNA or ribozymes are potentially capable of reducing the stability of the galactosyltransferase mRNA. These tools are not reliable and they have little chance to diminish the α; 13-Gal epitope to a sufficiently low level to prevent complement activation. Alternatively, the galactosyltransferase gene can be knocked out using homologous recombination and cloning. Encouraging experiments are in progress to reach this goal.

Most likely, these approaches will meet at best a partial success. Indeed, multiple independent antigens recognized by human antibodies are present on the surface of pig cells and it seems unrealistic to identify and eliminate all those molecules. On the other hand, it is known that the alternative pathway of complement which does not involve the recognition by antibodies, is strongly induced by xenografts. Thus, the withdrawal of the most intensively recognized antigens might at best partially abrogate the humoral and cellular rejections.

Another approach which met a spectacular although partial success came from the expression of genes exhibiting anticomplement effects. This was achieved by preparing transgenic pigs harbouring and expressing the human DAF (Decay Accelerating Factor or CD55) or CD59 genes. Those proteins are known to inactivate human complement in vitro. Their action is species-specific. This is at least the case for the DAF gene. CD55 and CD59 inhibit the complement by acting at two diffrent levels of the activation cascade (Cozzi and White 1995) and have additional effects in vitro. Interestingly, kidneys and hearts from transgenic pigs were maintained up to several days and weeks after being grafted to primate recipients (Bach et al, 1997; Schmoeckel et al 1998; Cowan et al 1999). At the end of the experiments, the grafted had not been attacked by complement. Their progressive rejection was rather due to the delayed acute rejection and to the induced cytotoxic T cell rejection.

The rejection of xeno and allografts implies the activation of cytotoxic T cells which specifically recognize the non-self xenoantigens. One theoretical possibility to prevent this rejection is to induce a specific tolerance by injecting the recipient with autologous hematopoietic stem cells transfected with genes from the donor, for example the gene encoding galactosyl transferase which directs the synthesis of the antigenic gal α-1,3 motif. Preliminary experiments suggest that this approach is not impossible (Bracy et al 1998; Platt 1999; Persidis 1999; Ohdan et al 1999). Patients might thus be prepared by a specific gene therapy in hematopoïetic stem cells to tolerate pig organs. This rather heavy protocol might be a severe limitation to achieve xenografting.

Preliminary experiments have shown that the injection of pig splenocytes by mice reduces the aggressiveness of spleen cells towards pig islet cells (You et al 1998). This simple approach might contribute to alternate the rejection mecanisms in patients transplanted with pig cells or organs.

This first success, although limited, inclines to consider that the addition of foreign genes to pig cells, but also the inhibition of some pig gene expression might finally allow an acceptable adaptation of pig cells and organs to human recipients. It is clear that many studies remain to be done before this becomes a reality. Even if pig is expected to be the essential, if not the only donor, other transgenic animals are and still have to be used. Indeed, generating transgenic pigs can be a routine work but it remains a laborious and costly task. Gene addition is possible in pig by microinjection into pronuclei in early embryo. Gene replacement by homologous recombination to inactivate endogenous genes has been performed essentially in mouse so far. The classical technique implies the use of totipotent embryonic stem cells (ES cells). The genetic modifications are carried out in the cultured ES cells which can be selected and finally used to generate chimeric animals harbouring the genetic alteration. Several vectors for homologous recombination can replace a host gene by a non functional version. Alternatively, with this method, a given gene can be replaced by an allele carrying a single point mutation only or by a quite different functional gene (Viville 1996). For still unknown reasons, reliable ES cells lines have been obtained from only a few mice strains and those other species allow the generation of chimeric animals but not the transmission of the mutation to progeny. The ES cells are then probably no more totipotent and they have become unable to participate to the generation of gonades and gametes. Recent experiments have shown that some totipotent cell lines established from pig primordial germ cells can give birth to highly chimeric animals (Mueller et al 1999). It remains to prove that the totipotent cell lines were able to participate to the generation of germ cells in the chimeric pig and transmit their genotype to progeny.

An alternative and quite promising method consists in using the embryo cloning technique by nuclear transfer. Fœtal cells can receive foreign gene during culture and further be used to regenerate living animals (Schniecke et al 1997). It is expected that the same method can be followed to replace endogenous pig genes using homologous recombinaison before embryo cloning. This protocol has recently met a success in sheep. Fœtal fibroblasts in which a gene was replaced by homologous recombination could give birth to living lambs after having been transferred to enucleated oocytes (Ayares 1999). It should be kept in mind however, that embryo cloning by nuclear transfer has proved to be possible only in ruminants and mouse so far. Improvement of the techniques is required before it can be applied to other species and namely to pig. Experiments in progress indicated that pig cloned embryos in which gene replacement was previously achieved could develop for several weeks if development but not reach the end of pregnancy.

Gene replacement implies a homologous recombination between the foreign DNA fragment and the targeted host gene. The classical methods used are based on the construction of vectors carrying two long DNA fragments strictly homologous to the targeted gene. The sequence containing the mutated region is inserted between the homologous regions. The use of restriction sites such as I-Sce 1 and of the corresponding restriction enzyme which cleaves both DNA strands considerably enhances the rate of homologous recombinations even when short DNA sequences are used (Cohen-Tannoudji et al 1998). Oligonucleotides composed of ribo- and deoxyribonucleotides and named RDO for this reason are strong inducers of homologous recombinations (Alexeev et al 1998; Barlett 1998). Double stranded RNAs kept in the nucleus have a strong and specific capacity to inhibit the expression of the corresponding gene by unknown mechanisms which might involve homologous recombination (Sharp, 1999). The additon of single stranded DNA carrying a mutation and associated with the bacteria recombination enzyme RecA to a cell or to an early embryo induces the mutation of the corresponding gene by a homologous recombination process with a high efficiency (Pati 1998). All these quite new and not fully validated methods can potentially be applied to cultured cells further used to regenerate embryos by nuclear transfer. Alternatively, these new tools could be directly used by microinjection into one cell embryos. These methods would greatly facilitate targeted gene mutation in various species including pig.

The inhibition of the expression of a given gene may be obtained by different methods not altering the structure of the gene. Oligonucleotides or RNA forming a triple helix in polypyrimidine rich region can specifically block transcription of a gene. Antisense or ribosymes may also specifically inactivate a mRNA (Vasquez and Wilson, 1998). Recent experiments indicate that double strand RNA are quite potent and specific tools to inactivate a mRNA in a specific manner (Wianny and Zernicka-Goetz, 2000; Bosher and Labouesse, 2000). Alternatively, transdominant negative protein can compete a cellular protein and reduce its activity to a low background (Bach et al, 1997). All these inhibitory molecules can be synthesized from transgenes.

The action of several genes must be studied before being tentatively exploited to prevent xenograft rejection. Animal models other than pigs are being used for this purpose. Mouse is a satisfactory model in many cases (Bach et al 1997; French et al 1998). Yet, the grafting of mouse organs or tissues to primates or other species is uneasy. For these reasons, transgenic rats (Charreau et al 1996) and rabbits (Taboit-Dameron et al 1999) are being used as models. Another interesting experimental approach consists in using adenoviral vectors carrying a gene to be tested. This tool may provide interesting informations on the action of a gene without being obliged to generate transgenic animals and it may incline or not to use transgenesis to study the gene more extensively.

It is of course presently impossible to predict how many genes will have to be added or inactived in the pig genome. Data shown in Fig.1 summarize the different mechanisms which should be studied and possibly controlled by transgenesis. It is conceivable that the number of genetic modifications in the pig genome will not exceed ten. If this is the case, the use of transgenic pigs as organ and cell donors appears accessible. The major hurdle still is the understanding of the rejection mechanisms and the identification of the genes to be used to generate the appropriate transgenic pigs.

5. The biosafety of xenografting

The first problem to solve with xenografting is certainly the biological compatibility of pig organs with the human organism. Pig hearts are expected to generate a sufficient blood supply to replace a human heart. The pig kidneys are different from human organs and the use of drugs controlling water and ion filtration in human recipients may be required.The treatment of patients with human erythropoietin is also expected to be required to compensate the lack of erithropoietin secretion by the grafted pig kidneys. The metabolism of glucose in pig and human is somewhat different. The grafting of pig Langerhans islets may lead to unsatisfactory situations for patients suffering from diabetes. Further studies are urgently needed to estimate the real capability of pig cells and organs to replace their human counterparts.

Another category of problems inherent of the grafting of pig cells or organs to human is the possible contamination by porcine pathogens. It is generally admitted that the classical pathogen-free conditions used to breed pigs should be sufficient to eliminate the common bacteria and parasites. This will require strictly controlled and standardized conditions of breeding. The same may not be true for other pathogens such as retroviruses or prions. Pig is generally considered as not suffering from diseases induced by retroviruses. However, several infectious retroviruses have been identified (Brown et al 1998). Recent experiments have shown that at least two retroviruses are secreted by pig cells (Patience et al 1997). These retroviruses are able to infect some human cell lines in culture. These observations suggest that additional experiments are required before being sure that pig can be use as safe donors of cells and organs. A moratorium in the USA and the UE regulates xenografting experiments in humans. Rules are being searched to define acceptable conditions for experimentation (Fox 1998). Observations done in several countries on 160 patients having been in contact in some way with living pig material indicated that none of them were stably seropositive for the porcine viruses tested (Paradis et al 1999). These data indicate that the known porcine retroviruses are not hightly infections in humans. Although relevant, the data reported by Paradis et al (1999) do not exclude that infection of patients cannot take place. This may be particularly true in patients treated by immunosuppressors. It should also be kept in mind that pig organs in which the α-Gal epitote has been deleted might generate retroviral particles much less readily attacked by the defense mechanisms of the human host. The same may be true to some extent with the organs expressing the human anticomplement proteins DAF or CD59 (Breun et al 1999). The real risk of human infection by various pig pathogens remains a matter of controversy (Weiss 1999; Hunkeler et al 1999; Collignon 1999; Paradis and Langford 1999). The risks of infection seem too low to dissuade experimenters to procede to the xenografting of pig cells or organs to humans. Nonetheless, if patients were to receive pig cells or organs, their immune reactions to porcine viruses and those of their relatives should be carefully followed thereafter. The problem is complicated by the fact that if the retroviruses are not highly infectious for humans at the time of grafting they might mutate or recombine to generate a population of viruses more aggressive and pathogenic for both humans and pigs.

An examination of the pig genome revealed that, as other species, it contains many retrovirus sequences. This was observed with all available breeds of pigs including wild animals. A systematic survey of the retroviral sequences should indicate which of them are still functional and directing the synthesis of infectious particles. Using those data, a selection of the animals expressing a minimal number of retroviral sequences could be done. In addition, the tools described above to mutate genes specifically by homologous recombination might be used to inactivate the functional retroviral sequences at the DNA or the RNA level.

6. The acceptability of xenografting

It is by no means certain that people will easily accept xenografting if it happens that the technique is available. For some people, the idea of carrying an animal organ cannot be tolerated. For others, pigs is particularly repulsive. Indeed, pig has not a good image for some reasons and it is also a strong symbol of physical and moral dirtiness. It should be mentioned that the persons who do not eat pork for religious reasons have not rejected the use of pig organs per se for xenografting. The use of transgenic animals, including pigs for medical purposes, is questionned by some people. Rules are being defined to determine optimal conditions in which transgenesis in pig could be acceptable (Houdebine 1999). It is clear that education is needed to inform patients and the public opinion of the real benefit that could be expected from xenografting. This should lead, in time, to a general acceptation of the technique. Inquiries have been done in several countries. They indicate that xenografting is not rejected per se by a majority of people. As could be expected this is particularly the case for those who are candidates to recieve cells or organs from pigs (Deschamps et al 1999).

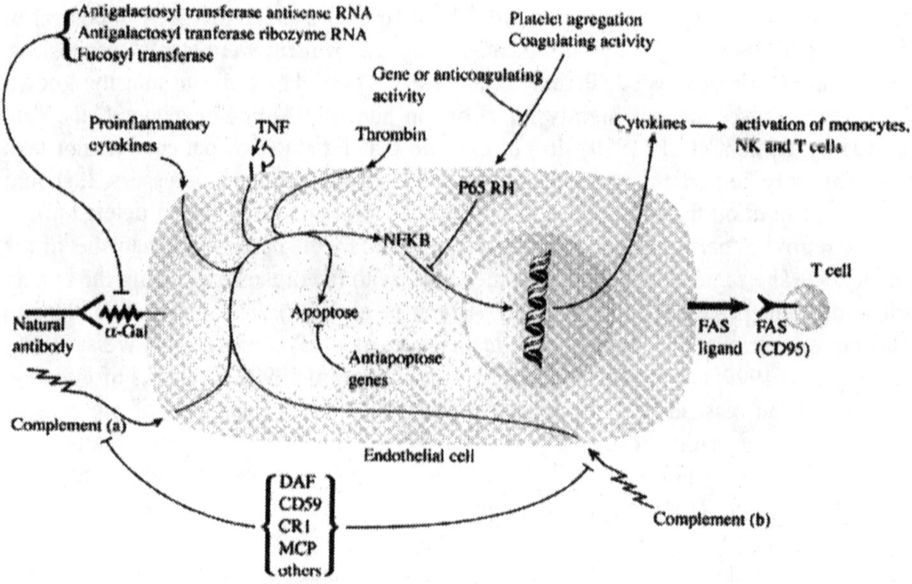

Figure 1 Schenatic representation of some of the mechanisms responsible for the rejection of xenografts. The recognition of xenoantigens such as the gal αl,3 gal motif at the surface of pig endothelial cells by natural antibodies induces the direct activation of complement (a). The alternative complement pathway (b) is unduced by the direct recognition of pig cells. Several human genes transferred into cultured cells or into animals (DAF, CD59; CR1, MCP and others) can inhibit the activation of human complement. Epitopes like the gal αl,3 gal motif may be eliminated to some degree by antisense RNA or ribozymes directed against the galactosyl transferase mRNA. The addition of fucose by fucosyl transferase transforms the foreign gal α-1,3 gal into a human structure. The action of cytokines which generates a proinflammatory state and induces apoptosis of endothelial cells can be counteracted by several anti-apoptotic genes and by P65 RH, a transdominant negative analogue of the transcription factors NFKB. The loss of coagulation factor by the activated endothelial cells can be compensated by the expression of transgenes coding for coagulation factors. The late rejection mechanisms involves the secretion of cytokines and the activation of cytotoxic T cells. The expression of the human FAS ligand at the surface of endothelial cells might induce the apoptosis of T cells after Fas-Fas ligand interaction. The rejection mechanisms described in this figure and their potential inhibition by transgenes only summarize the major types of experiments currently performed to tentatively adapt pig cells to human hosts.

7.Conclusion and perspectives

It is clear that xenotransplantation has raised number of important and new questions which have recieved so far a few answers only (Auchincloss and Wood 1998). The major unresolved problems still come from an insufficient knowledge of the rejection mechanisms. Technical problems, mamely the generation of transgenic pigs, have not recieved yet satisfactory solutions. The progress recently made in other species inclines to think that transgenesis efficiency will not be anymore a real limitation to xenografting in the coming years. The biosafety problems due to pig pathogens are the matter of serious concern. Yet, it seems reasonnable to consider that a limited number of xenograftings to humans might be performed under strickly controlled conditions. Although several independent and complementary directions have been taken to control rejection mechanisms, others have been only superficially explored so far. Genes involved in the natural tolerance of foreign organisms by hosts still have to be found and studied. This is the case for the mechanisms which allow a fœtus not to be destroyed by his mother or a parasite to be tolerated by its host (Malassagne et al unpublished data).

The complete sequencing of the pig histocompatibility genes which is in course should contribute to define lines of animals more easily tolerated by human hosts.

Whatever happens, it is clear that xenografting will not be the solution for cell or organ transplantation to humans but only an appropriate solution in some cases. It should be kept in mind that xenografting could be more and more in competition with bioartificial organs (Hinkeler 1999) artificial organs or human stem cells (Thomson and Odorico 2000).

References

Alexeev V., Yoon K. (1998) Stable and inheritable changes in genotype and phenotype of albino melanocytes induced by an RNA-DNA oligonucleotide. *Nature Biotech.* 16, 1343-1346.

Annas G.J., Caplan A., Elias S. (1999) Stem cell politics, ethics and medical progress. *Nature Med.* 5: 1339-1341.

Auchincloss H., Wood KJ (1998) Transplantation more questions than answers. *Cur. Opin. in Immun.* 10: 505-506.

Azizi S.A., Stokes D., Augelli B.J., DiGirolamo C., Prockop D.J. (1998) Engraftment and migration of human bone marrow stromal cells implanted in the brains of albino rats-similarities to astrocyte grafts. *Proc. Natl. Acad. Sci. USA* 95: 3908-3913.

Bach F.H., Ferran C., Soares M., Wrighton C.J., anrather J., Winkler H., Robson S.C. & Hancock W.W. (1997). Modification of vascular responses in xenotransplantation: Inflammation and apoptosis. *Nature Med.*, 9: 944-948.

Barlett R.J. (1998) Long-lasting gene repair *Nature Biotech.* 16, 1312-1313.

Bjornson C.R.R., Rietze R.L., Reynolds B.A., Magli M.C., Vescovi A.L. (1999) Turning brain into blood: a hematopoietic fate adopted by adult neural stem cells in vivo. *Science* 283: 234-537.

Bloom F.E. (1999) Breakthroughs 1999. *Science* 286: 2267.

Bosher J.M., Labouesse M. (2000) RNA interference: genetic wand and genetic watchdog. *Nature Cell Biol.* 2: E31-E36.

Bracy J.L., Sachs D.H., Lacomini J. (1998) Inhibition of xenoreactive natural antibody production by retroviral gene therapy. *Science* 281: 1845-1847.

Breun S., Salmons B., Günzburg W.H., Baumann J.G. (1999) Protection of MLV vector particles from human complement. *Biochem. Biophys. Res. Comm.* 264: 1-5.

Brown J., Matthews A.L., Sandstrom P.A., Chapman L.E. (1998) Xenotransplantation and the risk of retroviral zoonosis. *Trends Microbiol.* 6: 411-415.

Charreau B., Tesson L., Buscail J., Soulillou J-P, Anegon I. (1996). Analysis of human CD59 tissue expression directed by the CMV-IE-1 promoter in transgenic rats. *Trans. Res.* 5, 443-450.

Cohen-Tannoudji M., Robine S., Choulika A., Pinto D. El Marjou F., Babinet C., Louvard D., Jaisser F. (1998). I-SceI-Induced gene replacement at a natural locus in embryonic stem cells. *Mol. Cell. Biol.* 18, 1444-1448.

Collignon P. (1999) Transplants from pigs. *Science* 286: 1855

Cowen P., Goodman D., Aminian A., Brown A. Barlow H., Shinkel T., Kurek M. Francis D., Han W., Stewart A., Nottle M., Chen C.G., Salvaris E., Pearse M. and D'Apice A.J.F. (1999). Renal xenografts from triple transgenic pigs show prolonged survival in non-immunosuppressed baboons. *Transgenic Animal Research Conference Tahoe City USA.*

Cozzi E. & White D.J.G. (1995). The generation of transgenic pigs as potential organ donors for humans. *Nature Med.*, 9: 964-966.

Deschamps J., Chaillous L., Gouin E, Saï P. Acceptability of pig xenografts by patients with Type 1 diabetes mellitus and the general population. *Diabetes Care* (in press).

Dominko T., Mitalipova M., Haley B., Beyhan Z., Memili E., McKusick B., First N.L. (1999) Bovine oocyte cytoplasm supports development of embryos produced by nuclear transfer of somatic cell nuclei from various mammalian species. *Biol. Reprod.* 60: 1496-1502.

Fodor W. (1999) The development of transgenic livestock for biomedical purposes. *Transgenic Animal Research Conference Tahoe City USA.*

Fox J.L. (1998) FDA seeks "comfort factors" before removing hold on porcine xenotransplantation trials. *Nature Biotech.* 16: 224.

Fox J.L. (1999) Stem cell hearing stirs bioethics debate. *Nature Biotech.* 17: 11.

French A.J., Greenstein J.L., Loveland B.E., Mountford P.S. (1998) Current and future prospects for xenotransplantation. *Reprod. Fertil. Dev.* 10: 683-696.

Gearhart J. (1998) New potential for human embryonic stem cells. *Science* 282: 1061-1062.

Gibbons R.D., Meltzer D., Duan N. (2000) Waiting for organ transplantation. *Science* 287: 237-238.

Gussoni E., Soneoka Y., Strickland C.D., Buzney E.A., Khan M.K., Flint A.F., Kunkel L.M., Mulligan R.C. (1999) Dystrophin expression in the mdx mouse restored by stem cell transplantation. *Nature* 401: 390-394.

Gurdon J.B., Colman A. (1999) The future of cloning. *Nature* 402: 743-746.

Heath C.A. (2000) Cells for tissue engineering. *Trends Biotech.* 18: 17-19.

Hunkeler D. (1999) Bioartificial organs transplantated from research to reality. *Nature Biotech.* 17: 335-336.

Houdebine L.M. (1999) The ethical problems of transgenesis in pigs. Symposium on Bioethics of the European Council Oviedo (in press).

Hunkeler D. (1999) Bioartificial organs and acceptable risk. *Nature Biotech.* 17: 1045.

Jackson K.A., Mi T., Goodell M.A. (1999) Hematopoietic potential of stem cells isolated from murine skeletal muscle. *Proc. Natl. Acad. Sci.USA* 96: 14482-14486.

Lanza R.P., Cibelli J.B., West M.D. (1999) Human therapeutic cloning. *Nature Med.* 5: 975-977.

Lemischka I. (1999) The power of stem cells reconsidered? *Proc. Natl. Acad. Sci.USA* 96: 14193-14195.

Malassagne B, Taboit F, Conti F, Batteux F, Atia N, Chereau C, Conjeaud H., Theron MC, Attal J. Braet F, Houdebine LM, Calmus Y, Houssin D, Weill B 1998. A newly established porcine aortic endothelial cell line: characterization and application to the study of human to swine graft rejection. *Exp. Cell Res.* 238: 90-100.

Miyagawa S., Tanemura M., Koyota S., Koma M., Ikeda Y., Shirakura R., Taniguchi N. (1999) Masking and reduction of the galactose-α1,3-galactose (α-Gal) epitope, the major xenoantigen in swine, by the glycosyltransferase gene transfection. *Biochem. Biophy. Res. Commu.* 264: 611-614.

Ohdan H., Yang Y., Shimizu A., Swenson K.G., Sykes M. (1999) Mixed chimerism induced without lethal conditioning prevents T cell- and anti-Galα1,3Gal-mediated graft rejection. *J. Clin. Invest.* 104: 281-290.

Osman N., McKenzie I.F.C., Ostenried K., Ioannou Y.A., Desnick R.J., Sandrin M.S. (1997) Combined transgenic expression of α-galactosidase and α1,2-fucosyltransferase leads to optimal reduction in the major xenoepitope Galα(1,3)Gal. *Proc. Natl. Acad. Sci.USA* 94: 14677-14682.

Pati S. 1998. Genetically engineering and cloning animals, Parl city Deer Valley, USA, June 21-23.

Paradis K., Langford G. (1999) Response. *Science* 286: 1856-1857

Paradis K., Langford G., Long Z., Heneine W., Sandstrom P., Switzer W.M., Chapman L.E., Lockey C., Onions D. (1999) Search for cross species transmission of porcine endogenous retrovirus in patients treated with living pig tissue. *Science* 285: 1236-1241.

Patience C., Takeushi Y. & Wiess R.A., 1997. Infection of Human Cells by an Endogenous Retrovirus of Pigs, *Nature Med.* vol. 3, 3: 282-286.

Persids A. (1999) Xenotransplantation. *Nature Biotech.* 17, 205-206.

Platt J.L. (1998) New directions for organ transplantation. *Nature* 392: 11-17.

Platt J.L. (1999) Tolerance by transplantation: how much is enough, how much is too much? *Journal Clin. Investi.* 104: 227-228.

Schmoeckel M., Bhatti F.N., Zaidi A., Cozzi E., Waterworth P.D., Tolan M.J., Goddard M., Warner R.G., Langford G.A., Dunning J.J., Wailwork J., White D.J. (1998) Orthotopic heart transplantation in a transgenic pig-to-primate model. *Transplantation* 65, 1570-1577.

Schniecke A., Kind A.J., Ritchie W.A., Mycok K., Scott A.R., Ritchie M., Wilmut I., Colman A. and Campbell K.H.S. (1997). Human factor IX transgenic sheep produced by transfer of nuclei from transfected fetal fibroblasts. *Science* 278: 2130-2133.

Shamblott M.J., Axelman J., Wang S., Bugg E.M., Littlefield J.W., Donovan P.J., Blumenthal P.D., Huggins G.R., Gearhart J.D. (1998) Derivation of pluripotent stem cells from cultured human primordial germ cells. *Proc. Natl. Acad. Sci. USA* 95: 13726-13731.

Sharp P.A. (1999). RNAi and double-strand RNA. *Genes & Develop.* 13, 139-141.

Solter D., Gearhart J. (1999) Putting stem cells to work. *Science* 283: 1468-1470.

Taboit-Dameron F., Malassagne B., Viglietta C., Puissant C., Leroux-Coyau M., Chereau C., Attal J., Weill B., Houdebine L-M.. (1999). Association of the 5'HS4 sequence of the chicken ß-globin locus control region with human EF1α gene promoter induces ubiquitous and high expression of human DAF and CD59 cDNAs in transgenic rabbits. *Trans. Res.* 8, 223-236.

Thomson J.A., Itskovitz-Eldor J., Shapiro S.S., Waknitz M.A., Swiergiel J.J., Marshall V.S., Jones J.M. (1998) Embryonic stem cell lines derived from human blastocyts. *Science* 282, 1145-1147.

Thomson J.A., Odorico J.S. (2000) Human embryonic stem cell and embryonic germ cell lines. *Trends Biotech.* 18: 53-57.

Vasquez K.M., Wilson J.H. (1998) Triplex directed modification of genes and gene activity. *Science* 4-9.

Viville S. 1997. Mouse genetic manipulation via homologous recombination. In transgenic Animal Generation and Use Edited by L.M. Houdebine, Harwood Academic Publishers Amsterdam, pp 307-323.

Vogel G. (1999) NIH sets rules for funding embryonic stem cell research. *Science* 286: 2050-2051.

Wakayama T., Rodriguez I., Perry A.C.F., Yanagimachi R., Mombaerts P. (1999) Mice cloned from embryonic stem cells. *Proc. Natl. Acad. Sci.USA* 96: 14984-14989.

Weiss R.A. (1999) Xenografts and retroviruses. *Science* 285: 1221-1222.

Weiss S. (1999) Pathways for neural stem cell biology and repair. *Nature Biotech.* 17: 850-851.

Wianny F., Zernicka-Goetz M. (2000) Specific interference with gene function by double stranded RNA in early mouse development. *Nature Cell Biol.* **2**: 70-75.

Wilmut I., Schnieke A.E., Mc Whir J., Kind A.J. Campbell K.H.S. (1997). Viable offspring derived from fetal and adult mammalian cells. *Nature* 385, 810-813.

You S., Gouin E., Saï P. (1998) Spleen cells of non-obese diabetic mice fed with pig splenocytes display modified proliferation and reduced aggressiveness in vitro against pig islet cells. *Diabetologia* 41: 955-962.

Zigova T., Sanberg P.R. (1998) The rising star of neural stem cell research. *Nature Biotech.* 16: 1007-1008.

REINFORCED BIOARTIFICIAL SKIN IN THE FORM OF COLLAGEN SPONGE AND THREADS

EUN KYUNG YANG, YOUNG KWON SEO AND JUNG KEUG PARK
Department of Chemical Engineering, Dongguk University, Seoul 100-715, Korea fax 82 2 2271 3489 mailto jkpark@dgu.ac.kr

Abstract

Bioartificial skin requires high mechanical strength-scaffold to overcome the problem of easily being torn during handling and suturing. In addition, the scaffold should not even have potential toxicity to cell culture and host tissue. Therefore, in this study, we suggest how to make a stronger sponge without using any cross-linking treatments, which will show the latent cytotoxicity on the implant material. We made sponge type of bioartificial skin using its three major components, collagen, dermal fibroblasts, and epidermal keratinocytes. The sponge-type collagen scaffold for skin cell culture was prepared by freeze-drying of 7.5 mg/ml of type I rat tail collagen solution. We reinforced collagen sponge by incorporation of collagen mesh which was made by stacking five layers of 9 ×9 collagen threads. By this method we could increase tensile strength by as much as three times higher compare to the collagen sponge cross-linked using glutaraldehyde. The ultimate tensile strength of collagen sponges reinforced with collagen threads, uncross-linked collagen sponges, and collagen sponges cross-linked by glutaraldehyde were 0.086, 0.0038, and 0.029 MPa respectively.

1. Introduction

The engineering of skin tissue has been studied from a variety of approaches. Bell *et al.* devised a bilayered model of skin using a contracted collagen gel lattice containing living dermal fibroblasts which was then overlaid with epidermal cells, forming an organotypic model of skin as cited in [1]. However, the severe contraction of this skin model, occurring very early in culture, restricted its use for wound coverage in vivo. In deed, living fibroblasts contract a floating collagen gel matrix, in vitro, at a rate that depends on the number of cells, collagen population seeded in the gel, as is the case in

A. Van Broekhoven et al. (eds.), Novel Frontiers in the Production of Compounds for Biomedical Use, 365–380.
© 2001 *Kluwer Academic Publishers. Printed in the Netherlands.*

wounds in vivo. We also have examined the contraction behaviour of human dermal equivalent in the form of collagen gel as cited [2].

Another types different from gel type of bioartificial skin have been investigated as various types of freeze-dried sponges. Yannas *et al.* [3,4] developed the sponge type of collagen-chondroitin sulphate material. The collagen sponges would appear to be useful supports for various cell types used in tissue engineering as cited in [5,6]. However, biomaterials made of collagen generally stimulate a mild inflammatory response, which results in degradation of the implant by host enzymes. In order to control the rate of biodegradation of collagen and to improve its tensile properties, physical and chemical cross-linking techniques are used.

The method and extent of cross-linking profoundly influence the strength, biocompatibility, resorption rate, and antigenicity of collagen-based materials [7,8]. Physical and chemical methods for the treatment of collagenous tissue are available. The predominant chemical agents that have been investigated for the cross-linking treatment of collagenous tissue are glutaraldehyde [9], carbodiimides [10], hexamethylene diisocyanate [11], ascorbic acid [12], and acyl azide [13].

Glutaraldehyde is now most widely used reagent, however, as glutaraldehyde cross-linked biomaterials induce local cytotoxicity [14], other physical treatments have been evaluated. The primary advantage of physical methods is that they do not introduce chemicals that may cause potential harm. Typical physical modification with ultraviolet (UV) irradiation of collagen materials has been carried out to improve their physical, chemical and biological properties [15,16]. The limitation of physical methods appears to be in trying to achieve a balance between attaining the desired amount of cross-linking but preventing the onset of degradation of tissue scaffold as a consequence of the long exposure times inevitable for physical processes.

Bioartificial skin requires high mechanical strength-scaffold to overcome the problem of easily being torn during handling and suturing. In addition, the scaffold should not even have potential toxicity to cell culture and host tissue. Therefore, in this study, we suggest how to make a stronger sponge without using any chemical cross-linking treatments, which will show the latent cytotoxicity on the implant material. In this work, we made collagen mesh and incorporated it into collagen sponge in order to improve both tensile strength and biocompatibility. To determine the reinforcement efficiency, we compared the strength of sponges reinforced by collagen mesh to cross-linked sponges by glutaraldehyde, hexamethylene diisocyanate (HMDI), ascorbic acid, UV irradiation, and dehydrothermal treatment (DHT). Also we cultured skin fibroblasts in our collagen scaffold and examined the internal morphology and strength. Finally we made the bioartificial skin using its three major components, collagen, dermal fibroblasts, and epidermal keratinocytes. The collagen based dermal equivalent provides a three-dimensional framework for the fibroblasts and provides biological support both for the fibroblasts and the overlying keratinocytes. We used reinforced collagen sponge incorporated with threads as skin substitute, and cultured the reinforced living skin substitute.

2. Materials and Methods

2.1 COLLAGEN SCAFFOLDS

2.1.1 Extraction of Type I Collagen Solution

Type I collagen was extracted from rat-tail tendons according to the procedure described by Yannas *et al.* [3]. Tendons were manually dissected out from unskinned tails that were stored at –20°C. Acid-extractable collagen was dissolved from the solid material by stirring in 0.05 M acetic acid at 4°C for 72 h. The crude collagen solution was separated from the insoluble residue by centrifugation at 12,000 × g (10,000 rpm) for 1 h in a centrifuge (HA-500, Hanil Industrial Co., Korea). Type I collagen was purified by repeated precipitation in 5% NaCl and redissolution in 0.05 M acetic acid as follows. The pellet was discarded and crystalline NaCl was added to the supernatant to a concentration of 5 wt. The resulting precipitate was centrifuged at 2,000 × g over 10 min and the supernatant was discarded while the pellet was redissolved in 0.05 M acetic acid. The precipitation-centrifugation-redissolution cycle was repeated twice more. The collagen solution was dialysed at 4°C in dialysis tubing, MW cut-off 12,400 (D-9402, Sigma Co. U.S.A.) against 0.02 M Na_2HPO_4 over 48 h and was then centrifuged at 2,000 × g over 10 min. The supernatant was discarded while the pellet was lyophilised and stored.

2.1.2 Fabrication of Macroporous Collagen Sponge

The process of collagen sponge production is shown in Figure 1. A type I collagen solids was dissolved in 0.05% acetic acid at a concentration of 7.5 mg/ml. The resulting collagen solution was placed in a freezer (MDF-U2086S, Sanyo, Japan) at -80°C for 16 h, after which it was lyophilised by a freeze dryer (SFDSM06, Samwon Freezing Engineering Co., Korea) at -80°C for 24 h to obtain a sponge matrix as cited in [17]. In order to preserve the triple helical structure of collagen it is essential to freeze-dry the specimens prior to heating at 105°C [4]. We placed collagen sponges in a vacuum oven (Isotemp Model 282A, Fisher Scientific, Pennsylvania, USA.) and subjected to a vacuum of 1.3 in. of Hg at 105°C for 24 h. The sponges each were treated by various cross-linking protocols. In addition, these cross-linked materials were rinsed by 70% ethanol for 18 h and then stored in 70% ethanol.

To provide a nonporous surface as a substrate for culture epidermal cells, a thin film of collagen was laminated to one surface of the dermal substitute as cited in [18]. Films were prepared from collagen (0.3%) containing 3% dimethylsulphoxide (DMSO). The solution was distributed (50 $\mu l/cm^2$) onto the dish surface. The one side of prepared collagen sponge was touched to the solution, and wet collagen sponge was then frozen and lyophilised overnight.

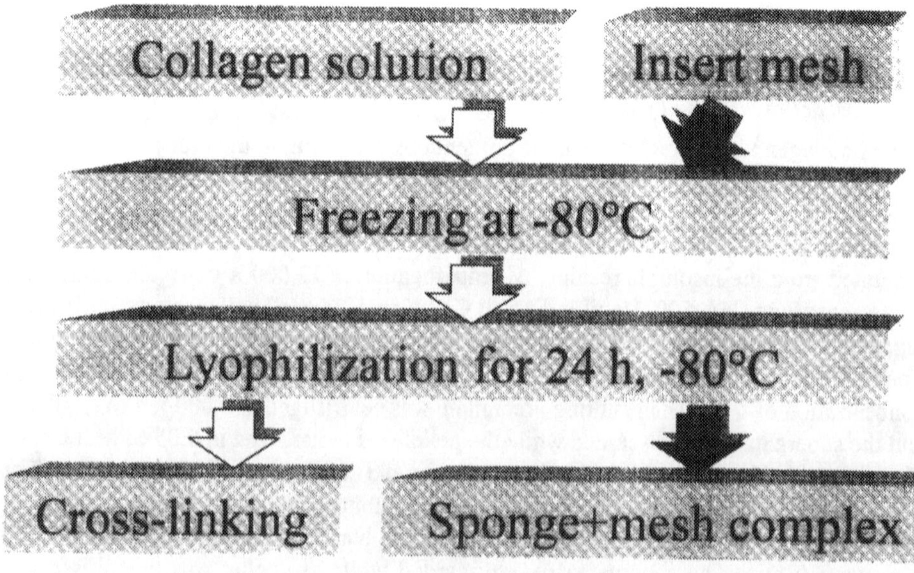

Figure 1. Diagram of the process of collagen sponge production. The sponge was made by freezing and lyophilisation of 7.5 mg/ml of type I collagen.

2.1.3 Fabrication of Collagen Threads and Mesh

Collagen threads were made from type I collagen according to a modification of the method described by Cavallaro and Kemp [19]. Our thread-making process is shown schematically in Figure 2. A 5.0 mg/ml collagen solution in 0.05 M acetic acid was loaded in a peristaltic pump (Haake Bucher instrument Inc., No. 426-2000, U.S.A.) set to infuse at 2.0 ml/min. Silicone tubing (2.0 mm I.D.) connected to an 20-gauge blunt stainless-steel needle immersed in coagulation solution containing 20% polyethylene glycol (PEG), MW 8000 (Duksan Co., Korea) in 94 mM sodium phosphate dibasic and 24 mM sodium phosphate monobasic at pH 7.55. The needle tip was moved along at 4 cm/s for thread formation. The collagen gelled on contact with the neutral pH solution, and the nascent thread began to dehydrate due to the osmotic pressure gradient formed between the collagen and the PEG solution. The coagulation had residence time of thread in the bath approximately 4 min. As the thread accumulated, it was transferred to rinse solution bath filled with 5.5 mM sodium phosphate dibasic, 0.5 mM potassium phosphate monobasic, and 75 mM NaCl, at pH 7.10; and remained there for 5 to 10 min. The threads were then partially dehydrated in isopropanol at room temperature for 16 h, and dried under tension by their own weight at 40°C for 30 min inside an incubator heated with air blowers. We fabricated one layer of the mesh by joining cross nine threads to the other nine threads. Then we made the mesh by stacking five layers of the 9 × 9 threads before they have been dried out.

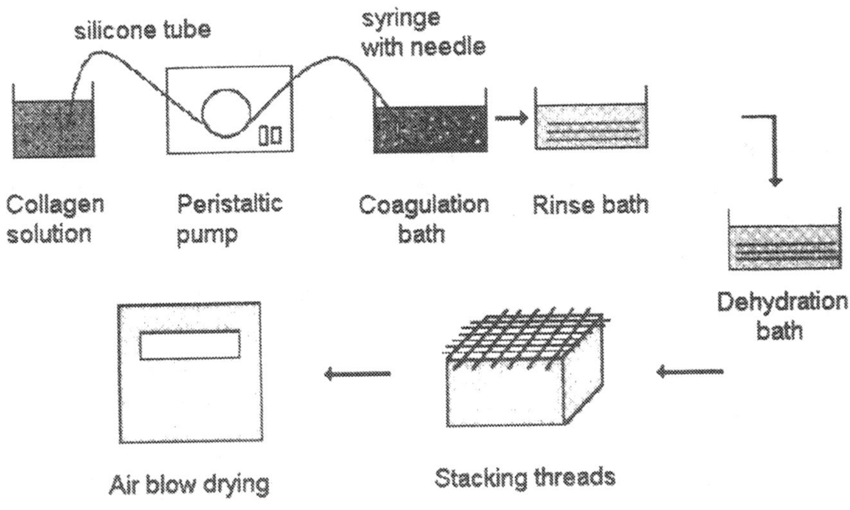

Figure 2. Schematic representation of the process of collagen thread-making. Collagen was extruded into a coagulation bath containing 20% polyethylene glycol 8000 in phosphate buffer at pH 7.55. The thread was then rinsed in a dilute phosphate-buffered saline, partially dehydrated in 70% isopropanol, and dried under tension by its own weight at 40 ℃ for 30 min inside a incubator heated with air blowers.

2.1.4 Reinforcement by Cross-linking Treatments

We examined three chemical cross-linking treatments as follows. Collagen sponges were cross-linked in 0.25% glutaraldehyde for 24 h, washed exhaustively with running tap water for 24 h, and were rinsed three times with distilled water to remove unreacted aldehyde. Other collagen sponges were cross-linked in 1% of hexamethylene diisocyanate (HMDI, Aldrich Co.) by methanol for 2 h at room temperature. We also applied ascorbate-copper ion system to cross-link collagen sponges. The sponges were immersed in 10 μM of cupric chloride in the 100 ml of 0.05 M sodium phosphate buffer (pH 6.0), and they were incubated at 37°C (water bath) for 5 min and then 1 ml of 1 μM ascorbate solution was added.

Two physical modifications of collagen sponge were examined as follows. Collagen sponges irradiated with a 254 nm of UV lamp for 8 h or 24 h at a distance of 20 cm in 4 °C chamber. For dehydrothermal treatment (DHT) cross-linking, collagen sponges were placed in a vacuum oven at room temperature and subjected to a vacuum of 1.3 in. of Hg for 1 h. The temperature within the vacuum oven was then increased to 110°C for 72 h.

2.1.5 Reinforcement by Incorporating with Collagen Mesh

To reinforce the collagen scaffold by biocompatible method, collagen mesh was incorporated in the sponges as shown in Figure 3. The complex of sponge and mesh was made by immersing the mesh into homogenised collagen solution before freezing to form sponge structure. At this time, the collagen sponge was not treated by any method.

2.1.6 Measurement of Mechanical Strength

We examined the mechanical property of collagen sponges and ultimate tensile strength was measured in the state of wet. The sponges in the form of rectangular strips were tested using a texture analyser (TA-XT2 stable microsystem, UK) and its X-TRAD program. The gage length (starting sample length) was set to 10 mm and specimen width and thickness used were 5.0 mm and 5.0 mm respectively. The crosshead speed was set at 0.1 mm/s. Breaking loads were measured directly from the force-displacement graph. Ultimate tensile strength (UTS) was determined by dividing breaking load by the cross sectional area.

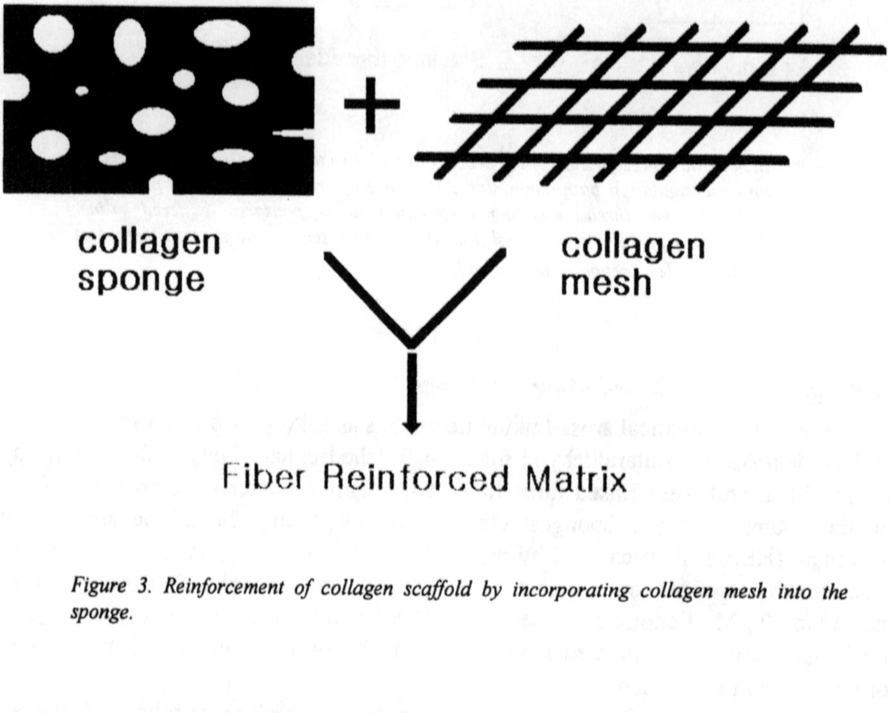

Figure 3. Reinforcement of collagen scaffold by incorporating collagen mesh into the sponge.

2.1.7 Morphological Analysis

We fixed the 7 d-cultured sponge type dermal substitute using 10% neutral buffered formalin for 2 h at 4 °C. Fixed samples were embedded in paraffin, and the 5 μm-sectioned paraffin ribbons were stained with haematoxylin and eosin.

For scanning electron microscopy (SEM), specimens were fixed for 2 h in a mixture of 4% glutaraldehyde and 2% formaldehyde in a phosphate buffer (a mixture of monosodium phosphate and dipotassium phosphate, 0.2 M, pH 7.4) at room temperature. After thorough rinsing with 0.175 M phosphate buffer, they were immersed in 2% osmium tetroxide buffered with 0.2 M phosphate for 2 h. The samples were dehydrated in acetone, subjected to critical point drying (E-3100, Bio-Rad, UK), sputter-coated with gold-palladium (E-5400, Bio-Rad, UK), and observed under a scanning electron microscope (JSM-840A, Jeol Inc., Japan) at 15 kV.

2.2 BIOARTIFICIAL SKIN

2.2.1 Primary Cell Culture

Human normal skin cells were aseptically isolated from a circumcised neonatal foreskin at the Cha General Hospital (Bundang, Korea) as cited in [20]. The epidermis and dermis came loose by incubation in 0.9 units/ml dispase (Gibco BRL, Grand Islands, N.Y., U.S.A.) in culture medium without serum for 16 h at 4°C. After two layers stripped off mechanically and the normal fibroblasts were isolated from that dermis by 2 mg/ml type I collagenase (Sigma Chemical, U.S.A.). Fibroblasts were cultured routinely in Dulbecco's modified Eagle medium (DMEM) supplemented with 10% foetal bovine serum (FBS) at 37 °C, 5% CO_2 incubator. Keratinocytes were isolated from that epidermis by 0.25% trypsin solution. The cells were grown in Keratinocyte-Serum Free Medium (Keratinocyte-SFM, Gibco BRL) consisting of MCDB 153 medium supplemented with insulin (0.005 g/l), hydrocortisone (0.074 mg/l), triiodothyronine (0.0067 mg/l), bovine pituitary extract (50 mg/l) and epidermal growth factor (EGF, 5 μg/l) at 37°C, 5% CO_2 incubator.

2.2.2 Culture of Bioartificial Skin

One millimetre of a harvested suspension containing 2×10^5 cells/cm^2 was absorbed into a collagen sponge with a diameter of 35 mm and thickness of 3 mm. The sponge was left for 24 h at 37°C, overlaid with 2 ml of DMEM containing 10% FBS. The 6 d-cultured sponge type dermal substitutes were fixed with 10% neutral buffered formalin (Sigma Chemical) and processed according to standard techniques for histological observations. The sections were stained by haematoxylin and eosin.

To count the viable cells in dermal equivalent, the cultured gels were digested by treatment for 2 h at 37°C with 3 mg/ml collagenase (type I, 1.6 units/mg solid, Sigma Chemical). The collagenase was dissolved in calcium acetate buffer consisting of 130 mM NaCl, 10 mM Ca acetate, and 20 mM Hepes at pH 7.2. Single cell preparations were obtained by incubating the samples for an additional 20 min at 37°C with 0.25% trypsin solution. Aliquots of the samples were mixed with trypan blue and the viable cell number was measured with a haemocytometer.

To assay cell viability, the culture medium was aspirated from the cultures and 1.0 ml PBS (phosphate buffered saline) containing 0.33 mg MTT/ml (3-[4,5-dimethylthiazole-2-yl]-2,5-diphenyltetrazolium bromide, Sigma Chemical Co.) was dispensed into each well. Cultures were then incubated for 4 h, and then the PBS solution was aspirated, and monolayers were extracted in 1.0 ml isopropanol (Oriental Chemical Co., Seoul, Korea), acidified with 0.04 N HCl (Junsei Chemical Co., Tokyo,

Japan) for 5 min at room temperature. Absorbencies of 3.0 ml aliquots were read at 570 nm in an UV-VIS spectrophotometer (Smart plus 2605, Young Hwa Co., Seoul, Korea). To production of a bioartificial skin, the dermal equivalent was transferred onto the 3 μm porous polycarbonate membrane of the culture insert (Nunc TC Insert, Nunc, Denmark). After the culture of dermal equivalent for 7 days, the culture medium was aspirated to leave the gel surface exposed; epidermal keratinocytes were applied to the surface at initial density of 2×10^5 cells/cm^2. The construct was submerged under keratinocyte culture medium inside and outside of the culture insert to allow these keratinocytes to spread and cover the surface of the dermal equivalent, which is accomplished in about 7 days of incubation. The developing multilayered bioartificial skin was then cultured at the air-liquid interface by removing the inside medium of the insert, and the underlying culture medium was switched to the keratinocyte culture medium containing 10% FBS. Since the nutrients and growth factors are diffused to the keratinocytes layer through the prepared dermal equivalent, the in vitro bioartificial skin is grown in the environment that is similar to in vivo epithelial tissue development. It was designed to enhance maturation and cornification of the stratified epidermal layer and support long-term maintenance of the fibroblasts [21].

2.2.3 Morphology Examination

Dermal equivalent samples for microscopic analysis were fixed for 2 h in a mixture of 4% glutaraldehyde and 2% formaldehyde in a phosphate buffer (a mixture of monosodium phosphate and dipotassium phosphate, 0.2 M, pH 7.4) at room temperature. After through rinsing with 0.175 M phosphate buffer, they were immersed in 2% osmium tetroxide buffered with 0.2 M phosphate for 2 h. For transmission electron microscopy (TEM), dehydrated specimens were embedded in Epon 812. Thin sections were observed and photographed with a Jeol JEM-1200EXII electron microscope.

We fixed the bioartificial skins using 10% neutral buffered formalin (Sigma Chemical) for 2 h at 4 °C. Fixed samples were embedded in paraffin, and the 5 μm-sectioned paraffin ribbons were stained with haematoxylin and eosin. The sections after implantation were examined with Masson's trichrome stain also.

3. Results and Discussion

3.1 COLLAGEN SCAFFOLDS

3.1.1 Preparation of Collagen Threads

Figure 4(a) shows a SEM (×1500) of a collagen thread, Figure 4(b) shows a light microphotograph (×40) of mesh made of the collagen threads. Diameters of collagen threads were measured microscopically (×40) using a measuring eyepiece, averaging the readings taken on more than one hundred threads samples.

Distribution of threads diameter is shown in Figure 5, and the average diameter was about 50 μm. Figure 6 shows mechanical strength of collagen threads produced by the method described. The results are similar to published tensile strength of Cavallaro *et al.* [19]. Both UV irradiation and glutaraldehyde cross-linked threads have dry ultimate tensile strength (UTS) of around 220 MPa, while uncross-linked threads have dry UTS of around 340 MPa as shown in Figure 6(a). The production of living skin substitute requires thread that is not only strong enough, but also flexible enough without breaking. To examine the wet UTS, the collagen threads were immersed in phosphate buffered saline (PBS) for 10 d before UTS measurement. In the hydrated wet state, UTS of collagen threads treated with glutaraldehyde was about 49 MPa that is larger value than that of uncross-linked wet thread, about 20 MPa as shown in Figure 6(b). In the dry state, collagen threads have high tensile strength up to about 340 MPa. The large decreases in these properties in the hydrated wet state suggest that water molecules act to break down hydrogen and electrostatic bonds that hold collagen fibrils together [22]. In addition, the UTS of uncross-linked collagen threads in the wet state was less than 20 MPa, suggesting that hydrogen and electrostatic bonding between molecules plays a critical role in the load-bearing capacity of this material. The role of cross-linking appears to be that of minimising the distances between neighbouring molecules. During the cell culture period, threads and mesh in the wet state should be used and required for about four weeks. Therefore the measurement of mechanical strength of collagen scaffold in the wet state necessary to compare and study in the various modifications and cross-link treatments.

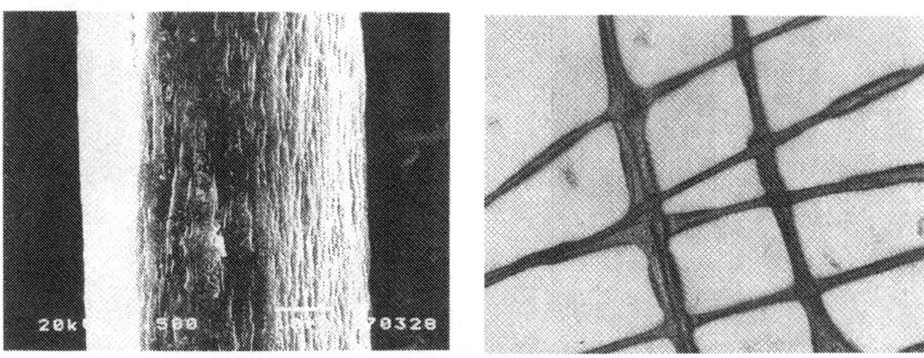

Figure 4. Collagen threads and mesh produced by the method described. (a) Scanning electron micrograph of a collagen thread made from type I calf skin collagen solution (5.0 mg/ml). Magnification × 1500. Bar = 10 μm. (b) Light micrograph of collagen mesh made of the collagen threads. Magnification × 40.

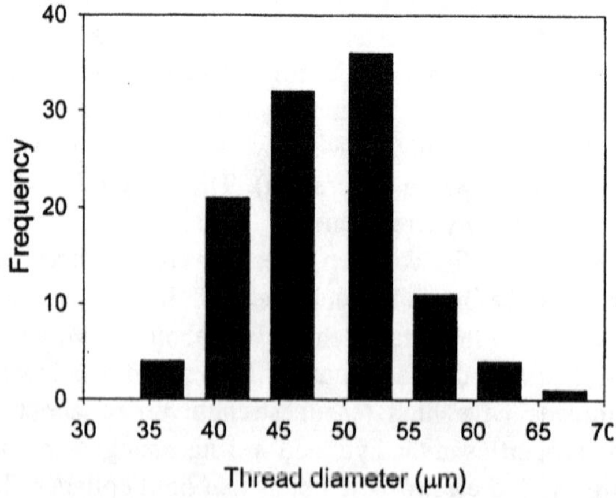

Figure 5. Histogram of diameter distribution of collagen threads. Average diameter is about 50 μm, which were measured microscopically using a measuring eyepiece, averaging the readings taken on more than one hundred threads.

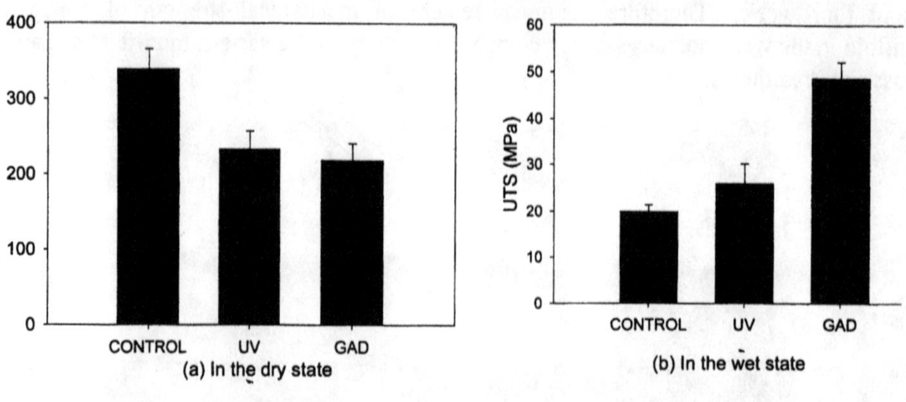

Figure 6. Ultimate tensile strength (UTS) of collagen threads. The threads were cross-linked by 254 nm ultraviolet irradiation (UV) or 2.5% glutaraldehyde treatment (GAD). These error bars show mean standard deviations for five determinations. (a) is the UTS in the dry state and (b) shows UTS in the hydrated wet state in which the threads were immersed in PBS solution for 10 days before UTS measurement.

3.1.2 Ultimate Tensile Strength of Collagen Sponges

The mechanical strength of sponges in the wet state by immersing in culture medium during ten days after cross-linking treatment show contrary results as shown in Figure 7. To prolong the mechanical integrity of the wet collagen sponge, we used chemical and physical treatment for cross-linking. For chemically modified collagenous tissues, physical treatment for cross-linking. For chemically modified collagenous tissues, following three procedures were applied. Glutaraldehyde is able to interact chemically with amino groups on collagen to form bifunctional cross-links as cited in [23].

Glutaraldehyde treated sponges were become rigid to 0.0072 MPa from 0.0037 MPa of control in the wet state. HMDI is another bifunctional molecule where the terminal isocyanate groups can react with amines of lysine on collagen to form the urea bond [8]. UTS of wet sponge after cross-linked with HMDI were measured about 0.0039 MPa. Other chemical treatment, ascorbate-copper ion system was executed to cross-link collagen sponge structure. It was indicated the average strength of 0.0045 MPa as shown in Figure 7. Chemical cross-linking agents such as glutaraldehyde, hexamethylene diisocyanate, or ascorbate-copper provide high strength, but may result in poor biocompatibility and protracted resorption. Therefore we tried physical cross-linking protocols as UV irradiation and dehydrothermal treatment. It was reported previously that UV cross-linked fibres may retain more native structure and should exhibit greater resistance to non-specific proteases in vivo. Formation of cross-links during UV irradiation (254 nm) is thought to be initiated by free radical formation on aromatic amino acid residues such as tyrosine and phenylalanine. Bonding between these radicals then forms cross-links [15].

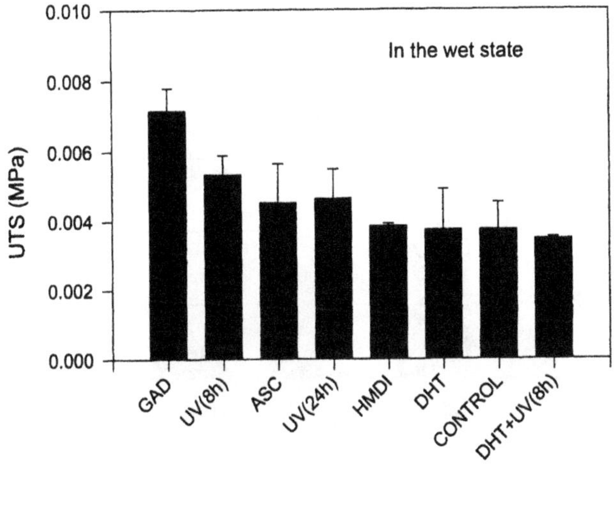

Figure 7. UTS of collagen sponges wetted during 10 days after modify with various cross-linking methods. GAD (glutaraldehyde), ASC (ascorbate-copper), UV (254 nm, ultraviolet irradiation), HMDI (hexamethylene diisocyanate), DHT (dehydrothermal treatment), control (uncross-linked sponges). These error bars indicate mean standard deviations (n = 5, p < 0.005).

The strength of UV cross-linked sponges indicates about 0.0053 MPa for 8 h irradiation, and although we increase the irradiation time to 24 h the strength didn't increase to about 0.0047 MPa as shown in Figure 7. Using higher irradiation times is restricted cross-linking collagen fibres that avoids potentially cytotoxic reaction products and provides moderate strength and resorption rate. However, this treatment requires 3 ~ 5 days to complete and results in partial degradation of the collagen. Formation of cross-links during DHT treatment is dependent on the exhaustive removal of bound water from collagen molecules. The removal of water results in condensation reactions between the carboxyl and amino groups on adjacent amino acid side chains. The process is time-consuming and may be incomplete even after 3 days. Our results of DHT at 110°C for 72 h indicate about 0.0038 MPa, and so they have showed no effect on strengthening. These results are shown in Figure 7 that these error bars indicate mean standard deviations for more than five determinations. *t*-Tests were performed comparing each sponge with its control, and those probabilities results significant at the $p < 0.005$ level are indicated on the Figure 7. The primary advantage of physical method is that they do not introduce chemicals that may cause potential harm. The limitation of physical methods appears to be in trying to achieve a balance between attaining the desired amount of cross-linking, and also preventing the onset of degradation of material as a consequence of the long exposure time inevitable for physical process [8].

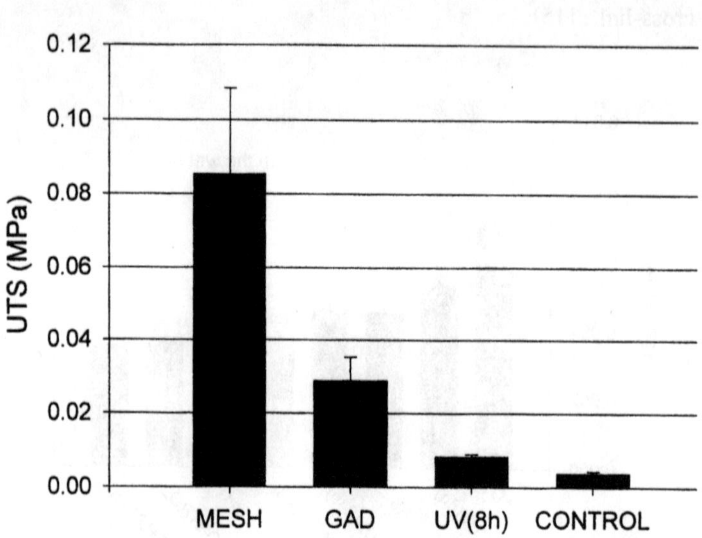

Figure 8. Comparison of reinforcement effects on ultimate tensile strength (UTS) of collagen sponges in the wet state immersed in serum containing culture medium during 3 days after modify with cross-linking. MESH indicates uncross-linked collagen sponge incorporated with collagen mesh. GAD indicates collagen sponge cross-linked by chemical treatment using glutaraldehyde, and UV(8h) indicates collagen sponge cross-linked by physical treatment with UV irradiation for 8 h. CONTROL indicates uncross-linked sponge without mesh. These error bars show mean standard deviations (n > 10, p < 0.005).

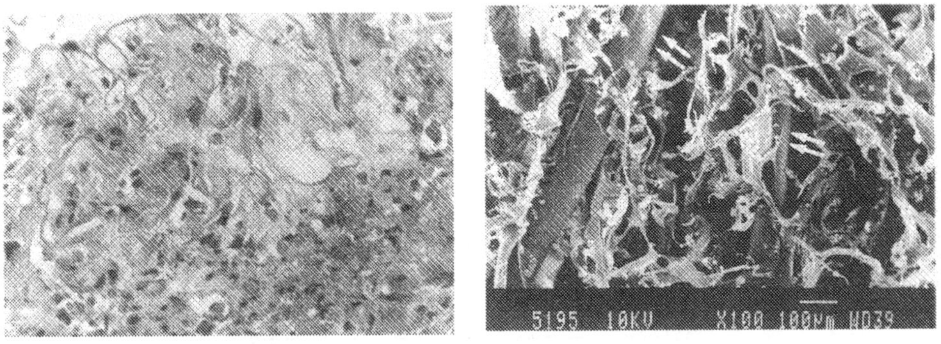

Figure 9. Micrograph of skin fibroblasts cultured in collagen sponge reinforced by incorporating with collagen mesh. (a) Light micrograph of the cultured dermal tissue stained with haematoxylin and eosin. Magnification × 200. (b) SEM of sponge structure and internal distribution of threads and cells. Arrows indicate collagen threads. Magnification, × 100.

We also examined the strength of uncross-linked collagen sponge incorporated with collagen mesh. All the sponges of Figure 8 were measured after wetted in serum containing culture medium during 72 h. According to the result of UTS measurement, sponges incorporated with mesh show about 0.086 MPa, and that is three times as stronger as glutaraldehyde treated sponges of 0.029 MPa. In Figure 8, uncross-linked control collagen sponges showed about 0.0038 MPa and UV irradiated sponges showed about 0.0083 MPa. In this procedure, we could reinforce collagen sponge without using cross-linking treatments. Reinforcement by threads and mesh had an effect on the implant material without latent cytotoxicity through chemical cross-linking procedure. In addition to the biocompatibility, we will be able to control the degree of mechanical tensile strength efficiency through changing the diameter of thread, a quantity of threads, thickness of mesh, and different way of weaving threads into mesh. By this procedure, it will be possible to control the tensile strength of the matrix efficiently.

3.2 BIOARTIFICIAL SKIN

To grow a three-dimensional dermal tissue, rat dermal fibroblasts were seeded into the reinforced collagen sponge incorporated with collagen mesh. These cells adhere rapidly to the sponge and mesh scaffold and attach completely in less than 24 h. Unlike monolayer cultures, where contact inhibition limits cell growth, the porous structure provides an environment for cell growth and extracellular matrix deposition to occur in a three-dimensional array. Histological examination (× 200) in Figure 9(a) shows that cultured skin fibroblasts and tissue associated with porous scaffold of collagen sponge. Figure 9(b) shows SEM (× 100) of sponge structure and internal distribution of mesh and fibroblasts. Homogeneous distribution of cells into pores and the surface of collagen sponge and threads can be seen clearly. Arrows indicate collagen threads that were incorporated into the sponge.

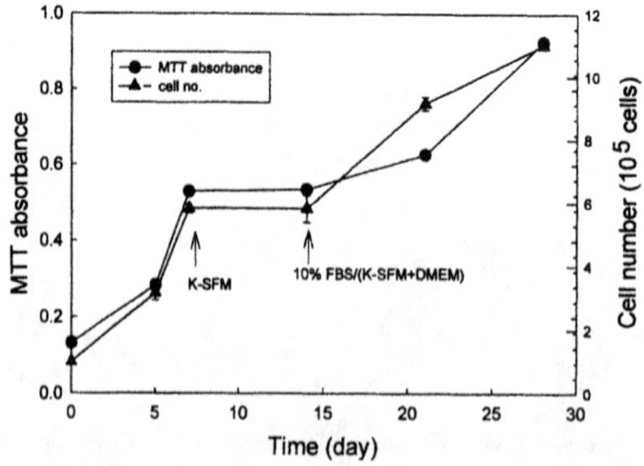

Figure 10. The cell growth curve of dermal fibroblasts is shown. 1×10^5 cells/ml were inoculated into a collagen sponge and were cultured in DMEM containing with 10% FBS. At 7th day, the medium was changed to Keratinocyte-SFM (K-SFM) from DMEM containing 10% FBS. At 14th day, the medium was changed to K-SFM plus DMEM containing 10% FBS. Cell counting and MTT assay was tested by the method described.

The fibroblasts growth in the collagen scaffold is shown in Figure 10. At 7th day, the medium was changed to Keratinocyte-SFM (K-SFM) from DMEM containing 10% serum. At 14th day, the medium was changed to K-SFM plus DMEM containing 10% serum.

Lamination of a smooth, continuous surface to the porous collagen sponge prevents passive invasion of the dermal sponge's interior by cultured keratinocytes. The scaffold surfaces of dermal collagen sponge are shown in Figure 11 with (a) or without (b) lamination. The lamination is very thin and does not affect the interior porosity of the sponge for dermal substitute. Inoculation and incubation of cultured epidermal keratinocytes results in complete coverage of the laminated surface with cells as shown in Figure 12.

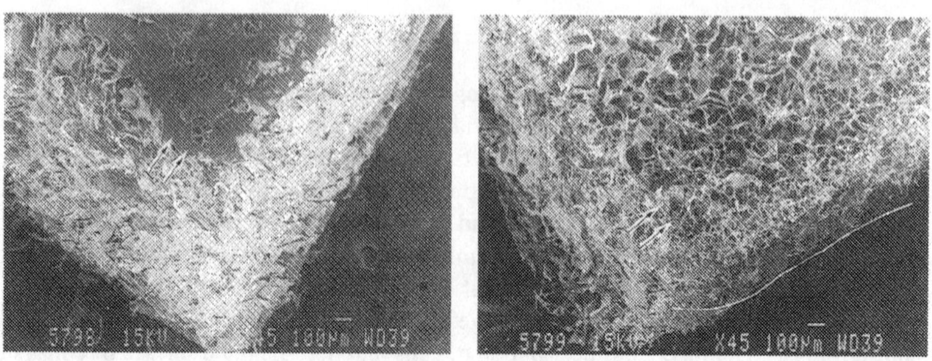

Figure 11. The scaffold surfaces of dermal collagen sponge are shown with (a) and without (b) production of laminated membrane. Magnification, × 45, bar = 100 μm.

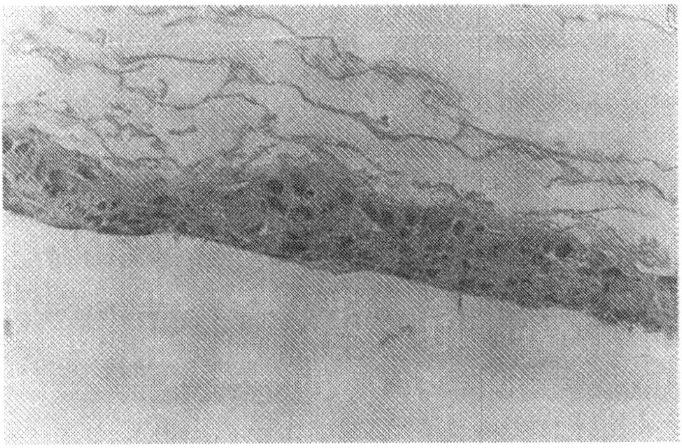

Figure 12. The morphology of bioartificial skin composed of keratinocytes and collagen scaffold containing fibroblasts stained with haematoxylin and eosin.

4. Conclusion and Future work

Bioartificial skin consisting of primary cells could be three-dimensionally differentiated to full-thickness skin tissue. The attached gel culture had a jumping effect of growth factor on cell growth at the point of lag phase. Our results indicate that the method of dermal matrix organisation influenced not only cell growth but also cell responsiveness. To develop the bioartificial skin 2×10^5 cells/cm^2 of keratinocytes were cultured on the dermal equivalent at air-liquid interface.

We made collagen scaffold in the form of sponge used for skin fibroblast culture. It was prepared by freeze-drying of 7.5 mg/ml of type I collagen solution. We tried cross-linking treatments by glutaraldehyde and ultraviolet irradiation. In addition, to improve biocompatible characteristics as well as mechanical strength, we incorporated mesh of collagen threads into the sponge and obtained much stronger than that of glutaraldehyde treatment. According to the result of UTS measurement, sponges incorporated with mesh showed about 0.086 MPa that is three times as stronger as glutaraldehyde treated sponges at 0.029 MPa. We will be able to control the degree of strength through changing the diameter of thread, quantity of thread, thickness of mesh, and different way of weaving threads into mesh. By this procedure, it will be possible to control the tensile strength of the matrix and bioartificial skin without using any toxic cross-linking reagents or additional physical treatments in the future.

Acknowledgement

This study was supported by a grant No. 961-1105-035-2 and 981-1105-020-2 from the Korea Science and Engineering Foundation.

References

Bell, E., Ivarsson, B., and Merrill C.: Production of a tissue-like structure by contraction of collagen lattices by human fibroblasts of different proliferative potential in vitro, *Proc. Natl. Acad. Sci. USA 76 (1979), 1274-1279.*

Yang, E.K., Lee, D.H., Park, S.N., Choe, T.B., and Park, J.K.: Contraction behavior of collagen gel and fibroblasts activity in dermal equivalent model, *J. Microbiol. Biotechnol. 7 (1997) 267-271.*

Yanas, I.V., Burke, J.F., Gordon, P.L., Huang, C., and Rubenstein, R.H.: Design of an artificial skin. II. Control of chemical composition, *J. Biomed. Mater. Res. 14 (1980) 107-131.*

Burke, J.F. Yannas, I.V., Quinby, W.C., Bondoc, C.C., and Jung, W.K.: Successful use of a physiologically acceptable artificial skin in the treatment of extensive burn injury, *Ann. Surg. 194 (1981) 413-428.*

Suzuki. S., Matsuda K., Maruguchi, T., Nishima, TY., and Ikada, Y.: Further application of bilayer artificial skin, *Br. J. Plast. Surg. 48 (1995) 222-229.*

Koide, M., Osaki, K., Konishi, J., Oyamada, K., Katakura, T., and akahashi, A.: A new type of biomaterial for artificial skin: Dehydrothermal cross-linked composites of fibrillar and denatured collagens, J. Biomed. Mater. Res. 27 (1993) 79-87.

Nimni, M.E., Cheung, D., Strates, B., Kodama, M., and Sheik, K.: Bioprosthesis derived from cross-linked and chemically modified collagenous tissues, in Nimni, M.E. (ed.), Collagen, Vol.III: Biotechnology, CRC Press, Florida, 1988, pp.1-37.

Khor, E.: Methods for the treatment of collagenous tissues for bioprostheses, Biomaterials 18 (1997) 95-105.

Bomalaski, M.D., Bloom, D.A., McGuire, E.J., and Panzl, A.: Glutaraldehyde cross-linked collagen in the treatment of urinary incontinence in children, J. Urology 155 (1996) 699-702.

Weadock, K., Olson, R.M., and Silver, F.H.: Evaluation of collagen crosslinking techniques, Biomater. Med. Dev. Artif. Organs 11 (1983/84) 293-318.

Chvapil, M., Speer, D.P., Holubec, H., Chvapil, T.A., and King, D.H.: Collagen fibers as a temporary scaffold for replacement of ACL in goats, J. Biomed. Mater. Res. 27 (1993) 313-325.

Kano, Y., Sakano, Y., and Fujimoto, D.: Cross-linking of collagen by ascorbate-copper ion system, J. Biochem.102 (1987) 839-842.

Rault, I., Frei, V., Herbage, D., Malak, N.A., and Huc, A.: Evaluation of different chemical methods for cross-linking collagen gel, films and sponges, J. Mater. Sci:Mater. in Med. 7 (1996) 215-221.

Huang-Lee, L.L.H., Cheung, D.T., and Nimni, M.E.: Biochemical changes and cytotoxicity associated with the degradation of polymeric glutaraldehyde derived cross-links, J. Biomed. Mater. Res. 24 (1990) 1185-1201.

Weadock, K.S., Miller, E.J, Bellincamp, L.D., Zawadsky, J.P., and Dunn, M.G.: Physical crosslinking of collagen fibers: Comparison of ultraviolet irradiation and dehydrothermal treatment, J. Biomed. Mater. Res. 29 (1995) 1373-1379.

Vizarova, K., Bakos, D., Rehakova, M., and Macho, V.: Modification of layerd atelocollagen by ultraviolet irradiation and chemical cross-linking: structure stability and mechanical properties, *Biomaterials 15 (1994) 1082-1086.*

Yang, E.K., Seo, Y.K., Lee, D.H., Lee, J.H., Kim, J., Yook, J.I., Park, J.H., and Park, J.K.: Fabrication of collagen sponge incorporated with collagen threads, *Biomaterials Research 3 (1999) 60-65.*

Boyce, S.T., Christianson, D.J., and Hansbrough, J.F.: Structure of a collagen-GAG dermal skin substitute optimized for cultured human epidermal keratinocytes, J. Biomed. Mater. Res. 22 (1988) 939-957.

Cavallaro, J.F., Kemp, P.D., and Kraus, K.H.: Collagen fabrics as biomaterials, *Biotech. Bioeng. 43 (1994) 781-791.*

Yang, E.K., Seo, Y.K., Youn, H.H., Lee, D.H., Park, S.N., and Park, J.K.: Tissue engineered artificial skin composed of dermis and epidermis, *Artificial Organs 24 (2000) 7-17.*

Yang, E.K., Jin, S.W., Kang, B.T., Kim, I.H., Park, J.K., Lee, S.S., Kim, J.W., and Park, S.N.: Construction of artificial epithelia tissues prepared from human normal fibroblasts and C9 cervical epithelia cancer cells carrying human papillimavirus type 18 genes, Biotechnol. Bioproc. Eng. 3 (1998) 1-5.

Wang, M.C., Pins, G.D., and Silver, F.H.: Collagen fibres with improved strength for the repair of soft tissue injuries, *Biomaterials 15 (1994) 507-512.*

Jayakrishnan, A., and Jameela, S.R.: Review: Glutaraldehyde as a fixative in bioprostheses and drug delivery matrices, Biomaterials 17 (1996) 471-484.

Part 6

ANTITUMOUR COMPOUNDS

TOWARDS THE GENERATION OF NOVEL ANTITUMOUR AGENTS FROM ACTINOMYCETES BY COMBINATORIAL BIOSYNTHESIS

JOSE A. SALAS, GLORIA BLANCO, ALFREDO F. BRAÑA, ERNESTINA FERNANDEZ, Mª JOSE FERNANDEZ, JOSE GARCIA BERNARDO, ANA GONZALEZ, FELIPE LOMBO, LAURA PRADO, LUIS M. QUIROS, CESAR SANCHEZ AND CARMEN MENDEZ

Departamento de Biología Funcional e Instituto Universitario de Oncología del Principado de Asturias (I.U.O.P.A). Universidad de Oviedo. Spain.

1.Introduction

From the introduction of antibiotics for the treatment of bacterial infectious diseases, search for new drugs has been mainly carried out through the development of screening programmes with the isolation of different microorganisms from various sources and the analysis of the compounds produced by these microorganisms. This has been combined with chemical modification of lead compounds to introduce specific changes in these molecules, thus generating families of semisynthetic or synthetic compounds, many of them clinically useful. Most of these screening programmes have used as raw material terrestrial macro- and micro-organisms. However, in the past two decades, great progress has been made in natural drug discovery from marine organisms through the isolation of many unique bioactive substances, some of which are now in advanced clinical trials (Fenical, 1997). These studies have been usually complemented by improving the yield of production through mutagenesis programmes and selection for high-level production strains. However, the search for novel useful drugs through screening programmes, although still clearly useful, require thousands of compounds to be tested before a novel interesting compound appears. Consequently new approaches for generating novel compounds are needed.

Many bioactive compounds are produced by species of the Actinomycetes family. Actinomycetes are filamentous gram-positive bacteria and they are commercially important microorganisms. They are producers of approximately two thirds of all known bioactive compounds and within them some have clinical application on the basis of their activity against different kinds of microorganisms. They include compounds presenting antibacterial activity, antifungal, antiparasite and antiviral activity. Furthermore, they synthesise compounds with immunosuppressant activity and others showing activity

A. Van Broekhoven et al. (eds.), Novel Frontiers in the Production of Compounds for Biomedical Use, 383–399.
© 2001 *Kluwer Academic Publishers. Printed in the Netherlands.*

against different cellular tumour lines and some of them have clinical application in the chemotherapeutic treatment of several cancer diseases (Fig. 1).

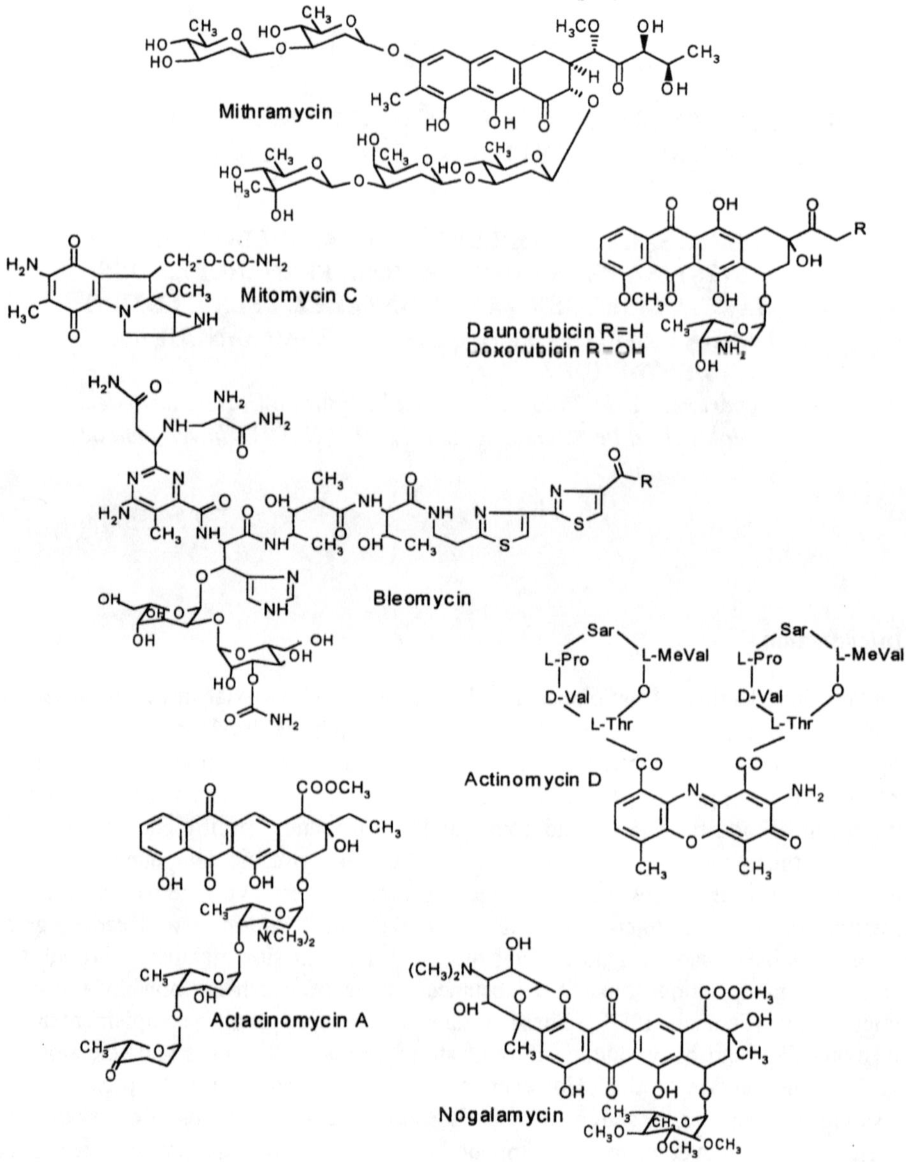

Figure 1. Structure of different anticancer agents produced by Actinomycetes

Structurally, many of these bioactive compounds belong to the polyketide group. Polyketides constitute one of the largest families of secondary metabolites and show a remarkable variety of chemical structures. However, all of them are synthesised through the successive condensation of short-chain carboxylic acids in a series of reactions catalysed by the polyketide synthase (for reviews see Hopwood and Sherman, 1990; Katz and Donadio, 1993; Hutchinson and Fujii, 1995; Hopwood, 1997; Bechthold and Salas, 1999). This condensation process will generate carbon chains of varying length, different side chains and a reduction pattern that will be differentially cycled and subsequently modified to give the mature polyketides. The final consequence is a great variety of different chemical structures.

The development of recombinant DNA technology has opened a new and exciting field of research for the generation of new bioactive compounds through the genetic manipulation of the biosynthetic pathways. The enormous capability of Actinomycetes to produce bioactive compounds is being now considered by the researchers as a great potential for such genetic manipulation.

2. Anticancer biosynthetic gene clusters

The isolation and cloning of genes involved in the biosynthesis of anticancer drugs produced by streptomycetes has been mainly possible by the pioneer work by researchers at the John Innes Institute (Norwich, U.K.) on the development of *Streptomyces* genetic tools. Two basic strategies have been used for the cloning of anticancer biosynthetic gene clusters. As mentioned above many anticancer drugs belong to the polyketide family and it has been extensively used the cloning by genetic homology as an approach for the isolation of polyketide gene clusters. This strategy is based in the use of gene probes from related biosynthetic pathways for the screening of gene libraries of polyketide producers. Particularly, the *actI-actIII* genes have resulted to be very useful probes when using this cloning strategy. The *actI* region encodes part of a type II polyketide synthase involved in the biosynthesis of the isochromanoquinone actinorhodin by *Streptomyces coelicolor* (Fernández-Moreno et al., 1992) and the *actIII* gene encodes a polyketide ketoreductase in the same pathway (Hallam et al., 1988). Using these two gene probes several anticancer (and many antibiotic) gene clusters have been isolated. A second probe that has been widely used are two genes from the streptomycin pathway by *Streptomyces griseus*, *strD* and *strE* genes (Stockmann and Piepersberg, 1992; Decker et al., 1996) that encode the earliest enzymatic steps in 6-deoxyhexose biosynthesis. A second complementary strategy widely used is the cloning of resistance genes and their use as gene probes to isolate biosynthetic genes. This approach is based in the frequent existence of chromosomal linkage between resistance genes and biosynthetic genes. Many anticancer drugs have also antibacterial activity and therefore the producer organisms have developed specific resistance mechanisms to survive during drug biosynthesis. These resistance genes can be cloned in another host and then used as probes for the screening of gene libraries. The first gene cluster cloned for an anticancer drug produced by Actinomycetes was that for tetracenomycin C biosynthesis in *Streptomyces glaucescens* (Motamedi and Hutchinson, 1987). Using the *actI* probe, these authors cloned the entire gene cluster for tetracenomycin C and they expressed this pathway into different streptomycetes. Since then, different laboratories have reported the isolation and characterisation of biosynthetic genes for anticancer agents from various Actinomycetes.

3. The aureolic acid group

(a)

(b)

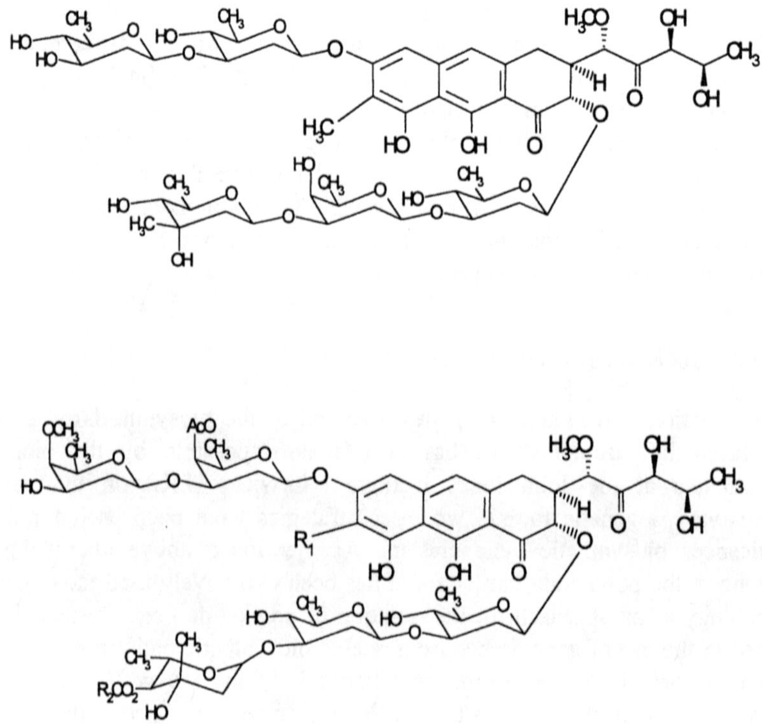

Figure 2. Structures of the family of the aureolic acid. A: mithramycin . B: chromomycin A3 (R_1=R_2=Me) and olivomycin (R_1=H; R_2=isopropanol)

Mithramycin, chromomycin and olivomycin (Fig. 2) belong to the family of the aureolic acid group of drugs. These compounds present antibacterial activity against Gram-positive bacteria but not against Gram-negative bacteria because of a permeability barrier and they also show cytostatic activity. Their mode of action is not well understood although they are not intercalative agents and they interact with nucleic acids by interacting with the minor groove of DNA in GC-rich regions and with an absolute requirement for magnesium ions for activity. Mithramycin shows a remarkable cytotoxicity against a variety of tumour cell lines, including brain tumours and experimental animal tumours, and has been clinically used for the treatment of certain tumours, such as disseminated embryonic cell carcinoma as well as for Paget's bone disease (Remers, 1979). It also finds use for control of hypercalcemia in patients with malignant disease. Structurally the aureolic acid family are aromatic polyketides and the chromophoric ring derives from the condensation of ten acetates (Montanari and Rosazza, 1988, 1990; Rohr et al., 1999). This polyketide moiety (aglycon) is further glycosylated with a trisaccharide and a disaccharide consisting of 2,6-dideoxyhexoses. In our group we were interested in the isolation and characterisation of genes involved in the biosynthesis of anticancer drugs of the aureolic acid group and the generation of novel drugs by genetic manipulation. Using the two cloning strategies mentioned above we isolated a series of overlapping clones in the bifunctional cosmid pKC505 (*Streptomyces-Escherichia coli*) that contain the mithramycin gene cluster from *Streptomyces argillaceus* ATCC 12956. In this region we have sequenced approximately 50 kb and we have identified 34 genes encoding the functions required for mithramycin biosynthesis (Fig. 3). Assignment of the functions to the different genes was made on the basis of three different approaches: (i) comparison of the deduced product of each gene with proteins in databases; (ii) insertional inactivation by gene replacement of most of the genes and analysis of the structure of the accumulated products and (iii) heterologous expression of some of the genes in another producer strains and analysis of the new compounds.

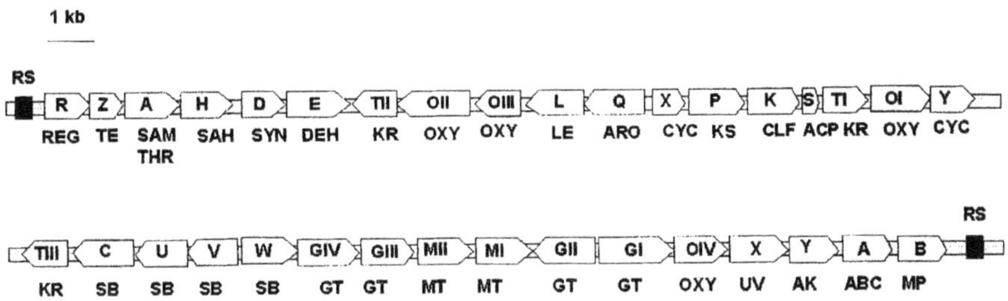

Figure 3. Gene organisation of the mithramycin cluster. RS, repeated sequences. REG, regulation. TE, thioesterase. SAM, S-adenosylmethionine-synthase. THR, tetrahydrofolate reductase. SAH, S-homocysteine hydrolase. SYN, dTDP-glucose synthase. DEH, dTDP-glucose 4,6-dehydratase. KR, ketoreductase. OXY, oxygenase. LE, loading enzyme. ARO, aromatase. CYC, cyclase. KS, ketosynthase. CLF, chain length factor. ACP, acyl carrier protein. SB, sugar biosynthesis. GT, glycosyltransferase. MT, methyltransferase. UV, DNA excision repair enzyme. AK, unknown function. ABC, ATP-binding protein. MP, membrane protein.

The different genes identified and proposed functions in mithramycin biosynthesis are as follows:

3.1. GENES INVOLVED IN THE BIOSYNTHESIS OF THE POLYKETIDE MOIETY.

A set of genes constituting a minimal type II polyketide synthase is represented by *mtmP* (β-ketoacylsynthase), *mtmK* (chain length factor) and *mtmS* (acyl carrier protein) (Blanco et al., 1996; Lombó et al., 1996). In addition, an aromatase (*mtmQ*) and two possible cyclases (*mtmX* and *mtmY*) would be responsible for generation of the tetracyclic nonglycosylated intermediates. Interestingly a gene, *mtmZ*, was found within the cluster which would encodes a thioesterase enzyme. As far as we knows such enzyme has not yet been reported for an aromatic polyketide cluster but it is forming part of type I polyketides. It is possible that this thioesterase could be involved in releasing the polyketide from the multienzymatic polyketide synthase. Also a gene, *mtm*L, was found that encodes an acyl-CoA ligase that could be responsible for channelling acetyl-CoA units for polyketide biosynthesis.

3.2. GENES ENCODING ENZYMES MODIFYING THE POLYKETIDE SKELETON.

Four genes encoding oxygenases were found within the mithramycin cluster (*mtmOI*, *mtmOII*, *mtmOIII* and *mtmOIV*). Two of these genes (*mtmOI* and *mtmOIII*) do not seem to be essential for mithramycin biosynthesis since insertional inactivation of these genes does not affect mithramycin biosynthesis (Prado et al., 1999b). The *mtmOII* product would catalyse the hydroxylation of the aglycon (Prado et al., 1999b) and the *mtmOIV* product is responsible for the opening of the fourth ring of a mithramycin intermediate, premithramycin B, by oxidative cleavage (Prado et al., 1999a). Two methyl groups are incorporated into the aglycon after its biosynthesis by the action of two S-adenosyl-methionine (SAM)-dependent methyltransferases. The *mtmMI* and *mtmMII* genes would code for a O- and C-methyltransferases, respectively, acting on C-4 and C-9 positions of the aglycon (Fernández Lozano et al., 2000).

3.3. GENES ENCODING AN ACTIVATED METHYL CYCLE

Methylation requires a methyl group donor. When S-adenosyl methionine (SAM) is the cofactor this can be probably provided by primary metabolism. However, in the mithramycin gene cluster we found two genes encoding an incomplete activated methyl cycle. The first gene (*mtmA*) would code for a protein with two domains. The first half resembles SAM-synthetases that catalyse the formation of SAM from methionine and ATP. The second half of the protein shows clear similarities with 5,10-methylenetetrahydrofolate reductases which convert N^5,N^{10}-methylene-tetrahydrofolate into N^5-methyl-tetrahydrofolate. The second gene (*mtmH*) encodes a S-adenosyl-L-homocysteine hydrolase responsible for the conversion of S-adenosylhomocysteine into methionine and adenosine. S-Adenosylhomocysteine is a potent inhibitor of SAM-dependent methyltransferases and it has therefore to be removed from the cytoplasm by the action of an S-adenosylhomocysteinyl hydrolase, generating homocysteine that now can be transformed into methionine by a methionine synthase incorporating a methyl group from N^5-methylene-tetrahydrofolate. The gene coding this enzyme activity is the only one not present within the mithramycin gene cluster. However, by Southern hybridisation it was found to be present in other region of the chromosome. The simultaneous deletion of the *mtmA* and *mtmH* genes from the chromosome of the mithramycin producer did not affect mithramycin biosynthesis but, interestingly, the levels of mithramycin production in this mutant were much higher than in the wild type strain (Mª J. Fernández Lozano, unpublished results).

3.4. GENES ENCODING SUGAR BIOSYNTHETIC ENZYMES

The disaccharide and trisaccharide attached to the mithramycin aglycon are formed by 6-deoxyhexoses (6DOH): D-olivose and D-olivose in the disaccharide and D-olivose, D-oliose and D-mycarose in the trisaccharide. 6DOH are normally synthesised through dTDP-activated intermediates. Two enzymatic steps in this pathway are common to the biosynthesis of all 6DOH. These are the activation of glucose-1-phosphate into dTDP-glucose by a glucose-1-phosphate:TTP thymidylyl transferase and the dehydration of dTDP-glucose by a TDP-D-glucose 4,6-dehydratase. These two enzymes are encoded in the mithramycin gene cluster by the *mtmD* and *mtmE* genes. Both genes have been expressed in *E. coli* and shown by *in vitro* assays that the corresponding enzymes catalyse the proposed enzyme reactions, respectively (Lombó et al., 1997). The result of the action of these two enzymes is the formation of TDP-D-4-keto-6-deoxyglucose, a key and crucial intermediate in the biosynthesis of all 6DOH. Insertional inactivation of *mtmD* generated a nonproducing mutant that accumulates a tetracyclic intermediate in mithramycin biosynthesis, premithramycinone (Lombó et al., 1997; Rohr et al., 1998), the last nonglycosylated intermediate in mithramycin biosynthesis. Several other genes encoding 6DOH enzymes have been also identified. The *mtmV* would code for a 2,3-dehydratase and would be involved in the biosynthesis of three 6DOH in mithramycin. *mtmC* would codes for a C-methyltransferase involved in D-mycarose biosynthesis; this has been confirmed by insertional inactivation and the accumulation of intermediates lacking the D-mycarose moiety (A.González, unpublished results). The *mtmU* gene would code a ketoreductase probably catalysing one of the last steps in D-oliose biosynthesis (F. Lombó, unpublished results).

3.5. GENES ENCODING GLYCOSYLTRANSFERASES

Four glycosyltransferase genes have been identified in the mithramycin cluster. Two of them (*mtmGI* and *mtmGII*) are involved in the formation and transfer of the diolivosyl disaccharide (Fernández et al., 1998). Most of the sugars attached to different antibiotic aglyca, are transferred as monosaccharides by glycosyltransferases. However, it is worth to mention that, in the case of mithramycin, the disaccharide is first formed by the action of a glycosyltransferase and then the second glycosyltransferase transfer this disaccharide to the aglycon (Fernández et al., 1998). This disaccharide is incorporated to the aglycon once the trisaccharide is already attached. The *mtmGIV* and *mtmGIII* are responsible for the incorporation of the first two sugars in the trisaccharide, D-olivose and D-oliose respectively, into the nonglycosylated aglycon premithramycinone (Blanco et al., 2000). However, it remains to be elucidated which glycosyltransferase is responsible for the attachment of the third sugar, D-mycarose.

3.6. GENES RESPONSIBLE FOR RESISTANCE AND SECRETION.

The mithramycin producer *Streptomyces argillaceus* is highly resistant to mithramycin but sensitive to the closely related chromomycin and olivomycin. Two genes (*mtrA* and *mtrB*) were identified that confer resistance to mithramycin when expressed in another host (Fernández et al., 1996). They encode an ABC transporter system composed of two components: a hydrophilic ATP-binding protein encoded by *mtrA* and a hydrophobic membrane component encoded by *mtrB*. Both components are necessary to confer resistance to mithramycin and possibly represents the secretion system to secrete mithramycin to the extracellular medium. Immediately upstream of *mtrA*, two other genes (*mtrX* and *mtrY*) have been found to which no clear function has been assigned. The *mtrX* gene product shows similarity with excision repair systems and it has been proposed that it could represent a secondary defence mechanism to repair some minor DNA damage caused by free mithramycin molecules accidentally not removed from the cytoplasm by the secretion system (Fernández et al., 1996).

4. Generation of novel compounds

Several approaches can be attempted for the generation of novel compounds through the genetic manipulation of anticancer biosynthetic pathways: insertional inactivation, tailoring modification and combinatorial biosynthesis.

4.1. INSERTIONAL INACTIVATION

Targeted gene disruption of selected and specific genes may produce the accumulation in the correspondent mutants of novel biosynthetic intermediates. In this approach, the normal expression of the selected gene is altered either by creating an *in frame* mutation or deletion or by the insertion of an antibiotic-resistance cassette. The new compounds generated can present activity or they can be useful lead compounds for further modifications by chemical means or biotransformation.

In the mithramycin pathway, *mtmD* encodes a dTDP-glucose synthase responsible for the activation of glucose-1-phosphate into dTDP-glucose. Inactivation of this gene is predicted to completely block the biosynthesis of all 6DOH in mithramycin. This gene was knocked out by insertional inactivation and the mutant strain generated accumulated a novel mithramycin intermediate, premithramycinone (Fig. 4) (Lombó et al., 1997; Rohr et al., 1998). This is a tetracyclic compound lacking sugars and its isolation was the first proof indicating that biosynthesis of mithramycin proceeds through tetracyclic intermediates. It is worth to mention that premithramycinone structurally resembles two compounds from *Aspergillus niger* that have been shown to be inhibitors of neuropeptide-Y receptors (Shu et al., 1995). A non-methylated derivative from premithramycinone (designated as 4-demethyl-premithramycinone) was also generated by deleting the four glycosyltransferases and two methyltransferases (Prado et al., 1999a). The same compound was accumulated by a mutant in the *mtmMI* methyltransferase gene (Fernández Lozano et al., 2000). During mithramycin biosynthesis, premithramycinone is first glycosylated with a trisaccharide and then with a disaccharide. It was tentatively assumed that the resulting compound, a tetracyclic fully glycosylated intermediate, would be the substrate for an oxygenase that would cause the oxidative cleavage of the fourth ring generating the final tricyclic structure with a lateral chain. Insertional inactivation of the *mtmOIV* oxygenase gene allowed to demonstrate this hypothesis and the isolation of a novel compound, premithramycin B (Fig. 5) (Prado et al., 1999a).

Interesting glycosylated compounds differing in the number of sugars attached to the aglycon were also generated in the mithramycin producer by insertional inactivation of two genes (*mtmGI* and *mtmGII*) encoding glycosyltransferases (Fernández et al., 1998). Independent inactivation of both genes produced the accumulation of several intermediates. Interestingly, both mutants accumulated the same compounds. Isolation of these compounds by preparative HPLC and elucidation of the respective structures showed that they were tetracyclic intermediates containing one, two or three sugars in the trisaccharide chain and no sugars at the position of the disaccharide (Fig. 6) (Fernández et al., 1998). The explanation to the fact that the same compounds were accumulated by both mutants was that one of the genes was responsible for the formation of the disaccharide and the other for the transfer of the disaccharide to the aglycon.

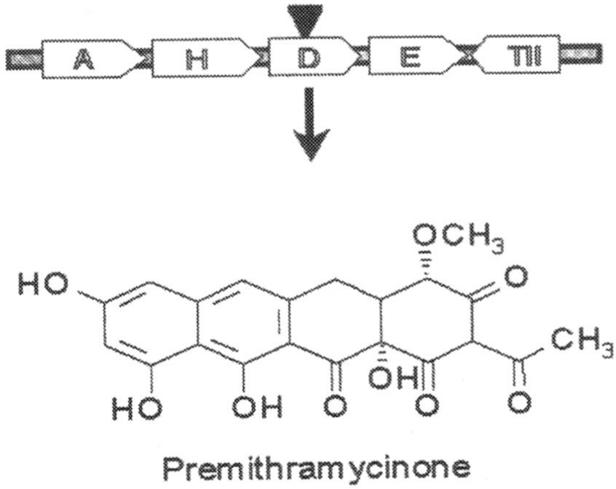

Figure 4. Generation of a novel tetracyclic compound by insertional inactivation of the dTDP-glucose synthase encoding gene of the mithramycin gene cluster. The mtmD *gene was inactivated by inserting an apramycin resistance cassette (black triangle) within the gene coding region (Lombó et al., 1997).*

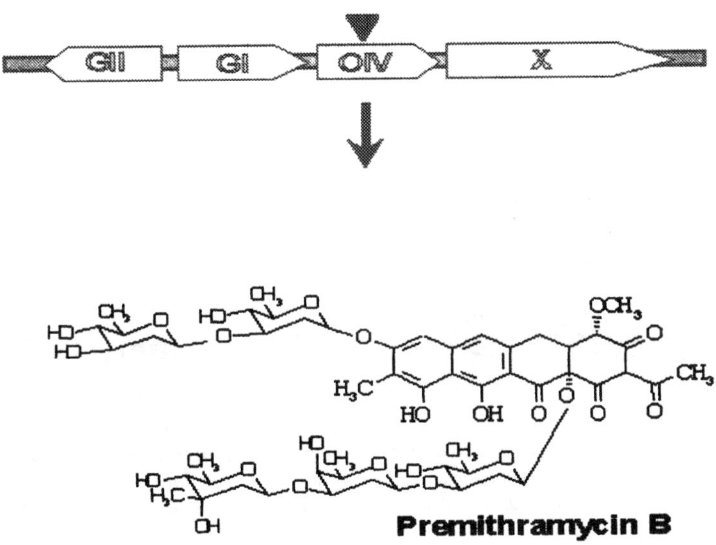

Figure 5. Generation of a novel tetracyclic glycosylated compound by insertional inactivation of the mtmOIV *oxygenase gene of the mithramycin gene cluster. The* mtmOIV *gene was inactivated by deleting most of the gene and replacing it by an apramycin resistance cassette (black triangle) (Prado et al., 1999a)*

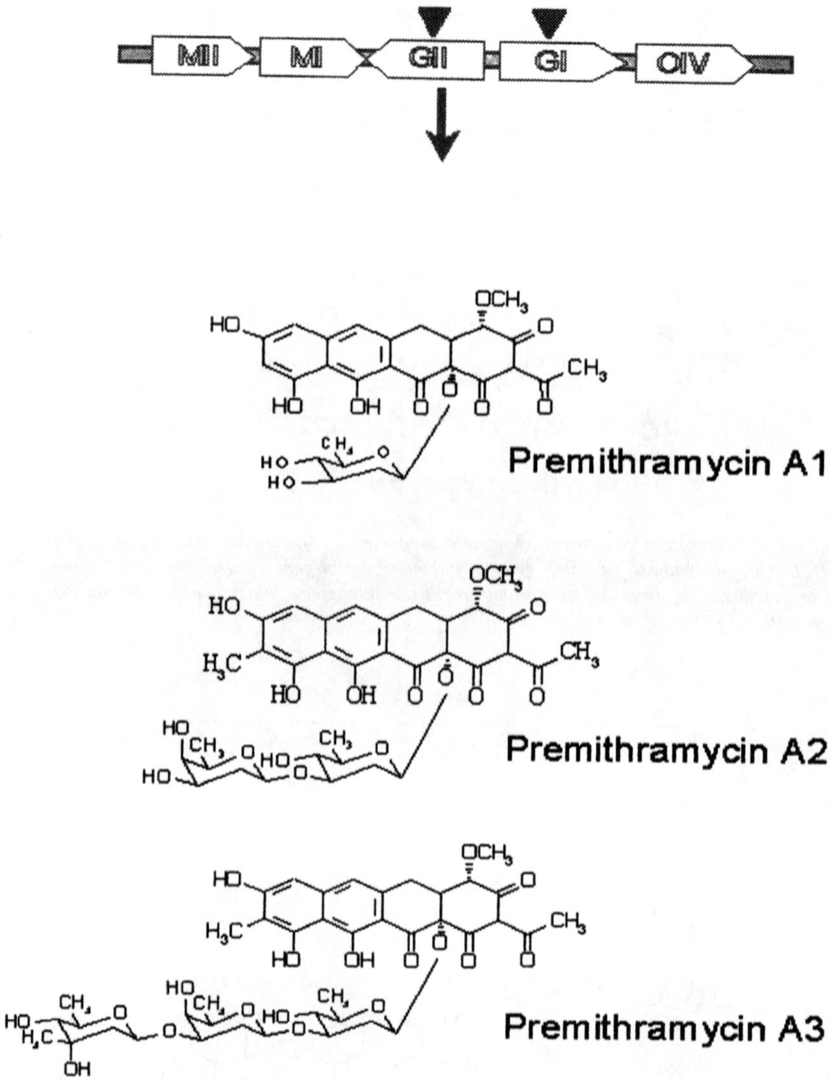

Figure 6. Generation of novel glycosylated compounds by insertional inactivation of the mtmGI and mtmGII genes from the mithramycin gene cluster. The three compounds shown were accumulated by mutants in which either mtmGI or mtmGII were independently inactivated (Fernández et al., 1998)

4.2. TAILORING MODIFICATION

Tailoring modification of specific compounds by expression of a gene from a biosynthetic pathway in another host can produce single and specific chemical modifications thus generating novel compounds. Urdamycin (an angucycline) and tetracenomycin C (an anthracycline) are two aromatic anticancer polyketides. The *urdE* gene from *Streptomyces fradiae* encodes an oxygenase involved in the introduction of a hydroxyl group in the urdamycin molecule (Decker and Haag, 1995). This gene was expressed in *Streptomyces glaucescens* (tetracenomycin C producer) under the control of the erythromycin resistance promoter *ermE*, resulting in the production of a new hybrid compound, 6-hydroxy-tetracenomycin C (Fig. 7). Another hybrid anticancer drug (11-hydroxyaclacinomycin A) was produced by the expression of the *dnrF* gene from *Streptomyces peucetius* encoding an aklavinone-11-hydroxylase (from a doxorubicin producer) in the aclacinomycin producer *Streptomyces galilaeus* (Fig. 8) (Hwang et al., 1995). This hydroxylated anthracycline was more active against leukaemia and melanoma cell lines that the parental aclacinomycin A.

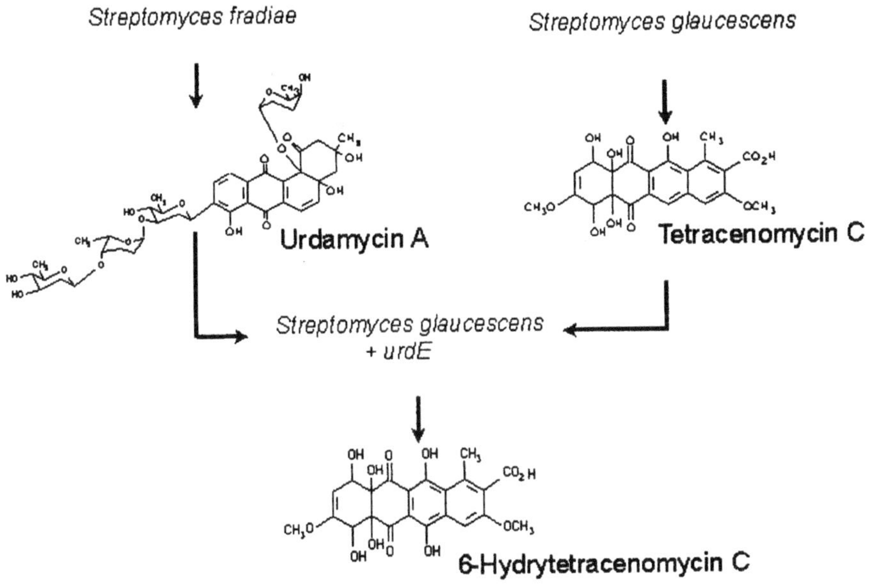

Figure 7. Novel hydroxylated tetracenomycin derivatives generated by tailoring modification.

Epidoxorubicin (epirubicin) is a clinically useful antitumor drug that until now has been produced semisynthetically following several chemical modification steps of doxorubicin. Recombinant DNA technology has been now applied to generate this compound in a less time-consuming and less expensive way. A mutant was generated in *Streptomyces peucetius* by inactivating the gene *dnmV* which encodes a 4-ketoreductase involved in the biosynthesis of daunosamine, the deoxysugar component of daunorubicin and doxorubicin. By expressing the *avrE* gene from *S. avermitilis* in this mutant epirubicin was obtained. The *avrE* gene encodes a deoxyhexose 4-ketoreductase with different stereospecificity than *dnmV* and therefore complementation of the *dnmV* mutant changed the stereospecificity of the 4-hydroxyl group (Madduri et al., 1998)

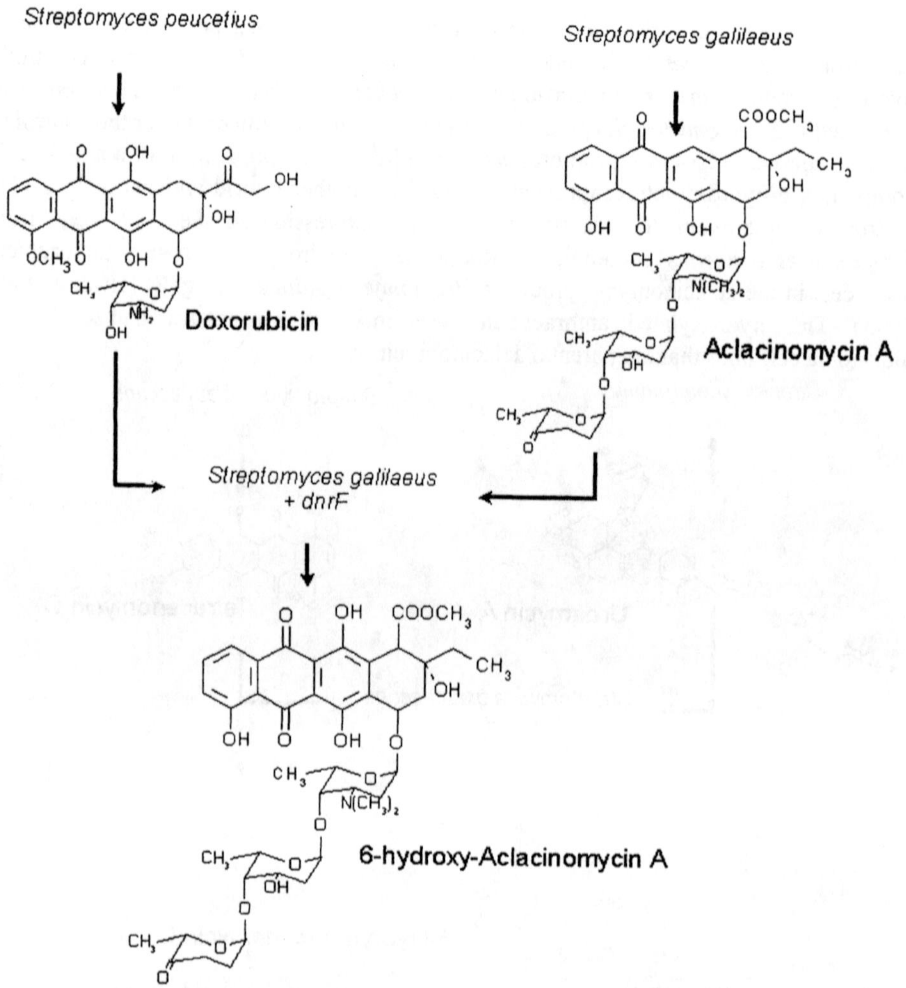

Figure 8. Novel hydroxylated anthracycline derivative generated by tailoring modification.

4.3. COMBINATORIAL BIOSYNTHESIS

Novel compounds can be also generated by recombining naturally occurring biosynthetic pathways. This means the coexistence in host of genes from different, but related, biosynthetic pathways. The concerted interaction of enzymes from both pathways may give rise to the generation of novel compounds.

Two different compounds were produced by expression of several genes from the mithramycin gene cluster into different hosts (Fig. 9). A DNA fragment containing several genes encoding a cyclase (*mtmX*), a ketoacylsynthase (*mtmP*), a chain length factor (*mtmK*) and an acyl carrier protein (*mtmS*) was expressed into *Streptomyces galilaeus* (aclacinomycin producer) and into *Streptomyces glaucescens* (tetracenomycin producer). The *S. galilaeus* transgenic strain was able to produce auramycinone as a result of the interaction between genes from the mithramycin and aclacinomycin pathways (Kantola et al., 1997). This is an intermediate in the biosynthesis of some anthracyclines and is not produced by the host strain. Expression of the same genes in the tetracenomycin producer *S. glaucescens* originated a novel compound, tetracenomycin M, resulting from the interaction between genes from the mithramycin and tetracenomycin pathways (Künzel et al., 1997). In *S. lividans*, a compound, SEK15, was produced through spontaneous cyclisations of the decaketide produced by the mithramycin polyketide synthase genes.

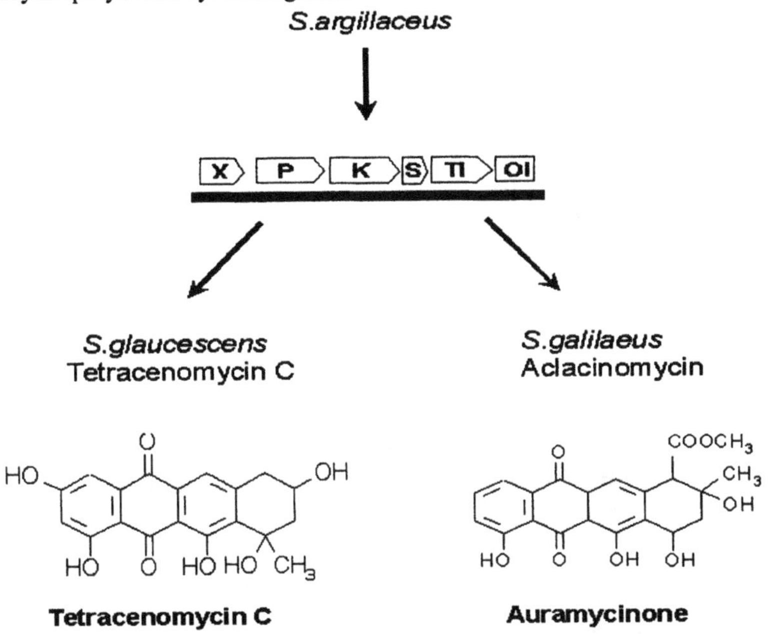

Figure 9. Novel compounds generated by expression of the mithramycin minimal polyketide synthase into different hosts.

Elloramycin is an anthracycline structurally quite similar to tetracenomycin C. It is a tetracyclic compound with a permethylated L-rhamnose attached to carbon 8 of the elloramycin polyketide aglycon. From a cosmid library of chromosomal DNA from the elloramycin producer, *Streptomyces olivaceus,* a clone was isolated that contains most (but not all) of the elloramycin biosynthetic gene cluster (Decker et al., 1995). When this cosmid clone, cos16F4, was transformed into a streptomycetes host it gave rise to the biosynthesis of an elloramycin biosynthetic intermediate, 8-demethyl-tetracenomycin C. This compound lacks the L-rhamnose moiety at carbon 8 and therefore is susceptible to glycosylation by an appropriate glycosyltransferase. Two novel glycosylated elloramycin derivatives were produced by transforming cosmid 16F4 into the urdamycin producer, *Streptomyces fradiae*: one containing D-olivose and the other L-rhodinose attached to the aglycon (Decker et al., 1995). In a similar experiment cosmid 16F4 was introduced into the mithramycin producer and three new compounds were produced (Fig. 10): D-olivosyl- (already mentioned above), D-mycarosyl- and D-diolivosyl-tetracenomcyin C (Wohlert et al., 1998). These novel compounds are the consequence of combinatorial biosynthesis based on the coexistence of genes from two anticancer biosynthetic pathways in a producer strain: the elloramycin genes encoding the aglycon (8-demethyl-tetracenomycin C) and genes from the urdamycin or mithramycin pathways

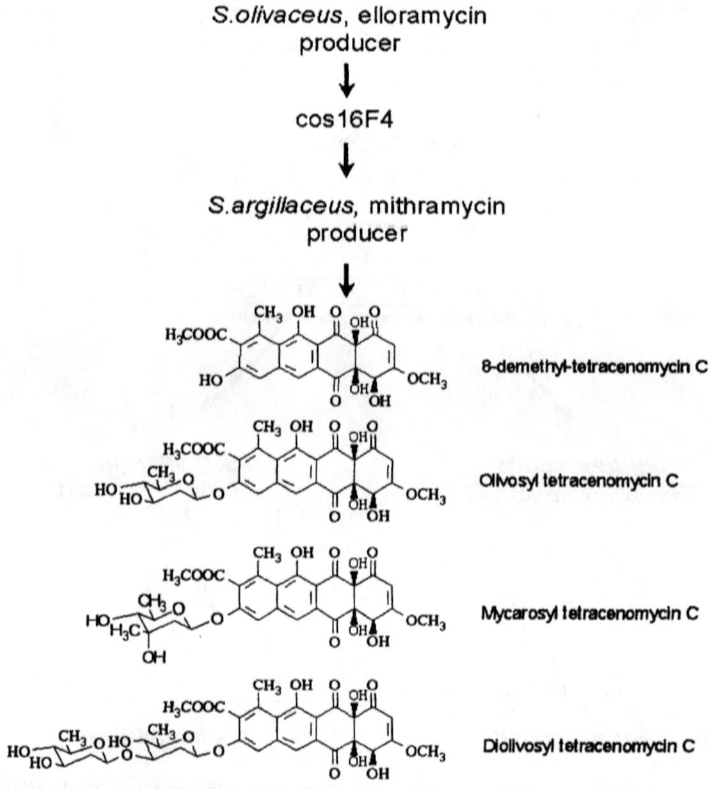

Figure 10. Novel hybrid glycosylated tetracenomycins by combination of elloramycin and mithramycin biosynthetic genes encoding the biosynthesis of the 6DOH transferred. Two possibilities exist to answer the question concerning the origin of the "substrate flexible glycosyltransferase". One would be that the flexible glycosyltransferase might be encoded by the elloramycin cluster and it would be flexible with respect to the sugar transferred to the aglycon. The alternative would be that the urdamycin or mithramycin glycosyltransferases might be flexible with respect to the aglycon that they glycosylate. Several lines of experimental evidence pointed out to the fact that the former hypothesis was the correct one: (i) they were also produced in a mithramycin nonproducer deletion mutant lacking all the mithramycin glycosyltransferases and (ii) they were produced in a Streptomyces lividans *strain in which a plasmid was introduced containing the mithramycin sugar biosynthetic genes but lacking any other mithramycin gene (Wohlert et al., 1998). All this experimental evidence clearly support the existence of a glycosyltransferase in cosmid 16F4 having a certain degree of flexibility with respect to the sugar to be transferred and able to accept at least L-rhodinose, D-olivose, D-mycarose and L-rhamnose. A gene encoding this glycosyltransferase,* elmG, *has been now isolated and is being characterised to define its sugar flexibility (G.Blanco, unpublished results).*

5. Concluding remarks

Current methods of obtaining novel compounds through the screening of biological samples, although clearly useful, are costly and time-consuming methods and frequently require thousands of compounds to be tested before a novel and promising lead compound appears. Recombinant DNA technology offers alternative and interesting possibilities for the generation of novel antitumor drugs in the near future. Combinatorial biosynthesis may be a powerful tool to create structural biodiversity as a complement to combinatorial chemistry. Many bioactive compounds (including antitumor drugs) are glycosylated compounds being, in many cases, the presence of the sugars essential for bioactivity. Sugar flexible glycosyltransferases might contribute to the generation of novel compounds with different glycosylation patterns, either containing different sugars to the ones normally attached to a position in the molecule or glycosylating a different position of the aglycon or changing the chain length in an oligosaccharide. In addition, heterologous gene expression in selected producer hosts may contribute to the introduction of specific and single chemical modifications into lead compounds which might be difficult to achieve by chemical means.

Acknowledgements

Research in the authors laboratory has been supported by grants of the Plan Nacional en Biotecnologia (BIO94-0037 and BIO97-0771) and from the European Union (Biotech BIO4-CT96-0068). Structure elucidation of the compounds generated in the authors laboratory was carried out in Dr. Jürgen Rohr`s laboratory (formerly Universität Göttingen, Germany and now Medical University of South Carolina, USA).

References

Bechthold, A. and Salas, J.A. (1999) Combinatorial biosynthesis of microbial metabolites. In: Jung, G. Combinatorial organic chemistry, pp. 381-407. Wiley-VCH, Weinheim.

Blanco, G., Fu, H., Méndez, C., Khosla, C. and Salas, J.A. (1996) Deciphering the biosynthetic origin of the aureolic acid group of antitumor agents. *Chemistry and Biology 3,* 193-196

Blanco, G., Fernández, E., Fernández, Mª J., Braña, A.F., Weißbach, U., Künzel, E., Rohr, J., Méndez, C. and Salas, J.A. (2000). Characterization of two glycosyltransferases involved in early glycosylation steps during biosynthesis of the antitumor polyketide mithramycin by *Streptomyces argillaceus. Molecular and General Genetics 262*, 991-1000

Decker, H. and Haag, S. (1995) Cloning and characterization of a polyketide synthase gene from *Streptomyces fradiae* Tu2717, which carries the genes for biosynthesis of the angucycline antibiotic urdamycin A and a gene probably involved in its oxygenation. *Journal of Bacteriology 177*, 6126-6136

Decker, H., Rohr, J., Motamedi, H., Zahner, H., Hutchinson, C.R. (1995) Identification of *Streptomyces olivaceus* Tü2353 genes involved in the production of the polyketide elloramycin. *Gene 166*, 121-126

Decker, H., Gaisser, S., Pelzer, S., Schneider, P., Westrich, L., Wohlleben, W. and Bechthold, A. (1996) A general approach for cloning and characterizing dNDP-glucose dehydratase genes from actinomycetes. *FEMS Microbiology Letters 141*, 195-202

Fenical, W. (1997) New pharmaceuticals from marine organisms. *Trends in Biotechnology 15*, 339-341

Fernández, E., Lombó, F., Méndez, C. and Salas, J.A. (1996) An ABC transporter is essential for resistance to the antitumor agent mithramycin in the producer *Streptomyces argillaceus. Molecular and General Genetics 251*, 692-698

Fernández, E., Weiβbach, U., Sánchez Reillo, C., Braña, A.F., Méndez, C., Rohr, J. and Salas, J.A. (1998) Identification of two genes from *Streptomyces argillaceus* encoding two glycosyltransferases involved in the transfer of a disaccharide during the biosynthesis of the antitumor drug mithramycin. *Journal of Bacteriology 180*, 4929-4937

Fernández Lozano, Mª J., Remsing, L.L., Quirós, L.M., Braña, A.F., Fernández, E., Sánchez, C., Méndez, C., Rohr, J. and Salas, J.A. (2000). Characterisation of two polyketide methyltransferases involved in the biosynthesis of the antitumor drug mithramycin by Streptomyces argillaceus. *Journal of Biological Chemistry 275*, 3065-3074.

Fernandez-Moreno, M.A., Martinez, E., Boto, L., Hopwood, D.A. and Malpartida, F. (1992) Nucleotide sequence and deduced functions of a set of cotranscribed genes of *Streptomyces coelicolor* A3(2) including the polyketide synthase for the antibiotic actinorhodin. *Journal of Biological Chemistry 267*, 19278-19290

Hallam, S.E, Malpartida, F. and Hopwood, D.A. (1988) Nucleotide sequence, transcription and deduced function of a gene involved in polyketide antibiotic synthesis in *Streptomyces coelicolor. Gene 74*, 305-320

Hopwood, D.A. (1997) Genetic contribution to understanding polyketide synthases. *Chemical Reviews 97*, 2465-2497

Hopwood, D.A. and Sherman, D.H. (1990) Molecular genetics of polyketides and its comparison to fatty acid biosynthesis. *Annual Review of Genetics 24*, 37-66

Hutchinson, C.R. and Fujii, I. (1995) Polyketide synthase gene manipulation: a structure-function approach in engineering novel antibiotics. *Annual Review of Microbiology 49*, 201-238

Hwang, C.K., Kim, H.S., Hong, Y.S., Kim, Y.H., Hong, S.K., Kim, S.J. and Lee, J.J. (1995) Expression of *Streptomyces peucetius* genes for doxorubicin resistance and aklavinone 11-hydroxylase in *Streptomyces galilaeus* ATCC 31133 and production of a hybrid aclacinomycin. *Antimicrobial Agents and Chemotherapy 39*, 1616-1620

Kantola, J., Blanco, G., Hautala, A., Kunnari, T., Hakala, J., Méndez, C., Ylihonko, K., Mantsala, P. and Salas, J.A. (1997) Folding of the polyketide chain is not dictated by minimal polyketide synthase in the biosynthesis of mithramycin and anthracycline. *Chemistry & Biology 4* , 751-755

Katz, L. and Donadio, S. (1993) Polyketide synthesis: prospect for hybrid antibiotics. *Annual Review of Microbiology 47*, 875-892

Künzel, E., Wohlert, S.E., Beninga, C., Haag, S., Decker, H., Hutchinson, C.R., Blanco, G., Mendez, C., Salas, J.A. and Rohr, J. (1997) Tetracenomycin M, a novel genetically engineered tetracenomycin resulting from a combination of mithramycin and tetracenomycin biosynthetic genes. *Chemistry European Journal 3*, 1675-1678

Lombó, F., Blanco, G., Fernández, E.F., Méndez, C. and Salas, J.A. (1996) Characterization of *Streptomyces argillaceus* genes encoding a polyketide synthase involved in the biosynthesis of the antitumor mithramycin. *Gene 172*, 87-91.

Lombó, F., Siems, K., Braña, A.F., Méndez, C., Bindseil, K. and Salas, J.A. (1997) Cloning and insertional inactivation of *Streptomyces argillaceus* genes involved in earliest steps of sugar biosynthesis of the antitumor polyketide mithramycin. *Journal of Bacteriology 179*, 3354-3357

Lombó,F., Braña,A.F., Méndez,C. and Salas,J.A. (1999) The mithramycin gene cluster of *Streptomyces argillaceus* contains a positive regulatory gene and two DNA repeated sequences that are located at both ends of the cluster. *Journal of Bacteriology 181*, 642-647.

Madduri, K., Kennedy, J., Rivola, G., Inventi-Solari, A., Filippini, S., Zanuso, G., Colombo, A.L., Gewain, K.M., Occi, J.L., MacNeil, D.J. and Hutchinson, C.R. (1998) Production of the antitumor drug epirubicin (4'-epidoxorubicin) and its precursor by a genetically engineered strain of *Streptomyces peucetius. Nature Biotechnology 16*, 69-74

Montanari, A. and Rosazza, J.P. (1988) The biosynthesis of chromomycin A3. *Tetrahedron Letters 29*, 5513-5516

Montanari, A. and Rosazza, J.P. (1990) Biogenesis of chromomycin A3 by *Streptomyces griseus. Tetrahedron Letters 43*, 883-889

Motamedi, H. and Hutchinson, C.R. (1987) Cloning and heterologous expression of a gene cluster for the biosynthesis of tetracenomycin C, the anthracycline antitumor antibiotic of *Streptomyces glaucescens. Proceedings of the National Academy of Sciences (USA) 84*, 4445-4449

Prado, L., Fernández, E., Weiβbach, U., Blanco, G., Quirós, L.M., Braña, A.F., Méndez, C., Rohr, J. and Salas, J.A. (1999a) Oxidative cleavage of premithramycin B is one of the last steps in the biosynthesis of the antitumor drug mithramycin. *Chemistry & Biology 6*, 19-30

Prado, L., Lombó, F., Braña, A.F., Méndez, C., Rohr, J. and Salas, J.A. (1999b) Analysis of two chromosomal regions adjacent to a type II polyketide synthase involved in the biosynthesis of the antitumor polyketide mithramycin in *Streptomyces argillaceus*. *Molecular and General Genetics 261*, 216-225

Remers, W.A. (1979) The chemistry of antitumor antibiotics. Vol. 1: 133-175. Wiley-Interscience. New York.

Rohr, J., Weiβbach, U., Beninga, C., Künzel, E., Siems, K., Bindseil, K., Prado, L., Lombó, F., Braña, A.F., Méndez, C. and Salas, J.A. (1998) The structure of premithramycinone and demethylpremithramycinone, early intermediates of the aureolic acid group antibiotic mithramycin. *Chemical Communications* 437-438

Rohr, J., Méndez, C. and Salas, J.A. (1999) New aspects of the biosynthesis of aureolic acid group antibiotics. *Bioorganic Chemistry 27*, 41-54

Shu, Y-Z., Cutrone, J.Q., Klohr, S.E. and Huang, S. (1995) BMS-192548, a tetracyclic binding inhibitor of neuropeptide Y receptors, from *Aspergillus niger* WB2346. II. Physico-chemical properties and structural characterization. *Journal of Antibiotics 48*, 1060-1065

Stockmann, M. and Piepersberg,W. (1992) Gene probes for the detection of 6-deoxyhexose metabolism in secondary metabolism-producing streptomycetes. *FEMS Microbiology Letters 90*, 185-190

Wohlert, S.-E., Blanco, G., Lombó, F., Fernández, E., Braña, A.F., Reich, S., Udvarnoki, G., Méndez, C., Decker, H., Salas, J.A. and Rohr, J. (1998) Novel hybrid tetracenomycins through combinatorial biosynthesis using a glycosyltransferase encoded by the elm-genes in cosmid 16F4 which shows a broad sugar substrate specificity. *Journal of the American Chemical Society 41*, 10596-10601.

CELL IMMOBILISATION OF *TAXUS MEDIA*

CHI WAI TANG AND FERDA MAVITUNA
Department of Chemical Engineering, University of Manchester Institute of Science and Technology, PO Box 88, Manchester M60 1QD, UK
E-mail: chi.w.tang@stud.umist.ac.uk *Fax No: 44 161 200 4399*

Summary

Taxol is a cytotoxic diterpene initially isolated from the bark of *Taxus brevifolia* (Pacific yew). It has been approved for use in cancer treatment. Since total synthesis is uneconomical, plant biotechnology can provide an alternative source for the drug. Callus was initiated from needles of *Taxus media* on different media under different conditions of light. Cell immobilisation in polyurethane foam particles showed improved growth.

1. Introduction

Paclitaxel (Taxol) and the related taxane compounds mainly occur in the plants of *Taxus spp.* (Strobel *et al.*, 1993). The bark of *T. brevifolia* was reported to have the highest content of Taxol (Fang *et al.*, 1993; Vidensek *et al.*, 1990; Witherup *et al.*, 1990), but the concentration was very low, only about 0.007-0.04% of the dry weight (Wheeler *et al*, 1992; Vidensek *et al.*, 1990).

We have experimented with conditions to allow initiation and growth of calli from needles of *T. media*. When considering a continuous production system using whole-cell biocatalysts, cell immobilisation is one of the most effective ways to maintain higher productivity of target metabolites. In plant cell culture, the actual productivity often changes due to the effects of intraparticle mass transfer (Furusaki *et al.*, 1988), redifferentiation and physicochemical interaction between the immobilising materials and cells (Haldimann and Brodelius, 1987). Polyurethane foam particles provided an economical and easy method to immobilise plant cells (Mavituna and Park, 1985; Lindsey *et al.*, 1983).

A. Van Broekhoven et al. (eds.), Novel Frontiers in the Production of Compounds for Biomedical Use, 401–407.
© 2001 *Kluwer Academic Publishers. Printed in the Netherlands.*

2. Materials and Methods

2.1 PLANT MATERIAL AND CALLUS INDUCTION

Callus was induced from the needles of *Taxus media.* For sterilisation, plant materials were washed and immersed in 70% ethanol for 2 min. Then, they were immersed in 2% sodium hypochlorite solution for 20 min, rinsed with sterile water, aseptically cut into explants of *ca.* 10 mm in length, then placed on 0.35% Phytagel (Sigma) culture media at 25°C in the complete dark and 16 hour light / 8 hours dark cycle photo-regime. Different basal media, MS (Murashige and Skoog, 1962), Gamborg's B5 (Gamborg, 1970) and woody plant (WP) media (Lloyd and McCown, 1980) supplemented with different plant growth regulators were used (Table 1).

***Table 1** The basal media and different plant growth regulators used in callus initiation.*

Medium No	Basal salt	2,4-D[1]	Kinetin[1]	NAA[1]
1	MS	1	1	0
2		1	0	0
3		0	1	0
4	B5	0	1	1
5		1	0	0
6		1	0	1
7	WP	1	0	0
8		1	1	0
9		0	1	1

[1] mg/L

***Table 2** The composition of the new medium (CWT) used for suspension culture Additions to B5 basal salt*

	mg/L		μg/L
Inositol	300	Thiamine HCl	6900
Glycine	75	Nicotinic acid	6250
Proline	115	Pyridoxine HCl	1750

***Table 2** The composition of the new medium (CWT) used for suspension culture Additions to B5 basal salt*

	mg/L		μg/L
Aspartic acid	133	Folic acid	500
Arginine	175	Biotin	50
Ascorbic acid	5		
Kinetin	0.1		g/L
ABA	0.1	Sucrose	7.5

2.2 CALLUS GROWTH MEASUREMENT

Callus growth was monitored by measuring the increase in fresh weight. To account for conditioning effects due to variation in inoculum size, a growth index was calculated as [(final wt - initial wt)/initial wt].

2.3 SUSPENSION CULTURE AND CELL IMMOBILISATION:

Callus was introduced into the new liquid medium (CWT) to form the suspension cultures (Table 2). Polyurethane foam (Declon, UK) particles in shake flasks and sheets in a bioreactor were used for cell immobilisation.

2.4 BIOREACTOR:

Four types of bioreactors were used; air lift, bubble, stirred tank and immobilised cell reactor. 6 sheets of 3cm X 15cm polyurethane foam were arranged vertically like baffles and used for cell immobilisation in a 4L stirred tank bioreactor.

3. Results and Discussion

3.1 EFFECT OF MEDIA ON CALLUS INITIATION FROM EXPLANTS:

Table 3 shows the result of callus initiation. Hormone regimes played a key factor in callus formation. 1 mg/L 2,4-D had the best effect on callus formation. Kinetin and cycled light photo-regime had detrimental effects on callus initiation. Less number of explants produced callus when cultured under cycled light. The callus was harder, slow to grow and light green to green. Fine, light yellow or yellow white callus could be induced from Gamborg's B5 medium with 1mg/L 2,4-D. Most of the callus induced under cycled light photo-regime did not proliferate when subcultured and turned brown. Only the calli induced from woody plant medium could be continuously subcultured.

Effect of light on callus growth: Callus induced under cycled light was hard, green and grew very slowly compared to the yellow-white callus induced in complete darkness. Figure 1 shows the growth index of callus culture under cycled light photo-regime and complete darkness. Fine, light yellow callus was subcultured every 4 weeks. This could also prevent browning of the culture.

Fig. 1 The growth index of *T. media* callus on CWT medium.

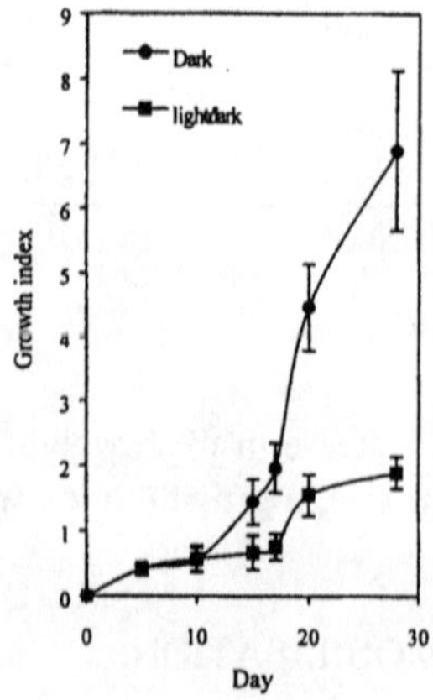

Fig.2 Growth of *T. media* suspension culture in shake flasks using CWT medium.

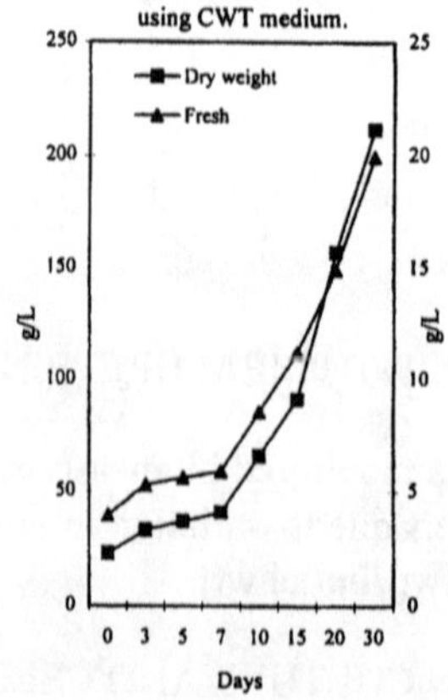

Fig. 3 The effect of cell immobilization on biomass dry weight per flask.

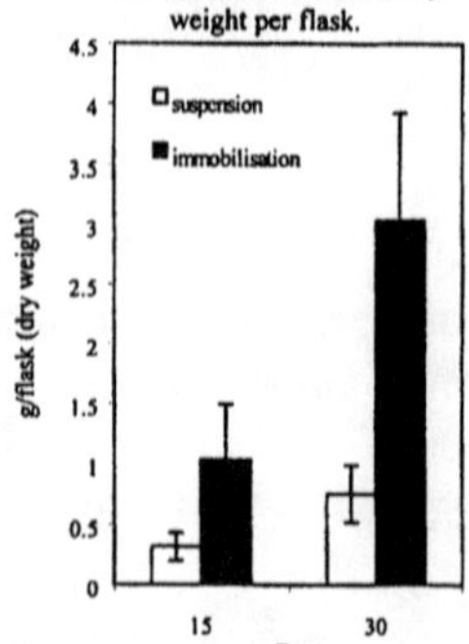

Fig. 4 The growth index in different bioreactors after 30 days.

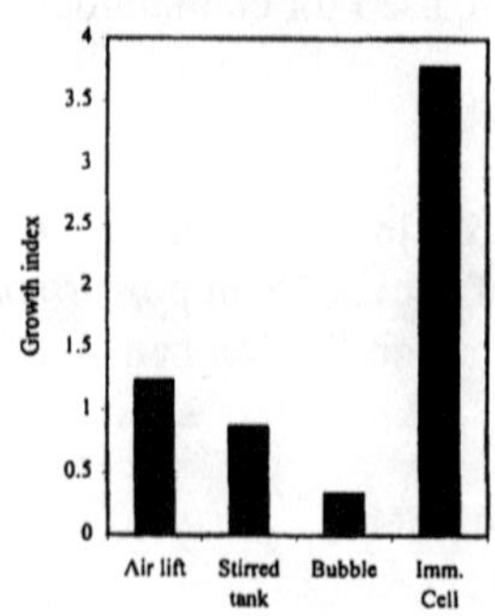

Table 3 *Percentage of callus initiation on different media and photo-regime after 30 days.*

Medium		Part 1 Photoregime = Dark
No	% induced	Callus formation
1	54	Yellow, fine

***Table 3** Percentage of callus initiation on different media and photo-regime after 30 days.*

Medium No	% induced	Part 1 Photoregime = Dark Callus formation
2	68	Yellow, fine, slow growth
3	12	Yellow, fine, very slow growth
4	28	Light yellow, fine, slow growth
5	88	Yellow-white, fine
6	72	Yellow-white, fine
7	64	White, fine, slow growth
8	14	White, fine, very slow growth
9	18	White, fine, very slow growth

***Table 3** Percentage of callus initiation on different media and photo-regime after 30 days.*

Medium No	% induced	Part 2 Cycled Light Callus formation
1	36	Light green, hard, very slow growth
2	62	Light green, hard, very slow growth
3	0	
4	4	Yellow-green, hard, very slow growth
5	62	Yellow-green, hard, very slow growth
6	56	Yellow-green, hard, very slow growth
7	32	Green, hard, very slow growth
8	6	Green, hard, very slow growth
9	22	Green, hard, very slow growth

3.2 SUSPENSION CULTURE:

Figure 2 shows the growth curve in terms of both fresh and dry weights during 30 days of cultivation in complete darkness.

3.3 IMMOBILISATION:

T. media cells were immobilised in the open pore network of the reticulated foam cubes initially by the process of filtration brought about by agitation in the bulk liquid. The initial entrapment through filtration led to the adhesion of cells to foam fibres and subsequent growth of aggregates around foam fibres. The majority of cells immobilised were starting to grow from the internal region of the foam cubes. The cell aggregate size started to increase with time. As the foam cube was filled up, the cells started to grow on the exterior surface of the foam cubes. Figure 3 shows that cell immobilisation can promote cell growth. The biomass yield per flask was about 4 times the freely suspended culture starting with the same inoculum concentration.

3.4 BIOREACTORS:

The cells in suspension culture were small and tended to float to the top of the liquid medium. Therefore, cells aggregated easily to form a “meringue” on the liquid surface in the headspace in the reactors. There was no indication of growth in the bubble reactor. The air lift reactor had the best growth, but it also formed a “meringue”. The stirred tank reactor had two impellers and the cells settled on the top impeller and again formed a “meringue”. Immobilised cell reactor provided the best result in terms of increased growth yield and cell viability. Figure 4 shows the results after 30 days of cultivation in terms of growth index.

4. Conclusion

Gamborg's B5 medium supplemented with 1mg/L 2,4-D gave the best result for callus initiation. Callus had better growth in complete darkness. Immobilisation enhanced biomass growth by over 4-fold in both shake flasks and bioreactor compared with suspension cultures.

References

Fang W, Wu Y, Zhou J, Chen W and Fang Q (1993) Phytochem. Anal. 4:115.
Furusaki S, Nozawa T, Isohara T and Furuya T (1988) Appl. Microbiol. Biotechnol. 29:437.
Gamborg OL (1970) Plant Physiol. 45:372.
Haldimann D and Brodelius P (1987) Phytochemistry 26:1431.
Lindsey K, Yeoman MM, Black GM and Mavituna F (1983) FEBS Letters 155:143.
Lloyd. G and McCown B (1980) Comb. Pro. Int. Plant Prop. Soc. Part 30:421.
Mavituna F and Park JM (1985) Biotechnol. Letters 7:637.
Murashige T and Skoog F (1962) Physiol. Plant 15:473.
Strobel GA, Stierle A and Hess WM (1993) Plant Science 92:1-12.
Vidensek N., Lim P., Campbell A. and Carlson C (1990) J. Natural Product 53:1609.

Wheeler NC, Jech KS, Masters S, Brobst SW, Hoover AJ, Snader KM and Alvarado AB (1992) J. Natural Product 55:432.
Witherup KM, Look SA, Stasko MW, Ghiorzi TJ and Muschik GM (1990) J. Natural Product 53:1249.

Part 7

PRE AND PROBIOTICS

THE ROLE OF PREBIOTICS IN HUMAN GUT MICROBIOLOGY

Prebiotic oligosaccharides

CATHERINE E. RYCROFT*, ROBERT A. RASTALL AND GLENN R. GIBSON
*Department of Food Science and Technology, The University of Reading, Reading, RG6 6BZ, UK *Corresponding author*

Abstract

The human large intestine is an intensely colonised area. This is because the colon contains a vast array of microorganisms, thought to represent the majority of all cells in the body. The nutritional, biological and clinical importance of bacteria resident in the gut is becoming increasingly recognised. Although it is known that many disease states involve bacterial metabolism, the human gut microflora may also be considered as relevant to an improvement in host health and welfare. Gut bacteria carry out a multi-disciplinary process known as fermentation, where dietary and indigenously produced residues are metabolised to a variety of different end products. The numerically predominant (culturable) anaerobes in the gut are Gram-negative rods belonging to the genus *Bacteroides*. Other groups which have been identified as significant include bifidobacteria, clostridia, eubacteria, lactobacilli, Gram-positive cocci, coliforms, methanogens and dissimilatory sulphate-reducing bacteria. It is thought that several hundred different bacterial species are present in the human large intestine.

The principal substrates for colonic bacterial growth are dietary carbohydrates which have escaped digestion in the upper gastrointestinal tract. These may be derived from either the diet or by endogenous secretions and include both carbohydrates and proteins. Whilst the products of gut proteolysis may be generally thought of as toxic towards host health, those of saccharolytic digestion may be considered to be benign and in some cases can contribute positively.

In humans, there are positive aspects to the gut fermentation which may improve certain aspects of host health. The microflora contains certain bacteria that can be perceived as health promoting, as well as pathogenic. For instance, bifidobacteria and lactobacilli may help to improve resistance to gut infections by inhibiting the growth of harmful bacteria, reduce blood lipid levels, improve the immune response and be

A. Van Broekhoven et al. (eds.), Novel Frontiers in the Production of Compounds for Biomedical Use, 411–428.
© 2001 *Kluwer Academic Publishers. Printed in the Netherlands.*

involved in protection against gut cancers. Whilst the definitive health outcomes are not clearly defined, there is currently much interest in increasing numbers and activities of these bacteria in the large gut, preferably at the expense of more harmful species. The manner in which this can be achieved is through dietary supplementation.

The use of probiotics has been widely supported. In this case, foodstuffs such as fermented milk products containing viable cultures perceived as beneficial (e.g. lactobacilli, bifidobacteria) are used to proliferate populations in the colon. Probiotics are defined as live microbial feed supplements which beneficially affect the host animal by improving its intestinal microbial balance. To be effective, probiotics must be capable of being prepared in a viable manner and on large scale (e.g. for industrial purposes), whilst during use and under storage the probiotic should remain viable and stable, be able to survive in the intestinal ecosystem and the host animal should gain beneficially from harbouring the probiotic. Some of these requirements may be difficult to attain.

An alternative, or additional, approach is the prebiotic concept. A prebiotic is a non digestible food ingredient that beneficially affects the host by selectively stimulating the growth and/or activity of one or a limited number of bacteria in the colon, that can improve the host health. Thus, the prebiotic approach advocates the administration of non viable entities. Dietary carbohydrates are candidate prebiotics but most promise has been realised with oligosaccharides. In particular the ingestion of fructooligosaccharides has been shown to stimulate bifidobacteria in the lower gut. As prebiotics exploit non-viable food ingredients, their applicability in diets is wide ranging.

1. The Human Large Intestine

The large intestine harbours a very complex and diverse bacterial microflora. Bacterial populations may reach up to 10^{11}-10^{12} bacteria for every gram of gut contents in the colon (Gibson *et al.* 1996). Bacteroides, bifidobacteria, clostridia, eubacteria, lactobacilli, Gram-positive cocci, coliforms, methanogens and sulphate-reducing bacteria seem to be of numerical significance (Gibson *et al.* 1996), and are all capable of growth under strictly anaerobic conditions.

A range of substrates of dietary origin, or produced by the host, are available for fermentation by the colonic microflora. Through diet, resistant starch (RS), non-starch polysaccharides (NSP) such as pectin, cellulose, hemicellulose, guar and xylan, sugars and oligosaccharides like lactose, lactulose, raffinose, stachyose and fructo-oligosaccharides escape absorption in the small intestine and are metabolised by colonic bacteria. Fermentable endogenous substrates include mucin glycoproteins, produced by goblet cells in the colonic epithelium, mucopolysaccharides such as chondroitin sulphate and heparin, and pancreatic and bacterial secretions (Salyers 1979, Cummings and Macfarlane 1986). Proteins and peptides are also available in the colon (Macfarlane *et al.* 1986), although to a lesser extent than the principal carbohydrates. These may originate from the diet, from pancreatic secretions or are produced by bacteria.

Major end-products of colonic fermentation are short-chain fatty acids (SCFA) such as acetate, propionate and butyrate, fermentation intermediates like succinate, lactate and ethanol, and gases such as hydrogen, methane, carbon dioxide and hydrogen sulphide.

Protein and peptide hydrolysis produces amines, ammonia, phenols, indoles and branched-chain fatty acids (BCFA) such as isovalerate and isobutyrate (Gibson and Roberfroid 1995).

2. Beneficial and Pathogenic Bacteria

Colonic bacteria are sometimes categorised as being either beneficial or potentially pathogenic due to their metabolic activities and type of fermentation end-product. Bifidobacteria and lactobacilli are considered to be health-promoting organisms. Lactobacilli may aid digestion of lactose, reduce constipation and infantile diarrhoea, help resist infections such as salmonellae, prevent traveller's diarrhoea and help in irritable bowel syndrome (IBS) (Salminen *et al.* 1993). Bifidobacteria are thought to stimulate the immune system, produce B vitamins, inhibit pathogen growth, reduce blood ammonia and blood cholesterol levels, and help to restore the normal flora after antibiotic therapy (Gibson and Roberfroid 1995). Such 'beneficial' bacteria have been added to yoghurts and fermented milks to be ingested and exhibit such desirable effects in the colon. This is the **probiotic** approach described as "live microbial feed supplements which beneficially affect the host animal by improving its intestinal microbial balance" (Fuller 1989).

3. The Prebiotic Concept

An alternative approach has been proposed where commensal bifidobacteria and lactobacilli are selectively promoted by the intake of certain non-viable substrates, known as **prebiotics**. Gibson and Roberfroid (1995) described a prebiotic as a "non-digestible food ingredient that beneficially affects the host by selectively stimulating the growth and/or activity of one or a limited number of bacteria in the colon, and thus improves host health." For such substrates to be classed as a prebiotic, four criteria should be satisfied (Gibson and Roberfroid 1995):- 1) the substrate must not be hydrolysed or absorbed in the stomach or small intestine, 2) it must be selective for beneficial commensal bacteria in the colon by encouraging the growth/metabolism of the organisms, 3) it will alter the microflora to a healthy composition, 4) the substrate must induce beneficial luminal/systemic effects within the host. Non-digestible oligosaccharides (NDOs) are dietary substrates which meet these criteria and possess prebiotic potential.

4. Methods for Evaluating Prebiotics

Both in vitro and in vivo methods have been used to determine the effects of different substrates on the human faecal microflora.

4.1. IN VITRO METHODS

The simplest in vitro fermenters are static batch cultures where the substrate is added at a known concentration to a vessel. These may contain pure cultures, defined mixed inocula or a faecal suspension, are incubated anaerobically at 37°C and sampled at intervals (Wang and Gibson 1993). As batch fermenters are closed systems where the substrate is limited, the culture follows typical bacterial growth curves and can only be used for short time course experiments. Moreover, the use of pure cultures does not allow competitive interactions to be investigated. Steady state continuous culture systems can also be used to simulate the intestinal ecosystem. These may range from single to multiple-stage chemostats. The latter can be useful gut models, in that physicochemical conditions imposed on each vessel can be made to represent a different region of the intestine. However, these models do not fully simulate the situation in vivo.

4.2. IN VIVO METHODS

Various in vivo approaches can be used. Animals, usually rats or mice, have been used to determine the effect of substrates on the faecal microflora. Three types of animal have been used; those with their own microflora (Morishita and Konishi 1994, Yamada et al 1993); gnotobiotic (germ-free), or inoculated with one type of organism (Valette et al 1993, Djouzi et al 1995); and those associated with a human faecal flora (HFA) (Rowland and Tanaka 1993, Mallett *et al.* 1987). Obviously however, the ultimate test for the effectiveness of a prebiotic is a human volunteer trial. In this case the trial should be placebo controlled, samples blind coded and a good spread of volunteers used. One drawback is the use of faeces as test material to determine fermentative interactions.

4.3. USE OF MOLECULAR METHODS

In most studies investigating prebiotics, changes in intestinal microflora have been detected using classical phenotypic and biochemical methods of identification. With the advent of a range of molecular techniques, this situation is changing. Pure colonies obtained by classical culturing techniques can be identified, to species level, by sequence analysis. DNA is extracted from colonies, amplified by the polymerase chain reaction (PCR) and then the 16S rRNA gene is sequenced and compared to an information database (Wilson and Blitchington 1996, O'Sullivan 1999). However, the gut microflora consists of a large proportion of unculturable cells which are viable organisms that can be seen under the microscope but do not form colonies on agar plates (Liesack and Stackebrandt 1992). Routine microbiological methods, as well as being time-consuming, labour-intensive and unable to account for these unculturable organisms, are also of low resolution. To overcome this, a range of species and genus-specific oligonucleotide probes have been developed, using information obtained by sequence analysis. Such approaches can be used in situ where total cells are extracted from a faecal sample and hybridised using labelled probes. Perhaps the most common method is fluorescent in situ hybridisation (FISH) whereby genus-specific fluorescently-labelled probes hybridise bacterial rRNA and fluorescent cells enumerated (Langendijk *et al.* 1995, O'Sullivan 1999).

To gain information on the full species diversity of a sample, which does not rely on culturing, direct community analysis (DCA) can be used. This technique involves extracting total DNA or RNA from faeces, amplifying by the PCR, cloning into a vector resulting in many gene clones and sequencing these to identify the range of species present in the faecal sample (O'Sullivan 1999). The use of FISH in conjunction with DCA allows quantitative and qualitative analysis of the bacterial composition (including unculturable organisms) of the intestinal microflora in response to certain substrates.

5. Bifidogenic Factors

Much of the early work was carried out in Japan, e.g. Yazawa *et al.* (1978). A range of oligosaccharides and polysaccharides were screened for their ability to promote *Bifidobacterium infantis* and *Bifidobacterium breve* in pure culture and compared to the effect on *Escherichia coli*, *Streptococcus faecalis* and *Lactobacillus acidophilus* as representatives of other intestinal bacteria. Of all the substrates tested, *B. infantis* utilised a range of mono- and disaccharides, maltotriose (Glu α1-4 Glu α1-4 Glu), raffinose (Gal α1-6 Glu α1-2β Fru), stachyose (Gal α1-6 Gal α1-6 Glu α1-2β Fru), inulin (Glu α1-2[β Fru 1-2]$_n$ where n>10) and oligosaccharides from cellulose, amylose, dextran and inulin. Of these, raffinose, stachyose, inulin of low molecular weight (<4500) and tri- to penta-saccharides from dextran were useful substrates for bifidobacteria.

Yazawa and Tamura (1982) suggested that the selectivity for bifidobacteria may be improved by a high molecular weight and a fructose molecule at the reducing end of the sugar. It has been concluded that bifidobacteria need sugar sources to supply energy and cellular components, and this results in the production of lactic and acetic acids which can inhibit the growth of potential pathogens (Tamura 1983). Generally, di-, tri- and tetrasaccharides consisting of glucose, galactose and fructose monomers were well utilised by bifidobacteria (Tamura 1983). Yazawa and Tamura (1982) concluded that raffinose was a good bifidobacterial-promoting factor.

Minami *et al.* (1983) used a similar method to evaluate the fermentation of enzymatically synthesised isogalactobiose (Gal β-β Gal), galsucrose (Gal α-β Fru) and lactosucrose (Gal β1-4α Glu α1-2β Fru) in comparison to raffinose, lactose and glucose. Lactose and glucose were utilised by *B. infantis*, *Bifidobacterium longum*, *Bifidobacterium adolescentis*, *L. acidophilus*, *E. coli* and *S. faecalis* but the other four sugars were used only by the bifidobacteria. Whilst such techniques are useful baseline studies, it is the selectivity of fermentation that is a major prerequisite for prebiotics. In this case, mixed culture experiments are required.

6. Oligosaccharides as Prebiotics

A range of oligosaccharides have been assessed for their prebiotic potential using in vitro methods, animal models and human clinical trials.

6.1 LACTULOSE

Lactulose is a synthetic disaccharide in the form Gal β1-4 Fru. Lactulose was originally used as a laxative as it is not hydrolysed or absorbed in the small intestine (Saunders and Wiggins 1981). Lactulose has also received attention as a bifidogenic factor and has been administered as such (Tamura 1983, Modler *et al.* 1990, Modler 1994).

Lactulose increased lactobacilli and bifidobacteria and significantly decreased bacteroides in a mixed continuous faecal culture (Fadden and Owen 1992), although total bacterial numbers decreased. A 10% (w/v) lactulose diet fed to rats significantly increased bifidobacteria with little change in total bacteria or lactobacilli (Suzuki *et al.* 1985). As no other bacterial groups were enumerated, the full microbial diversity was not represented.

Bifidobacteria also significantly increased while *Cl. perfringens*, bacteroides, streptococci and *Enterobacteriaceae* significantly decreased on feeding lactulose to eight humans at 3g/day for 14 days (Terada *et al.* 1992). In addition, decreases in the detrimental metabolites ammonia, indole, phenol, *p*-cresol and skatole, and enzymes β-glucuronidase, nitroreductase and azoreductase supported beneficial claims of lactulose. Although lactulose possesses prebiotic activity, it is not yet widely distributed as such.

6.2. INULIN AND FRUCTO-OLIGOSACCHARIDES

Inulin is a polysaccharide of the form Glu α1-2[β Fru 1-2]$_n$ where n > 10 (Crittenden and Playne 1996). The structural relatives of inulin, fructo-oligosaccharides (FOS, a lower molecular weight version) have been the best documented oligosaccharides for their effect on intestinal bifidobacteria and are considered important prebiotic substrates. They are produced in large quantities in several countries and are added to various products such as biscuits, drinks, yoghurts, breakfast cereals and sweeteners (Mizota 1996). The term "fructo-oligosaccharides" may refer to the inulin hydrolysis product oligofructose (OF) or the FOS products known as "Neosugar" or "Meioligo". The latter are mixtures of 1-kestose (Glu-Fru$_2$), 1-nystose (Glu-Fru$_3$) and 1^F-β-fructofuranosylnystose (Glu-Fru$_4$) (Hidaka *et al.* 1986).enzymatically synthesised from sucrose (Hidaka *et al.* 1986).

Batch culture studies where faecal slurries were incubated with inulin, OF, starch, polydextrose, fructose and pectin for 12h (Wang and Gibson 1993) showed the greatest increase in bifidobacteria with OF and inulin, indicating the prebiotic nature of these substrates. Continuous culture systems inoculated with faecal slurries were later used to investigate FOS fermentation (Gibson and Wang 1994). In accordance with earlier studies, bifidobacteria, and to a lesser extent lactobacilli preferred OF and inulin to glucose, whereas bacteroides could not grow on OF. By varying parameters in the chemostat, optimum conditions for growth of bifidobacteria but inhibition of bacteroides, clostridia and coliforms were concluded to be low pH (pH 5.5), high culture dilution rate ($0.3h^{-1}$) and 1% (w/v) concentration of carbohydrate, i.e. similar to the physicochemical environment of the proximal colon. Three-stage chemostats, have confirmed the enhanced proliferation of bifidobacteria by OF in conditions resembling the proximal colon (Gibson and Wang 1994, McBain and Macfarlane 1997).

A later single-stage chemostat study, with FOS, (Sghir *et al.* 1998) demonstrated discrepancies between classical microbiological techniques and molecular approaches. Agar plate counts showed an increase in the combined populations of bifidobacteria and lactobacilli to reach 98.7% of the total bacterial flora by steady state. However, 16S rRNA genus-specific probes indicated an initial increase in the bifidobacterial population which decreased after 6 days, whilst lactobacilli thrived in the low pH fermenter (pH 5.2-5.4) maintaining a high population at steady state. Changes observed in the SCFA profile corresponded well with the population data obtained through probe methods.

Rats that were previously fed tyrosine and tryptophane (capable of producing putrefactive products) were administered a 10% (w/v) Neosugar diet, and this resulted in increased SCFA, decreased faecal pH, and significantly decreased concentrations of the tyrosine derivatives phenol and p-cresol (Hidaka *et al.* 1986).

Several studies have been conducted using human subjects although the dose, substrate, duration and volunteers vary (Table 1). There were large variations between the subjects in their microflora compositions and response to the substrates, particularly between Western and Eastern subjects (Buddington *et al.* 1996). Also, a greater bifidogenic effect was seen in subjects with a low initial bifidobacterial count (10^7/g of faeces) than in those with high initial numbers ($10^{9.5}$/g of faeces) (Hidaka *et al.* 1986).

Table 1. *In vivo studies with inulin and fructo-oligosaccharides (FOS)*

Subjects	Substrate	Dose	Duration	Results	Reference
23 Senile adults	Neosugar	8g/d	14d	Significantly increased bifidobacteria	Hidaka *et al.* (1986)
2 adults	Neosugar	8g/d	2 months	Increase in bifidobacteria, Reduction in SCFA and putrefaction	Hidaka *et al.* (1986)
10 adults	Neosugar	4g/d	14d	Increased bifidobacteria and lactobacilli	Williams *et al.* (1994)
10 adults	FOS	12.5g/d	12d	Significant increase in bifidobacteria	Bouhnik *et al.* (1994)
8 adults	Oligofructose (OF), and inulin	15g/d (in controlled diet)	15d	Both test substrates significantly increased bifidobacteria. OF significantly reduced bacteroides, clostridia and fusobacteria	Gibson *et al.* (1995)

Table 1. *In vivo studies with inulin and fructo-oligosaccharides (FOS)*

Subjects	Substrate	Dose	Duration	Results	Reference
12 adults	Neosugar	4g/d (in controlled diet)	25d	Significant increase in aerobes, enterobacteria and bifidobacteria. Only the latter decreased after feeding stopped	Buddington *et al.* (1996)
10 senile adults	Inulin	20g/d, then 40g/d	8d, then 11d	Significant increase in bifidobacteria at 40g/d, significant reduction in enterococci, bacteroides and enterobacteria	Kleesen *et al.* (1997)

6.3. GALACTO-OLIGOSACCHARIDES

Galacto-oligosaccharides are galactose-containing oligosaccharides of the form Glu α1-4[β Gal 1-6]$_n$ where n= 2-5, and are produced from lactose syrup using the transgalactosylase activity of the enzyme β-galactosidase (Crittenden 1999). Three products are available with slightly differing compositions. Transgalactosylated oligosaccharides (TOS), are produced using β-galactosidase from *Aspergillus oryzae* (Tanaka *et al.* 1983), and consist of tri-, tetra-, penta- and hexa-galacto-oligosaccharides. Oligomate 55, is prepared using β-galactosidase from *A. oryzae* and *Streptococcus thermophilus* (Ito *et al.* 1990) and contains 36% tri-, tetra-, penta- and hexa-galacto-oligosaccharides, 16% disaccharides galactosyl glucose and galactosyl galactose, 38% monosaccharides and 10% lactose. Finally, a transgalactosylated disaccharide (TD) preparation is produced using β-galactosidase from *S. thermophilus* (Ito *et al.* 1993).

Adding 20g/day TOS to a continuous culture (Durand *et al.* 1992) and 10g/d TOS to a semi-continuous culture containing human faecal bacteria (Bouhnik *et al.* 1997) increased gas and SCFA. Although bacteria were not enumerated, increased lactate and acetate was suggested to be result of a proliferation of lactic acid bacteria (lactobacilli and bifidobacteria) in response to TOS addition.

Oligomate, TOS and TD have been studied in rats, HFA rats and humans to demonstrate their prebiotic effects (Table 2). All three gave a prebiotic effect at doses of 10g/d for Oligomate (Ito *et al.* 1990) and TOS (Tanaka *et al.* 1983), and 15g/d for TD (Ito *et al.* 1993). The prebiotic effect was clear, although Teuri *et al.* (1998) found no significant change in bifidobacteria but a significant increase in total bacteria on MRS, a medium purportedly selective for lactic acid bacteria.

Unlike TOS, Oligomate (10g/d) increased gas production and bloating in humans according to symptoms recorded by the subjects. It is possible that a higher proportion of monosaccharides and lactose, and reduced content of higher oligosaccharides may have led to elevated gas production.

Table 2. *Rat and human studies with galacto-oligosaccharide preparations*

Subjects	Substrate	Dose	Duration	Results	Reference
Rats	TOS	5% (w/v)	7 weeks	Significant increase in bifidobacteria, SCFA and acetate. Significant decrease in staphylococci and streptococci	Morishita and Konishi (1994)
Human flora associated (HFA) rats	TOS	5% or 10% (w/v)	Unknown	Significant increase in SCFA and H_2. For the10% diet, significant decrease in lactate and succinate, decreased butyrate, valerate and isoacids	Andrieux and Szylit (1992)
HFA rats	TOS	5% (w/v)	4 weeks	Significant increase in bifidobacteria and lactobacilli, significant decrease in enterobacteria. Significant decrease in nitrate reductase, β-glucuronidase and conversion of IQ[a] to 7-OHIQ[b]	Rowland and Tanaka (1993)
5 men	TOS	3g/d, then 10g/d	1week, then 1week	3g/d - little effect 10g/d – significant increase in bifidobacteria and lactobacilli and significant decrease in bacteroides	Tanaka *et al.* (1983)
12 men	Oligomate	0, 2.5, 5 then 10g/d	1 week for each dose	Bifidobacteria increased with dose. For 10g/d ,significant increase in bifidobacteria and lactobacilli	Ito *et al.* (1990)
12 men	TD	15g/d	6d	Significant increase in bifidobacteria and lactobacilli, significantly lower bacteroides. Significant decrease in indole, p-cresol, NH_3, propionate, valerate, isovalerate and isobutyrate	Ito *et al.* (1993)
8 adults	TOS	10g/d	21d	Significantly increased bifidobacteria, significant reduction in breath H_2	Bouhnik *et al.* (1997)

Table 2. *Rat and human studies with galacto-oligosaccharide preparations*

Subjects	Substrate	Dose	Duration	Results	Reference
6 adults	TOS	15g/d	14d	Significant increase in total count on MRS[c]. No change in bifidobacteria	Teuri *et al.* (1998)

[a] 2-amino-3-methyl-3H-imidazo[4,5-*f*]quinoline
[b] Hydroxy derivative of IQ, a carcinogen stimulating DNA damage in the colonic mucosa, resulting in colon cancer
[c] Agar for enumeration of lactic acid bacteria and bifidobacteria

6.4. SOYBEAN OLIGOSACCHARIDES

The soybean oligosaccharides are raffinose and stachyose which promote growth of *B. infantis* but not *E. coli, S. faecalis* or *L. acidophilus* (Tamura 1983). Soybean oligosaccharide extract (SOE) consists of 23% (w/v) stachyose and 7% (w/v) raffinose; and when purified by activated charcoal chromatography contains 71% stachyose, 20% raffinose and 2% other sugars, i.e. a refined soybean oligosaccharide product (SOR). In pure culture studies, SOR were fermented to a far greater degree by bifidobacteria than any other organisms tested (Hayakawa *et al.* 1990).

The addition of a low concentration (0.1% (w/v)) of SOR to a two-stage continuous culture of faecal bacteria (Saito *et al.* 1992) resulted in a three-fold increase in the proportion of bifidobacteria in the total bacterial count. As only these bacterial groups were enumerated, any other changes that occurred in the microflora were overlooked. A significant decrease in azoreductase activity was recorded as well as large decreases in β-glucosidase and β-glucuronidase. These results do not correspond fully with those from in vivo trials however (Wada *et al.* 1992).

Human trials have been carried out to assess the prebiotic activity of soybean oligosaccharides (Table 3). Again variation was observed between volunteers but overall, raffinose and SOE both showed prebiotic activity although less of the latter was needed to induce a response.

Table 3. *Human studies with soybean oligosaccharides*

Subjects	Substrate	Dose	Duration	Results	Reference
7 adults	Raffinose	15g/d	4 weeks	Significant increase in bifidobacteria, significant decrease in bacteroides and clostridia	Benno *et al.* (1987)
6 adults	SOE	10g/d	3 weeks	Significant increase in bifidobacteria and	Hayakawa *et*

Table 3. Human studies with soybean oligosaccharides

Subjects	Substrate	Dose	Duration	Results	Reference
				lactobacilli, large decrease in clostridia and peptostreptococci	*al.* (1990)
7 adults	SOE	10.6g/d	3 weeks	Significant increase in bifidobacteria, slight decrease in bacteroides, significant decrease in NH_3 and β-glucosidase, large decrease in indole, skatole, phenol and p-cresol	Wada *et al.* (1992)

6.5. LACTOSUCROSE

Lactosucrose is produced from a mixture of lactose and sucrose using the enzyme β-fructofuranosidase (Playne and Crittenden 1996) and has been found to be bifidogenic in pure culture studies (Tamura 1983; Fujita *et al.* 1991).

A later pure culture study compared lactosucrose with lactulose, FOS, SOR, raffinose and glucose for its utilisation by various intestinal bacteria (Hara *et al.* 1994). Six bifidobacteria and three lactobacilli strains grew to the same extent (comparable end pH) on lactosucrose and glucose, whereas all the other organisms tested preferred glucose. Contrary to the data of Fujita *et al.* (1991), lactosucrose did not appear to favour bifidobacterial growth. However, as the end pH was used to represent growth, this does not take into account the rate of utilisation or the fact that the amount or type of acid produced varies with different microorganisms.

Studies in cats and humans (Table 4) have demonstrated the prebiotic potential of lactosucrose and associated desirable effects such as increased SCFA, lactate and acetate, decreases in putrefactive metabolites, reduction in detrimental enzymes known to produce toxins/carcinogens and lowering of faecal odour in cats. Although lactosucrose appears to have some prebiotic effects even at 3g/d, there is a need for larger studies with adults, with particular regard to the minimum effective dose.

Table 4. In vivo studies with lactosucrose

Subjects	Substrate	Dose	Duration	Results	Reference
2 adults	55 % lactosucrose	20g/d	10d	Significant increase in bifidobacteria, decreased bacteroides and clostridia	Fujita *et al.* (1991)
13 senile	55%	Approx.	3 weeks	Significant increase in	Kumemura *et*

Table 4. In vivo studies with lactosucrose

Subjects	Substrate	Dose	Duration	Results	Reference
adults	lactosucrose	10.3g/d		bifidobacteria, significant increase in bacteroides, eubacteria and peptococci. Significant decrease in NH_3, skatole, indole and p-cresol. Significant increase in acetate and significant decrease in butyrate	*al.* (1992)
8 cats	95% lactosucrose	175mg/d	2 weeks	Significant increase in bifidobacteria and lactobacilli, significant decrease in clostridia, enterobacteria, fusobacteria, staphylococci and *Spirochaetae*. Significant decrease in indole, NH_3, ethylphenol and significant decrease in phenol, skatole and p-cresol. Decrease in faecal odour	Terada *et al.* (1993)
8 adults	95% lactosucrose	3g/d, then 6g/d	1 week, then 1 week	Significant increase in bifidobacteria, significant decrease in clostridia and bacteroides. Significant increase in NH_3, sulphide, phenol, and indole, decreased p-cresol, skatole and ethylphenol. Significantly decreased β-glucuronidase, azoreductase and nitrate reductase. Significant increase in SCFA, lactate and acetate	Hara *et al.* (1994)

6.6. ISOMALTO-OLIGOSACCHARIDES

Isomalto-oligosaccharides (IMO) are composed of glucose monomers linked by α1-6 glucosidic linkages. The commercial mixture, Isomalto-900 produced by incubating α-amylase, pullulanase and α-glucosidase with cornstarch (Kohmoto *et al.* 1988) contains isomaltose (Glu α1-6 Glu), isomaltotriose (Glu α1-6 Glu α1-6 Glu) and panose (Glu α1-6 Glu α1-4 Glu).

Pure culture studies showed panose, isomaltose, isomaltotriose and Isomalto-900 to be utilised as well as raffinose by all the bifidobacteria tested, with the exception of *B. bifidum* which gave no growth on any of the substrates (Kohmoto *et al.* 1988). Bacteroides species utilised all the sugars but fewer clostridia grew on the IMO than the raffinose. Overall, IMO appeared at least as selective, if not more so, for bifidobacteria than raffinose.

Human studies showed the substrates to possess prebiotic potential (Table 5) with a minimum effective dose of 8-10g IMO (13-15g/d Isomalto-900) (Kohmoto *et al.* 1991). The IMO3 fraction had a greater prebiotic effect than IMO2 (Kaneko *et al.* 1994) but further studies are needed to compare the effects of IMO2 and IMO3 on bacteria other than bifidobacteria. Our own (unpublished) studies with a 3-stage continuous culture model of the gut have shown that IMO fermentation maintained a lactic acid flora whilst also allowing the generation of butyrate. As this is thought to be a desirable metabolite of colonic function, it may be that IMO's are effective prebiotics.

Table 5. *Human studies with isomalto-oligosaccharides (IMO)*
[a] Isomaltose, isomaltotriose and panose.
[b] Fraction of Isomalto-900, containing 63.8% isomaltose and 22.6% other disaccharides (nigerose and kojibiose).
[c] Fraction of Isomalto-900, containing 27.7% panose, 12.1% isomaltotriose and 50.1% tetra-, penta- and hexasaccharide components.

Subjects	Substrates	Dose	Duration	Results	Reference
6 men	Isomalto-900 (64% IMO)	20g/d	10d	Significantly increased bifidobacteria, lesser increase in bacteroides	Kohmoto *et al.* (1988)
18 senile adults	Isomalto-900	20g/d	14d	Significantly increased bifidobacteria, esp. with low initial bifidobacteria counts, lesser increase in bacteroides	Kohmoto *et al.* (1988)
12 men	IMO[a]	4.3, 6.5, 8.2 or 9.8g/d	10d for each dose	Significantly increased bifidobacteria for all doses but significant increase only at 9.8g/d	Kohmoto *et al.* (1991)
9 men	IMO	10g/d	14d	Significantly increased bifidobacteria Tendency for most other bacteria to increase	Kohmoto *et al.* (1991)
2 groups of 7 men	IMO2[b] or IMO3[c]	10, 20 then 5g/d	12d for each dose	IMO2-significantly increased bifidobacteria at 20g/d, no effect at 5 or 10g/d. IMO3- significantly increased bifidobacteria with all three doses	Kaneko *et al.* (1994)

6.7. GLUCO-OLIGOSACCHARIDES

Gluco-oligosaccharides (GOS) are enzymatically synthesised, using a glucosyl-transferase from *Leuconostoc mesenteroides,* which transfers glucose molecules from the sucrose donor to the acceptor, maltose (Valette *et al.* 1993). The resulting mixture is composed of 18% mono-, di- and trisaccharides, 18% tetrasaccharides, 33% pentasaccharides and 31% hexa- and heptasaccharides comprising glucose units linked by α1-6 and α1-2 glycosidic bonds. GOS was poorly hydrolysed and digested in the intestinal tract of gnotobiotic rats (Valette *et al.* 1993).

A defined mixed culture of *Bact. thetaiotamicron*, *B. breve* and *Cl. butyricum* incubated with 0.5% (w/v) GOS (Djouzi *et al.* 1995) resulted in no change in the *Bact. thetaiotamicron* population but increased the *B. breve* count, and reduced *Cl. butyricum* numbers. However, such data should be interpreted with caution, as the use of defined mixed culture does not allow the full diversity and complexity of the intestinal microflora to be determined.

The same three organisms were inoculated separately (monoxenic) or jointly (trixenic) into gnotobiotic rats, half of which were fed a control diet and the other half a 2% (w/v) GOS diet for four weeks (Djouzi *et al.* 1995, Table 6). In the trixenic rats, decreased butyrate and hydrogen production suggested that clostridia did not degrade the GOS to a great extent, as observed in co-culture study (Djouzi *et al.* 1995). GOS was degraded equally by *B. breve* and *Bact. thetaiotamicron* in monoxenic rats, but not to the same extent by *Cl. butyricum*. In the trixenic rats, GOS had no prebiotic effect in terms of bacterial populations.

Table 6. *Effects of gluco-oligosaccharides GOS administration on monoxenic and trixenic rats*[a]

Changes after 2% (w/v) GOS administration[c]				
Rats[b]	**Bacterial populations**	**Hydrogen production**	**SCFA profile**	**GOS digestibility (%)**[d]
Bacteroides thetaiotaomicron	No change	Low (for control or GOS diet)	No change	82
Bifidobacterium breve	Slight increase	None (for control or GOS diet)	No change	80
Clostridium butyricum	No change	High, especially on GOS diet	Significant increase in SCFA, slight increase in acetate	58
Trixenic	No change in any organism	Decreased on GOS feeding	Significant increase in SCFA, decreased succinate and lactate	93

[a] *Data from Djouzi* et al. *(1995)*
[b] *Gnotobiotic rats inoculated with one (monoxenic) or the three organisms (trixenic).*
[c] *Changes compared to control rats which were not fed GOS.*
[d] *GOS digestibility was 21% in gnotobiotic rats.*

Controlled human studies are needed to elicit the response of faecal bacteria, metabolites, SCFA, enzyme activity and gas production when GOS is administered.

6.8. XYLO-OLIGOSACCHARIDES

Xylo-oligosaccharides (XOS) are chains of xylose molecules linked by β1-4 bonds and mainly consist of xylobiose, xylotriose and xylo-tetraose (Hopkins *et al.* 1998). They are produced enzymatically by hydrolysis of xylan from birch wood (Campbell *et al.* 1997), oats (Jaskari *et al.* 1998) or corncobs (Playne and Crittenden 1996). Few studies have been conducted on XOS fermentation by gut bacteria, although Okazaki *et al.* (1990) carried out a small trial in which five male subjects were fed 1g and five were fed 2g/day XOS for 3 weeks. The volunteers showed some variation in their response to the XOS but overall, both doses significantly increased faecal bifidobacteria and decreased bacteroides, the changes being more pronounced for the higher dose. When XOS administration ceased, the proportion of bifidobacteria decreased to its original level. A dose of 1g/d XOS was sufficient to effect a bifidogenic response, although larger controlled human studies are needed to study a range of doses, to compare XOS to other prebiotics such as FOS, and observe the effect of XOS administration on normal gut function.

7. Conclusions

This review has suggested that many potential prebiotic agents exist. However, there is now a clear need for large comparative studies with a range of potentially prebiotic oligosaccharides using a thorough methodology, with molecular procedures directed towards the human gut flora being an important development. It is equally critical that appropriate food vehicles for use be determined, as well as the minimum effective dose and preferred target population groups. Moreover, the realistic health consequences need to be determined.

References

Andrieux, C. and Szylit, O. (1992) Effects of galacto-oligosaccharides (TOS) on bacterial enzyme activities and metabolite production in rats associated with a human faecal flora, *Proceedings of the Nutrition Society 51,* 7A.

Benno, Y., Endo, K., Shiragami, N., Sayama, K. and Mitsuoka, T. (1987) Effects of raffinose intake on human fecal microflora, *Bifidobacteria Microflora 6*, 59-63.

Bouhnik, Y., Flourie, B., Ouarne, F., Riottot, M., Bisetti, N., Bornet, F. and Rambaud, J.C. (1994) Effects of prolonged ingestion of fructo-oligosaccharides (FOS) on colonic bifidobacteria, fecal enzymes and bile acids in humans, *Gastroenterology 106,* (4) A598.

Bouhnik, Y., Flourie, B., D'Agay-Abensour, L., Pochart, P., Gramet, G., Durand, M. and Rambaud, J.C.(1997) Administration of transgalacto-oligosaccharides increases fecal bifidobacteria and modifies colonic fermentation metabolism in healthy humans, *Journal of Nutrition 127,* 444-448.

Buddington, R.K., Williams, C.H., Chen, S. and Witherly, S.A. (1996) Dietary supplement of neosugar alters the fecal flora and decreases activities of some reductive enzymes in human subjects, *American Journal of Clinical Nutrition 63,* 709-716.

Campbell, J.M., Fahey, G.C. and Wolf, B.W. (1997) Selected indigestible oligosaccharides affect large bowel mass, cecal and fecal short-chain fatty acids, pH and microflora in rats, *Journal of Nutrition 127,* 130-136.

Crittenden, R.G. and Playne, M.J. (1996) Production, properties and applications of food-grade oligosaccharides, Trends in Food Science and Technology 7, 353-361.

Crittenden, R.G. (1999) Prebiotics. In : Probiotics: a critical review (G.Tannock, ed.) Wymondham, Horizon Scientific Press.

Cummings, J.H. and Macfarlane, G.T. (1986) The control and consequences of bacterial fermentation in the human colon, Journal of Applied Bacteriology 70, 443-459.

Djouzi, Z., Andrieux, C., Pelenc, V., Somarriba, S., Popot, F., Paul, F., Monsan, P. and Szylit,O. (1995) Degradation and fermentation of α-gluco-oligosaccharides by bacterial strains from human colon:in vitro and in vivo studies in gnotobiotic rats, Journal of Applied Bacteriology 79, 117-127.

Durand, M., Cordelet, C., Hannequart, G. and Beaumatin, P. (1992) In vitro fermentation of a galacto-oligosaccharide by human bacteria in continuous culture, Proceedings of the Nutrition Society 51, 6A.

Fadden, K. and Owen, R.W. (1992) Faecal steroids and colorectal cancer: the effect of lactulose on faecal bacterial metabolism in a continuous culture model of the large intestine, European Journal of Cancer Prevention 1, 113-127.

Fujita, K., Hara, K., Sakai, S., Miyake, T., Yamashita, M., Tsunetomi, Y. and Mitsuoka, T. (1991) Effect of 4^G-β-D-galactosylsucrose (lactosucrose) on intestinal flora and its digestibility in humans, Denpun Kagaku 38, 249-255.

Fuller, R. (1989) Probiotics in man and animals, Journal of Applied Bacteriology 66, 365-378.

Gibson, G.R. and Wang, X. (1994) Enrichment of bifidobacteria from human gut contents by oligofructose using continuous culture, FEMS Microbiology Letters 118, 121-128.

Gibson, G.R., Beatty, E.R., Wang, X. and Cummings, J.H. (1995) Selective stimulation of bifidobacteria in the human colon by oligofructose and inulin, Gastroenterology 108, 975-982.

Gibson, G.R. and Roberfroid,M.B. (1995) Dietary modulation of the human colonic microbiota : introducing the concept of prebiotics, Journal of Nutrition 125, 1401-1412.

Gibson, G.R., Willems, A., Reading, S. and Collins, M.D. (1996) Fermentation of non-digestible oligosaccharides by human colonic bacteria, Proceedings of the Nutrition Society 55, 899-912.

Hara, H., Li, S., Sasaki, M., Maruyama, T., Terada, A., Ogata, Y., Fujita, K., Ishigami, H., Hara, K., Fujimori, I. and Mitsuoka, T. (1994) Effective dose of lactosucrose on fecal flora and fecal metabolites of humans, Bifidobacteria Microflora 13, 51-63.

Hayakawa, K., Mizutani, J., Wada, K., Masai, T., Yoshihara, I. and Mitsuoka, T. (1990) Effects of soybean oligosaccharides on human faecal flora, Microbial Ecology in Health and Disease 3, 293-303.

Hidaka, H., Eida, T., Takizawa, T., Tokunaga, T. and Tashiro, Y. (1986) Effects of fructooligosaccharides on intestinal flora and human health, Bifidobacteria Microflora 5, 37-50.

Hopkins, M.J., Cummings, J.H. and Macfarlane, G.T. (1998) Inter-species differences in maximum specific growth rates and cell yields of bifidobacteria cultured on oligosaccharides and other simple carbohydrate sources, Journal of Applied Microbiology 85, 381-386.

Ito, M., Deguchi, Y., Miyamori, A., Matsumoto, K., Kikuchi, H., Matsumoto, K., Kobayashi, Y., Yajima, T. and Kan, T. (1990) Effects of administration of galactooligosaccharides on the human fecal microflora, stool weight and abdominal sensation, Microbial Ecology in Health and Disease 3, 285-292.

Ito, M., Kimura, M., Deguchi, Y., Miyamori-Watabe, A., Yajima, T. and Kan, T. (1993) Effects of transgalactosylated disaccharides on the human intestinal microflora and their metabolism, Journal of Nutritional Science and Vitaminology 39, 279-288.

Jaskari, J., Kontula, P., Siitonen, A., Jousimies-Somer, H., Mattila-Sandholm, T. and Poutanen,K. (1998) Oat β-glucan and xylan hydrolysates as selective substrates for Bifidobacterium and Lactobacillus strains, Applied Microbiology and Biotechnology 49, 175-181.

Kaneko, T., Kohmoto, T., Kikuchi, H., Shiota, M., Iino, H. and Mitsuoka, T. (1994) Effects of isomaltooligosaccharides with different degrees of polymerisation on human fecal bifidobacteria, Bioscience and Biotechnological Biochemistry 58, 2288-2290.

Kleessen, B., Sykura, B., Zunft, H-J. and Blaut, M. (1997) Effects of inulin and lactose on fecal microflora, microbial activity and bowel habit in elderly constipated persons, American Journal of Clinical Nutrition 65, 1397-1402.

Kohmoto, T., Fukui, F., Takaku, H., Machida, Y., Arai, M. and Mitsuoka, T. (1988) Effect of isomalto-oligosaccharides on human fecal flora, Bifidobacteria Microflora 7, 61-69.

Kohmoto, T., Fukui, F., Takaku, H. and Mitsuoka, T. (1991) Dose-response test of isomaltooligosaccharides for increasing fecal bifidobacteria, Agricultural and Biological Chemistry 55, 2157-2159.

Kumemura, M., Hashimoto, F., Fujii, C., Matsuo, K., Kimura, H., Miyazoe, R., Okamatsu, H., Inokuchi, T., Ito, H., Oizumi, K. and Oku, T. (1992) Effects of administration of 4^G-β-D-galactosylsucrose on fecal microflora, putrefactive products, short-chain fatty acids, weight, moisture and pH, in the subjective sensation of defecation in the elderly with constipation, Journal of Clinical and Biochemical Nutrition 13, 199-210.

Langendijk, P.S., Schut, F., Jansen, G.J., Raangs, G.C., Kamphuis, G.R., Wilkinson, M.H.F. and Welling, G.W. (1995) Quantitative fluorescence in situ hybridisation of Bifidobacterium spp. with genus-specific 16S rRNA-targeted probes and its application in fecal samples, Applied and Environmental Microbiology 61, 3069-3075.

Liesack, W. and Stackebrandt, E. (1992) Unculturable microbes detected by molecular sequences and probes, Biodiversity and Conservation 1, 250-262.

Macfarlane, G.T., Cummings, J.H. and Allison, C. (1986) Protein degradation by human intestinal bacteria, Journal of General Microbiology 132, 1647-1656.

Mallett, A.K., Bearne, C.A., Rowland, I.R., Farthing, M.J.G., Cole, C.B. and Fuller, R. (1987) The use of rats with a human faecal flora as a model for studying the effects of diet on the human gut microflora, Journal of Applied Bacteriology 63, 39-45.

McBain, A.J. and Macfarlane, G.T. (1997) Investigations of bifidobacterial ecology and oligosaccharide metabolism in a three-stage compound continuous culture system, Scandinavian Journal of Gastoenterology 32, 32-40.

Minami, Y., Yazawa, K., Tamura, Z., Tanaka, T. and Yamamoto, T. (1983) Selectivity of utilisation of galactosyl-oligosaccharides by bifidobacteria, Chemical and Pharmaceutical Bulletin 31, 1688-1691.

Mizota, T. (1996) Functional and nutritional foods containing bifidogenic factors, Bulletin of the International Dairy Foundation 313, 31-35.

Modler, H.W., McKellar, R.C. and Yaguchi, M. (1990) Bifidobacteria and bifidogenic factors, Canadian Institute of Food Science and Technology Journal 23, 29-41.

Modler, H.W. (1994) Bifidogenic factors - sources, metabolism and applications, International Dairy Journal 4, 383-407.

Morishita,Y. and Konishi,Y. (1994) Effects of high dietary cellulose on the large intestinal microflora and short-chain fatty acids in rats, Letters in Applied Microbiology 19, 433-435.

O'Sullivan, D.J. (1999) Methods for analysis of the intestinal microflora. In : Probiotics : a critical review (G. Tannock, ed.) Wymondham, Horizon Scientific Press.

Okazaki, M., Fujikawa, S. and Matsumoto, N. (1990) Effects of xylooligosaccharide on growth of bifidobacteria, Journal of Japanese Society of Nutrition and Food Science 43, 395-401.

Playne, M.J. and Crittenden, R. (1996) Commercially available oligosaccharides, Bulletin of the International Dairy Foundation 313, 10-22.

Rowland, I.R. and Tanaka, R.(1993) The effects of transgalactosylated oligosaccharides on gut flora metabolism in rats associated with a human faecal microflora, Journal of Applied Bacteriology 74, 667-674.

Saito, Y., Takano, T. and Rowland, I. (1992) Effects of soybean oligosaccharides on the human gut microflora in in vitro culture, Microbial Ecology in Health and Disease 5, 105-110.

Salminen, S., Ramos, P. and Fonden, R. (1993) Substrates and lactic acid bacteria. In : Lactic acid bacteria (S.Salminen and A.von Wright, eds.) New York, Marcel Dekker, Inc.

Salyers, A.A. (1979) Energy sources of major intestinal fermentative anaerobes, American Journal of Clinical Nutrition 32, 158-163.

Saunders, D.R. and Wiggins, H.S. (1981) Conservation of mannitol, lactulose and raffinose by the human colon, American Journal of Physiology 241, G397-G402.

Sghir, A., Chow, J.M. and Mackie, R.I. (1998) Continuous culture selection of bifidobacteria and lactobacilli from human faecal samples using fructooligosaccharide as selective substrate, Journal of Applied Microbiology 85, 769-777.

Suzuki, K., Endo, Y., Uehara, M., Yamada, H., Goto, S., Imamura, M. and Shioza, S. (1985) Effect of lactose, lactulose and sorbital on mineral utilisation and intestinal flora, *Journal of Japanese Society of Nutrition and Food Science 38,* 39-42.

Tamura, Z. (1983) Nutriology of bifidobacteria, *Bifidobacteria Microflora 2,* 3-16.

Tanaka, R., Takayama, H., Morotomi, M., Kuroshima, T., Ueyama, S., Matsumoto, K., Kuroda, A. and Mutai, M. (1983) Effects of administration of TOS and *Bifidobacterium breve* 4006 on the human fecal flora, *Bifidobacteria Microflora 2,* 17-24.

Terada, A., Hara, H., Kataoka, M. and Mitsuoka, T. (1992) Effect of lactulose on the composition and metabolic activity of the human faecal flora, *Microbial Ecology in Health and Disease 5,* 43-50.

Terada, A., Hara, H., Kato, S., Kimura, T., Fujimori, I., Hara, K., Maruyama, T. and Mitsuoka, T. (1993) Effects of lactosucrose (4^G-β-D-galactosylsucrose) on fecal flora and fecal putrefactive products of cats, *Journal of Veterinary and Medical Science 55,* 291-295.

Teuri, U., Korpela, R., Saxelin, M., Montonen, L. and Salminen, S. (1998) Increased fecal frequency and gastrointestinal symptoms following ingestion of galacto-oligosaccharide-containing yoghurt, *Journal of Nutritional Science and Vitaminology 44*, 465-471.

Tomomatsu, H. (1994) Health effects of oligosaccharides, *Food Technology 48,* 61-65.

Valette, P., Pelenc,V., Djouzi, Z., Andrieux, C., Paul, F., Monsan, P. and Szylit, O.(1993) Bioavailability of new synthesised glucooligosaccharides in the intestinal tract of gnotobiotic rats, *Journal of Science of Food Agriculture 62,* 121-127.

Wada, K., Watabe, J., Mizutani, J., Tomoda, M., Suzuki, H. and Saitoh, Y. (1992) Effects of soybean oligosaccharides in a beverage on human fecal flora and metabolites, *Journal of Agricultural Chemical Society of Japan 66,* 127-135.

Wang, X. and Gibson, G.R.(1993) Effects of the in vitro fermentation of oligofructose and inulin by bacteria growing in the human large intestine, *Journal of Applied Bacteriology 75,* 373-380.

Williams, C.H., Witherly, S.A. and Buddington, R.K. (1994) Influence of dietary neosugar on selected bacterial groups of the human faecal microbiota, *Microbial Ecology in Health and Disease 7,* 91-97.

Wilson, K.H. and Blitchington, R.B. (1996) Human colonic bacteria studied by ribosomal DNA sequence analysis, *Applied and Environmental Microbiology 62,* 2273-2278.

Yamada, H., Itoh, K., Morishita, Y. and Taniguchi, H. (1993) Structure and properties of oligosaccharides from wheat bran, *Cereal Foods World 38,* 490-492.

Yazawa, K., Imai, K. and Tamura, Z. (1978) Oligosaccharides and polysaccharides specifically utilisable by bifidobacteria, *Chemical and Pharmaceutical Bulletin 26,* 3306-3311.

Yazawa, K. and Tamura, Z. (1982) Search for sugar sources for selective increase of bifidobacteria, *Bifidobacteria Microflora 1,* 39-44.

THE INFLUENCE OF INTESTINAL MICROFLORA ON MUCOSAL AND SYSTEMIC IMMUNE RESPONSES

STEPHANIE BLUM, DIRK HALLER, SUSANA ALVAREZ, PABLO PEREZ AND EDUARDO J. SCHIFFRIN*
Nestlé Research Center, Nestec LTD, Vers-chez-les-Blanc, 1000 Lausanne 26, Switzerland, Tel. + 41 21 785 8671, Fax + 41 21 785 8925, e-mail:eduardo.schiffrin@rdls.nestle.com Corresponding author*

Summary

The intestinal mucosa, colonised by commensal microorganisms, constitutes the interface with the external environment, through which most pathogens initiate infectious processes in mammals. Therefore, intestinal mechanisms of defence need to discriminate between the commensal, symbiotic microflora, including food ingested probiotic bacteria and exogenous pathogens. To date, we do not fully understand the mechanisms of discrimination. However, there is increasing evidence that innate and adaptive immune responses participate in this process.

To understand the molecular basis of differences in the mucosal response to non-pathogenic microorganisms, different human *in vitro* models were established. Using these *in vitro* models, we studied the capacity of mucosal immunocompetent cells to react to bacterial signals. A characteristic response to different commensal bacteria was detected, discriminating between gram-positive and gram-negative bacteria. Furthermore, differences in cellular activation by food-derived *Lactobacillus* species were observed.

In this chapter, we discuss the importance of non-pathogenic microorganisms, including selected probiotic bacteria, in modulation of the host gut mucosal defences and maintenance of homeostasis and integrity. The understanding of the molecular basis of these modulatory functions will provide a unique opportunity to the food industry to prevent or treat, by nutritional means, intestinal disorders associated with food allergy, intestinal infections, inflammatory bowel disease and autoimmunity.

A. Van Broekhoven et al. (eds.), Novel Frontiers in the Production of Compounds for Biomedical Use, 429–445.
© 2001 *Kluwer Academic Publishers. Printed in the Netherlands.*

Abbreviations

FAE, follicular associated epithelium; GALT, gut associated lymphoid tissue; GIT, gastrointestinal tract; GC, germinal centre; HLA, histocompatibility leukocyte antigen; IBD, inflammatory bowel disease; ICAM-1, intercellular adhesion molecule–1; IEC, intestinal epithelial cell; IEL, intraepithelial lymphocyte; IFNγ, interferon-gamma; IL, interleukin; LAB, lactic acid bacteria; LPL, lamina propria lymphocytes; MCP-1, monocyte chemoattractive protein-1; MHC, major histocompatibility complex; PBMC, peripheral blood mononuclear cells; PRR : Pattern recognition receptor , TGF-β, tumour growth factor-beta; TNFα, tumour necrosis factor-alpha

1.Introduction

Mucosal surfaces represent large areas of interface between the host and the external environment. Physiologically, they can be sterile, i.e. the distal pulmonary track, or colonised as the distal gastrointestinal tract (GIT). Mucosal mechanisms of defence have evolved with common strategies for all mucosal surfaces but, in the case of a colonised mucosa, additional characteristics occur. Whereas a strong response against invasive pathogens has to be mounted, non-responsiveness or hypo-responsiveness to food antigens or indigenous bacteria has to be guaranteed (Underdown & Schiff, 1986). This lack of immunological response is an active process, based on various mechanisms, which are globally called oral tolerance (Garside, Mowat & Khoruts, 1999). Moreover, the intestinal immune system remains non-reactive to the resident microflora, which has also been explained on the basis of immunological tolerance. This phenomenon is of major importance for preservation of gut integrity, which is a basic requirement for the host survival, both from the nutritional and defensive point of view. It is suggested that failure of immunological tolerance towards commensal microflora can be one mechanism in inflammatory bowel disease (IBD) (Duchmann, Kaiser & Hermann, 1995). Under physiological conditions, mucosal defences are able to cope with environmental antigens or infectious agents, without triggering constant and severe inflammation, which would result in tissue damage. Both endogenous mediators and luminal factors, including those derived from bacteria, seem to be implicated in this homeostasis.

Recent work conducted in our group was aimed at the understanding how non-pathogenic, commensal and food derived microorganisms can modulate the mucosal immune response and contribute to the preservation of gut integrity.

2.The innate immune system

In vertebrates two types of immune responses can be distinguished: innate (non-clonal) and adaptive (clonal). The receptors of the innate immune system, called pattern recognition receptors (PRR), are germ line-encoded and recognise conserved molecular structures shared by a large variety of pathogens (Lipford, Heeg, & Wagner,1998; Stahl

& Ezekowictz, 1998). The adaptive immune system uses somatically generated receptors which are clonally distributed between T and B cells. These receptors are highly specific, as they undergo gene rearrangements and somatic mutations generating an adapted specificity for the antigen.

The traditional distinction between innate and adaptive immunity has become obsolete, as new results clearly indicate a fundamental role of innate components of immunity in adaptive immune responses. In this context, macrophages, which are widely distributed beneath epithelial surfaces, play a keyrole, as they guard the sites of antigen entry. Many of the receptors expressed on macrophages are currently considered as pattern recognition receptors (PPR). It was demonstrated that PPR participate in the delivery of signals that will activate the adaptive immune system. Furthermore, this activation strongly depends on the interaction of bacteria or bacterial products with the receptor (Kirschning,Wesche, Ayres & Rothe, 1998).

Both innate and adaptive immune responses are of major importance for host defence. In particular, the intestinal mucosa is the interface to a highly variable bacterial environment, composed of the commensal microflora, potential pathogenic components, food derived microorganisms and, occasionally, obligate pathogens. Thus, a fine tuning of responses to maintain a continuous low-grade activation of the mucosal immune system are mandatory. This will allow immune activation without inflammatory tissue damage, but also rigorous immune responses upon the encounter with harmful pathogens. Endogenous mediators and luminal factors are thought to play major roles in the orchestration of discriminative immune responses (Klapproth, Donnenberg, Abraham & Mobley, 1995).

3.Adaptive immunity at mucosal sites

3.1 THE MUCOSAL SECRETORY IMMUNE SYSTEM

The mucosal secretory immune response has been extensively studied as a mechanism of defence against enteropathogens. The humoral mucosal immune response starts in a well defined anatomical compartment where antigens have a facilitated access to the host. The induction takes place at lymphoid aggregates, i.e. Peyer's patches (Abreu-Martin & Targan, 1996). Peyer's patches are covered by a specialised epithelium that contains the M cell, adapted for sampling the intestinal content (Neutra, Frey & Kraehenbuhl, 1996). The underlying lymphocytes are arranged in follicles containing T and B cell compartments. The prominent germinal centres (GC) of the gut associated lymphoid tissue (GALT) are the main lymphopoietic site for mucosal B cells, with a preferential commitment to immunoglobulin A (IgA) production. GC development depends upon antigenic challenge, mainly of microbial origin. B cells migrate into the follicles and GC according to their affinity for specific antigens. B cells then undergo somatic mutation of their antigen receptors (immunoglobulins) which leads to increased affinity for the specific epitopes. Follicular dendritic cells in the GC retain immune complexes on their surfaces exposed in such a way that B cells, with the correct mutations, are stimulated and, consequently, rescued from deletion. In contrast, low affinity B cells are deleted by apoptosis. Immunoglobulin isotype switching occurs predominantly towards the IgA isotype. $CD4^{+}$ T cells (T helper lymphocytes), expressing CD40 ligand (CD40L) and producing IL-4 , IL-10, IL-5, co-localise in the B cell zone of the GC and provide help for B cell isotype switching (Liu & Arpin, 1997).

Antigen-specific B and T lymphocytes leave the inductive sites, travel through lymph and the blood stream and home back into the mucosa. The specific homing of mucosal primed lymphocytes is directed, in part, by the expression of α4 β7 integrin, which recognises the mucosal addressing cell surface molecule (MAdCAM-1) expressed on mucosal blood vessels (Berlin, Berg, Briskin, & Butcher 1993). Terminally differentiated B cells will migrate to the lamina propria (LP) compartment, where IgA is secreted and transported through the epithelial layer towards the intestinal lumen by the polymeric Ig receptor or secretory component. When secretory IgA (sIgA) reaches the intestinal lumen, it reacts with specific antigens preventing the physical interaction of noxious agents with the mucosal surface. This process is called immune exclusion and does not imply activation of inflammatory processes. Production and secretion of IgA is further regulated at the LP by (i) endogenous mediators, such as TGF-β and IL-5, mainly produced by regulatory T cells (Lebman, Lee & Coffman, 1990) (ii) intestinal bacterial colonisation (Kett, Baklien, Kral, 1995). .

3.2 STIMULATION OF IGA PRODUCTION BY THE PROBIOTIC MICROORGANISM L.JOHNSONII LA1

Fermented milk products containing probiotic bacteria, such as *L. johnsonii* La1 (10^{10} cfu/ml, daily dose), were shown to have immune adjuvant effects, as demonstrated by a significant increase of total serum IgA in human adult volunteers. Furthermore, consumption of *L. johnsonii* La1 in conjunction with an attenuated oral *Salmonella typhi* vaccine (Vivotif), promoted the specific immune response, as assessed by a significant increase of Ty21a specific serum IgA (Link-Amster, Rochat, Saudan, 1994).

It has been reported that the human indigenous microflora is only partially covered by IgA specific antibodies and even less so by IgG and IgM (van der Waaij, Limburg, Mesander, 1994). An important proportion of the microflora (close to 50%) is not covered by antibodies. These findings seem to show that the partial unresponsiveness to the autochthonous microflora may appear after a transient immune response took place, which, in fact, is suggested by the gnotobiotic animal model. On the other hand, the effect of ingested bacteria, such as probiotics, for maintaining activation at the germinal centre (GC) level is not known. However, they could contribute to it and thereby promote an IgA response which is not only specific against bacterial antigens but also against bystander antigens sampled through the follicular associated epithelium (FAE) containing the M cells.

4. The epithelial compartment

4.1 INTRAEPITHELIAL LYMPHOCYTES

The epithelial compartment, consisting mainly of intestinal epithelial cells (IEC) and intraepithelial lymphocytes (IEL), has now been recognised as a central player in mucosal defences and tissue homeostasis (Eckmann,Jung, Schürer-Maly, Panja & Kagnoff, 1993; Jung, Eckmann, Yang, Panja & Kagnoff, 1995). IEL are located within the intestinal compartment in close association to the basal and lateral membrane of the IEC. Constitutively or upon stimulation, they are capable to modify the epithelial immune-phenotype and its immunological or defensive functions (Schiffrin, Borel & Donnet-Hughes, 1995). These effects could be mediated by surface molecules, such as αE β7/E-cadherin interactions, and cytokines, mainly IFN-γ. Human and murine IELs are enriched for T cells that express $\gamma\delta$ TCR. In humans, IEL express a limited array of TCR $\alpha\beta$ and to a lesser extend $\gamma\delta$, indicating that they recognise a restricted range of antigens in the context of MHC class I related molecules (Blumberg, Terhorst, Bleicher, 1991; Tanaka, Morita, Nieves, 1995; Panja, Blumberg, Balk, 1993). $\gamma\delta$ TCR+ IEL is a population that seems to be less dependent upon bacterial antigenic challenge than the $\alpha\beta$ TCR IEL, since colonisation of germ-free mice preferentially induce the appearance of the later.

Most of the IELs are mature T cells ($CD3^+$) of the $CD8^+$ (suppressor/cytotoxic) phenotype, which express the homodimeric form of the CD8 molecule (CD8$\alpha\alpha$). The predominant expression of CD8$\alpha\alpha$ is considered as an evidence of their extrathymic origin, because peripheral blood $CD8^+$ T cells exclusively express the CD8 $\alpha\beta$ heterodimer (Guy-Grand, Cerf-Bensussan, Malissen, 1991). It is commonly accepted that TCR/CD3 mediated signals in IELs are diminished compared to peripheral T lymphocytes. This functional characteristic is discussed in regard to the specific bi-directional cross-talk with neighbouring IECs and LP-T cells and the control of immune stimulation within this microenvironment. Moreover, IEL may influence functional aspects of adjacent IEC, such as barrier functions and ion transport, probably by secretion of soluble factors.

The specific homing of IEL into mucosal sites seems to depend on the expression of αE β7 integrin found on all IELs. It's ligand, E-cadherin, is expressed on the basolateral membrane of the enterocyte (Cepek, Shaw, Parker, 1994). These integrins can function as accessory molecules in TCR-mediated responses (Sarnacki, Begue, Buc, 1992; Lefrancois, Barrett, Havran, 1994). Since αE β7 expression is up-regulated by TGF-β (Parker, 1992), mainly derived from enterocytes, it will be of great interest to understand how luminal signals can modulate IEC dependent TGF-β production. Therefore, nutrients with the capacity to modulate TGFβ production could have a major impact on mucosal mechanisms of defence.

IEL show high cytolytic (CTL) activity *in vitro*, suggesting that one of their main function *in vivo* could be the elimination of infected or highly damaged cells. It was shown that a fraction of IELs contains intracytoplasmic perforin and granzyme-containing granules and exert immediate cytolytic activity upon TCR stimulation.

Furthermore, IELs express Fas-ligand and are able to kill Fas-bearing targets (Rocha, Guy-Grand & Vasalli, 1995). To date it is not clear which cytolytic pathway, Fas or perforin-mediated, is preferentially used by IEL (Gelfanov, Lai & Liao, 1996).

4.3 . INTESTINAL EPITHELIAL CELLS AS ACTIVE PARTNERS IN MUCOSAL IMMUNE DEFENSES

Intestinal epithelial cells (IEC) are considered to be a constitutive component of the mucosal immune system. They participate in the initiation and regulation of the mucosal immune response to bacteria by interacting with immune cells of the GALT, lamina propria lymphocytes (LPL) and intraepithelial lymphocytes (IEL). Activated IECs express higher levels of HLA class II molecules, classical class I and nonclassical HLA class Ib molecules, such as CD1d, the adhesion molecule ICAM-1, complement factors and cytokine receptors (Eckmann, Jung, Schürer-Maly, 1993). Upon stimulation they are able to produce a wide range of immunomodulatory cytokines (Jung, Eckmann, Yang, 1995). In addition, they can actively participate in the local reaction against pathogens, exerting a form of innate immunity. Moreover, the endogenous microflora seems to have a modulatory effect on the mucosal immune homeostasis and therefore on the mucosal mechanisms of defence. The importance of microflora-derived host protection is evident by the higher susceptibility of germ-free animals to intestinal infections. To date, IECs are thought to be implicated in the recognition of components of the intestinal microflora, including food derived probiotic bacteria, and the transduction of bacteria derived signals to resident mucosal immune cells.

The use of specific bacterial strains in foods (functional nutrients), which provide the consumer with health benefits, particularly for the control of gastrointestinal infections, makes this a field of high interest (Brassart & Schiffrin, 1997).

In the next paragraphs, we summarise data obtained in our group on the molecular mechanisms of bacterial interaction with intestinal epithelial cells using different human *in vitro* models.

5. Modulation of the mucosal immune response by commensal bacteria

5.1 REGULATION OF THE IMMUNE PHENOTYPE OF INTESTINAL EPITHELIAL CELLS IN VITRO

At the colonised mucosal surfaces the host entertains a symbiotic relation with the commensal microflora that implies both nutritional and defensive aspects. With regard to the defensive function, the resident microflora prevents colonisation of the gut by pathogens, the so-called "barrier effect" of the microflora or colonisation resistance. The exclusion of exogenous microorganisms by commensals involves the competence for mucosal specific niches, nutrients, steric hindrance and the production of bactericidal or bacteriostatic products. These aspects will not be discussed in this chapter.

There is increasing evidence that the endogenous microflora has a modulatory effect on the mucosal immune homeostasis and thereby increases mucosal mechanisms of defence. To better understand the functional role of bacteria in gut homeostasis, the molecular mechanisms have to be elucidated at different compartmental levels.

Non-pathogenic bacteria normally do not invade the host. Therefore, the signal for modulation of mucosal immune homeostasis has to be "processed" via the IECs by

- the release of soluble mediators that will translocate through the epithelial layer to neighbouring immune cells
- the modification of the luminal ecology due to their metabolic activity
- changes in epithelial phenotype and function.
- Relevant IEC immune markers can be grouped into molecules involved in (i) antigen presentation: MHC class II, Cd1, which is an early event of the immune response, (ii) cross-talk between IECs and lymphocytes: ICAM-1, Fas, IFN-γ receptor, and (iii) soluble mediators such as cytokines and chemokines: IL-8, TNF-α, MCP-1, which promote the recruitment and activation of immune cells in the different intestinal micro-environments.

Stimulation of the human intestinal cell line HT-29 *in vitro* using a non-pathogenic *E. coli* increased the expression of ICAM-1 (CD54) and IFN-γR (CD119). Furthermore, when *E. coli* was combined with IFN-γ, constitutively produced by IELs, a synergistic effect was detected for the expression of MHC class II molecules (HLA-DR), ICAM-1, Fas and IFN-γR (Table 1). In addition, the pro-inflammatory cytokine/chemokines TNFα and IL-8 were induced by the gram-negative bacteria and the induction was significantly increased when the *E. coli* was combined with IFN γ (Figure 1). Lactic acid bacteria (*L. johnsonii* La1) in contrast, did not show any agonistic effect with respect to the on-set of pro-inflammatory cytokines (Delneste, Donnet-Hughes & Schiffrin, 1998).

Table 1. Bacterial modulation of HLA-DR,CD54 and CD95 Expression on HT-29 cells HT-29 cells (60-80% confluence) were cultured for 24 h with bacteria (50 bacteria/HT-29 cell) with or without 25 U / ml IFNg. HLA-DR, CD54 (ICAM-1) and CD95 (Fas) expression was determined by flowcytometry. Results are presented in mean fluorescence intensity (MFI) minus the MFI of the isotype control. Significant change (p< 0.05) compared to IFN□ single treatment. ** Significant change (p< 0.05) compared to basal expression.*

Stimulus	HLA -DR	CD54	CD95
No stimulus	0	5	5
L .johnsonii La1	0	5	4
E. coli	0	26**	5
IFNγ	9	20**	10
LPS	0	6	5
La1 + IFNγ	15*	27*	14*
E. coli + IFNγ	26*	80*	16*
LPS + IFNγ	19*	60*	16*

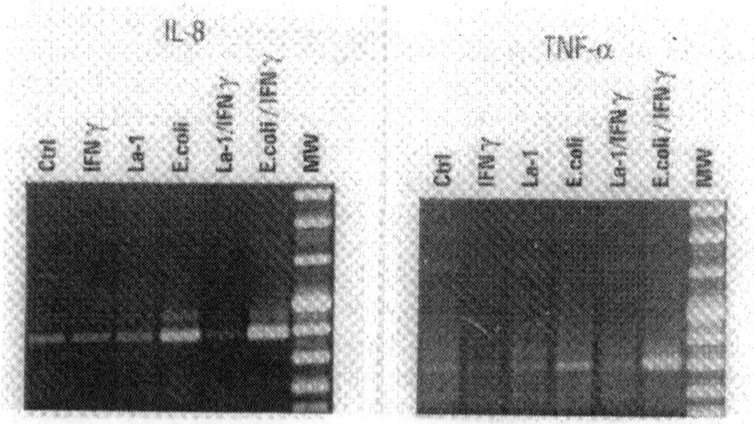

Figure 1 Potentiating effect of IFNγ to induce pro-inflammatory cytokines in HT-29 cells after stimulation with non-pathogenic bacteria. Absence of pro-inflammatory cytokines TNFα, IL-8 and MCP-1 after stimulation of HT-29 cells with L. johnsonii La1. RT-PCR analysis of TNFα, IL-8 and MCP-1 mRNA expression in undifferentiated HT-29 cells after bacterial challenge. E. coli, L. johnsonii La1 and L. sakei (24 h, 10^6 cfu/ml), IFN[illegible] (25 U/ml), (ctrl), no treatment.

The fact that LAB alone or in combination with IFNγ did not induce any of the pro-inflammatory cytokines suggests that LAB could participate in tissue protection against the deleterious effect of an ongoing inflammatory process. Bacterial - epithelial cell contact was mandatory for induction of IEC phenotypic changes, giving support to the importance of adherence to intestinal epithelium as a selection criteria for probiotic bacteria. The underlying physical interaction between prokaryote and eukaryotic cells has not been elucidated, but these observations could imply that either specific receptors or physical surface properties, such as hydrophobicity (Perez, Minaard, Disalvo & De Antoni, 1998), are important to cell-to-cell interactions.

The regulation of the immune phenotype with regard to specific molecules involved in cell-to-cell interactions, playing a key role in the homeostasis of the immune system, seems to suggest a role for IEC in the regulation of the cellular environment at the intra-epithelial compartment and at the lamina propria.

5.2 INTERACTION OF NON-PATHOGENIC BACTERIA WITH MIXED MUCOSAL CELL POPULATIONS:

HUMAN CACO-2/LEUKOCYTE CO-CULTURES IN VITRO

There is increasing evidence that bacterial signals to the host need to be processed by a network of different mucosal cells, resulting in an integrated response that dictates the host reaction against a constantly changing microbial environment in the intestine.

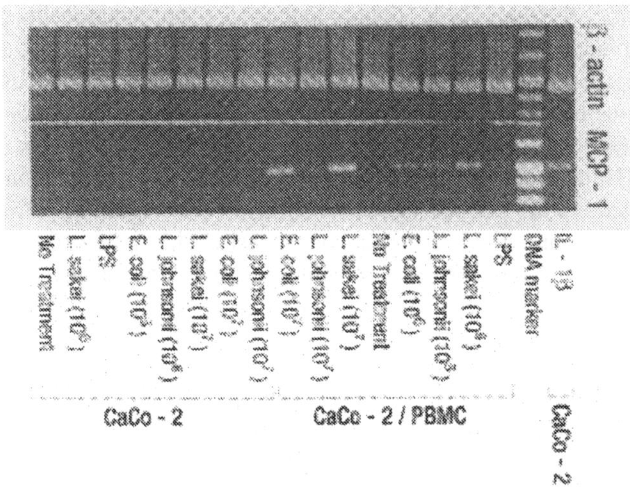

Figure 2: Differential chemokine expression in CaCO-2 cells. RT-PCR analysis was used to determine IL-8 (A) and MCP-1 (B) mRNA expression in CaCO-2 cells upon stimulation of CaCO-2/leukocyte co-cultures or CaCO-2 cells alone with non-pathogenic E. coli, L. johnsonii *and* L. sakei *(16 h, 10^6 and 10^7 cfu/ml), respectively. LPS (1 μg/ml), IL-1β (10 ng/ml) and culture medium (no treatment) were used as controls. Results represent one of three independent experiments.*

To investigate such interactions, a human *in vitro* model was established with CaCo-2 cells and peripheral blood mononuclear cells (PBMC) using transwell culture technique (Haller, Bode, Hammes, Pfeifer, Schiffrin & Blum, 2000). The immune response to different non-pathogenic bacteria was assessed by the determination of cytokine expression in IECs and leukocytes. The pro-inflammatory cytokines TNF-α, IL-8 and MCP-1 were induced in CaCo-2 cells upon challenge with non-pathogenic *E. coli* and *Lactobacillus sakei* (Figure 2). In contrast, *L. johnsonii* La1 did not stimulate the production of these cytokines, but upregulated the expression of TGF-β. Responsiveness of IECs to non-pathogenic bacterial signals was dependent on the presence of PBMCs. In addition, the underlying immune cells also responded in a discriminative manner to different bacteria, although the bacteria had no direct access to this compartment. *E. coli* and *L. sakei* exclusively induced TNFα and IL-10 protein secretion from leukocyte-sensitised co-cultures, whereas no induction of these pro-inflammatory cytokines occurred with *L. johnsonii* La1. These results strengthen the hypothesis that bacterial signalling at the mucosal surface is dependent on epithelial-immuno crosstalk, which seems responsible for the innate reaction that can, in fact, distinguish between different non-pathogenic microorganisms. This discriminative response occurred in both compartments, probably orchestrated by cell secretory products, which are not entirely identified yet. These results also indicate, that, depending on the *lactobacillus* strain, a more pro-inflammatory (*L. sakei*) or more immune-regulatory (*L. johnsonii* La1) type of immune response might be stimulated at the mucosal site. A comparison of non pathogenic and enteropathogenic E. coli revealed that i) induction of TNFα was only attained with live bacteria and ii) TNFα mRNA was expressed earlier (6h) and sustained until 36 h after infection of CaCO-2/leukocyte co-cultures with the EPEC wildtype strain E2348/69 and the intimin$^{-/-}$ mutant CVD206. In contrast, non-pathogenic bacteria only induced a transient expression of TNFα which was downregulated in the course of stimulation (Figure 3).

The presented results provide direct evidence on the beneficial effect of specific probiotic strains on intestinal immune homeostasis. This knowledge gives unique possibilities to the food industry to improve gut homeostasis by nutritional interventions.

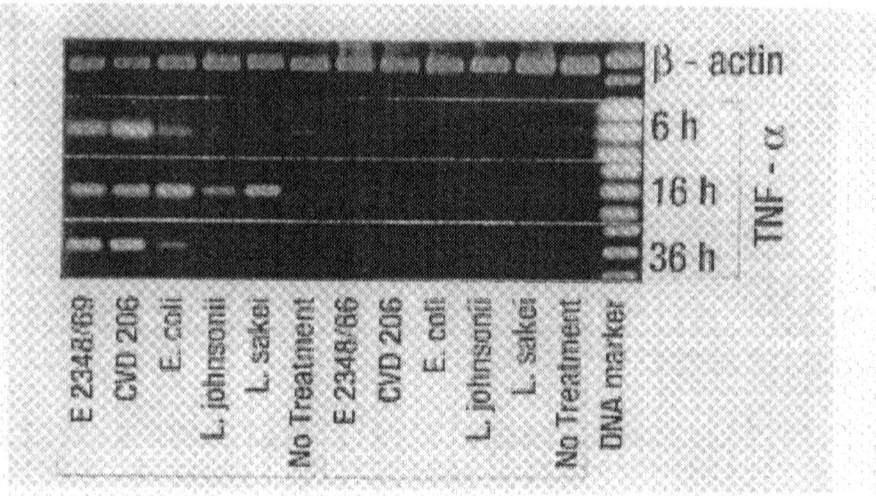

Figure 3: Differential induction of TNF-α mRNA by leukocyte sensitised CaCO-2 cells stimulated with enteropathogenic E. coli *and non-pathogenic bacteria. Stimulation of CaCO-2/leukocyte co-cultures with live or heat killed (95°C/30 min) EPEC strain E2348/69 and its* eaeA *deletion mutant CVD206, non-pathogenic* E. coli, L. johnsonii *La1 and* L. sakei *(10^7 cfu/ml) for 4 h in the absence of gentamycin (150 µg/ml). Culture medium (no treatment) was used as a control. The expression of TNF-α mRNA in CaCO-2 cells was determined after 6, 16, 36 h of incubation by RT-PCR analysis. Results represent one of three independent experiments.*

6. Modulation of the systemic immune response

6.1 INTERACTION OF NON-PATHOGENIC BACTERIA AND BLOOD LEUKOCYTES

Although commensal bacteria do not massively invade the host, it is possible that in defined mucosal environments, like the Peyer's patches, they may interact directly with immune cells and induce their activation.

Fluorescent-labelled bacteria (*Lactobacillus* spp., *Bifidobacteria* spp.) were used to assess their capacity to directly interact with human blood leukocyte subpopulations using flow-cytometry. Notably, differences in the percentage of association between different strains and subpopulation were detectable. As shown in Table 2 *Lactobacillus johnsonii* La1 showed the strongest interactions with lymphocytes, monocytes and granulocytes.

Table 2. Percentage of leukocyte subpopulations showing association of fluorescent labelled bacteria Human unstimulated PBMC were incubated with BCEF-AM fluorescent labelled bacteria in a ratio 20:1 (bacteria:leukocytes), washed three times and analysed by flowcytometry. The interaction of BCEF-AM-bacteria with leukocyte subpopulations is demonstrated as the percentage of fluorescence positive cells within the gated population.

Strain	lymphocytes	monocytes	granulocytes
***Bifidobacterium bifidum* CIDCA 536**	1.48 ± 0.04	6.43 ± 0.01	13.8 ± 0.64
***Bifidobacterium infantis* CIDCA 538**	0.90 ± 0.01	9.33 ± 0.32	32.1 ± 0.76
***Lactobacillus johnsonii* La1**	4.64 ± 0.05	25.05 ± 2.14	46.89 ± 1.99
Lactobacillus acidophilus La10	2.41 ± 0.67	11.59 ± 1.48	27.83 ± 0.56

Assuming that bacteria/leukocyte cell-to-cell interactions could lead to the stimulation of immune cells, an assessment of cell proliferation and cytokine production was performed. ^{3}H-thymidine incorporation by proliferating leukocytes was determined at 3 and 5 days after incubation with bacterial cells. Although all bacteria induced proliferation at day 5, *L. johnsonii* La1 was the strongest effector. When $CD4^+$, $CD8^+$, $CD19^+$, $CD56^+$ purified leukocyte subpopulations were stimulated with non-pathogenic bacteria, only $CD3^-CD56^+$ NK cells showed expression of the activation antigens CD69 and CD25 and proliferated significantly. Further analysis demonstrated that the activation of human NK cells by non-pathogenic bacteria required direct contact with the bacteria and was increased in the presence of accessory cells, such as primed macrophages, which could provide the necessary co-stimulatory signals (Haller, Blum, Bode, Hammes & Schiffrin, 2000).

The various bacterial strains also induced a differential cytokine response. Whereas *L. johnsonii* La1 and *L .sakei* strongly induced the secretion of IFNγ and IL-12, *E. coli* preferentially induced IL-10 (Figure 4 and 5).

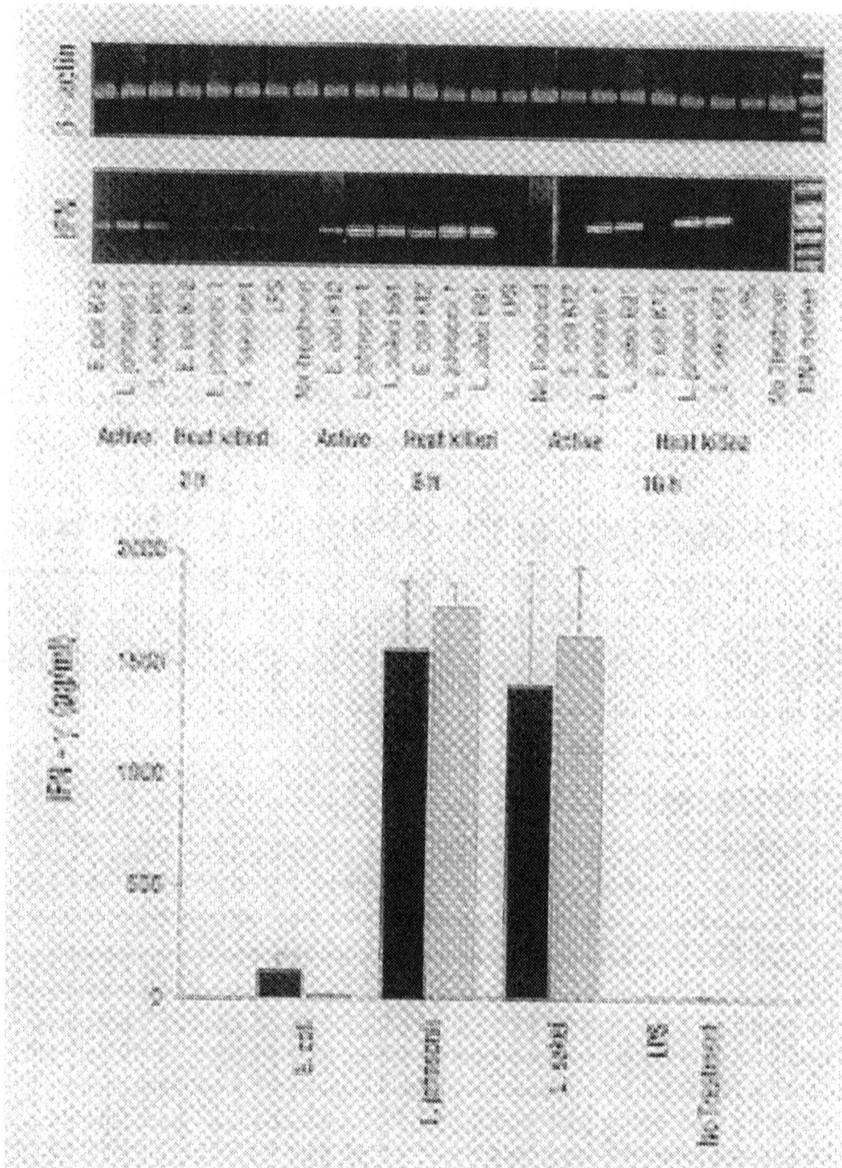

Figure 4. Expression of IFN-γ by PBMC upon the stimulation with non-pathogenic bacteria. RT-PCR and ELISA analysis was used to determine IFN-γ expression by PBMC (106/ml) upon the stimulation with heat killed and live bacterial cells (10^6 cfu/ml) of E. coli, L. johnsonii *La1,* L. sakei *or LPS (1 μg/ml). Gene transcription (IFN-γ, IL-12 p40) was determined after 2, 6, and 16 h. Protein secretion was analysed after 16 h of stimulation. No antibiotics were added to the cultures. Values are given as mean ± SD of triplicates and represent one of three independent experiments.*

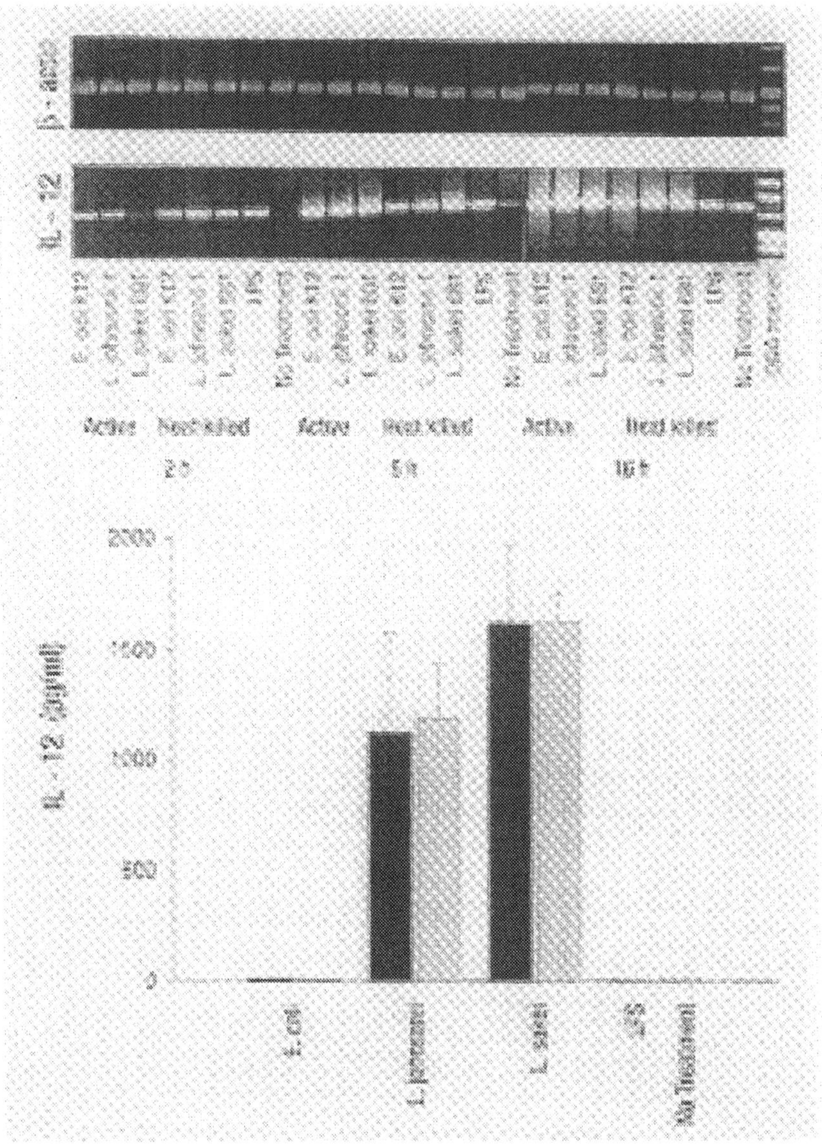

<u>Figure 5</u>. Expression of IL-12 by PBMC upon the stimulation with non-pathogenic bacteria. RT-PCR and ELISA analysis was used to determine IL-12 expression by PBMC (106/ml) upon the stimulation with heat killed and live bacterial cells (10^6 cfu/ml) of E. coli, L. johnsonii *La1,* L. sakei *or LPS (1 µg/ml). Gene transcription (IL-12 p40) was determined after 2, 6, and 16 h. Protein secretion (IL-12 p70) was analysed after 16 h of stimulation. No antibiotics were added to the cultures. Values are given as mean ± SD of triplicates and represent one of three independent experiments.*

7. Regulation of the mucosal immune response by luminal factors. New perspectives

It has been shown that IEC may be participants in the initiation of the mucosal immune response as suggested by antigen presentation *in vitro*. On the other hand, antigen presentation by IEC could involve, in addition to MHC class II molecules, specific restriction molecules such as CD1d or other nonpolymorphic class-1 molecule, such as thymus leukaemia (TL) antigens. In addition to the IEC accessory function, they produce cytokines under physiological conditions, and in response to pathogenic bacteria. The differentially expressed cytokines then promote the recruitment and activation of effector immune cells for clearance of the pathogens.

Results obtained in our group using different human IEC *in vitro* models put forward further support for both functions. Moreover, the regulation of the immune phenotype with regard to specific molecules involved in cell-to-cell interactions, seems to suggest a role for the regulation of the cellular environment at the intra-epithelial compartment and at the lamina propria. Notably, the IEC are in permanent interactions with the luminal content, including commensal microflora, and the endogenous cellular network of professional immune cells. Thus, the appropriate function of IEC can adapt the physiological reactivity of the host tissues to a highly changing intestinal content. A dysfunction of this interphase could promote a discordance between the luminal signal and the initiated response. This could be the origin of pathological conditions due to exaggerated responses to non-dangerous signals, like food antigens, resulting in food allergy or in chronic inflammation, i.e. inflammatory bowel disease (IBD). Both conditions could result from the breakdown of oral tolerance, for which one of the underlying mechanisms is the secretion of inhibitory cytokines such as TGF-β. It is notable to emphasise that the probiotic strain *L. johnsonii* La1 selectively induced the production of TGF-β. In fact, the specific systemic unresponsiveness to oral antigens, namely oral tolerance, is an important physiological characteristic of the mucosal immune system.

On the other hand, the unresponsiveness to a harmful luminal signal could be associated with a lack of appropriate response for controlling potential or obligate pathogens, as it can be observed in some cases of intestinal immunological disorders. The fine tuning of this dynamic interphase is probably not only depending on the IEC function but on an intricate cell-to-cell crosstalk, where IEL and LP immune cells are further participants. However, there is a strong evidence that the IEC plays a central role in delivering early signals to neighbouring cells. The knowledge about the downstream cascade of events will give us unique possibilities to modify the gut homeostasis by nutritional interventions. Finally, it is possible that the IEC play an effector role in the clearance or very basic defence mechanisms against pathogens. These functions may be, in turn, regulated by the immune cells in the vicinity.

Our current research interest is focused on the identification of the molecules and receptors implicated in the sensitising effect of leukocytes on intestinal epithelial cells. Therefore, new molecular techniques, such as gene expression profiling of bacterial activated and non-activated IEC using DNA arrays will be applied. Combined with functional proteomics of IEC this will give us new opportunities to dissect the molecular basis of immuno-modulation by non-pathogenic bacteria in the human intestine.

8. Conclusion

Intestinal bacterial signalling of the mucosal mechanisms of defence can cover a wide range, from the physiological interactions between the resident microflora and the host cells in the superficial mucosal micro-environment, until the inflammatory, exudative reaction promoted by invasive pathogens, which implies the recruitment of blood leukocytes. Both types of reactions are adaptive mechanisms of defence adequate to the triggering signal. Nevertheless, it is obvious that the second type of response implies inflammatory tissue damage of the mucosal barrier and a response that can lead to a systemic inflammatory reactions. Therefore, the understanding of how commensal bacteria or nutrients, may enhance defences without inflammation is of major interest. The strengthening of this mechanisms will diminish the interactions between virulent micro-organisms and the mucosal surface.

The first level of defence takes place at the interface intestinal lumen/ mucosa and is represented mainly by secretory immunity (sIgA), and possibly other components of innate immunity secreted into the intestine. There is growing experimental evidence that the epithelial layer also participate in local mechanisms of defence as an effector cell by altering it's immune-phenotype and as part of a local control against infections.

In this chapter, we demonstrated the importance of non-pathogenic microorganisms, including selected probiotic bacteria, in modulation of the host gut mucosal defences and maintenance of homeostasis and integrity. Human *in vitro* models, followed by studies in experimental animals and human studies, will allow a better understanding of the 'mode of action' of distinct lactic acid bacteria. This will lead to the selection of probiotics optimally tailored for the application in infant nutrition, nutrition for the elderly or clinical nutrition.

References

Abreu-Martin MT, Targan SR. (1996) Regulation of immune responses of the intestinal mucosa. *Critical Reviews in Immunology* 16 :277-309.

Berlin C, Berg EL, Briskin MJ, Andrew DP, Kilshaw PJ, Holzmann B, Weismann IL, Hamann, A, Butcher EC. (1993) Alpha-4 beta-7 intefrin mediates lymphocyte binding to the mucosal addressin MAdCAM-1.*Cell* 74:158-195.

Blumberg RS, Terhorst C, Bleicher P, McDermott FV, Allan CH, Landau SB, Trier JS, Balk S. (1991) Expression of a nonpolymorphic MHC class I-like molecule, CD1d, by human intestinal epithelial cells. *J.Immunol.*147:2518-2524

Bland PW, Warren LG. (1986) Antigen presentation by epithelial cells of the rat small intestine. *Immunology* 58: 1-7.

Barclay AN, Mason DW. (1982) Induction of Ia antigen in rat epidermal cells and gut epithelium by immunological stimuli. *J.Exp.Med.*156: 1665-1669.

Brassart D, Schiffrin EJ. (1997) The use of probiotics to reinforce mucosal defence mechanisms. *Trends in Food Science and Technology* 8: 321-326

Coffman RL, Lebman DA, Shrader B. (1989) Transforming growth factor beta specifically enhances IgA production by liposaccharide-stimulated murine B-lymphocytes. *J.Exp.Med.* 170: 1039-1044.

Cerf-Bensussan N, Quaroni A, Kurnick JT, Bhan AK. (1984) Intraepithelial lymphocytes modulate Ia expression by intestinal epithelial cells. *J.Immunol.* 132: 2244-2252.

Cepek KL, Shaw SK, Parker CM, Russell GJ, Morrow JS, Rimm DL, Brenner MB. (1994) Adhesion between epithelial cells and lymphocytes is mediated by E-cadherin and the alpha E beta-7 integrin. *Nature* 372: 190-193.

Delneste Y, Donnet-Hughes A, Schiffrin EJ. (1998) Functional Foods: mechanisms of action on immunocompetent cells. *Nutrition Reviews* 56:S93-S98.

Duchmann R, Kaiser I, Hermann E, (1995) Tolerance exists towards resident intestinal flora but is broken in active inflammatory bowel disease (IBD). *Clin Exp Immunol* 102 :448-455

Eckmann L, Jung HC, Schürer-Maly C, Panja A, Morzycka-Wrobleska E, Kagnoff MK. (1993) Differential cytokine expression by human intestinal epithelial cell lines: regulated expression of Interleukin 8. *Gastroenterology* 105:1689-1697.

Garside P, Mowat A McI, Khoruts A. (1999) Oral tolerance and disease. *Gut*.44 :137.142

Guy-Grand D, Cerf-Benussan N, Malissen B, Malasssis-Seris M, Briottet C, Vasalli P. (1991) Two gut intraepithelial CD8+ lymphocyte populations with different T cell receptors: a role for the gut epithelium in T cell differentiation. *J.Exp.Med.* 173: 471-478.

Gelfanov V, Lai YG, Liao NS. (1996) Activated $\alpha\beta$-CD8$^+$ but not $\alpha\alpha$-CD8$^+$, TCR-$\alpha\beta^+$ murine intestinal intraepithelial lymphocytes can mediate perforin-based cytotoxicity whereas both subsets are active in Fas-based cytotoxicity. *J.Immunol.* 157: 35-41.

Haller D, Bode Ch, Hammes WP, Pfeifer AMA, Schiffrin EJ, Blum S. (2000) Non-pathogenic bacteria elicit a differential cytokine response by intestinal epithelial cell/leukocyte co-cultures. *Gut, in press*

Haller D, Blum S, Bode Ch, Hammes WP, Schiffrin EJ. (2000) Activation of human peripheral blood mononuclear cells by nonpathogenic bacteria in vitro: evidence of NK cells as primary targets. *Infection & Immunity* 68 (2): 752-759.

Hughes A, Block KJ, Bhan AK, Gillen D, Giovino VC, Harmatz PR. (1991) Expression of MHC class II (Ia) antigen by the neonatal enterocyte: the effect of treatment with IFNγ. *Immunology* 72: 491-496.

Jung HC, Eckmann L, Yang SK, Panja A, Fierer J, Morzycka-Wrobleska E, Kagnoff MK. (1995) A distinct array of pro-inflammatory cytokines is expressed in human colon eoithelial cells in response to bacterial invasion. *J.Clin.Invest.* 95: 55-65.

Kett K, Baklein K, Bakken A, Kral JG, Fausa O, Brandtzaeg P. (1995) Intestinal B-cell isotype response in relation to bacterial load: evidence for immunoglobulin A subclass adaptation. *Gastroenterology* 109: 819-825.

Klapproth JM, Donnenberg MS, Abraham JM, Mobley HLT, James SP. (1995) Products of enteropathogenic *Escherichia coli* inhibit lymphocyte activation and lymphokine production. *Infection & Immunity* 63: 2248-2254.

Kirschning CJ, Wesche H, Ayres TM, Rothe M. (1998) Human toll-like receptor 2 confers responsiveness to abcterial lipopolysaccharide. *J. Exp. Med.* 188: 2091-2097.

Kotler DP, Giang TT, Thiim M, Nataro JP, Sordillo EM, Orenstein JM. (1995) Chronic bacterial enteropathy in patients with AIDS. *The Journal of Infectious Diseases* 171: 552-558

Lebman DA, Lee FD, Coffman RL. (1990) Mechanism for transforming growth factor beta and IL-2 enhancement for IgA expression in liposaccharide-stimulated B cell cultures. *J.Immunol.* 144: 952-959.

Link-Amster H, Rochat F, Sauda KY, Mignot O, Aeschlimann JM. (1994) Modulation of the humoral immune response and changes in the intestinal microflora through fermented milk intake. *FEMS Immunol Med Microbiol* 10(1):55-63.

Lipford GB, Heeg K, Wagner H. (1998) Bacterial DNA as immune activator. *Trends in Microbiology;* 6 :496-500

Liu YJ, Arpin C. (1997) Germinal centre development. *Immunological Reviews* 156:11-126

Neutra MR, Frey A, Kraehenbuhl JP. (1996) Gateways for mucosal infection and immunization. *Cell* 86: 345-348.

Panja A, Blumberg RS, Balk S, Mayer L. (1993) Cd1d is involved in T cell-intestinal epithelial cell interactions. *J.Exp.Med.* 178: 1115-1119.

Parker CM. (1992) A familiy of b7 integrin on human lymphocytes. *PNAS* 89:1924-28

Perez PF, Minaard J, Disalvo EA, De Antoni GL. (1998) Surface properties of bifidobacterial strains of human origin. *Appl. Environ. Microbiol.* 64: 21-26

Rocha B, Guy-Grand D, Vasalli P. (1995) Extrathymic T cell differentiation. *Current Opinion in Immunology* 7: 235-242.

Sarnacki S, Begue B, Buc H, Le Deist F, Cerf-Benussan N. (1992) Enhancement of CD3 induced activation of human intestinal intraepithelial lymphocytes by stimulation of the β7-containing integrin defined by HLM-1 monoclonal antibody. *Eur.J.Immunol.* 22: 2887-2892.

Schiffrin EJ, Borel Y, Donnet-Hughes A. (1995) Modulation of the MHC class I and II molecules by bacterial products on intestinal epithelial cells. *Adv Exp Med Biol* 371A:195-196.

Smart CJ, Trejdosiewicz LK, Badr-el-Din S, Heatley RV. (1988) T lymphocytes of the human colonic mucosa: functional and phenotypic analysis. *Clin.Exp.Immunol.* 73: 63-67.

Stavnezer J, Shocket P. (1991) Effects of cytokines on switching to IgA and alpha germline transcripts in the B loymphoma 1.29. Transforming growth factor-beta activates transcription of the unrearranded C alpha gene. *J.Immunol.* 147: 4374-4383.

Stahl PD, Ezekowitz AB. (1998) The mannose receptor is a pattern recognition receptor involved in host defense. *Current Opinion in Immunology* 10: 50-55

Tanaka Y, Morita CT, Nieves E, Brenner MB, Bloom BR. (1995) Natural and synthetic non peptide antigens recognized by human γδ T cells. *Nature* 375: 155-158.

Underdown BJ, Schiff JM. (1986) Strategic defense initiative at the mucosal surfaces. *Annu Rev Immunol* 4: 389-417.

van der Waaij LA, Limburg PC, Mesander G, van der Waaij D. (1996) In vivo IGA coating of anaerobic bacteria in human feaces. *Gut* 38(3):348-54

Wicox CM, Schwartz DA, Cotsonis G, Thompson SE. (1996) Chronic unexplained diarrhea in human immunodeficiency cirus infection: determination of the best diagnostic approach. *Gastroenterology* 110: 30-37

INDEX

GPSR Compliance
The European Union's (EU) General Product Safety Regulation (GPSR) is a set of rules that requires consumer products to be safe and our obligations to ensure this.

If you have any concerns about our products, you can contact us on

ProductSafety@springernature.com

In case Publisher is established outside the EU, the EU authorized representative is:

Springer Nature Customer Service Center GmbH
Europaplatz 3
69115 Heidelberg, Germany

www.ingramcontent.com/pod-product-compliance
Lightning Source LLC
Chambersburg PA
CBHW060748150926
51501CB00003B/123
* 9 7 8 0 7 9 2 3 6 7 4 7 5 *